Mathematical Modeling of Water Quality: Streams, Lakes, and Reservoirs

Wiley IIASA International Series on Applied Systems Analysis

1 CONFLICTING OBJECTIVES IN DECISIONS
Edited by David E. Bell, *University of Cambridge*,
Ralph L. Keeney, *Woodward-Clyde Consultants, San Francisco*,
and Howard Raiffa, *Harvard University*.

2 MATERIAL ACCOUNTABILITY
Rudolf Avenhaus, *Nuclear Research Center Karlsruhe,*
and University of Mannheim.

3 ADAPTIVE ENVIRONMENTAL ASSESSMENT AND MANAGEMENT
Edited by C. S. Holling, *University of British Columbia.*

4 ORGANIZATION FOR FORECASTING AND PLANNING: EXPERIENCE IN THE SOVIET UNION AND THE UNITED STATES
Edited by W.R. Dill, *New York University*,
and G. Kh. Popov, *Moscow State University.*

5 MANAGEMENT OF ENERGY/ENVIRONMENT SYSTEMS: METHODS AND CASE STUDIES
Edited by Wesley K. Foell, *University of Wisconsin, Madison.*

6 SYSTEMS ANALYSIS BY MULTILEVEL METHODS
Yvo M. I. Dirickx, *Twente University of Technology*,
and L. Peter Jennergren, *Odense University.*

7 CONNECTIVITY, COMPLEXITY, AND CATASTROPHE IN LARGE-SCALE SYSTEMS
John Casti, *New York University.*

8 PITFALLS OF ANALYSIS
Edited by Giandomenico Majone
and E. S. Quade, *International Institute for Applied Systems Analysis.*

9 CONTROL AND COORDINATION IN HIERARCHICAL SYSTEMS
W. Findeisen, *Institute of Automatic Control, Technical University, Warsaw*;
F. N. Bailey, *University of Minnesota*;
M. Brdyś, K. Malinowski, P. Tatjewski, *and* A. Woźniak, *Institute of Automatic Control, Technical University, Warsaw.*

10 COMPUTER CONTROLLED URBAN TRANSPORTATION
Edited by Horst Strobel, *University for Transportation and Communication "Friedrich List," Dresden.*

11 MANAGEMENT OF ENERGY/ENVIRONMENT SYSTEMS: NATIONAL AND REGIONAL PERSPECTIVES
Edited by Wesley K. Foell, *University of Wisconsin, Madison, and* Loretta A. Hervey, *International Institute for Applied Systems Analysis.*

12 MATHEMATICAL MODELING OF WATER QUALITY: STREAMS, LAKES, AND RESERVOIRS
Edited by Gerald T. Orlob, *University of California, Davis.*

12 International Series on
Applied Systems Analysis

Mathematical Modeling of Water Quality: Streams, Lakes, and Reservoirs

Edited by

Gerald T. Orlob

University of California, Davis

A Wiley–Interscience Publication
International Institute for Applied Systems Analysis

JOHN WILEY & SONS
Chichester–New York–Brisbane–Toronto–Singapore

Library of Congress Cataloging in Publication Data:

Mathematical modeling of water quality: streams, lakes, and reservoirs

(International series on applied systems analysis; 12)

'A Wiley–Interscience publication.'

Includes index.

1. Water quality—Mathematical models. I. Orlob, Gerald T. II. Series.

TD365.M27 1982 363.7′394′0724 82-17457

ISBN 0 471 10031 5 (U.S.)

British Library Cataloguing in Publication Data:

Mathematical modeling of water quality: streams, lakes, and reservoirs—(Wiley international series in applied systems analysis (IIASA)

1. Water quality—Mathematical models

I. Orlob, Gerald T.

ISBN 0 471 10031 5

Filmset at Composition House Limited, Salisbury. Wiltshire.

Printed and bound in Great Britain at the Pitman Press, Bath.

The Authors

GERALD T. ORLOB
Department of Civil Engineering
University of California
Davis, CA 95616
USA

M. BRUCE BECK
Department of Civil Engineering
Imperial College
London SW7 2BU
England

M. J. GROMIEC
Institute of Meteorology and Water Management
ul. Podlesna 61
01-673 Warsaw
Poland

DONALD R. F. HARLEMAN
Ralph M. Parsons Laboratory for Water Resources and Hydrodynamics
Department of Civil Engineering
Massachusetts Institute of Technology
Cambridge, MA 02139
USA

J. JACQUET
Electricité de France
64 Rue La Boëtie
75008 Paris
France

S. E. JØRGENSEN
The Royal Danish School of Pharmacy
Universitetsparken 2
2100 Copenhagen
Denmark

DANIEL P. LOUCKS
School of Civil and Environmental Engineering
Cornell University
Ithaca, NY 14853
USA

PETER MAUERSBERGER
Institute for Geography and Geoecology
Academy of Sciences of the GDR
Müggelseedamm 260
1162 Berlin
German Democratic Republic

OLEG F. VASILIEV
Institute of Hydrodynamics
Siberian Branch of USSR Academy of Sciences
630090 Novosibirsk
Prospect Nauki 15
USSR

MASATAKA WATANABE
National Institute for Environmental Studies
Yatabe
Tsukuba
Ibaraki 305
Japan

Preface

A capability to represent quantitatively the responses of the aquatic environment to actions of man is an essential part of resource management. It is manifest in the mathematical model, which, if correctly designed, will simulate real-world behavior under conditions yet to be experienced. Used with discretion and appreciation of its limitations, the model can become a helpful tool in the management process, enabling the user to explore new horizons of the imagination, to compare choices, and to identify pathways toward superior solutions to practical problems.

Mathematical modeling of water quality presents a special challenge to the systems analyst because it demands integration of so many disciplines. It is dependent upon hydrology and hydromechanics for description of the movement of water and the mechanisms of mixing. It calls upon climatology, meteorology, and atmospheric physics to specify conditions at the air–water interface. It draws on the chemistry of dilute solutions, chemical kinetics, and biochemistry for determination of the fate of substances dissolved or suspended in water. It requires knowledge of the interrelationships of aquatic life forms and their environment—an understanding of aquatic ecology. Perhaps it is this interdisciplinary aspect of water quality modeling that has attracted so many competent scientists to become practitioners of the art. Whereas a decade or so ago water quality models were novelties in the technical literature, today they are acknowledged with increasing frequency as necessary elements in environmental and resource management.

Yet, despite the enthusiasm with which modeling of aquatic systems has apparently been embraced, there exists a gap between conception of the model as an exercise of the mind and its use as a practical tool. One has only to examine the literature to see that comparatively few water quality models have attained a status that enables the technology they represent to be transferred to others.

It was this deficiency in the art that prompted me to propose early in 1977 that IIASA undertake to survey water quality modeling as a part of its State-of-the-Art Series. Professor Zdzislaw Kaczmarek, then in charge of IIASA's water program, invited me to visit Laxenburg for a two-week period and to outline in detail a program for the survey. The proposal I presented to IIASA for a monograph series on water quality modeling was accepted in principle and work on this volume on *Streams, Lakes, and Rservoirs* was initiated in September 1978. We were successful, as you will see, in securing the collaboration of an eminently qualified group of contributors to the modeling art, each individual preparing chapters for which his experience best qualified him. This book is the result of their combined efforts, stimulated from time to time by the editor, by Professor Kaczmarek, and by Professor Oleg Vasiliev, who served both as Deputy Director of IIASA and as a major contributor to the project.

The charge to our group was to capture as best we could the essence of water quality modeling—the basic principles upon which it is based, the practical problems in conceptualizing real-world phenomena in model form, and the use of models as aids in decision making. I believe we have met this charge and that you, the reader—likely a modeler yourself—will find that the examples we have chosen to illustrate water quality modeling of streams, lakes, and reservoirs are fairly representative of the state of our art.

G. T. Orlob
Benicia, California

Contents

Notation for Book

A	cross-sectional area; algal biomass
$\mathbf{A}$	eddy viscosity (tensor)
A_r	affinity of rth chemical reaction
B_k	ρ_k/ρ, mass fraction of biomass of kth biotic component
c_p	specific heat at constant pressure
C	concentration
E	exchange (bulk dispersion) coefficient
E_k	rate of excretion by kth biocomponent
Eu	Euler or Ruark number
f	Coriolis parameter
F	external force
Fr	Froude number
g	gravitational acceleration
G_e	Gibbs free enthalpy (per unit mass)
G_k	grazing rate of kth biocomponent
h	density of enthalpy
H	height; depth
I	light intensity
J	diffusion flow
k_B	Boltzmann constant
K	coefficient
$\dot{L}$	length; concentration of organic matter
$\mathcal{M}_j$	molar mass of jth chemical constituent
M_k	nonpredatory mortality rate of kth biocomponent
n_j	number of moles of jth chemical constituent
N	nitrogen concentration
N_j	ρ_j/ρ, mass fraction of jth chemical constituent

p	pressure
P	phosphorus concentration
$\mathbf{P}$	dissipative component of pressure tensor (viscosity)
P_k	rate of production of kth primary producer
Pr	Prandtl number
Q	flow rate (downstream)
$\boldsymbol{r}$	space vector
R_k	respiration rate of kth biocomponent
R_0	universal gas constant
Re	Reynolds number
Ri	Richardson number
Ro	Rossby number
St	Strouhal number
t	time
T	temperature
u	density of internal energy
u, v, w	velocity components
U	Q/A, downstream or wind velocity in one-dimensional flow
$\mathbf{v}$	velocity
V	volume
w_r	$\dot{\xi}_r$, rate of rth chemical reaction
$\boldsymbol{W}$	vector of heat flow
x, y, z	space coordinates
Z	zooplankton concentration
α	$1/\rho$, specific volume
Δ	Laplace operator
ζ	free elevation of surface
η	light extinction coefficient
θ	temperature coefficient
λ	wavelength of light
μ	growth rate
μ_j	chemical potential of jth chemical constituent (per unit mass)
ν	kinematic viscosity
ν_{jr}	stoichiometric coefficient of jth chemical constituent in rth chemical reaction
ξ_r	extent of rth chemical reaction
$\rho(\rho_j)$	mass density (of jth chemical constituent)
τ	shear stress
ϕ	potential of external forces (e.g. gravity)
Ψ	velocity potential ($\boldsymbol{v} = \text{grad}\,\Psi$)
ω	angular velocity.

Chapters 2–4 and 6–12 have individual lists of extra notation.

1 Introduction

G. T. Orlob

1.1. WATER QUALITY IN PERSPECTIVE

Water as a natural resource has been a primary concern of man throughout his existence. Only recently, however, has he learned that it is limited, not only in quantity, but also in its capacity to assimilate the waste products of human activity without endangering life itself.

Our earliest records acknowledge his requirements for a continuous source of fresh, drinkable water and his dependence on the natural hydrological cycle, which in its extremes could severely restrict, even preclude, life processes. Agricultural societies developed and flourished where the natural cycle of flooding brought new soil and nutrient to the land, but they also failed for want of water of sustained high quality. Salinization of ground- and surface-waters is believed to have been a primary cause of the demise of ancient agricultural centers in the Fertile Crescent of the Middle East, just as it threatens agricultural productivity in many parts of the world today.

While the gross differences in quality of natural waters—between sea- and rain-waters, for example—were appreciated by early civilizations, the processes resulting in changes of quality were not. It was not until the late nineteenth century that a reasonably correct representation of the hydrological cycle was advanced. A consciousness of *quality* was not developed until the connection between enteric diseases in man and water pollution was recognized.

The identification of disease-producing organisms and of their transport from source to receptor by water stimulated rapid development in the technology of water treatment and, as a matter of historical record, brought about a sharp decline in the incidence of enteric diseases in those parts of the world where this technology was practiced. At the same time, monitoring the quality of natural waters for evidence of pollution became an essential part of disease prevention. Analytical techniques were devised for measurement of specific constituents in solution or suspension. With these tools in hand, scientists and engineers began to inquire more deeply into the phenomenological behavior of natural water bodies, especially freshwater sources—the streams, lakes, and reservoirs with which this book is concerned.

1.2. THE BEGINNINGS OF MODELING

Stimulated by the need to control pollution of the major sources of freshwater supply and to ensure protection of public health, sanitary engineers were probably among the first to examine quantitatively the physical, chemical, and biological responses of streams to loadings of nutrients, either natural or man-induced. In the 1920s the Ohio River Commission in the United States began an intensive study of sources of pollution and their impacts on domestic water supply. From this investigation emerged one of the first mathematical models of an aquatic environment, if not *the* first—the Streeter–Phelps equation describing the balance of dissolved oxygen in a stream (Streeter and Phelps, 1925).

This comparatively simple formulation recognized that the change in the dissolved oxygen deficit below saturation in a stream with a steady discharge could be represented by the sum of two processes: uptake of oxygen from solution by the biochemical oxidation of dissolved and suspended organic matter, and mass transfer of oxygen from the air to the turbulent stream through the air–water interface. It was stated as a first-order ordinary differential equation that, when integrated, provided an analytical solution describing the dissolved oxygen deficit as a function of time (or days of flow) from the point of discharge of a steady source of biodegradable pollutant. This solution, called the "oxygen sag equation," was to command the attention (even to the extent of preoccupation) of sanitary engineers for the next several decades. Probably no single "model" in the history of our art† has been the subject of more intensive study. Investigators sought to determine biodegradation and reaeration coefficients, both in the laboratory and in the field, and to obtain even more elegant analytical solutions to the original differential equation embellished with a variety of conceptual complications that arose from an increased awareness of the behavior of the real system. A considerable amount of literature grew out of these efforts, which are summarized in Chapter 6. Yet, until the advent of modern computational hardware and techniques after the Second World War, the Streeter–Phelps model was seriously restricted in its usefulness. The necessary requirements for closed-form analytical solutions precluded treating many cases that were considered essential to realistic description of the water body.

---

† Or "science," if the reader prefers.

Development of the computer and the mathematical techniques that accompanied it, especially numerical methods for solving previously intractable problems, quickly effected, *inter alia*, a change in water resource technology. In the relatively short span of five years or so in the late 1950s, methods were developed for solving large sets of simultaneous algebraic equations and finite-difference representations of more complex linear and nonlinear differential equations. Members of the engineering professions quickly recognized the potential of the methods to reduce the burden of work and to expedite solutions to the more difficult problems posed by a society that was becoming exponentially more complex.

In this same period the rise in public concern for protection and enhancement of the aquatic environment resulted in large expenditures for pollution control through advanced wastewater treatment technology. Seeking to evaluate the benefits of such measures in terms of the quantitative response of the aquatic environment, engineers began to draw on the computer sciences for tools to solve traditional problems, like that posed in the Streeter–Phelps formulation.

Among the first models of this new era in water management was the Delaware Estuary Comprehensive Study model, formulated by Thomann (1963) and his associates for the Federal Water Pollution Control Administration in the United States. The DECS model was an important extension of the classical Streeter–Phelps equation to a case where multiple waste loads of varying strengths were distributed along a narrow estuary of nonuniform cross section, and where the rates of biodegradation and reaeration could be expected to vary spatially and temporally with hydrological conditions. The uniqueness of the model lay in its capability for rapid, quantitative assessment of alternative strategies for pollution control in terms of water quality (of dissolved oxygen in this example). This capability was only realizable because of advances in computer hardware, which have continued unabated.

1.3. GROWTH OF WATER QUALITY MODELING

In the late 1960s the mounting public pressure in the United States for control of pollution stimulated investment in a host of special studies to find the best alternative ways of protecting the aquatic environment. The computer and the computer sciences, especially mathematical modeling, were stressed by zealous practitioners of the modeling art as aids to decision makers. Encouraged by even the modest success of the DECS experience and a few others, governmental agencies began to invest heavily in this new technology.

During this period the comparatively simple Streeter–Phelps model of dissolved oxygen concentration and biochemical oxygen demand (DO–BOD) appeared in a variety of computerized forms. The model DOSAG developed by the Texas Water Development Board (1970) solved the steady state problem for a multisegment river system, providing flexibility in dealing with variable deoxygenation coefficients. QUAL I, also produced for the Texas agency (Masch and Associates, 1970), simulated stream temperature as well as DO and BOD, allowing temperature adjustments in rate coefficients to be made internally during simulation. An extended version, QUAL II, produced for the US Environmental Protection Agency (Water Resources Engineers, Inc., 1973), included the capability to simulate more complex stream systems for both steady and unsteady flow and to evaluate the impacts of nutrient loading on the oxygen resources of the stream. Ecological interactions, including photosynthesis and primary productivity, have been the subjects of the most recent work on deterministic modeling of streams (Kelly, 1975; Stehfest, 1977).

Mathematical modeling of lakes was stimulated by two environmental concerns: the impact of reservoir releases on downstream water quality, and the nutrient enrichment of impoundments by tributaries. Applying the principle of heat energy conservation, Water Resources Engineers, Inc. (WRE, 1968) and a group at the Massachusetts Institute of Technology (MIT: Huber *et al.*, 1972) developed models of the reservoir as a one-dimensional system of horizontal slices to simulate the vertical distribution of heat in the impoundment over an annual cycle. These models were extended to consider other quality constituents, including DO, BOD (Markofsky and Harleman, 1973), nutrients, and even biota (Chen and Orlob, 1975), and to predict the quality of downstream releases from operating reservoirs. To obtain better agreement between model and impoundment, various improvements were made: by representing inflow and withdrawal more realistically, by better formulation of heat exchange at the air–water interface (Tennessee Valley Authority, 1972), and by accounting for detention time within volume elements (Fontane and Bohan, 1974). In addition to the finite-difference explicit and implicit formulations used in the WRE and MIT models, the one-dimensional lake model was expressed in finite-element form by Baca and Arnett (1976).

The eutrophication of lakes such as Zurich See and Lake Erie drew the attention in the late 1960s of biological researchers, who proposed fairly simple models of nutrient enrichment of lakes (Vollenweider, 1969). They used the models to focus on gross nutrient balances in two-layer (epilimnion–hypolimnion) systems and to explore the cause–effect relationships between carbon, nitrogen, and phosphorus inputs and primary productivity (O'Melia, 1974; Imboden, 1974; Vollenweider, 1975). In a somewhat parallel modeling activity stimulated by the International Biological Program, more intricate ecological models of lakes, like CLEANER (Park *et al.*, 1974), were developed. Initially, these modeling efforts concentrated on ecological correctness and less attention was given to the physical behavior of the lake system, e.g. circulation, mixing, and stratification. More recently, this generation of models has been adapted, as in the one-dimensional temperature models and LAKECO (Hydrologic Engineering Center, 1974), to multisegment systems, for example, MS CLEANER (Leung *et al.*, 1978). The ecological conceptualizations of such models were the subject of continuing research and improvement throughout the 1970s (Straškraba, 1973; Jørgensen, 1976).

Modeling of large lakes, where transport processes are often governed by wind stresses, was necessarily concentrated first on representing circulation and mixing phenomena. The models evolved from the comparatively simple, single-layer storm surge predictors (Platzman, 1963) to multilayer, multisegment models based on the phenomenological equations of motion in two and three dimensions (Simons, 1973). Applications of these models to large wind-driven lakes like Lake Ontario and Lake Vänern (Simons *et al.*, 1977) have provided essential input for modeling the water quality of such systems.†

Notwithstanding the limits to simulation of circulation processes in large lakes, model developers have not hesitated to tackle problems of describing the water quality and ecological behavior of such water bodies. Given a capability to describe the principal components of wind-driven circulation in a compatible hydrodynamic model, Lam and Simons (1976) devised an advection–diffusion water quality model of Lake Erie. Patterson *et al.* (1975) adapted a link–node estuary model to Green Bay in northeastern Lake Michigan, driving it with circulation derived from a two-dimensional, orthogonal-mesh, finite-difference model patterned after that of Leendertse (1967). Di Toro *et al.* (1975) developed a seven-segment–seven-constituent Lotka–Volterra type of primary production model for western Lake Erie. A so-called "comprehensive" model for Lake Ontario was developed by Chen *et al.* (1975), who used an approach similar to that employed in LAKECO (Chen and Orlob, 1975). It was driven by a three-dimensional circulation derived from a finite-difference hydrodynamic model and employed 15 state variables ranging from nutrients to fish. The model conceptualized Lake Ontario as a system of seven layers over the maximum depth and included more than 200 segments.

1.4. PROLIFERATION OF MODELS AND THE NEED FOR TECHNOLOGY TRANSFER

In the late 1970s the technology of modeling, at least that of water quality, reached a point where advancement seemed to depend more on the availability of reliable data from the field than on the ingenuity of the modeler or on the

† Parallel development of models of the hydrodynamics of shallow estuaries and coastal seas has contributed greatly to the technology of modeling large and complex water bodies.

computer. The focus shifted toward implementation of the model as a tool, requiring that it be calibrated and validated against the water body. It was necessary to determine the sensitivity of the model in terms of realistic measures of the system response, e.g. dissolved oxygen and phytoplankton production, if the model was to be accepted as an adjunct to the decision-making process.

It was clear, also, that there had been a proliferation of models, many of which were merely instruments of research, often ends in themselves and of little value in the management of aquatic resources. A redundancy was apparent. While models were usually original, reflecting the "artistic" abilities of the designers, there was a tendency to "reinvent the wheel" rather than to make true advances in modeling. Moreover, there was a wide disparity in documentation, a step in the model development process that is absolutely essential if the utility of the model is to be realized by other than the developer. Transfer of modeling technology had been sadly neglected in the rush to develop.

It is in the nature of modeling, as it has been and is currently practiced, that the documentation necessary for effective technology transfer rarely appears in the literature. A survey of reference sources conducted by the author in 1976 revealed that, of some 400 documents dealing with mathematical modeling of surface water impoundments, about 90 percent were of limited circulation, e.g. technical reports, conference proceedings, and technical memoranda. In addition, the volume of information of relevance to a topic like water quality modeling is too large for one to digest without proper organization and dissemination.

In early 1976 the author proposed the development of a centralized clearing house, or "register," for the coordination and transfer of technology related to computer software for planning and managing water resources. The concept was presented to IIASA through Professor Dr. Z. Kaczmarek and developed further by the author while in Laxenburg as a visiting scientist in May 1976. On this occasion a project within the scope of the IIASA state-of-the-art survey was identified, outlines were prepared for a series of monograph on water quality modeling, and a committee was formed to prepare this volume on *Streams, Lakes, and Reservoirs.*

During late 1976, under the sponsorship of the Technology Transfer Program of the Office of Water Research and Technology (OWRT), US Department of the Interior, the author studied the feasibility of the proposed software register. The work included a survey, through 1976, of the state of the art of mathematical modeling of surface water impoundments, the development of a format for documentation of software for a national register, an assessment of user potentials, and a determination of the feasibility of an institutional structure for technology transfer. A two-volume report comprising a narrative description and review of capabilities of the principal useful models identified in the survey, a bibliography of some 400 pertinent references, and specifications for about 90 working models was presented to OWRT (Orlob, 1977). The report served as a source document for the present work, which was greatly expanded to provide the necessary breadth of international coverage.

1.5. SCOPE OF THIS VOLUME

The original prospectus for consideration by IIASA discussed a series of monographs dealing with water quality modeling, each encompassing a part, or several parts, of the aquatic environment. This first volume is limited to modeling water quality in streams, lakes, and reservoirs. However, simply because it is the introductory volume and should set the tone for future works in the same subject area, it gives special attention to certain topics of a general nature, e.g., those concerned with the modeling process itself and with the basic principles of modeling water quality. Overall, however, the emphasis is on models, especially those that are of demonstrable utility in the management of water quality and are well documented, i.e. in a suitable condition for effective technology transfer.†

Having summarized our work, it is as well to indicate how we expect the reader to use this book, before we proceed with the details. We have tried to assemble and to describe carefully the best examples of the modeling art. Aside from the chapters of general interest to modelers, i.e. the modeling process and basic principles, each chapter presents at the outset a summary of historical highlights. Each chapter tends to follow the general sequence by which models are developed, i.e. conceptualization, formulation, computational representation, solution, calibration, validation, and application. Because our bias is toward truly useful models, we have tried to give concrete examples of their applications, normally showing comparisons between simulated and observed measures of response. Each chapter concludes with a candid summary, by each author, of the capabilities and limitations of models in the category discussed.

Chapter 2, prepared by Dr. Bruce Beck, describes the modeling process in a stepwise fashion. It begins with some essential basic definitions of technical terminology used by modelers. The procedure whereby the essence of an aquatic environment is captured conceptually, described in mathematical form, and translated into the language of computation is outlined. The nature and importance of calibration, validation, and sensitivity testing are emphasized, although a detailed exposure of these topics is deferred to Chapter 11.

Chapter 3, by Professor Peter Mauersberger, lays the foundation for deterministic water quality modeling with a comprehensive treatment of the

† This work does not really present the "state of the art" as such, but merely the best effort to capture the essence of mathematical modeling of water quality as experienced by the individual authors.

general physical, chemical, and biological principles. The coverage is mathematically rigorous, yet the presentation emphasizes the general rather than the special case.

Chapter 4, written by Professor S. E. Jørgensen, follows somewhat the pattern of Chapter 3, in that basic ecological principles are identified, but it endeavors as well to provide a broad coverage of the bases for translating these principles into mathematical terms.

Chapter 5, by Dr. J. Jacquet, examines the modeling of the thermal regime in streams. It includes the effects of multiple heat sources, diurnal heating and cooling, longitudinal and lateral mixing, stratification, and the stochastic nature of the heat exchange process. The appendix includes the most current quantitative statement of the component heat fluxes at the air–water interface.

Chapter 6, prepared by Dr. M. J. Gromiec, Professor D. P. Loucks, and the editor, deals with the classic problem of simulation of water quality in streams, ranging from the historical foundation in the Streeter–Phelps equation to the more elaborate predictive models, like QUAL II, that deal with a full range of biological parameters. The appendices summarize the extensive research on formulation of deoxygenation and reaeration rates.

Chapter 7, by the editor, reviews the development of one-dimensional models of water quality in stratified impoundments. The chapter presents the basic formulation for the one-dimensional temperature model and the implicit solution technique for the resulting set of energy balance equations. Extension of the model to treat multiple quality and biological state variables is described. Illustrative examples are given of simulation of temperature and dissolved oxygen in stratified reservoirs.

Chapter 8, by Dr. M. Watanabe, Professor D. R. F. Harleman, and Professor O. Vasiliev, reviews the literature in modeling the hydrodynamic and water quality processes of two- and three-dimensional lake systems. Formulations based on the Navier–Stokes equations are given, solution techniques are described, and the interrelationships between hydrodynamics and water quality are emphasized. Illustrative examples of two- and three-dimensional modeling of water quality in large lakes are presented. A special section deals with the unique problem of thermally stratified plumes and cooling ponds.

Chapter 9, by Professor S. E. Jørgensen, provides a comprehensive examination of techniques for modeling the ecology of lakes and reservoirs. A discussion is presented of the many variants in the formulation of interactions between nutrients, biota, and environmental parameters that govern rates of growth, respiration, mortality, etc. The performance of ecological models is demonstrated by examples selected from actual case studies.

Chapter 10, also by Professor Jørgensen, presents the latest experience in modeling the fate of toxic substances and evaluating their effects on aquatic biota.

Chapter 11, prepared by Dr. Bruce Beck, is an extension of material presented in Chapter 2 on the modeling process, dealing specifically with sensitivity analysis, calibration, and validation. The theoretical bases for these essential steps in modeling the aquatic environment are outlined. Illustrative examples are presented.

Chapter 12, by Professor D. P. Loucks, describes the use of models in water quality management. Methods are presented for the evaluation of alternative wastewater treatment methods in relation to management objectives and quality standards. The role of deterministic water quality models of streams and lakes is discussed.

The final chapter, by the editor, summarizes the status of mathematical modeling of water quality of streams, lakes, and reservoirs as this group of authors has presented it. In particular, we try to highlight the major practical accomplishments of the modelers and to bring the real limitations of models to the attention of potential users. Some of the more notable deficiencies in water quality modeling at present are identified, and we attempt, with some trepidation, to identify the most promising avenues for future research and development.

REFERENCES

Baca, R. G. and R. C. Arnett (1976). A finite-element water quality model for eutrophic lakes. *Finite Elements in Water Resources* vol. 4, eds. W. G. Gray, G. Pinder, and C. Brebbia (Plymouth, Devon: Pentech), pp. 125–147.

Chen, C. W., M. Lorenzen, and D. J. Smith (1975). A comprehensive water quality–ecologic model for Lake Ontario. *Report to Great Lakes Environmental Research Laboratory, National Oceanic and Atmospheric Administration by Tetra Tech, Inc.*, 202 pp.

Chen, C. W., and G. T. Orlob (1975). Ecologic simulation of aquatic environments. *Systems Analysis and Simulation in Ecology* vol. 3, ed. B. C. Patten (New York, NY: Academic Press), ch. 12, pp. 475–528.

Di Toro, D. M., D. J. O'Connor, R. V. Thomann, and J. L. Mancini (1975). Phytoplankton–zooplankton–nutrient interaction model for western Lake Erie. *Systems Analysis and Simulation in Ecology* vol. 3, ed. B. C. Patten (New York, NY: Academic Press), ch. 11, pp. 423–474.

Fontane, D. G., and J. P. Bohan (1974). Richard B. Russell Lake water quality investigation. *Waterways Experiment Station, Vicksburg, MS Technical Report* H-74-14, 26 pp.

Huber, W. C., D. R. F. Harleman, and P. J. Ryan (1972). Temperature prediction in stratified reservoirs. *Proceedings of American Society of Civil Engineers, Journal of Hydraulics Division*, Paper 8839, pp. 645–666.

Hydrologic Engineering Center (1974) (revised 1978). WQRRS, generalized computer program for river–reservoir systems. *US Army Corps of Engineers, HEC, Davis, CA User's Manual* 401-100, 100A, 210 pp.

Imboden, D. M. (1974). Phosphorus model for lake eutrophication. *Limnology and Oceanography* 19:297–304.

Jørgensen, S. E. (1976). A eutrophication model for a lake. *Journal of Ecological Modeling* 2(2): 147–165.

Kelly, R. A. (1975). The Delaware Estuary Model. *Ecological Modeling in a Resource Management Framework*, ed. C. S. Russell. *Resources for the Future Working Paper* QE-1 (Baltimore, MD: Johns Hopkins University Press), pp. 103–134.

Lam, D. C. L., and T. J. Simons (1976). Numerical computations of advective and diffusive transports of chloride in Lake Erie, 1970. *Journal of Fisheries Research Board of Canada* 33: 537–549.

Leendertse, J. (1967). Aspects of a computational model for well mixed estuaries and coastal seas. *Rand Corporation, Santa Monica, CA Research Memorandum* 5294-PR.

Leung, K. K., R. A. Park, D. J. Desormeau, and J. Albanese (1978). MS CLEANER—An overview. *Proceedings of Conference on Modeling and Simulation, Pittsburg, PA* (Troy, NY: Center for Ecological Modeling, Rensselaer Polytechnic Institute), 8pp.

Markofsky, J., and D. R. F. Harleman (1973). Prediction of water quality in stratified reservoirs. *Proceedings of American Society of Civil Engineers, Journal of Hydraulics Division* 99(HY5), Paper 9730, pp. 729–745.

Masch, F. D., and Associates (1970). QUAL I, simulation of water quality in streams and canals, program documentation and user's manual. *Report to Texas Water Development Board, Austin, TX.*

O'Melia, C. R. (1974). Phosphorus cycling in lakes. *North Carolina Water Resources Research Institute, Raleigh Report* 97, 45 pp.

Orlob, G. T. (1977). Mathematical modeling of surface water impoundments, vols. I and II. *Report to Office of Water Research and Technology by RMA, Inc., Lafayette, CA*, 119 pp., appendix, and references.

Park, R. A., D. Scavia, and N. L. Clesceri (1974). CLEANER, the Lake George model. *US International Biological Program, Eastern Deciduous Forest Biome Contribution* 184, 13 pp.

Patterson, D. J., E. Epstein, and J. McEvoy (1975). Water pollution investigations: Lower Green Bay and Lower Fox River. *Report to US Environmental Protection Agency*, No. 68-01-1572, 371 pp.

Platzman, G. W. (1963). The dynamic prediction of wind tides on Lake Erie. *Meteorology Monographs* 4(26), 44 pp.

Simons, T. J. (1973). Development of three-dimensional numerical models of the Great Lakes. *Canada Center for Inland Waters, Burlington, Ontario, Science Series* No. 12, 26 pp.

Simons, T. J., L. Funkquist, and J. Svensson (1977). Application of a numerical model to Lake Vänern. *Swedish Meteorology and Oceanography Institute Report* RH09, 15 pp.

Stehfest, H. (1977). Mathematical modeling of self-purification of rivers. *Professional Paper* PP-77-11 (Laxenburg, Austria: International Institute for Applied Systems Analysis).

Straškraba, M. (1973). Limnological models of reservoir ecosystems. *Proceedings of International Symposium on Eutrophication and Water Pollution Control, Castle Reinhardsbrunn, DDR* (Biology Society, DDR).

Streeter, H. W., and E. B. Phelps (1925). A study of the pollution and natural purification of the Ohio River, III. Factors concerned in the phenomena of oxidation and reaeration. *US Public Health Service, Washington, DC Bulletin* 146, 75 pp.

Tennessee Valley Authority (1972). Heat and mass transfer between a water surface and the atmosphere. *TVA Water Resources Research Engineering Laboratory, Norris, TN Report* 14, 123 pp.

Texas Water Development Board (1970). DOSAG-I simulation of water quality in streams and canals. *Program documentation and user's manual.*

Thomann, R. V. (1963). Mathematical model for dissolved oxygen. *Proceedings of American Society of Civil Engineers, Journal of Sanitary Engineering Division* 89(SA5): 1–30.

Vollenweider, R. A. (1969). Möglichkeiten und Grenzen elementarer Modelle der Stoffbilanz von Seen. *Archiv für Hydrobiologie* 66: 1–36.

Vollenweider, R. A. (1975). Input–output models. *Schweizerische Zeitschrift für Hydrologie* 37(1): 53–82.

Water Resources Engineers, Inc. (1968). Prediction of thermal energy distribution in streams and reservoirs. *Report to California Department of Fish and Game*, 90 pp.

Water Resources Engineers, Inc. (1973). Computer program documentation for the stream quality model QUAL II. *Report to US Environmental Protection Agency*, 90 pp.

2 A Procedure for Modeling

M. B. Beck

2.1. INTRODUCTION

It is difficult to claim that one has available a universal systematic procedure for model development. This chapter certainly makes no such claim. We are concerned throughout this book with the state of the *art*, and art indeed is the operative word in the matter of characterizing the nature of the modeling procedure. Most analysts engaged in the development of models for the quality of water bodies have assembled their own unique collections of techniques for, and attitudes toward, model building. In this chapter we shall suggest a particular sequence in the modeling procedure; it is not expected that such a sequence is one to which rigid adherence is demanded or even possible. More important, rather, we shall thereby cover the majority of problems associated with model development. At this stage in the book our attention is merely to outline in qualitative and nonmathematical terms the principal features of each step in the modeling procedure. In Chapter 11, a more technical and detailed discussion of the same topics will be given: in particular, case studies will be reported as examples of the application of methods of sensitivity analysis, calibration, and parameter estimation. Chapter 11 will also illustrate the problem of model validation.

It is first necessary to define the overall problem that motivates the present discussion of the modeling procedure. This chapter intends, therefore, to answer the questions: how do we derive from the ensemble of general theory a set of relationships, i.e. the model, for a *specific* water body; and how do we demonstrate the good or poor approximation of the behavior of that model to the observed behavior of reality? The two questions also define the way in which the modeling procedure divides into two parts. Up to section 2.5 our procedure describes the selection and combination of the various expressions required for the specific model *before* any field experiment has been used to test the validity of the model. From section 2.6 onward the discussion attends to the matter of model modification *after* the collection of field data; in other words, the latter half of the chapter discusses system identification, model calibration, and validation.

The "before" and "after" character of the modeling procedure refers additionally to the way in which we might view the *a priori* and *a posteriori* knowledge of the behavior of a system. Figure 2.1, adapted from Eykhoff (1974), gives an example of the relationship between the two types of knowledge. The plentiful sources of error in both our structural and measured knowledge of a

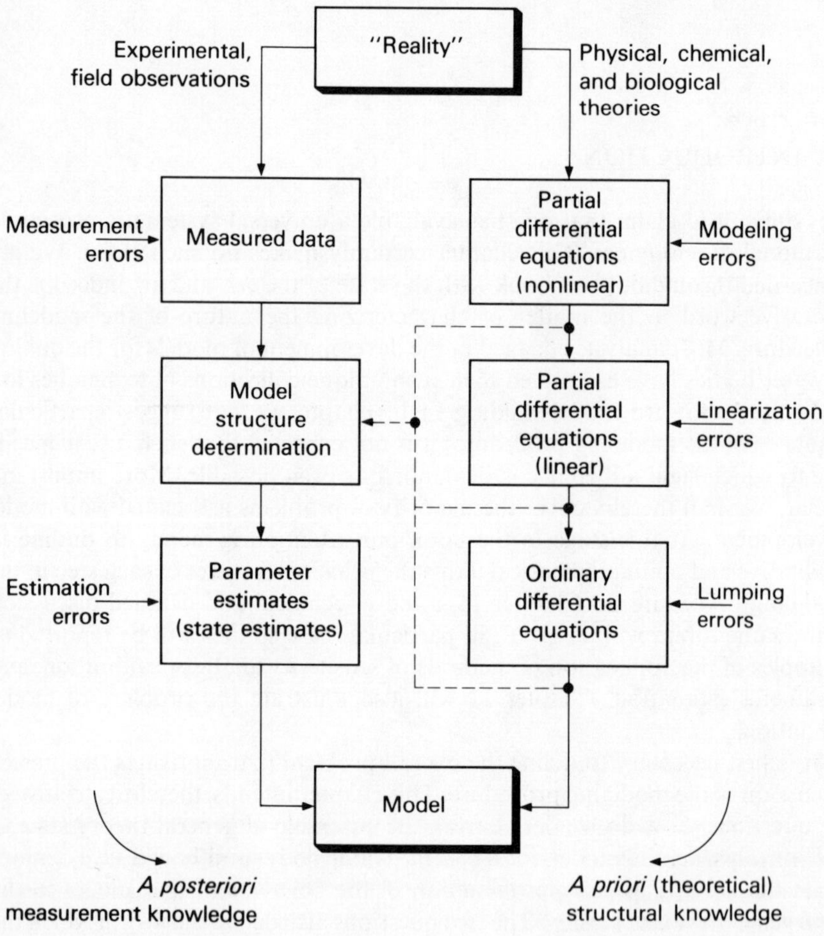

FIGURE 2.1 Combining *a priori* and *a posteriori* knowledge in the modeling procedure (adapted from Eykhoff, 1974).

process should emphasize the fact that models are at best working hypotheses about the nature of reality. There is never a good reason for overconfidence in the performance of a model. Again, we can use the dual nature of Figure 2.1 and the modeling procedure to illustrate two frequently polarized approaches to modeling. The first, associated with *a priori* structural knowledge, is a deductive reasoning approach: from an existing general theory we deduce the model relationships for the specific case study. The second, associated with *a posteriori* measurement knowledge, is an inductive reasoning approach: assuming no *a priori* knowledge (theory) of process behavior, we attempt to develop the specific information acquired from the particular sample set of data into a more general model. The only valid reason for relying upon one of these approaches alone may be the inevitable difficulty of obtaining experimental data. We shall see aspects of this duality reflected throughout the chapter; they have also been treated in a recent review by Somlyody (1981). Figure 2.2 shows the sequence of the modeling procedure.

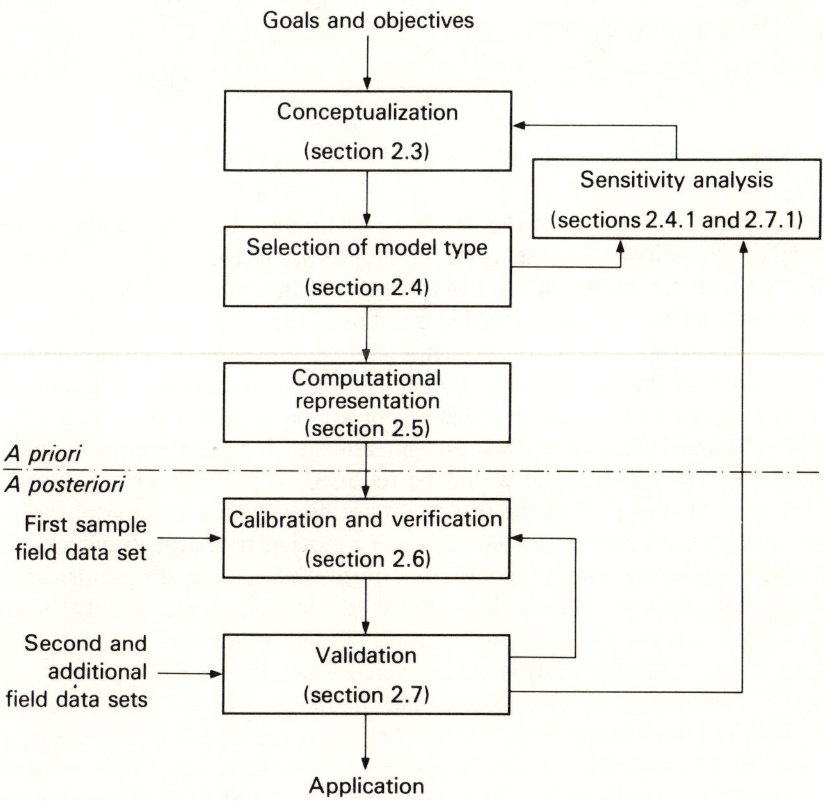

FIGURE 2.2 An outline of the modeling procedure and organization of the chapter.

There are five basic steps in the sequence:

(i)	conceptualization	(section 2.3)
(ii)	selection of model type	(section 2.4)
(iii)	computational representation	(section 2.5)
(iv)	calibration and verification	(section 2.6)
(v)	validation	(section 2.7).

Figure 2.2 indicates also some of the primary feedback, or iterative, loops in the modeling procedure. It will thus be observed that sensitivity analysis has been assigned a position of importance both at an early *a priori* stage and at a later *a posteriori* step in two of the feedback loops. Like the rest of the diagram, this location for sensitivity analysis does not have to be interpreted in any strict sense. There are as many variants on the theme of Figure 2.2 as there are authors on the subject of modeling. Much of what follows will be a distillate of ideas from the works of Orlob (1975), Jørgensen (1978), Young (1978), Rinaldi *et al.* (1979), and Beck (1979).

2.2. DEFINITION OF GOALS AND OBJECTIVES

Though perhaps not part of the modeling procedure itself, a common theme among the majority of statements on the subject of this chapter is that the goals and objectives for model application determine the nature of the model. As pointed out in Chapter 1, our attitude is not one of seeking a universal model to solve, in general, all manner of problems. For example, the definition of objectives might state that the model is intended as a guide to long-term planning by determining the year-to-year average response of a river system to patterns of population and industrial growth and movement. It is clearly not sensible to suggest that the same model would be required to fulfill these objectives as would be needed to predict the probability of intermittent stream deoxygenation resulting from the diurnal variations of a particular sewage discharge.

Two broad categories of goals and objectives can be distinguished. In a *research* context the model has to provide indicators for further fruitful directions of investigation. An awareness of the immediate use for the model is not necessary before the study is undertaken. Rather, the concise representation of *a priori* and measured information that the model offers and the possibility for a gain in comprehension (of system behavior) are of primary importance. An essentially research-oriented model may, nevertheless, be used to make forecasts about the probable future behavior of the system. In a *management* context the immediate application of the model must be known and carefully specified. "Management" of water quality has traditionally been understood to mean long-range planning, the design of treatment facilities, or the problem of legislation for discharge consents and standard setting. Applications of models to such problems will be discussed in Chapter 12. The term management, however, may also include the design and operation of real-time control and forecasting systems. Models used for this purpose in water quality management are at present of lesser significance and are, therefore, not treated in detail in this book. The reader will find more suitable accounts of these kinds of models by Young and Beck (1974), Rinaldi *et al.* (1979), and Beck (1980).

2.3. CONCEPTUALIZATION

Let us assume that we have chosen an objective and have a specific water body in mind. The first step of the modeling procedure is conceptualization. At this point the analyst is interested in how the physical system is represented in three-dimensional space. It is of some importance, for instance, to know the locations of control structures and tributaries along a river, or to know whether various portions of a lake can be considered to be essentially deep or shallow. Usually conceptualization will involve a choice regarding the possible (spatial) segregation of the water body into a number of discrete segments and layers. Besides a spatial separation of the water body it may be necessary to include a grouping and differentiation of biotic species according to how one visualizes their roles in the ecology of the water body. For example, a particular species of phytoplankton with high nuisance value might best be considered in terms of a component mass balance that is separate from a component mass balance for all other phytoplankton species. In addition, though perhaps less obvious, it might occasionally be appropriate to partition the temporal dimension into ranges of quickly changing and slowly changing variations. A classic problem of this kind is the matter of coupling, or decoupling, the hydrophysical and ecological sectors of a lake system model (Jørgensen and Harleman, 1978; Jørgensen, 1979). There is, of course, no guarantee that these choices of spatial, temporal, and ecological aggregation/decomposition will turn out to be correct at later stages of the modeling procedure. Moreover, as others have argued eloquently elsewhere (e.g. Young, 1978), we should bear in mind that the decomposition of a complex environmental system for individual analysis of its component parts does not always imply that subsequent reassembly will yield a characterization of the behavior of the whole.

With conceptualization of the modeling problem comes also model formulation. The relevant variables for description of the desired water quality characteristics and expressions for their interaction have to be chosen. The system definition will have been completed so that in addition we can define in abstract terms the groups of quantities shown by Figure 2.3. In physical terms we have the following groups of variables.

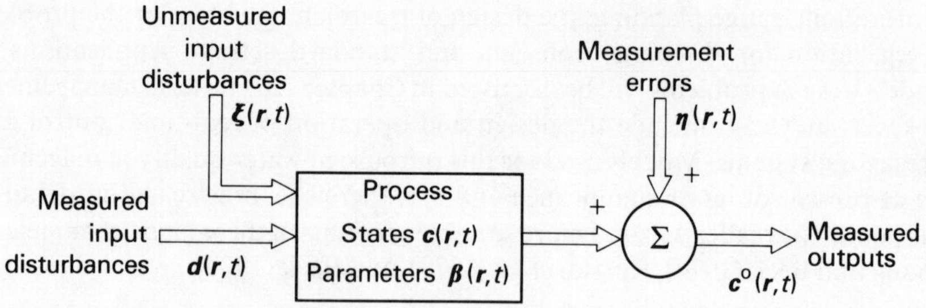

FIGURE 2.3 Definition of the system and variables.

(i) The group of variables denoted by the vector $\boldsymbol{d}$, *measured input disturbances* (or forcing functions), might comprise the recorded day-to-day variations of total biochemical oxygen demand (BOD), suspended solids, and ammonia nitrogen concentrations in a treated sewage discharge to a river.

(ii) The group of variables denoted by $\boldsymbol{\xi}$ represents *unmeasured* (*unknown*) *input disturbances*. These might include items such as yearly rate of organic phosphorus loading to a lake system from diffuse nonpoint sources. A predominant characteristic of the unmeasured inputs is that they will generally be expected to exhibit a random variability.

(iii) *The process state variables* $\boldsymbol{c}$ characterize the essential properties and behavior of a process (or system) as functions of space and time. Previously, in discussions of ecological models, the term compartment variable has been used frequently to identify state variables. We have adopted the notation $\boldsymbol{c}$ because for the most part in this book the state of water quality is described by a vector of component concentration variables.

(iv) The group of variables $\boldsymbol{c}^o$ is defined as *measured output variables*. In fact, by and large these variables are merely measurements of some of the process state variables. However, it is not difficult to imagine a situation in which $\boldsymbol{c}^o$ is not so straightforwardly related to $\boldsymbol{c}$: for example, a particulate phosphorus measurement would include phosphorus bound in algal cells, in zooplankton, and in detrital material, with all three types of phosphorus binding possibly being defined as model state variables.

(v) The last category of variables, $\boldsymbol{\eta}$, represents the random and systematic *measurement errors* that derive from process instrumentation and laboratory analysis; such errors are inherent in all measurements $\boldsymbol{c}^o$, thus prohibiting the possibility of $\boldsymbol{c}^o$ being an absolutely exact measure of $\boldsymbol{c}$.

One further group of variables in Figure 2.3 remaining to be discussed is the model *parameters* $\boldsymbol{\beta}$: for instance, the reaeration rate coefficient or chemical kinetic rate constants that appear in the equations of the system model. A highly desirable property of these parameters is that they should be invariant with respect to time and space, i.e. truly constant. This desirable property will be shown to be an extremely important feature of certain aspects of model development and analysis. Yet we know that the value of a reaeration rate constant varies with stream discharge and it is quite common that a BOD decay rate coefficient will also be different for different reaches of a river. One is thus led to question how constant a parameter ought to be. Indeed, the difference between a quantity that is a state variable and a quantity that is a parameter becomes almost negligible when one considers a state variable that does not vary with time, i.e. when part of the system is at steady state, or a parameter that displays seasonal, and therefore temporal, fluctuations. To attempt to preserve the difference between the notions of state and parameter is actually not particularly useful in the later discussion of the model calibration problem. Perhaps an ambivalent attitude toward the distinction is desirable: sometimes the difference between state and parameter is important, and sometimes it is not. For these kinds of reason the parameters are specified as functions of (three-dimensional) space $\boldsymbol{r}$ and time t in Figure 2.3; all other groups of variables discussed above are likewise defined. The parameters are, in fact, implicitly functions of other variables, such as temperature and stream flow, and hence they are functions of time and space.

The values of the parameters may not necessarily be known precisely at the stage of conceptualization and formulation; it is more likely that the values can only be said to lie within a certain range according to previously published results. The principal objective of formulation is the specification of causal relationships that describe the phenomena thought to govern the behavior of the system. In other words, this is what we might call *a priori model structure determination.* We therefore note the complementary activity of *a posteriori* model structure identification, which will be discussed in section 2.6.3.

2.4. SELECTION OF MODEL TYPE

We have distinguished between categories of variables. It is also useful to distinguish between various types of model and to discuss briefly their characteristics. In section 2.2 we have actually made a division of models into the groups of:

(i) "research" or "management" models.

Furthermore, Figure 2.1 allows us to classify the following pairs of model types:

(ii) distributed or lumped models;
(iii) nonlinear or linear models.

Some explanation is needed. A distributed model or, commonly, *distributed-parameter model*, is one in which variations of all the quantities in Figure 2.3 are considered to be continuous functions of time and space. This form of model arises rather naturally in the analysis of water bodies, the most familiar example being the advection–diffusion model for the transport of a dissolved substance along a stream. Several other examples of distributed-parameter models will be found in Chapter 8.

A distributed-parameter model accounting for variations in the three orthogonal directions (x, y, z) is probably the most accurate form of model that one might use to describe the behavior of quality in a water body. It is also the most difficult form of model to solve. Frequently, however, the analyst may decide on the basis of prior experimental observations that horizontal gradients of dissolved material in a lake, for example, are not sufficiently large to merit inclusion in the model. Thus let us suppose that we "lump" together parts of the system description so that for certain finite volumes of the water body, or within certain bounded spatial locations, water quality is assumed to be uniform and independent of position *within the defined volume*. A distributed-parameter model would then be reduced by that assumption to a *lumped-parameter model*. A typical example of this kind of model is the continuously stirred tank reactor (CSTR) idealization of lake water quality dynamics. Whereas the lumped-parameter model is frequently expressed in ordinary differential equation form, the distributed-parameter model is usually defined by partial differential equations.

In general, the distributed or lumped models are *nonlinear models*. A special case of the general class of nonlinear models is the *linear model*. Almost always the analyst strives to obtain a linear system model because of the many powerful techniques available for comprehensive analysis of such models. The great advantage of the linear model is that it obeys the principle of superposition. Hence it is possible to say that, for instance, if the model output response O_A is related to the input forcing function I_A, and likewise O_B is related to I_B, then the combination of inputs $(a'I_A + a''I_B)$ will produce a model response $(a'O_A + a''O_B)$, where a' and a'' are proportionality constants. Where appropriate we may exploit the linearity of a model in order to separate the *free response* of the system due only to the initial state of the system, i.e. the general solution of the differential equation, from the *forced responses* to the input disturbances, i.e. the particular solution of the differential equation. The Streeter–Phelps model of stream BOD–DO interaction (Streeter and Phelps, 1925) is a linear model; the Lotka–Volterra model of predator–prey interaction is a typical nonlinear model.

Another pair of complementary model types may be defined as:

(iv) stochastic or deterministic models.

A *stochastic model* of system behavior would reduce to a *deterministic model* provided that the stochastic input disturbances $\boldsymbol{\xi}$ and the random measurement errors $\boldsymbol{\eta}$ are assumed to be zero, i.e.

$$\boldsymbol{\xi}(\boldsymbol{r}, t) = \mathbf{0} \qquad \text{and} \qquad \boldsymbol{\eta}(\boldsymbol{r}, t) = \mathbf{0}$$

for all $\boldsymbol{r}$, t in Figure 2.3, and provided that the parameters are known exactly (as opposed to being estimates stated in terms of statistical distributions). It is worth remembering that the assumption of a deterministic model is tantamount to the assumption that one has perfect knowledge of the behavior of a system. In other words, the future response of the system is completely determined by a knowledge of the present state and future measured input disturbances. Alternatively, it has been said that (Papoulis, 1965) "probabilistic considerations are necessary only because of our ignorance," a profound statement indeed if we infer from it that all natural phenomena are deterministic. Stochastic models are not treated in depth in this book—except for their application in stream temperature prediction (Chapter 5)—and thus the reader is referred elsewhere for the development of such models (e.g. Loucks and Lynn, 1966; Thayer and Krutchkoff, 1967; Padget *et al.*, 1977; Tiwari *et al.*, 1978; Finney *et al.*, 1982).

Our next classification of models, that between

(v) dynamic or steady state models,

makes one of the most significant distinctions that can be drawn, since it often furnishes a dividing line in choosing the type of model best suited to particular problems. Usually the analyst will be interested in approximating a distributed-parameter description to a *steady state model* by assuming all variables and parameters to be independent of time t. A classical example of the use of a steady state model is that in which the average spatial variations of quality in a river system are computed for an average time-invariant set of wastewater discharge, temperature, and stream flow rate conditions. Strictly speaking, if all variables but the stream discharge are held constant with time, then one is dealing with a *dynamic model*; a time-varying discharge implies that water quality at any fixed spatial location is not at a steady state. Like the lumped-parameter model, the advantage of the steady state model is its potential for simplifying subsequent computational effort through the elimination of one of the independent variables in the model relationships. A typical dynamic model application is that of examining over a period of years the response of a lake ecological system after installation of nutrient removal treatment at an adjacent wastewater treatment plant.

In the approximation of a lumped-parameter model, where behavior is said to be uniform across a discrete portion of three-dimensional space, and in the approximation of a steady state model, where behavior is effectively said to be uniform over a discrete interval of time, the real questions of judgment are: how great are the variations in time and space of the water quality characteristics relevant to the problem; and how many of those individual variations can be approximated by constants or neglected?

So far we have discussed models that are understood to be characterized by a set of differential equations. Difference equation forms also arise readily in water quality modeling. For instance, in practice it is rather atypical to have available *continuous-time* (analog) measurements of process variables, just as it is improbable to imagine a *spatial continuum* of measurements. Partly for these reasons, namely that we shall often want to compare a model prediction with *discrete-time* (digital, sample) or *discrete-point* measurements, partly because numerical solutions to differential equations are obtained through difference equation approximations, and partly because analytical solutions are sometimes available that relate variables at discrete points of time and space to variables at other discrete points, we shall thus refer frequently to difference equation models.

Last, but not least, we introduce the classification of models into:

(vi) internally descriptive (mechanistic) or black box (input–output) models.

It would be incorrect to suggest that the analyst has to make a choice between one or the other of these model types; it is better to view them as defining the two ends of the spectrum of models (Karplus, 1976). Here again we see the duality of the modeling procedure. An *internally descriptive model* is closely associated with *a priori* information and with a deductive reasoning process; the *black box model* is much more naturally oriented toward *a posteriori* information and inductive reasoning. One might say that the internally descriptive model characterizes *how* the inputs are connected to the states and how, in turn, the states are connected to each other and to the outputs of the system. In contrast, the black box model reflects only *what* changes the input disturbances will effect in the output responses. Of course, the major difference between these two categories of models is that one of them, the internally descriptive model, provides a description of the internal mechanisms, i.e. $\boldsymbol{c}$ and $\boldsymbol{\beta}$, of process behavior. The black box model makes no such explicit reference to what is inside the process block of Figure 2.3; it deals only with what is measurable: the inputs and outputs.

The association of black box models with input–output relationships allows us also to mention the subject of *transfer function* models and *frequency response analysis*. Differential equations may be transformed into algebraic equations by the application of Laplace transforms. It is customary to refer to this as a transformation from time-domain analysis, in which time is the independent variable (in the differential equation), to analysis in the frequency domain. A transfer function model appears then as the ratio of the Laplace-transformed output variable to the Laplace-transformed input variable, i.e. a transfer function can be derived for each pairwise combination of an input and an output variable. When the input forcing variable of the system exhibits sinusoid-type variations (in time) it is possible to observe both input and system output response oscillations and hence determine experimentally the form of the transfer function model. Such a determination of the model is known as frequency response analysis, since in effect one is analyzing the amplitude attenuation and phase shift between input and output sinusoids at given frequencies. Frequency response analysis is in fact a classical form of system identification (section 2.6 and Chapter 11) typically applied in electrical engineering and control engineering systems. Its use in water quality modeling has been limited (e.g. Thomann, 1973), although as a component of systems analysis it offers a perspective that conveniently complements the view provided by the time-domain, state-space approach of this book. For example, it may be quite helpful to decompose the nature of the behavior of a system into responses resulting from essentially low-frequency (slowly changing) input disturbances and responses from essentially high-frequency (rapidly changing) disturbances (e.g. in the problem of experimental design, section 2.6.2).

Two examples will illustrate the use of black box models. Let us suppose that the system inputs and outputs are defined as time variations of quantities at two spatial locations. A model that *directly* relates chlorophyll *a* variations (output) at the center of a reservoir to variations in the phosphorus concentration (input) of the major feed river can be said to be a black box model. Alternatively, let us consider the case where inputs and outputs are defined as time variations of several quantities at a single spatial location. If we wish to predict day-to-day DO variations (output) from a multiple regression relationship using conductivity and temperature variations as inputs, this too would come under our definition of a black box model. We may thus conclude with the following remark. Much of the eventual character of a black box model depends upon how the analyst draws a line (conceptually) around the system—thus separating the system from its environment—and how he chooses to categorize the quantities of interest into groups of variable types (according to Figure 2.3).

2.4.1. *A Priori* Sensitivity Analysis

Even without experimental field data having been collected for model evaluation, certain important questions about the suitability of the model can now be posed. The answers to these questions—questions of *a priori* sensitivity analysis—may lead to a restructuring of the model at the conceptualization stage.

Figure 2.2 shows that sensitivity analysis is part of a feedback loop during both the *a priori* and *a posteriori* phases of the modeling procedure.

The major point that we wish to establish from sensitivity analysis is the relative sensitivity of the model predictions to changes in the values of the model parameters $\boldsymbol{\beta}$. From this analysis we might also be able to infer the sensitivity of the model performance to various modes of behavior associated with the various parameters and to those sectors of the model to which these parameters are attached. A *sensitivity coefficient* s_{ij} for the change Δc_i in the ith state variable of the model in response to a change $\Delta\beta_j$ in the value of the jth parameter may simply be defined as

$$s_{ij} = \frac{\Delta c_i/\bar{c}_i}{\Delta\beta_j/\bar{\beta}_j}. \tag{2.1}$$

We have normalized the relationship by inclusion of a nominal reference value $\bar{\beta}_j$ for the parameter, which would give a nominal reference value $\bar{c}_i$ for the predicted state variable response. The notion of sensitivity analysis has been well known in control engineering since the 1950s (with the work of Bode). Only relatively recently has it been applied to water quality and ecological systems, for example by Rinaldi and Soncini-Sessa (1978), Jørgensen *et al.* (1978), and van Straten and de Boer (1979). In general, Δc_i and $\Delta\beta_j$ are understood as small changes in the neighborhoods of $\bar{c}_i$ and $\bar{\beta}_j$. A definition of the type given by eqn. 2.1 enables the analyst to investigate whether a certain percentage change in a parameter has no real significance, $s_{ij} \approx 0$, whether β_j is a dominant parameter, or whether a small change in β_j induces instability in the model structure.

Two points can now be mentioned briefly in connection with sensitivity analysis. First, if the output response of the model is found to be insensitive to changes in the value of any parameter (i.e. $s_{ij} = 0$), that parameter is said to be *not identifiable*. Such a problem of *parameter identifiability*, which will be discussed further in section 2.6.2, means that it is not possible to estimate certain parameters unless the model relationships are specified in an alternative form. Conversely, if a parameter has a strong influence on a particular output variable, then measurements of that variable will influence the ability to estimate accurate values of the parameter. In Chapter 11 this link between sensitivity coefficients and parameter estimation will be made more explicit. Second, we may note that *stability analysis* and sensitivity analysis are closely related. Both are concerned with the nature of the behavior of the model in the region of some nominal or equilibrium solution. Conditions can be derived for which linear models are globally stable, but for nonlinear models—typical of those encountered in modeling ecological systems—it is usually only possible to state the constraints guaranteeing local, i.e. neighborhood, stability (e.g. Ikeda and Adachi, 1976; Adachi and Ikeda, 1978).

Clearly, sensitivity analysis can yield much insight both into the nature of the model and into the suitability of the model for calibration.

2.5. COMPUTATIONAL REPRESENTATION

It would be fortunate if all water quality models were sufficiently simple to yield either an analytical solution or a solution requiring nothing more than pencil and paper. Closed-form solutions to the model equations are partly becoming the exception rather than the rule, and in part they are becoming increasingly impractical. The discussion in section 2.4 reveals that our analyses are usually strongly tied to the solution of differential equations; and if not differential equations, then difference equations are employed that are readily amenable to the recursive function of the modern digital computer. Hence numerical solution of ordinary differential equations has become so commonplace that the techniques for such solution are a regular feature of standard mathematical texts for scientists and engineers (e.g. Kreyszig, 1972). Because of this facility seldom is little more demanded of the analyst than that he be able to write down the functional forms of the differential equations for subsequent solution by computer program library routines. The analyst would, naturally, be imprudent if he did not check the degree of numerical error in his model solutions.

The case of partial differential equations and their solution is not so straightforward. For this reason a great deal in the character of a model may depend upon how the differential equations are transformed into an approximate set of difference equations. Clearly such considerations of numerical solution are inherent in the earlier choices of lumping together certain groups of spatial variations in the distributed-parameter model and in the choices of specifying which parts of the dynamic behavior of the system can be said to be at a steady state. Due reference will therefore be made from time to time in later chapters to particular methods of distributed-parameter model solution.

The product from the modeling procedure at this point (Figure 2.2) is a model *potentially capable of simulating* a portion, perhaps even a major portion, or "reality." We have come thus to the boundaries of what is justified without using the *a posteriori* measured information of field data.

2.6. CALIBRATION AND VERIFICATION

Let us briefly assess our situation. The *raison d'être* of our book is a consequence of the immense possibilities for complex system simulation created by the advent of electronic computers. We may say that large computers have fostered large water quality models—just as they have fostered large models in every other field of technology and science. But the ability to conduct simulation exercises with large, complex models has in no way necessarily increased our understanding of actual behavior or strengthened the validity of the models as approximations of reality. In this section we shall discuss some procedures for formally testing our models; in Chapter 11 it will become evident that only *rather low-order* (*i.e. small*) *models are amenable to rigorous calibration and verification*. One does not have to search far for an answer to why it is not possible to verify many of the larger water quality models currently available, e.g. those of Chen and Orlob (1975); we simply do not have the facilities to gather all the field data that would ideally be required for model calibration (Beck, 1978a; Jørgensen and Harleman, 1978; Jørgensen, 1979). Yet the answer to the question is also partly bound to the fact that the behavior of water quality in streams and lakes is rarely well defined and is often quite uncertain (Young, 1978; Beck, 1981). We cannot therefore expect to place as much confidence in large water quality models as we might have in a model of a petrochemical plant. Another question is: how, for instance, do errors of estimation in the calibrated model parameter values affect the predictions obtained from the model about future behavior? This topic, which might be termed *a posteriori* sensitivity analysis, will be discussed briefly in section 2.7.1.

Figure 2.4 gives an outline of the calibration and verification phase of the modeling procedure; calibration is here considered as somewhat more complex than straightforward parameter estimation. For our subsequent discussion of

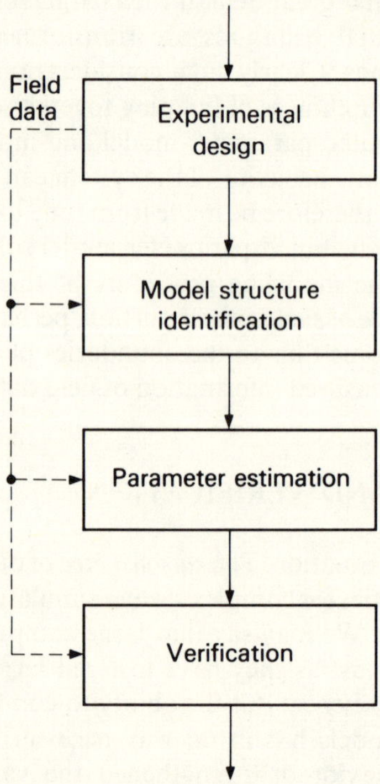

FIGURE 2.4 An outline of the calibration and verification phase of the modeling procedure.

this topic (in Chapter 11) we shall draw heavily upon ideas and algorithms from the fields of system identification (e.g. Eykhoff, 1974; Mehra and Lainiotis, 1976) and time-series analysis (e.g. Box and Jenkins, 1970). Before proceeding, however, let us make a short digression and consider in greater depth some of the implications of attempting to model poorly defined systems.

2.6.1. Observing the Behavior of the System

The central issue of model calibration and verification is that of obtaining estimates, $\hat{\boldsymbol{\beta}}$, for the model parameter values, and of comparing some prediction, $\hat{\boldsymbol{c}}^{o}$, of the outputs with the actually observed outputs $\boldsymbol{c}^{o}$. Predictions of the outputs, and not predictions, $\hat{\boldsymbol{c}}$, of the state variables, are required for the comparative aspects of model verification. In other words, if we recall Figure 2.3 the problem of calibration may be stated thus:

> *Given* a set of experimental field data comprising the measured inputs $\boldsymbol{d}$ and the measured outputs $\boldsymbol{c}^{o}$:
> *Determine* values for the parameters $\boldsymbol{\beta}$ and the states $\boldsymbol{c}$ of the model chosen to characterize the system behavior.

The theme of the *a posteriori* phase of the modeling procedure is centered upon the retrieval, manipulation, and *restructuring of measured information*: how can we translate information about the "external" description of the system, $\boldsymbol{d}$ and $\boldsymbol{c}^{o}$, into information about the "internal" description of the system, $\hat{\boldsymbol{\beta}}$ and $\hat{\boldsymbol{c}}$? Since restricted measurement facilities and considerable complexity are the dominant features of microbiological/ecological† systems, what is the likelihood of success in the application of techniques of calibration and verification?

To answer these questions it is instructive to recast Figure 2.3 into the representation of Figure 2.5. Block 1 includes the fundamental microbiology and biochemistry of the system, such as phytoplankton production or microorganism–substrate interaction. At this level a high degree of literally *microscopic* detail would be required to characterize, or model, all the phenomena in the process under study. Yet the structure of these relationships, and the changing patterns of dominant species in the ecological community, though microscopic in detail, cannot necessarily be ignored, for they may have gross macroscopic impacts on overall process conditions, such as algal blooms with the consequences of severe oxygen depletion.

For block 2 the more *macroscopic* features of the state behavior of the system, e.g. variations in pH and water temperature, will reciprocally influence what happens at the microscopic, biochemical level. In general, however, most of

---

† This is not to ignore the complexities of hydrodynamics; we merely use microbiological aspects of water quality as the illustrative example.

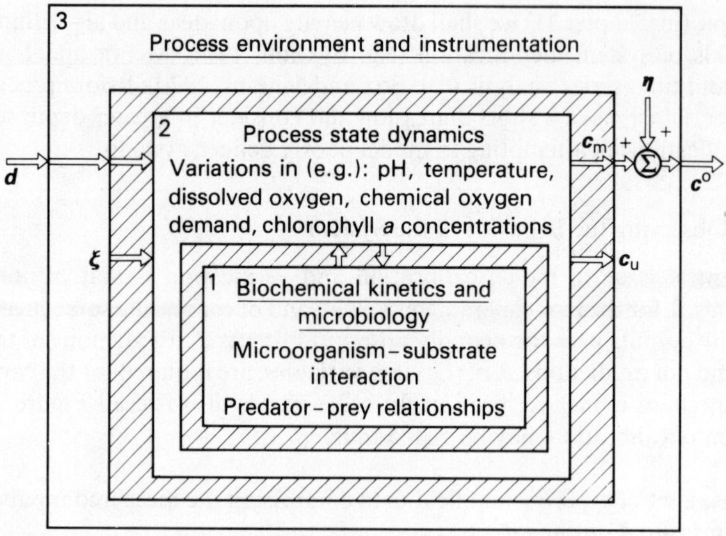

FIGURE 2.5 Observing the behavior of the system (to be compared with Figure 2.3).

the microscopic detail of block 1 belongs to the category of state variables that are not easily measured, namely $\boldsymbol{c}_u$. Hence this fine detail is "lost," as it were, to the process environment (block 3). That is to say, direct measurement of the variables characteristic of block 1, for instance the concentrations of nitrifying bacteria, is extremely difficult unless specialized experimental and analytical facilities are available to the investigator. The relatively few variables in block 2 that are easily measured, $\boldsymbol{c}_m$, amount only to the more macroscopic, sometimes crude, measurements of quantities like chemical oxygen demand, and concentrations of chlorophyll *a* and suspended solids.

Block 3 of Figure 2.5 represents in part the *system environment*, from which all manner of unobserved disturbances and unpredictable mechanisms of behavior, $\boldsymbol{\xi}$, will interact with the more deterministic features of the phenomena accounted for in blocks 1 and 2. Block 3 also represents the instrumentation and analytical procedures from which arise unavoidable components of measurement error, $\boldsymbol{\eta}$. Thus block 3 is intended to introduce elements of *uncertainty* into the picture of system behavior.

Therefore, the following can finally be stated in answer to our earlier question about the likelihood of success in the application of techniques for model calibration and verification. Clearly, if measurements of only some of the process inputs $\boldsymbol{d}$ and only some of the process outputs $\boldsymbol{c}^o$ can be obtained, then relatively little information is available with which to estimate the process states $\boldsymbol{c}$ and parameters $\boldsymbol{\beta}$. In particular, it is unlikely that there will be much information *directly* related to the microscopic detail of block 1 in Figure 2.5.

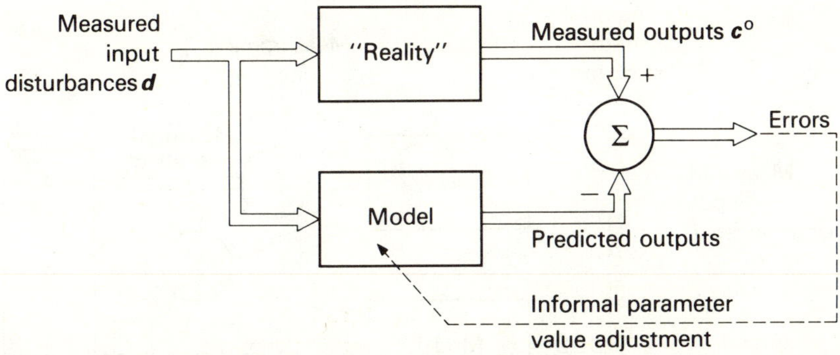

FIGURE 2.6 A rudimentary method of model calibration.

Moreover, the relationships between $\boldsymbol{d}$, $\boldsymbol{c}$, and $\boldsymbol{c}^{\mathrm{o}}$ are likely to be significantly obscured by the uncertainty originating from the process environment and instrumentation. One should therefore be modest in proclaiming the success of currently published calibration studies.

A large proportion of the procedures already applied to water quality model calibration might be classified as essentially procedures of "trial and error" simulation. In other words, this is the type of informal calibration procedure whereby, according to Figure 2.6, the analyst starts with some model structure and set of associated parameter values, so that the simulated performance of the model is compared with the observed behavior of the system under study. Then, if the model is found to be inadequate in its characterization of reality, the analyst may decide simply to adjust some of the parameter values on an *ad hoc* basis until the desired (i.e. observed) performance is obtained. On the other hand, the model may be so much in error that the analyst is required to alter the structure of the relationships between the variables accounted for in the model.

However, since reality is evidently subject to randomness in its observed behavior, the techniques of interest here will be required to recognize the presence of chance errors and disturbances (Figure 2.7). At the same time these *formal* methods of estimation and calibration should be able to discriminate effectively against such ever-present noise in the field data. It is not, in practice, a matter of the analyst being unaware of the stochastic aspects of the modeling problem, or of the informal trial-and-error simulation method being wholly inadequate; for whether one calibrates model performance along the lines of Figure 2.6 or 2.7, a major part of the calibration and verification exercise is devoted precisely to filtering out the uncertainty in the observed patterns of behavior. Moreover, these calibration methods should be able to assist one in making inferences about the behavior of the inaccessible (i.e. not easily measureable), microscopic portion of the state variables from information on the more

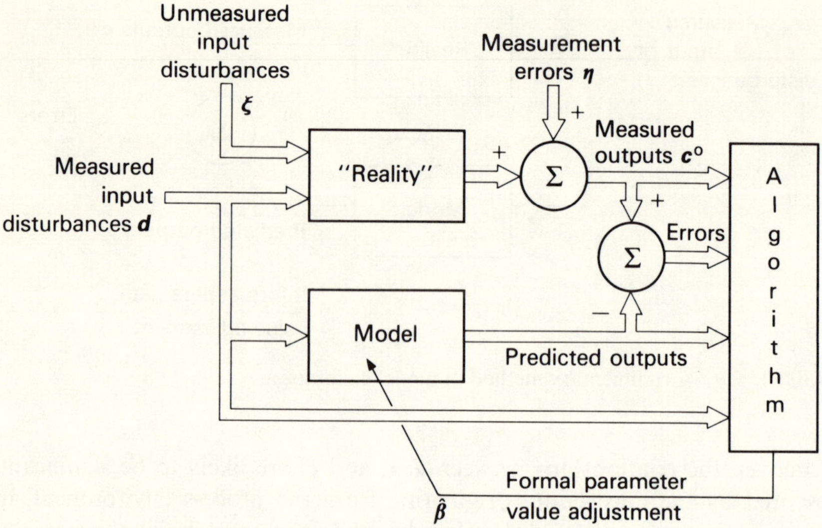

FIGURE 2.7 A more formal method of model calibration.

accessible, macroscopic sector of the state variables. If the algorithms of calibration can fulfill these functions, even in some small measure, then we might consider their application to have been beneficial.

2.6.2. Experimental Design

The success of any modeling procedure that sets itself the objective of demonstrating how well, or how badly, the model simulates reality is strongly dependent upon the quality of the field data available. The ideal would be the ability to make certain specialized and deliberate *in situ* experiments. Seldom, if ever, is the analyst permitted this luxury: it is difficult to imagine there being sufficient control over the input forcing functions of the system for a specified sequence of experimental events to be transmitted to the system. Thus the analyst is not free to experiment with his system, and any "experiments" as such reduce simply to the observation of behavior under "normal operating conditions" (Eykhoff, 1974). Given these constraints, it is useful to examine briefly how best to organize one's sampling effort in order to obtain pertinent observations of the behavior to be modeled. Since we are concerned with *in situ* experiments our remarks are not particularly relevant to those cases where the analyst undertakes a laboratory experiment, for example a chemostat or more complex hydrodynamic experiment, to examine hypotheses about the field system.

There are common features shared by the problems of determining the data requirements of parameter estimation algorithms and designing experimental and routine monitoring programs. Indeed, algorithms that we shall use later for illustrating parameter estimation in Chapter 11 are also algorithms that have been applied for the specification of water quality sampling programs (Moore, 1973; Lettenmaier and Burges, 1977; Bras, 1978; Kitanidis *et al.*, 1978; Canale *et al.*, 1980). The common theme of experimental design and parameter estimation is the determination of the number and kind of measurements required for distinguishing between "signal" and "noise" in the observed information. In other words, we wish to be reasonably confident that our results are not significantly impaired by measurement error and other sources of uncertainty (as discussed in section 2.6.1).

If it is assumed that the objective is the calibration of a dynamic model, the choice of sampling frequency, together with a decision regarding the duration of the data collection program, is of crucial importance (Gustavsson, 1975). Two very rough rules of thumb state that:

(i) *The sampling interval* should be at most as long as the minimum time constant (or response time) of interest; alternatively, the sampling interval should be one-sixth of the period of the fastest sinusoid-type variation expected in the behavior of the system (Shannon and Weaver, 1949).

(ii) *The length of the experiment* should ideally cover a period with a magnitude of at least ten times the magnitude of the largest time constant of interest; to some extent this kind of determination is related to the observation that the degree of subsequent parameter estimation error is inversely proportional to the length, i.e. number of samples, of the experiment.

Both rules are concerned with the speeds of response of the output variables to changes in the input variables. Thus, if the idea of a time constant is approximately interpreted as, say, the detention time for water in a lake, and if this same notion can also be crudely translated into the time scales for biological growth of a species and rates at which nutrients are cycled, one has, according to these rules, the basis of an experimental design.

Let us discuss the choice of sampling frequency in greater detail. For example, it may well be that a fixed sampling frequency, as in a routine monitoring program, is not always advisable. An "adaptive" strategy would be more appropriate in collecting data for a phytoplankton model, in which an important objective is the estimation of model parameters relating to the dynamics of a bloom condition. Provided that the analyst has prior knowledge of the timing of the algal bloom he may, for a short period of the year, allocate his sampling effort to high-frequency measurements of chlorophyll *a* concentration. Another view of this problem is that relatively rapid changes in water quality (i.e. relatively small time constants in the system behavior) are effective only during certain intervals and thus require only intermittent, intensive sampling effort. Alternatively, sensitivity analysis of the *a priori* model might indicate that the model responses are particularly sensitive to certain parameter values at certain points in time (and space). In Chapter 9 an adaptive experimental program will be reported for a lake modeling study.

For the calibration of a complex multivariable model having several inputs and several outputs, the determination of sampling frequency alone does not assist the analyst in making choices about which variables to measure. The problem is that a wrong choice of measured variable combinations may eventually lead to difficulties of *model structure and parameter identifiability*, as already noted in section 2.4.1. In general one has a problem of identifiability if the structure of the proposed model is such that the information contained in the field data cannot be translated into information about the values of certain parameters. An analysis of the identifiability of the *a priori* model may therefore reveal the following kinds of features (Cobelli *et al.*, 1979):

(i) those parameters in the model that can be uniquely estimated from an experiment with given input and output measurement combinations;
(ii) an appropriate combination of possible input and output measurements that will allow the unique estimation of all the model parameters.

We would, for instance, have a typical problem of identifiability in the model if it were possible to estimate only the sum of two rate constants, K_1 and K_2, as a value $\hat{\beta}$, i.e. $K_1 + K_2 = \hat{\beta}$. There is no unique solution to this relationship and many pairs of values of K_1 and K_2 will sum to the value $\hat{\beta}$. Since this kind of prior analysis requires a model, it will not have escaped notice that a good experimental design almost implies the end point of the exercise, i.e. a reasonably good model of reality.

2.6.3. Model Structure Identification

We come now to the problem of model structure identification. More specifically, let us say that *a posteriori model structure identification* attempts, by reference to the field data, to establish how the measured system input disturbances $\boldsymbol{d}$ are related to the system state variables $\boldsymbol{c}$, and how the state variables are in turn related both to each other and to the measured system outputs $\boldsymbol{c}^{o}$. This is, therefore, at least conceptually, a different problem from the straightforward matter of parameter estimation usually associated exclusively with the activity of model calibration. *A posteriori* model structure identification is technically a much more difficult problem to solve than the estimation of parameter values. In practice, as will be demonstrated later, the application of a parameter estimation algorithm is frequently implicit in the solution of the model structure identification problem (Beck, 1979).

It may be helpful to visualize model structure identification as analogous to the choice of whether to fit a straight line or a curve to a set of experimental data. Or again, within the above-mentioned broad definition of this problem, model structure identification is also concerned with identifying the correct form of the mathematical expressions that are contained in the model equations. Let us suppose that, as an example, we are examining the uptake or removal of a nutrient/substrate in a batch chemostat reaction, and that our first hypothesis is a linear kinetic model,

Model I

$$\frac{dc_1(t)}{dt} = \dot{c}_1(t) = -[\beta_1]c_1(t), \tag{2.2}$$

in which the dot notation refers to differentiation with respect to time t; c_1, the concentration of substrate, is the state variable, and β_1 is a parameter representing a first-order kinetic decay rate constant. For our second hypothesis about the observed system behavior we might propose a Monod-type kinetic expression together with the presence of a mediating microorganism in the reaction,

Model II

$$\dot{c}_1(t) = -\left[\frac{\beta'_1 c_2(t)}{\beta'_2 + c_1(t)}\right]c_1(t) \tag{2.3a}$$

$$\dot{c}_2(t) = -\left[\frac{\beta'_3 c_2(t)}{\beta'_2 + c_1(t)}\right]c_1(t) - \beta'_4 c_2(t), \tag{2.3b}$$

where the additional state variable c_2 is the microorganism concentration and we have a vector $[\beta'_1, \beta'_2, \beta'_3, \beta'_4]$ of associated model parameters. Now we recall that there are presumably some noise-corrupted measurements available from this experiment, but that we do not know which, if either, of Models I and II better characterizes the nature of the observed behavior. Model structure identification is then the problem of choosing, by reference to the *in situ* data, the number of state variables to be accounted for in the model, the problem of defining how these state variables depend upon each other, and the problem of identifying the correct form of the expression to go inside the square brackets of eqns. 2.2 and 2.3. If both models are thought, *a priori*, to be good approximations of reality, we might also call this a problem of model discrimination in the sense used by Shastry *et al.* (1973). However, if neither hypothesis is adequate and if a more complex pattern of behavior is suggested by the analysis of the data, the definition given above will be the most useful interpretation of model structure identification to be borne in mind here.

The matter of model structure identification will be discussed again in the next section and elsewhere. Its importance in the modeling procedure is partly related to the fact that water quality and ecological systems are not well defined, and therefore the investigator is frequently ignorant of the "true" observed relationships between the system variables. Thus let us return briefly to the discussion of internally descriptive and black box modeling approaches. Probably the best view of the two approaches is that internally descriptive and black box models represent complementary conceptual frameworks for system identification; some attempts have been made to illustrate this view (Beck, 1978b). Since the two approaches are complementary, more is to be gained from their joint application than from the exclusive use of either model type. For much of the time system identification, and model structure identification in particular, is confronted with the need to offer plausible hypotheses about "unexplained" relationships in a set of field data. It seems only prudent, therefore, to approach each such problem from a variety of angles, and to gather all the available evidence for synthesis of the next hypothesis.

2.6.4. Parameter Estimation

Parameter estimation deals with the computation of values for the parameters that appear in the model equations, *once the structure of these relationships has been properly identified.* A basic principle of parameter estimation, as illustrated by Figure 2.7, is that the estimates $\hat{\boldsymbol{\beta}}$, say, of the model parameters $\boldsymbol{\beta}$ are obtained by minimizing some function of the errors,

$$\boldsymbol{\epsilon}\{\hat{\boldsymbol{\beta}}\} = \boldsymbol{c}^{\text{o}} - \hat{\boldsymbol{c}}^{\text{o}}\{\boldsymbol{\beta}\}, \tag{2.4}$$

between the output response observations $\boldsymbol{c}^{\text{o}}$ and model predictions of those output variables, $\hat{\boldsymbol{c}}^{\text{o}}$. One of the simplest and most well known of parameter estimation schemes is that of least-squares estimation, where the loss (error) function to be minimized is

$$J = \sum \boldsymbol{\epsilon}^{\text{T}}\{\hat{\boldsymbol{\beta}}\}\boldsymbol{\epsilon}\{\hat{\boldsymbol{\beta}}\}. \tag{2.5}$$

In (2.5) the summation is carried out over all available observations; if $\boldsymbol{\epsilon}$ is assumed to be a column vector with p elements, then J is a scalar function, since

$$\boldsymbol{\epsilon}^{\text{T}}\boldsymbol{\epsilon} = \epsilon_1^2 + \epsilon_2^2 + \cdots + \epsilon_p^2, \tag{2.6}$$

where superscript T denotes the transpose of a vector or matrix.

At this point in the book it is not necessary to develop further the subject of parameter estimation; ample illustration of the application of techniques of parameter estimation will be provided in Chapter 11. It is useful, however, to draw a distinction between *off-line* schemes (block data processing) and *on-line* schemes of parameter estimation, since this leads to another view of the prior

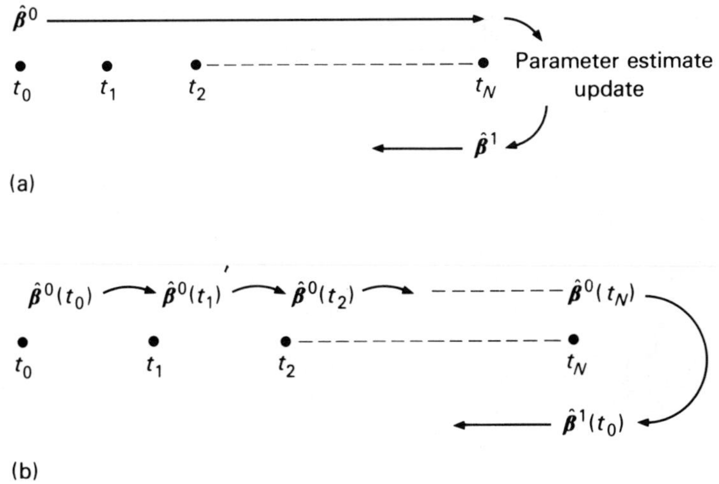

FIGURE 2.8 Methods of parameter estimation: (a) off-line; (b) recursive. The notation t_k in this example represents the kth discrete sampling instant in a time series with N samples.

problem of model structure identification. We shall deal first with a summary of the off-line scheme. As shown in Figure 2.8(a), an off-line procedure keeps the parameter estimates constant at their *a priori* values $\hat{\boldsymbol{\beta}}^{a}$, while the complete set of field data (the sampled measurements for the period from t_1 to t_N) is processed by the algorithm. A value for the loss function, such as that given by (2.5), is calculated at the end of each iteration. The algorithm attempts then to minimize the value of the loss function with respect to the model parameters and computes an updated set of parameter values, $\hat{\boldsymbol{\beta}}^1$, for substitution into the next iteration through the data.

In contrast to the off-line algorithms, a *recursive* estimation algorithm computes revised parameter estimates $\hat{\boldsymbol{\beta}}^0(t_k)$, say,† at each sampling instant t_k of the field data. Here, t_k denotes the time of the kth observation of system behavior. The data are therefore processed in a serial fashion, as shown by Figure 2.8(b). The minimization of the value of the error loss function is included implicitly, rather than explicitly, in the algorithms. At the end of the block of data the estimates $\hat{\boldsymbol{\beta}}^0(t_N)$ are substituted for the *a priori* parameter values $\hat{\boldsymbol{\beta}}^1(t_0)$ of the next iteration through the data. Further iterations through the set of field data are required since any initially incorrect estimates $\hat{\boldsymbol{\beta}}^0(t_0)$ contribute errors to the calibration loss function that are larger than the errors

† We use the argument of discrete time merely for illustration; otherwise, it is easy to imagine a recursive algorithm being applied in a stepwise manner to data obtained from a series of fixed spatial locations along a river.

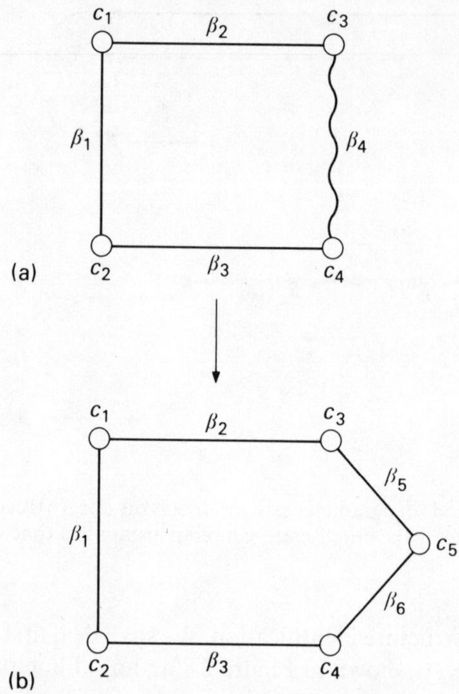

FIGURE 2.9 A conceptual picture of the problem of model structure identification.

contributed by initially correct estimates. By implication, therefore, the minimum of the loss function is unlikely to have been located after the first iteration.

Now let us reconsider model structure identification in the light of our knowledge of recursive parameter estimation. We suppose that the state variables **c** in a water quality model may be represented conceptually by the nodes of Figure 2.9(a), and that the parameter values are visualized as the "elastic" connections between the state variables. If the assumption has been made that all the parameter values are constant with time and yet a recursive algorithm yields an estimate of one or more of the parameters (β_4, say) that is significantly time-varying, one may question the correctness of the chosen model structure. The reason for this is as follows. The general nature of an estimation procedure is to fit the model (i.e. state variable) predictions to the field observations. Hence when any persistent *structural* discrepancy is detected between the model and observed patterns of behavior, this will manifest itself as an attempt by the estimation procedure to adapt the model, i.e. the parameter values, toward "reality." Such time variations of the parameter values can, of course, occur for different reasons; for instance, the parameter may be truly time-varying in accordance with some seasonal fluctuation. However, for the purposes of our example in Figure 2.9(a) we might suppose that the actual structure of the relationships underlying the observed system behavior would be better represented by the introduction of a new state variable and two new parameters, as shown in Figure 2.9(b). If this were indeed the correct model structure, recomputation of the parameter values should give recursive estimates that are essentially constant.

Our example has two objectives. First, it should emphasize the earlier statement that model structure identification and parameter estimation are closely related and that the former problem can sometimes be solved by recourse to a parameter estimation routine. Second, it should be apparent that an exercise in accurate parameter estimation is of dubious significance if the prior problem of model structure identification has not been satisfactorily resolved.

2.6.5. Verification

"Verification" and "validation" are easily confused and both have come to be interpreted in several different ways. However, since we have placed "verification" together with "calibration," thus implying association with the *a posteriori* phase of the modeling procedure, the following will serve as our working definition: verification is the determination of whether the "correct" model has been obtained from a given *single* set of experimental data. Findeisen *et al.* (1978), in their sample glossary of systems analysis, give a different definition: "a model is said to be verified if it behaves in the way the model builder wanted it to behave." Such a definition, in the present discussion, would be more relevant to the *a priori* stages of the modeling procedure, for instance at the stage of *a priori* sensitivity analysis (section 2.4.1). On reflection it must seem that both definitions suffer from being vague; and in fact the arguments for satisfying oneself that the model is verified, in an *a posteriori* sense, are also rather circular. Usually these arguments are as follows. We assume that a model structure has been identified, that the parameters have been estimated, and thus that a sequence of final model response errors can be computed, as illustrated by Figure 2.10. Almost inevitably it will have been necessary at some stage during the analysis of the field data to have made assumptions about the statistical properties of the noise sequences idealized in the model of reality, for example the processes ξ and η in Figure 2.3. If these assumptions are valid, the model response errors should also conform to certain statistical properties, and in particular to those of white noise sequences. To conform with the properties of white noise any error sequence should broadly satisfy the following constraints: that its mean value is zero; that it is not correlated with any other error sequence; and that it is not correlated with the sequences of measured input forcing functions. Evaluation of the error sequences in this fashion can, therefore,

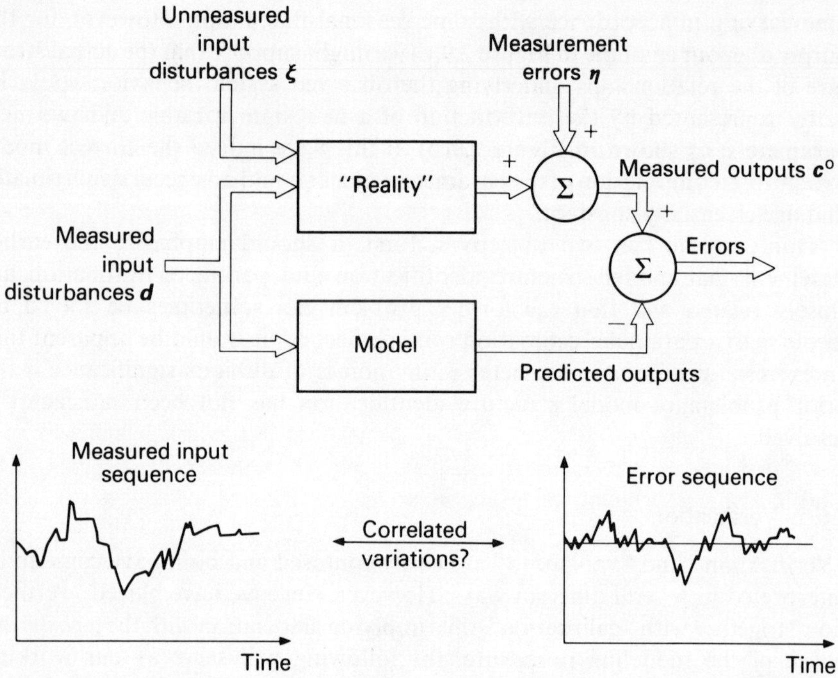

FIGURE 2.10 Model verification: computing the residual error sequences and checking their statistical properties.

provide a check essentially on whether the final model invalidates some of the assumptions inherent in its development.

Should the error sequences not conform to their desired properties, this suggests that the model does not characterize adequately all of the (relatively) more deterministic features of the observed dynamic behavior. A strong correlation between variations in a given input and the variations in the model response errors of a given output, for example, would indicate that the model structure should be modified to accommodate additional significant relationships between those two variables. Analysis of the model performance along these lines, therefore, directs attention once again to the problem of model structure identification.

2.7. VALIDATION

It is clear that calibration and verification represent the bulk of the procedure for model development and testing, once an experimental data set has been obtained. There is, however, no guarantee that the validity of the model extends beyond the sample data set against which it has been calibrated. Validation is, then, the testing of the adequacy of the model against a second, *independent* set of field data. Because validation thus entails the design and implementation of new experiments, it is unfortunately a step in the analysis that is all too rarely attempted. We may further observe that Findeisen *et al.* (1978) give the following as part of their definition of validation: "a model can never be completely validated; we can never prove that its results conform to reality in all respects; it can only be invalidated." Similar notions are evident in an article by Schweppe (1978), who discusses the use of statistical hypothesis testing for the determination of model validity. Statistical hypothesis testing, as he says, does not prove that a hypothesis (model) is true; it may merely indicate whether it is possible to reject the hypothesis on the basis of the available observations and the chosen set of test statistics.

Perhaps too much time should not be spent in lamenting any lack of interest shown by analysts in exhaustively validating their models. For example, Mankin *et al.* (1977) offer the following advice:

> Let us dismiss the question: Have you proven that your model is valid? with a quick **NO**. Then let us take up the more rewarding and far more challenging question: Have you proven that your model is useful for learning more about the ecosystem?

2.7.1. *A Posteriori* Sensitivity Analysis

Much of what has been said above implies that validation is not a "once and for all" step. Indeed, like the other topics of this chapter, validation defines only a location in the iterative procedure of model development. Figure 2.2 shows two feedback loops returning from the validation stage: one to the calibration and verification phase; and one to the block labeled "sensitivity analysis."

A priori sensitivity analysis (section 2.4.1) establishes the relative magnitudes of changes in the simulated model output responses to changes in the model parameter values. *A posteriori* sensitivity analysis, if we may call it such, examines the distribution of model responses that are possible, given the distributions of estimated parameter values. This requires some further explanation. Calibration of a model does not imply that all parameter values will be known exactly once the model is calibrated; there remain irreducible errors of estimation associated with the calibrated parameter values. In fact these estimation errors reflect a mixture of uncertainty and errors in the field data used for calibration, and uncertainty in the values initially specified for the parameters prior to calibration. If the calibrated model is now applied to the prediction of future events, the uncertainty in the parameter estimates (together with the uncertainty in the present state of water quality and uncertainty in the future input forcing functions, e.g. solar radiation) would be propagated forward as a degree of uncertainty, or error, in the predicted response of the water body. A simple example may illustrate the point. Let us suppose that, during model calibration, it turns out that one particular parameter has a large estimation error. Among several reasons, such as the problem of identifiability discussed in sections 2.4.1 and 2.6.2, this large error might result from the fact that the type of system behavior associated with that parameter is not exhibited in the available field data. In other words, there is no information in the observations with which to estimate the given parameter accurately. Thus, when the model is used for prediction, and if the assumed range of future forcing functions drives the model into a pattern of behavior that is sensitive to this particular parameter, then the following should occur. The distribution of future output responses should become wide, indicating both little confidence in model performance and great uncertainty about the pattern of responses. In effect, such analysis reveals that certain types of behavior cannot be predicted accurately because little previous empirical evidence is available for calibration of the model of that behavior.

The attention given so far to this kind of *a posteriori* sensitivity analysis has been modest, although current developments indicate a rapidly increasing interest in related topics. Typically, one might approach the problem under the heading of Monte Carlo simulation, as did Mankin *et al.* (1976) and Tiwari *et al.* (1978), although both of these studies lack the direct connection of such analysis with prior model calibration results. Some analytical results have also been obtained for a number of basic simple forms of ecological models by O'Neill and Gardner (1979) who, with co-workers, have undertaken a comprehensive examination of the effects of error and uncertainty on model predictions. Other approaches are suggested by Argentesi and Olivi (1976), Di Toro and van Straten (1979), Beck *et al.* (1979), Fedra *et al.* (1981), and Beck and van Straten (1983).

2.8. SUMMARY AND CONCLUSIONS

In this chapter a procedure for modeling has been introduced. Several salient problems within that procedure have been discussed in qualitative terms. The major objective of the chapter has been to show that the modeling procedure divides into two parts: that which can be accomplished on the basis of existing knowledge and theory; and that which ought to be accomplished when experimental field data are available. Much of the former part, the *a priori* phase of the modeling procedure, is largely intuitive and not methodological. The *a posteriori* stage of modeling covers more technical topics that may be less familiar to the reader. For that reason Chapter 11 has been designed as a complement to this chapter. Sensitivity analysis, system identification, time-series analysis, and parameter estimation will be the subjects of Chapter 11.

An important point emerging from the present chapter is that (*a posteriori*) model structure identification can be a particularly difficult problem to solve during calibration. Indeed, calibration as a whole is not a trivial exercise, not the least because good-quality field data are rarely available. Thus it may not always be possible to adhere to the calibration, verification, and validation phases of the modeling procedure, though such adherence would be highly desirable. Subsequent chapters of this book emphasize in general the development of specific models from basic theoretical principles (i.e. the *a priori* aspect of modeling). The reader should keep in mind the problems that each particular model might pose if it were to be calibrated and verified against field data.

REFERENCES

Adachi, N., and S. Ikeda (1978). Stability analysis of eutrophication models. *Research Memorandum* RM-78-53 (Laxenburg, Austria: International Institute for Applied Systems Analysis).

Argentesi, F., and L. Olivi (1976). Statistical sensitivity analysis of a simulation model for the biomass–nutrient dynamics in aquatic ecosystems. *Proceedings of 4th Summer Computer Simulation Conference, Simulation Councils, La Jolla, CA*, pp. 389–393.

Beck, M. B. (1978a). Mathematical modeling of water quality. *Collaborative Proceedings* CP-78-10 (Laxenburg, Austria: International Institute for Applied Systems Analysis).

Beck, M. B. (1978b). Random signal analysis in an environmental sciences problem. *Applied Mathematical Modelling* 2(1):23–29.

Beck, M. B. (1979). Model structure identification from experimental data. *Theoretical Systems Ecology*, ed. E. Halfon (New York, NY: Academic Press), pp. 259–289.

Beck, M. B. (1980). Forecasting and control of water quality. *Real-Time Forecasting/Control of Water Resource Systems*, ed. E. F. Wood (Oxford: Pergamon), pp. 179–202.

Beck, M. B. (1981). Hard or soft environmental systems? *Ecological Modelling* 11:233–251.

Beck, M. B., E. Halfon, and G. van Straten (1979). The propagation of errors and uncertainty in forecasting water quality—Part I: Method. *Working Paper* WP-79-100 (Laxenburg, Austria: International Institute for Applied Systems Analysis).

Beck, M. B., and G. van Straten (eds.) (1983). *Uncertainty and Forecasting of Water Quality* (Berlin: Springer) (in the press).

Box, G. E. P., and G. M. Jenkins (1970). *Time-series Analysis, Forecasting and Control* (San Francisco, CA: Holden–Day).

Bras, R. L. (1978). Sampling network design in hydrology and water quality sampling: A review of linear estimation theory applications. *Applications of Kalman Filter to Hydrology, Hydraulics and Water Resources*, ed. C.-L. Chiu (Pittsburgh, PA: University of Pittsburgh, Stochastic Hydraulics Program), pp. 155–200.

Canale, R. P., L. M. de Palma, and W. F. Powers (1980). Sampling strategies for water quality in the Great Lakes. *US Environmental Protection Agency, Washington, DC Report* EPA 600/3-80-055.

Chen, C. W., and G. T. Orlob (1975). Ecologic simulation of aquatic environments. *Systems Analysis and Simulation in Ecology* vol. 3, ed. B. C. Patten (New York, NY: Academic Press), pp. 476–588.

Cobelli, C., A. Lepschy, and G. Romanin-Jacur (1979). Structural identifiability of linear compartmental models of ecosystems. *Theoretical Systems Ecology*, ed. E. Halfon (New York, NY: Academic Press), pp. 237–258.

Di Toro, D. M., and G. van Straten (1979). Uncertainty in the parameters and predictions of phytoplankton models. *Working Paper* WP-79-27 (Laxenburg, Austria: International Institute for Applied Systems Analysis).

Eykhoff, P. (1974). *System Identification—Parameter and State Estimation* (Chichester: Wiley).

Fedra, K., G. van Straten, and M. B. Beck (1981). Uncertainty and arbitrariness in ecosystems modelling: a lake modelling example. *Ecological Modelling* 13:87–110.

Findeisen, W., A. Iastrebov, R. Lande, J. Lindsay, M. Pearson, and E. S. Quade (1978). A sample glossary of systems analysis. *Working Paper* WP-78-12 (Laxenburg, Austria: International Institute for Applied Systems Analysis).

Finney, B. A., D. S. Bowles, and M. P. Windham (1982). Random differential equations in water quality modeling. *Water Resources Research* 18(1):122–134.

Gustavsson, I. (1975). Survey of applications of identification in chemical and physical processes. *Automatica* 11:3–24.

Ikeda, S., and N. Adachi (1976). Dynamics of the nitrogen cycle in a lake and its stability. *Ecological Modelling* 2:213–234.

Jørgensen, S. E. (1978). Review of eutrophication models. *Modelling, Identification and Control in Environmental Systems*, ed. G. C. Vansteenkiste (Amsterdam: North-Holland), pp. 473–500.

Jørgensen, S. E. (ed.) (1979). Hydrophysical and ecological models of shallow lakes and reservoirs. *Collaborative Proceedings* CP-78-14 (Laxenburg, Austria: International Institute for Applied Systems Analysis).

Jørgensen, S. E., and D. R. F. Harleman (eds.) (1978). Hydrophysical and ecological modelling of deep lakes and reservoirs. *Collaborative Proceedings* CP-78-7 (Laxenburg, Austria: International Institute for Applied Systems Analysis).

Jørgensen, S. E., H. Mejer, and M. Friis (1978). Examination of a lake model. *Ecological Modelling* 4:253–278.

Karplus, W. J. (1976). The future of mathematical models in water resources systems. *System Simulation in Water Resources*, ed. G. C. Vansteenkiste (Amsterdam: North-Holland), pp. 11–18.

Kitanidis, P., C. S. Queiroz, and D. Veneziano (1978). Sampling networks for violation of water quality standards. *Applications of Kalman Filter to Hydrology, Hydraulics and Water Resources*, ed. C.-L. Chiu (Pittsburgh, PA: University of Pittsburgh, Stochastic Hydraulics Program), pp. 213–229.

Kreyszig, E. (1972). *Advanced Engineering Mathematics* (New York, NY: Wiley).

Lettenmaier, D. P., and S. J. Burges (1977). Design of trend monitoring networks. *Proceedings of American Society of Civil Engineers, Journal of Environmental Engineering Division* 103(EE5): 785–802.

Loucks, D. P., and W. R. Lynn (1966). Probabilistic models for predicting stream quality. *Water Resources Research* 2:593–605.

Mankin, J. B., R. H. Gardner, and H. H. Shugart (1976). The COMEX computer code: Monte Carlo analysis of ecosystem attributes. *Proceedings of 1976 Summer Computer Simulation Conference*, pp. 433–436.

Mankin, J. B., R. V. O'Neill, H. H. Shugart, and B. W. Rust (1977). The importance of validation in ecosystem analysis. *New Directions in the Analysis of Ecological Systems*, ed. G. S. Innis. *Simulation Council Proceedings Series* 5(1):63–72.

Mehra, R. K., and D. G. Lainiotis (eds.) (1976). *System Identification: Advances and Case Studies* (New York, NY: Academic Press).

Moore, S. F. (1973). Estimation theory applications to design of water quality monitoring systems. *Proceedings of American Society of Civil Engineers, Journal of Hydraulics Division* 99:815–831.

O'Neill, R. V., and R. H. Gardner (1979). Sources of uncertainty in ecological models. *Methodology in Modelling and Simulation*, eds. B. P. Zeigler *et al.* (Amsterdam: North-Holland), pp. 447–463.

Orlob, G. T. (1975). Present problems and future prospects of ecological modeling. *Ecological Modeling in a Resource Management Framework*, ed. C. S. Russell. *Resources for the Future Working Paper* QE-1 (Baltimore, MD: Johns Hopkins University Press), pp. 283–312.

Padgett, W. J., G. Schultz, and C. P. Tsokos (1977). A stochastic model for BOD and DO in streams when pollutants are discharged over a continuous stretch. *International Journal of Environmental Studies* 11:45–55.

Papoulis, A. (1965). *Probability, Random Variables and Stochastic Processes* (New York, NY: McGraw-Hill).

Rinaldi, S., and R. Soncini-Sessa (1978). Sensitivity analysis of generalized Streeter–Phelps models. *Advances in Water Resources* 1:141–146.

Rinaldi, S., R. Soncini-Sessa, H. Stehfest, and H. Tamura (1979). *Modeling and Control of River Quality* (New York, NY: McGraw-Hill).

Schweppe, F. C. (1978). Model identification problems. *Applications of Kalman Filter to Hydrology, Hydraulics and Water Resources*, ed. C.-L. Chiu (Pittsburgh, PA: University of Pittsburgh, Stochastic Hydraulics Program), pp. 115–133.

Shannon, C. E., and W. Weaver (1949). *The Mathematical Theory of Communication* (Urbana, IL: University of Illinois Press).

Shastry, J. S., L. T. Fan, and L. E. Erickson (1973). Nonlinear parameter estimation in water quality modeling. *Proceedings of American Society of Civil Engineers, Journal of Environmental Engineering Division* 99(EE3): 315–331.

Somlyody, L. (1981). Water quality modelling: A comparison of transport-oriented and biochemistry-oriented approaches. *Working Paper* WP-81-117 (Laxenburg, Austria: International Institute for Applied Systems Analysis).

van Straten, G., and B. de Boer (1979). Sensitivity to uncertainty in a phytoplankton–oxygen model for lowland streams. *Working Paper* WP-79-28 (Laxenburg, Austria: International Institute for Applied Systems Analysis).

Streeter, H. W., and E. B. Phelps (1925). A study of the pollution and natural purification of the Ohio River. *US Public Health Service, Washington, DC Bulletin* 146.

Thayer, R. P., and R. G. Krutchkoff (1967). Stochastic model for BOD and DO in streams. *Proceedings of American Society of Civil Engineers, Journal of Sanitary Engineering Division* 93(SA3): 59–72.

Thomann, R. V. (1973). Effect of longitudinal dispersion on dynamic water quality response of streams and rivers. *Water Resources Research* 9:355–366.

Tiwari, J. L., J. E. Hobbie, J. P. Reed, and M. C. Miller (1978). Some stochastic differential equation models of an aquatic ecosystem. *Ecological Modelling* 4:3–27.

Young, P. C. (1978). General theory of modelling for badly defined systems. *Modelling, Identification and Control in Environmental Systems*, ed. G. C. Vansteenkiste (Amsterdam: North-Holland), pp. 103–135.

Young, P. C., and M. B. Beck (1974). The modeling and control of water quality in a river system. *Automatica* 10:455–468.

CHAPTER 2: NOTATION

$\boldsymbol{c}, \hat{\boldsymbol{c}}$	vector of state variables, state estimates
$\boldsymbol{c}_{\mathrm{m}}, \boldsymbol{c}_{\mathrm{u}}$	vector of measurable, not easily measurable state variables
$\boldsymbol{c}^{\mathrm{o}}, \hat{\boldsymbol{c}}^{\mathrm{o}}$	vector of measured, model-predicted output response variables
$\bar{\boldsymbol{c}}$	nominal deterministic reference trajectory for the state vector
$\boldsymbol{d}$	vector of measured input (forcing) variables
J	(squared error) loss function
$\boldsymbol{r}$	vector representing three-dimensional space (x, y, z directions)
s_{ij}	sensitivity coefficient
t, t_k	independent variable of time, kth discrete instant of time
$\boldsymbol{\beta}, \hat{\boldsymbol{\beta}}$	vector of parameters, parameter estimates
$\bar{\beta}_j$	nominal reference value of parameter j
$\boldsymbol{\epsilon}$	predicted model response errors (or residual errors)
$\boldsymbol{\eta}$	vector of (stochastic) measurement errors
$\boldsymbol{\xi}$	vector of (stochastic) unmeasured input disturbances of system.

3 General Principles in Deterministic Water Quality Modeling

P. Mauersberger

3.1. INTRODUCTION

3.1.1. The Aim of this Chapter

The aim of this chapter is to present the theoretical background of deterministic water quality modeling. For any scientific discipline, as well as for management activities, it is of great importance to ensure against defects in the fundamental premises. Therefore, in this chapter the equations used in water quality modeling will be derived from fundamentals of macroscopic physics, chemistry, and biology. The resulting system of equations represents a complicated mathematical problem. For this reason, simplified versions of these equations and the assumptions under which they are valid will also be discussed. Anyone who is especially interested in the application of a model for a river reach or for a special lake may pass quickly through parts of this chapter to one of the later chapters.

Every deterministic water quality model must be based on the following general principles:

conservation of mass and elements
narrow bands of biomass composition
conservation of momentum
conservation of energy
boundary conditions and initial conditions
laws governing chemical, biochemical, and biological processes
the second law of thermodynamics.

Therefore, this chapter will discuss:

the main "components" of deterministic water quality models and numerical methods for solving the typical "balance equations" (section 3.1);

basic and specialized model equations derived from the laws of conservation of mass, momentum, and energy (sections 3.2–3.4);

mathematical descriptions of chemical reactions and biological processes (section 3.5);

the role of entropy in water quality modeling, and some relationships between deterministic, stochastic, and cybernetic modeling methods (section 3.6).

Figure 3.1 surveys the topics covered in the chapter. The formulation and application of special models will be described in later chapters.

It is necessary that a deterministic ecosystem model obey the second law of thermodynamics, according to which dissipation (or entropy production) must be positive. While the balance equations for mass, momentum, and energy play the roles of "book-keepers," the entropy principle, like the "director," determines the development of the ecosystem. The structure, state, and evolution of the aquatic ecosystem are regulated by the mutual effects of entropy-producing and entropy-reducing processes inside the water body and across its boundaries.

Furthermore, entropy and entropy production depend on practically all hydrophysical, hydrochemical, and hydrobiological variables and processes of the aquatic ecosystem. Therefore, by the entropy principle the many phenomenological relationships are reduced to a few important equations and basic principles. This is of significance for the unified treatment of physical, chemical, and (to a first approximation) biological processes in the ecosystem. It is also of importance if we are looking for "new" types of variables (especially "generalized" forces and potentials) suited for the description of complex ecosystems with the aid of a reduced number of variables.

Before we discuss these aspects in section 3.6.2, we must assemble the basic physical, chemical, and biological concepts of water quality modeling.

3.1.2. Aims and Types of Deterministic Water Quality Models

Macroscopic, deterministic models are widely used in studying the water quality of lakes, rivers, estuaries, and other types of water resources. These models, taking into account the physical, chemical, and biological processes inside the system as well as the fluxes of matter and energy across its boundaries, are intended to serve both as research tools and as bases for water resources management. Further improvement in our capability to model water quality is necessary, but may be seriously limited without better knowledge of the important processes taking place in aquatic systems and without the application of the most advanced modeling techniques.

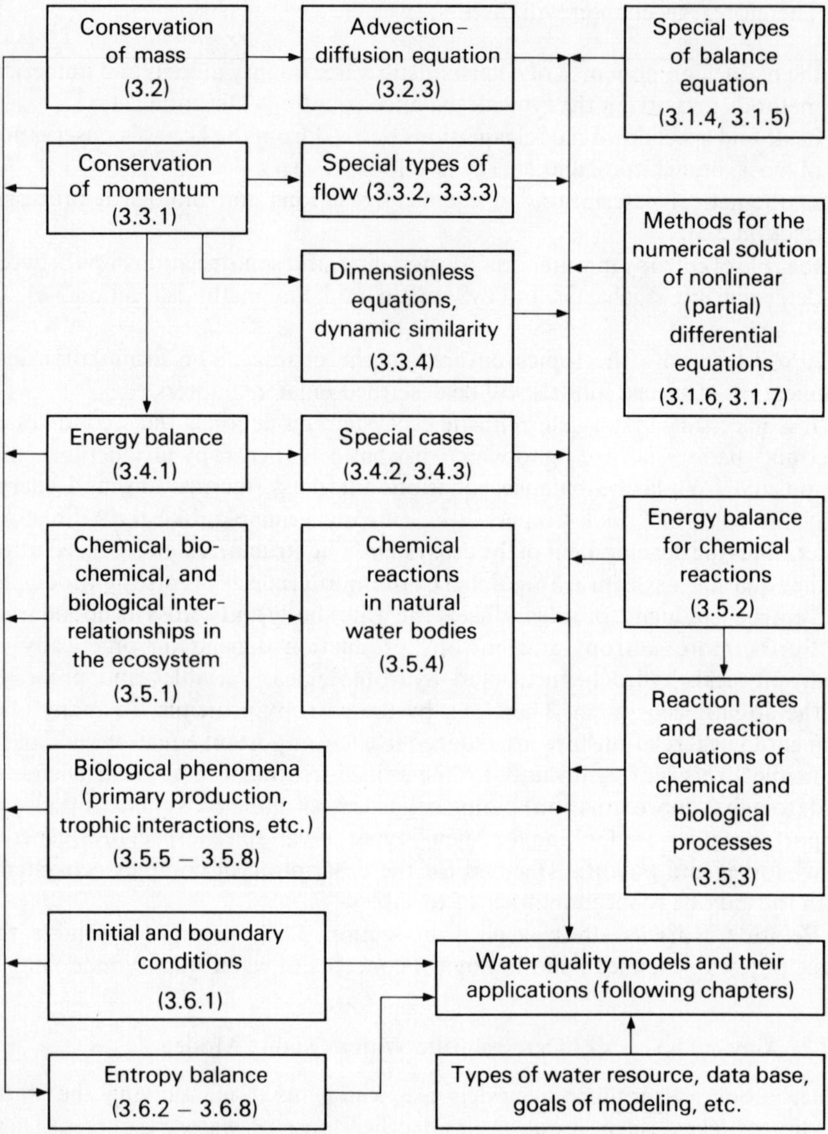

FIGURE 3.1 Survey of Chapter 3 (section numbers are in parentheses).

Since hydrophysical and ecological equations are invariably coupled, water quality models are usually characterized by high degrees of complexity and very difficult mathematical problems. Selecting the appropriate degree of complexity is a challenging task. Usually, models are generated in accordance with preset objectives of the modeling effort and with predetermined amounts and quality of field observations. For instance, the long-term ecological response of an entire lake, including the "hydrodynamics," can be modeled only by introducing the fluxes necessary to satisfy gross mass and energy balances, or by introducing fluxes related to spatial differences in concentration. However, if the highly transient response of a lake (e.g. in an extreme situation) is sought, it may be essential for dynamic models to be coupled to the ecological equations.

There is no "general" water quality model that can deal with the myriad problems of different types of streams, lakes, and reservoirs. The structure of a model is influenced by its particular goal. Scientific investigations may be structured by combining detailed measurements with elaborate submodels of the processes under consideration. Planning and management of water resources frequently require "engineer-developed models," derived through gross simplification of both the hydrodynamic behavior and the aquatic ecosystem. In such cases, scientific rigor is relaxed in favor of practicability. Each model represents a compromise between the complexity of the real water resources system and the simplicity of a well posed mathematical problem suited for the description and prediction of selected features of the real system.

Thus, in order to satisfy best the requirements of both science and practical water management, it is necessary to establish a *methodology* of water quality modeling. Basic concepts of this methodology will be stated in this chapter. We shall start with the fairly general "basic laws" of hydrothermodynamics and ecology and then proceed to simplified mathematical statements by which we can more easily make the transition to special models.

3.1.3. Physical, Chemical, and Biological Components and Processes

Important processes within the water body include chemical reactions, primary production, grazing, egestion, excretion, respiration, and nonpredatory mortality. Many of these processes are deeply influenced by variables describing the physical state of the system, e.g. temperature and turbidity, or by hydrodynamic characteristics, e.g. velocities, velocity gradients, shears, and turbulence.

The set of equations for a particular deterministic water quality model can be chosen to simulate the properties and interrelationships of any or all of the following components and processes:

Physical Solar radiation, temperature, pressure, density, external and frictional forces, flow velocity, diffusion flows, heat flow, kinetic and internal energy, entropy, sedimentation, etc.

Chemical Dissolved inorganic and organic chemicals, particulate organic material (detritus), inaccessible nutrients, heavy metals, complex synthetic and toxic compounds, etc.

Biological Primary producers (diatoms, green algae, blue-green algae, phytobenthos), zooplankton (raptors, selective filterers, nonselective filterers, etc.), zoobenthos, fishes, biota in different life stages, etc.

In the following sections we shall identify the basic concepts upon which models must be founded. However, we shall not deal with questions of selecting the variables and the equations best suited for a special type of water body. The development, calibration, validation, and application of specialized water quality models will be treated in subsequent chapters.

3.1.4. Balance Equations

In view of both the hydrodynamic and ecological aspects of natural water bodies, the system of equations for a water quality model must consist of at least the balances of mass, momentum, energy, and entropy and the equations describing chemical, biochemical, and biological processes. Appropriate initial values and boundary conditions must be included. From the mathematical point of view, a nonlinear initial-boundary-value problem of differential equations must be solved. A question arises if there is one and only one stable solution uniquely dependent on the initial and boundary values. Furthermore, stable, converging numerical methods are required for the solution of these equations. Here, we confine ourselves to some preliminary remarks on the general types of basic equations that may result and on some of the more popular numerical methods used for solution (sections 3.1.6 and 3.1.7).

The general type of balance equation for a scalar quantity F (which may be a function of space, time, and other parameters) is

$$\frac{\mathrm{d}}{\mathrm{d}t}\int_{V(t)} F(\boldsymbol{r}, t)\mathrm{d}V = \int_{V(t)} G(F, \ldots; \boldsymbol{r}, t)\mathrm{d}V - \oint_{A(t)} S^{\mathrm{n}}\,\mathrm{d}A. \tag{3.1}$$

G denotes the sources and sinks of the field F, and may depend on F in a nonlinear manner. S^{n} is the flux across the boundary $A(t)$ of the volume $V(t)$, directed outward and normal to $A(t)$. Table 3.1 shows some examples of sources and fluxes.

A deterministic water quality model generally consists of a set of interconnected balance equations. However, for simplicity we shall confine ourselves to only one equation of this type and assume that there is an independent determination of the flow field $\boldsymbol{v}(r, t)$, a fairly common approach. We shall leave discussion of the larger, more complex problems to later sections and chapters. By use of the relationship

$$\frac{\mathrm{d}}{\mathrm{d}t}\int_{V(t)} F\,\mathrm{d}V = \int_{V(t)}\left(\frac{\mathrm{d}F}{\mathrm{d}t} + F\,\mathrm{div}\,\boldsymbol{v}\right)\mathrm{d}V = \int_{V(t)}\left(\frac{\partial F}{\partial t} + \mathrm{div}(F\boldsymbol{v})\right)\mathrm{d}V,$$

TABLE 3.1 Examples of Sources G and Fluxes $\boldsymbol{S}$ of the Field F.

F	G	$\boldsymbol{S}$
ρ, mass density	0	0
ρ_j, mass density of the jth chemical component	$\sum_r \nu_{jr}\mathcal{M}_j w_r + Y_j$	$\boldsymbol{J}(\rho_j) = \rho_j(\boldsymbol{v}_j - \boldsymbol{v})$
u_e, electromagnetic energy density	$-\boldsymbol{j}\cdot\boldsymbol{E}$	$\boldsymbol{S}_e$, flux of electromagnetic energy
$u_t = u_e + (u + v^2/2 + \phi)\rho$, total energy density	$\sum_j \rho_j \boldsymbol{v}_j \cdot \boldsymbol{F}_j$	$\boldsymbol{W} + \boldsymbol{S}_e + \boldsymbol{v}\cdot(\mathbf{P} - \mathbf{I}p)$

$\boldsymbol{E}$, electric field; $\boldsymbol{F}_j$, external force; $\mathbf{I}$, unit tensor; $\boldsymbol{j}$, electric current density; $\boldsymbol{J}$, diffusion; $\mathcal{M}_j$, molar mass of the jth constituent; $\mathbf{P}$, dissipative part of the pressure tensor (viscosity); p, pressure; $\boldsymbol{v}_j$, velocity of the jth constituent; $\boldsymbol{v}$, velocity of center of mass; w_r, reaction rate of the rth chemical reaction; $\boldsymbol{W}$, heat flow density; Y_j, rate of biochemical reactions (cf. eqns. 3.85–3.86); ν_{jr}, stoichiometric coefficient of the jth constituent in the rth chemical reaction.

the following differential equation can be derived from the "integral" balance equation 3.1:

$$\frac{\partial F}{\partial t} + \operatorname{div}(\boldsymbol{v}F + \boldsymbol{S}) = G(F, \ldots; \boldsymbol{r}, t). \tag{3.2}$$

The local time variation $\partial F/\partial t$ results from convection $\boldsymbol{v}F$, from nonconvective transport $\boldsymbol{S}$, and from local sources and sinks G. Initial and boundary conditions must be added; for instance,

$$F(\boldsymbol{r}, 0) = F_0(\boldsymbol{r}), \qquad \boldsymbol{r} \in V, \tag{3.3}$$

$$v^{\mathrm{n}}F + S^{\mathrm{n}} = g(F, \boldsymbol{r}, t), \qquad \boldsymbol{r} \in A, t \geq 0. \tag{3.4}$$

3.1.5. Special Types of Balance Equation

Reduction to an Ordinary Differential Equation

In water quality modeling, very often the field variable F is considered to be constant within a finite time-independent volume V, or the mean values are used:

$$\hat{F} = \frac{1}{V}\int_{V(t)} F\,\mathrm{d}V, \qquad G^* = \int_{V(t)} G\,\mathrm{d}V, \qquad S^* = \oint_{A(t)} S^{\mathrm{n}}\,\mathrm{d}A. \tag{3.5}$$

The balance equation 3.1 then reduces to an ordinary differential equation,

$$\frac{\mathrm{d}(V\hat{F}(t))}{\mathrm{d}t} = G^*(\hat{F}, t) - S^*(t), \tag{3.6}$$

which, in general, is also nonlinear. G^* describes changes due to processes inside V, while S^* results from the exchange of matter and energy with the surroundings through the surface A (outflow, $S^* > 0$; inflow, $S^* < 0$). Appropriate initial conditions must be prescribed but, as a rule, in water quality modeling the exchange processes S^* govern the state and development of the system. The initial values are most often of minor consequence.

Averaging Along the Vertical

We derive a balance equation for the mean value $\tilde{F}$ of the field F, taken along the vertical by integrating (3.2) from the bottom of the basin or river to the free surface. Cartesian coordinates (x, y, z) are used, with z increasing toward the zenith. The flow velocity $\boldsymbol{v}$, flux $\mathbf{S}$, and ∇ operator are divided between the horizontal components $\boldsymbol{v}_h$, $\mathbf{S}_h$, ∇_h and the vertical components $\boldsymbol{v}_z$, $\mathbf{S}_z$, $\partial/\partial z$. Furthermore, we introduce: $z = \zeta_0(x, y)$ for the bottom of the water body; $z = \zeta(x, y, t)$, the equation of the free water surface; $H(x, y, t) = \zeta(x, y, t) - \zeta_0(x, y)$, the depth of the water body; and $\tilde{F}(x, y, t) = (1/H) \int_{\zeta_0}^{\zeta} F(x, y, z, t)\mathrm{d}z$, the mean value of F. Analogous definitions are valid for G, $\boldsymbol{v}_h$, and $\mathbf{S}_h$. Then, from (3.2) the following balance equation for $\tilde{F}$ can easily be derived:

$$\frac{\partial(H\tilde{F})}{\partial t} + \nabla_h \cdot (H\tilde{\boldsymbol{v}}_h\tilde{F} + \mathbf{S}^*) = HG^*, \qquad \textbf{(3.7)}$$

where

$$G^* = \tilde{G} + \frac{1}{H}\left(F(x, y, \zeta, t)\frac{\partial\zeta}{\partial t} + (\boldsymbol{v}_h F + \mathbf{S}_h) \cdot \nabla_h H - v_z F + S_z\right)_{z=\zeta_0}^{z=\zeta} \qquad (3.8)$$

and

$$\mathbf{S}^* = \tilde{\mathbf{S}}_h + \frac{1}{H}\int_{\zeta_0}^{\zeta} \boldsymbol{v}'_h F' \,\mathrm{d}z. \qquad (3.9)$$

Here, $\boldsymbol{v}'_h$ and F' denote the deviations of the local values $\boldsymbol{v}_h$ and F from the averages:

$$\boldsymbol{v}'_h = \boldsymbol{v}_h - \tilde{\boldsymbol{v}}_h, \qquad F' = F - \tilde{F}.$$

The source factor G^* in (3.7) consists of the mean value $\tilde{G}$ of the source in (3.2) and of contributions from fluxes through the air–water interface and across the bottom of the basin or river, which are determined by the boundary conditions. The horizontal flux $\mathbf{S}^*$ is determined not only by the average value $\tilde{\mathbf{S}}_h$ of the nonconvective flux $\mathbf{S}_h$, but also by the dispersion due to the deviations $\boldsymbol{v}'_h$ and F', which is very often the dominating process.

Averaging Over the Cross-Sectional Area of a River

In order to derive from (3.2) or (3.7) the type of balance equation used in one-dimensional models of river flow (section 3.2.3 and Chapter 7), we define:

(a) The equations describing the positions of the river banks:

$$y = Y_1(x, t), \qquad y = Y_2(x, t).$$

The water flows in the x direction. The positions of the banks can be functions of time, e.g. if the flow rate varies with time.

(b) The cross-sectional area $A(x, t)$ of the river:

$$A(x, t) = \int_{Y_1(x,t)}^{Y_2(x,t)} \int_{\zeta_0(x,y)}^{\zeta(x,y,t)} \mathrm{d}z\, \mathrm{d}y$$

$$= \int_{Y_1(x,t)}^{Y_2(x,t)} H(x, y, t)\mathrm{d}y. \tag{3.10}$$

(c) The mean value $\bar{F}$ of F across $A(x, t)$ (and analogously the mean values $\bar{G}$ and $\bar{S}_x$):

$$\bar{F}(x, t) = \frac{1}{A} \int_{Y_1}^{Y_2} \int_{\zeta_0}^{\zeta} F(x, y, z, t)\mathrm{d}z\, \mathrm{d}y$$

$$= \frac{1}{A} \int_{Y_1}^{Y_2} H(x, y, t)\tilde{F}(x, y, t)\mathrm{d}y. \tag{3.11}$$

(d) The net downstream velocity:

$$\bar{u}(x, t) = \int_{Y_1}^{Y_2} \int_{\zeta_0}^{\zeta} v_x(x, u, z, t)\mathrm{d}z\, \mathrm{d}y$$

$$= \frac{1}{A} \int_{Y_1}^{Y_2} H(x, y, t)\tilde{u}_x(x, y, t)\mathrm{d}y. \tag{3.12}$$

(e) The mean flow rate:

$$Q(x, t) = \bar{u}(x, t)A(x, t). \tag{3.13}$$

Integrating (3.2) over $A(x, t)$ or (3.7) over the river width, $Y_1 \leq y \leq Y_2$, yields the following "one-dimensional" river flow equation:

$$\frac{\partial(A\bar{F})}{\partial t} + \frac{\partial}{\partial x}(Q\bar{F} + AS^{**}) = AG^{**}, \tag{3.14}$$

where

$$G^{**} = \overline{G^*} + \left[H\left(\tilde{F}\frac{\partial Y}{\partial t} + \tilde{v}_x\tilde{F}\frac{\partial Y}{\partial x} + \tilde{v}_y\tilde{F} + \tilde{S}_y^*\right)\right]_{y=Y_1}^{y=Y_2} \tag{3.15}$$

and

$$S^{**} = \overline{S_x^*} + \frac{1}{A}\int_{Y_1}^{Y_2} H(\tilde{v}_x - \bar{u})(\tilde{F} - \bar{F})\mathrm{d}y. \tag{3.16}$$

In (3.14), the quantities A, $\bar{F}$, Q, S^{**}, and G^{**} may be functions of x and t. According to (3.15), the sources G^{**} include additional terms resulting from boundary effects. To discuss (3.16) we assume that the nonconvective flux $\mathbf{S}$ in (3.2) results from molecular diffusion and turbulent mixing. Then, $\mathbf{S}$ may be approximated by

$$\mathbf{S}(x, y, z, t) = -(D_{\mathrm{M}} + \epsilon)\mathrm{grad}\, F(x, y, z, t), \tag{3.17}$$

where the molecular diffusion coefficient D_{M} and the coefficient ϵ of turbulent mixing (eddy diffusivity) are functions of time and space. If, as usual, it is assumed that

$$S^{**}(x, t) = -E(x, t)\frac{\partial \bar{F}(x, t)}{\partial x}, \tag{3.18}$$

it follows from (3.9) and (3.16) that the effective diffusion coefficient E,

$$E(x, t) = \bar{D}_{\mathrm{M}} + \bar{\epsilon} + D', \tag{3.19}$$

includes effects of molecular and turbulent diffusion as well as the dispersion coefficient D' due to the deviations of the actual fields from their averages. The relative effect of the dispersion term depends on how well the net downstream velocity $\bar{u}$ is determined. If it is well done, dispersion can be minimized or even reduced to the level of computation errors.

When Q, A, E, G^{**}, and $\bar{F}$ are independent of time, (3.14) reduces to

$$\frac{\mathrm{d}}{\mathrm{d}x}\left(Q(x)\bar{F}(x) - A(x)E(x)\frac{\mathrm{d}\bar{F}(x)}{\mathrm{d}x}\right) = A(x)G^{**}(x), \tag{3.20}$$

the solution of which is

$$\begin{aligned}\bar{F}(x) = {} & \bar{F}(x_0)\exp\left(\int_{x_0}^{x}\frac{U(\xi)}{E(\xi)}\,\mathrm{d}\xi\right)\\ & - \int_{x_0}^{x}\frac{Q(x_0)\bar{F}(x_0) - A(x_0)E(x_0)\bar{F}^{*}(x_0)}{A(\xi)E(\xi)}\exp\left(\int_{\xi}^{x}\frac{U(\zeta)}{E(\zeta)}\,\mathrm{d}\zeta\right)\mathrm{d}\xi\\ & - \int_{x_0}^{x}\frac{\int_{x_0}^{\xi}G^{**}(\zeta)A(\zeta)\mathrm{d}\zeta}{A(\xi)E(\xi)}\exp\left(\int_{\xi}^{x}\frac{U(\zeta)}{E(\zeta)}\,\mathrm{d}\zeta\right)\mathrm{d}\xi,\end{aligned} \tag{3.21}$$

where $\bar{F}(x_0)$ is the value of $\bar{F}$ at an arbitrarily chosen point x_0,

$$\bar{F}^{*}(x_0) = \lim_{x\to x_0}\frac{\mathrm{d}\bar{F}(x)}{\mathrm{d}x},$$

$Q(x_0)\bar{F}(x_0)$ is the transport by downstream flow at $x = x_0$,

$-A(x_0)E(x_0)\bar{F}^*(x_0)$ is the downstream transport caused by the effective diffusion at x_0, and

$$U(x) = \frac{Q(x)}{A(x)}$$

is the mean flow rate.

The one-dimensional river flow model will be used in the following chapters of this book.

3.1.6. The Finite-Difference Method

We cannot present the details of mathematical methods for solving the non-linear balance equations 3.1, 3.2, or 3.6. We only wish to give a short outline that will serve to illustrate the application of the finite-difference method.

Applying (3.1) to a time-independent finite volume element V_i, the center (node) of which is situated at $\boldsymbol{r}_i$, and integrating (3.1) over the interval from t_μ to $t_{\mu+1}$, we obtain

$$\int_{V_i} F(\boldsymbol{r}, t_{\mu+1})\mathrm{d}V - \int_{V_i} F(\boldsymbol{r}, t_\mu)\mathrm{d}V$$
$$= \int_{t_\mu}^{t_{\mu+1}} \int_{V_i} G(F, \ldots; \boldsymbol{r}, t)\mathrm{d}V\,\mathrm{d}t - \int_{t_\mu}^{t_{\mu+1}} \sum_m \int_{A_{i,m}} (v^{\mathrm{n}}F + S^{\mathrm{n}})\mathrm{d}A\,\mathrm{d}t. \quad (3.22)$$

$A_{i,m}$ is one of the surface elements of V_i, and Σ_m is the sum over all the elements (Figure 3.2).

Let us introduce the following quantities, abbreviations, and approximations:

$$h_\mu = t_{\mu+1} - t_\mu, \qquad h_{ik} = |\boldsymbol{r}_i - \boldsymbol{r}_k| \quad (3.23a)$$

$$F_{i,\mu}V_i = F(\boldsymbol{r}_i, t_\mu)V_i = \int_{V_i} F(\boldsymbol{r}, t_\mu)\mathrm{d}V \quad (3.23b)$$

$$\int_{t_\mu}^{t_{\mu+1}} \int_{V_i} G\,\mathrm{d}V\,\mathrm{d}t = \tfrac{1}{2}(G_{i,\mu+1} + G_{i,\mu})V_i h_\mu \quad (3.23c)$$

$$E = v^{\mathrm{n}}F + S^{\mathrm{n}} \quad (3.23d)$$

$$\int_{t_\mu}^{t_{\mu+1}} \sum_m \int_{A_{i,m}} E\,\mathrm{d}f\,\mathrm{d}t = \tfrac{1}{2}\sum_m \int_{t_\mu}^{t_{\mu+1}} (E(\boldsymbol{r}_m, t) + E(\boldsymbol{r}_i, t))A_{i,m}\,\mathrm{d}t$$
$$= \tfrac{1}{4}\sum_m (E_{m,\mu+1} + E_{m,\mu} + E_{i,\mu+1} + E_{i,\mu})A_{i,m}h_\mu. \quad (3.23e)$$

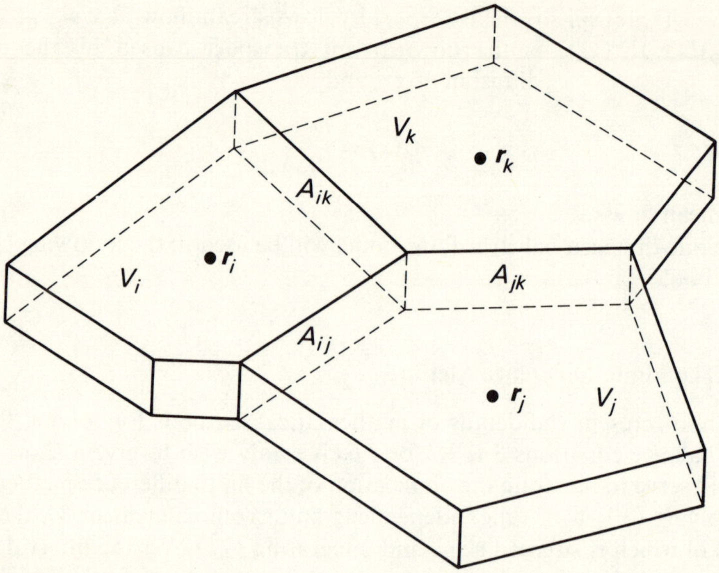

FIGURE 3.2 Example of a simple system of finite volume elements V_i, V_j, V_k suitable for the application of finite-difference methods for solving the balance equation.

Using (3.23) in (3.22), we arrive at

$$\frac{1}{h_\mu}(F_{i,\mu+1} - F_{i,\mu}) = \tfrac{1}{2}(G_{i,\mu+1} + G_{i,\mu})$$
$$- \tfrac{1}{4}\sum_m \frac{A_{i,m}}{V_i}(E_{m,\mu+1} + E_{m,\mu} + E_{i,\mu+1} + E_{i,\mu}), \qquad (3.24)$$

where

$$E_{k,\nu} = v^{\mathrm{n}}_{k,\nu} F_{k,\nu} + S^{\mathrm{n}}_{k,\nu}.$$

If $G = G(F, \ldots; \boldsymbol{r}, t)$, then $G_{k,\nu}$ depends on $F_{k,\nu}$. Thus, eqns. 3.24 are, in general, also nonlinear. These equations are the basis for the solution of (3.1) and (3.2) by the finite-difference method. The volume V of the water body must be divided into elements V_i ($i = 1, 2, 3, \ldots$) and the resulting system of equations 3.24 must be solved step by step ($\mu = 1, 2, 3, \ldots$). One possibility for linearization consists in using $F_{k,\mu}$ instead of $F_{k,\mu+1}$ on the right-hand side of (3.24), especially if it is assumed that $G_{i,\mu+1} \approx G_{i,\mu}$.

Equations 3.24 may easily be transformed into the following system of equations:

$$B_{ii}F_{i,\mu+1} + \sum_m B_{im}F_{m,\mu+1} = P_{i,\mu} \qquad (i \neq m), \qquad \mathbf{(3.25)}$$

where

$$B_{ii} = 1 + \sum_j \frac{h_\mu A_{ij}}{4V_i} v^{\mathrm{n}}_{i,\mu+1}$$

$$B_{im} = \frac{h_\mu A_{im}}{4V_i} v^{\mathrm{n}}_{m,\mu+1} \qquad (i \neq m) \tag{3.26}$$

and

$$P_{i,\mu} = F_{i,\mu} + \tfrac{1}{2}h_\mu(G_{i,\mu+1} + G_{i,\mu}) - \sum_m \frac{h_\mu A_{im}}{4V_i}(S^{\mathrm{n}}_{m,\mu+1} + S^{\mathrm{n}}_{i,\mu+1} + E_{m,\mu} + E_{i,\mu}). \tag{3.27}$$

Here, $i, j, m = 1, 2, 3, \ldots$ are the numbers of the space elements. Written in matrix form, (3.25) becomes

$$[\mathbf{B}]\{F\} = \{P\}. \tag{3.28}$$

It should be pointed out that $\{P\}$ comprises $\{F\}$ if G depends on F. If this is so, an iteration method must be used for solving (3.28). An example of the application of the finite-difference method is given, for instance, in Chapter 7. Finite-difference methods are widely treated in the literature, e.g. by Collatz (1959), Richtmyer and Morton (1967), and Marchuk (1975).

3.1.7. The Finite-Element Method

The initial-boundary-value problem represented by (3.2)–(3.4) can be transformed into an initial-value problem of an ordinary differential equation by the Galerkin method. It is supposed that the volume V of the water body is divided into space elements V_k ($k = 1, 2, 3, \ldots$) and that a set of weighting functions $w_k(\boldsymbol{r})$ ($k = 1, 2, 3, \ldots$) is given that fulfills the conditions

$$w_k(\boldsymbol{r}_k) = 1, \qquad w_k(\boldsymbol{r}_i) = 0 \text{ if } i \neq k$$

$$w_k(\boldsymbol{r}) \neq 0 \text{ inside } V_k \tag{3.29}$$

$$w_k(\boldsymbol{r}) = 0 \text{ outside } V_k.$$

At least the first derivatives of $w_k(\boldsymbol{r})$ are square-integrable. Figure 3.3 shows a special type of two-dimensional finite element V_i. The two-dimensional weighting function $w_k(x, y)$ identically vanishes outside V_k. We are seeking the solution $F(\boldsymbol{r}, t)$ of (3.2)–(3.4) in the form of a spline interpolation:

$$F(\boldsymbol{r}, t) = \sum_{j=1}^{N} a_j(t) w_j(\boldsymbol{r}) \tag{3.30}$$

$$F(\boldsymbol{r}, 0) = \sum_{j=1}^{N} a_j(0) w_j(\boldsymbol{r}), \qquad F(\boldsymbol{r}_k, t) = a_k(t). \tag{3.31}$$

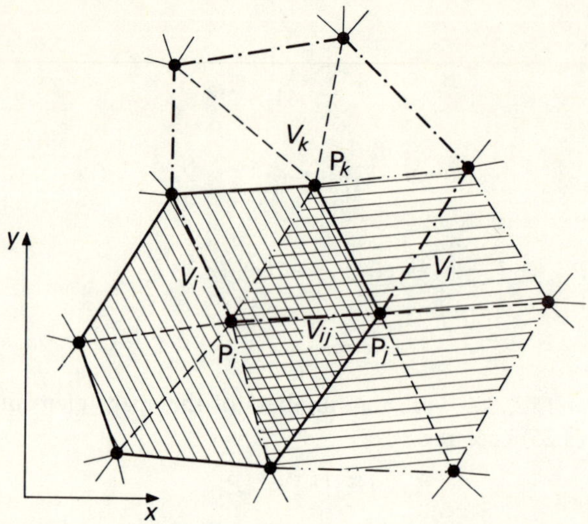

FIGURE 3.3 Three two-dimensional finite elements V_i, V_j, V_k, each consisting of six triangles. V_{ij} is common to V_i and V_j. $V_{ij} \neq 0$ only if P_i and P_j are neighboring points.

Multiplying (3.2) by $w_k(\boldsymbol{r})$, integrating over V, and using Green's theorem, the boundary conditions (3.4), and the expansion (3.30), we produce the following system of equations ($k = 1, 2, 3, \ldots$), from which the unknown functions $a_j(t)$ can be derived:

$$\sum_{j=1}^{N} \left(\frac{\mathrm{d}a_j(t)}{\mathrm{d}t} A_{jk} - a_j(t) B_{jk} \right) = C_k, \qquad \textbf{(3.32)}$$

where

$$A_{jk} = \int_V w_j(\boldsymbol{r}) w_k(\boldsymbol{r}) \mathrm{d}V = A_{kj} \qquad (3.33)$$

$$B_{jk} = \int_V w_j(\boldsymbol{r}) \boldsymbol{v} \cdot \mathrm{grad}\, w_k(\boldsymbol{r}) \mathrm{d}V \neq B_{kj} \qquad (3.34)$$

$$C_k(t) = \int_V (G(F, \ldots; \boldsymbol{r}, t) w_k(\boldsymbol{r}) + \boldsymbol{S} \cdot \mathrm{grad}\, w_k(\boldsymbol{r})) \mathrm{d}V$$
$$- \oint_A w_k(\boldsymbol{r}) g(F, \ldots; \boldsymbol{r}, t) \mathrm{d}A. \qquad (3.35)$$

On account of the conditions (3.29), the integrals over V in (3.33)–(3.35) are reduced to integrals over the small volume V_{jk} common to V_j and V_k. The matrix elements A_{jk} and B_{jk} are equal to zero if the finite elements V_j and V_k do not intersect. Thus, the matrices are sparse (with diagonal predominance and banding). A_{jk} and B_{jk} can be evaluated for very complicated geometries of V if triangular, rectangular, or curvilinear volume elements V_k and suitable functions $w_k(\boldsymbol{r})$ are defined. Elements can be varied in size, shape, and orientation to give the modeler the maximum opportunity to describe correctly the geometrical properties of the water body.

With the help of a difference method (e.g. the Crank–Nicolson method), we can integrate (3.32) with respect to time, starting from the values

$$a_j(0) = F(\boldsymbol{r}_j, 0), \qquad j = 1\,(1)\,N, \tag{3.36}$$

which result from (3.31). In general, the right-hand side, C_k, of (3.32) depends on the unknowns $a_k(t)$; therefore, iteration procedures are to be applied.

Finite-difference and finite-element methods are powerful tools for the numerical solution of the much more complicated system of differential equations characterizing a deterministic water quality model. The reader should find the following literature relevant: Collatz (1959), Richtmyer and Morton (1967), Zienkiewicz and Cheung (1967), Strang and Fix (1973), Marchuk (1975), Mitchel and Wait (1977), and Pinder and Gray (1977). We now turn to the basic equations of water quality models.

3.2. CONSERVATION OF MASS—ADVECTION AND DIFFUSION

3.2.1. Continuity Equation, Mass Fractions

In water quality modeling we use a description in terms of a limited number of macroscopic observables, such as the density of water, ρ_0 [ML^{-3}], the mass density ρ_j of the jth chemical component ($j = 1, 2, 3, \ldots$), and the mass densities ρ_k of the biotic components of the aquatic ecosystem ($k = 1, 2, 3, \ldots$). We also introduce the mass fractions

$$N_j = \frac{\rho_j}{\rho}, \qquad B_k = \frac{\rho_k}{\rho} \tag{3.37}$$

and the total mass density ρ:

$$\rho = \rho_0 + \sum_j \rho_j + \sum_k \rho_k = \sum_i \rho_i, \tag{3.38}$$

which fulfills the continuity equation

$$\frac{\partial \rho}{\partial t} + \operatorname{div}(\rho \boldsymbol{v}) = 0 \qquad \text{or} \qquad \frac{d\rho}{dt} + \rho \operatorname{div} \boldsymbol{v} = 0. \tag{3.39}$$

3.2.2. Chemical Reactions

Chemical reactions (with reaction rate w_r [mol L^{-3} T^{-1}] for the rth reaction) and biochemical processes Y_j [ML^{-3} T^{-1}] are to be taken into account in the mass balances of the chemical components:

$$\underset{}{\frac{\partial \rho_j}{\partial t}} + \underset{\text{convection}}{\operatorname{div}(\rho_j \boldsymbol{v})} + \underset{\text{diffusion}}{\operatorname{div} \boldsymbol{J}(\rho_j)} = \underset{\substack{\text{chemical}\\\text{reactions}}}{\sum_r \nu_{jr} \mathscr{M}_j w_r} + \underset{\substack{\text{biochemical}\\\text{reactions}}}{Y_j} \qquad \textbf{(3.40)}$$

or†

$$\rho \frac{\mathrm{d}N_j}{\mathrm{d}t} + \operatorname{div} \boldsymbol{J}(N_j) = \sum_r \nu_{jr} \mathscr{M}_j w_r + Y_j. \qquad \textbf{(3.41)}$$

Equations 3.41 follow from (3.40), by substitution of (3.37) and (3.39). The essential parts of (3.40) and (3.41) are:

convective displacement:

$$\rho_j \boldsymbol{v};$$

velocity of center of mass:

$$\boldsymbol{v} = \sum_i \frac{\rho_i \boldsymbol{v}_i}{\rho}; \qquad (3.42)$$

and effective diffusion consisting of molecular diffusion, turbulent "eddy diffusivity," and other types of random mixing processes:

$$\boldsymbol{J}(\rho_j) = \rho_j(\boldsymbol{v}_j - \boldsymbol{v}) = \rho N_j(\boldsymbol{v}_j - \boldsymbol{v}) = \boldsymbol{J}(N_j). \qquad (3.43)$$

The scales of all the mixing phenomena and diffusion processes are of particular importance for the validity of assumptions like

$$\boldsymbol{J}(\rho_j) = -D \operatorname{grad} \rho_j \qquad \text{or} \qquad \boldsymbol{J}(N_j) = -D^* \operatorname{grad} N_j, \qquad (3.44)$$

where D [L^2 T^{-1}] and D^* are scalar diffusion coefficients or diffusion tensors of rank 2. Furthermore, in (3.40) and (3.41) we use:

- ν_{jr} the stoichiometric coefficient of the chemical reaction ($\nu_{jr} > 0$ for products, but $\nu_{jr} < 0$ for reactants),
- $\mathscr{M}_j$ the molar mass of the jth chemical component.

Even in simple cases the reaction rates w_r are nonlinear functions of the densities ρ_j of the reacting chemicals. Consequently, the mass balances (3.40) and (3.41)

† $\frac{\partial}{\partial t}(\rho N_j) + \operatorname{div}(\rho N_j \boldsymbol{v}) = \rho \frac{\mathrm{d}N_j}{\mathrm{d}t} + N_j\left(\frac{\partial \rho}{\partial t} + \operatorname{div}(\rho \boldsymbol{v})\right) = \rho \frac{\mathrm{d}N_j}{\mathrm{d}t}.$

are nonlinear partial differential equations. These "diffusion equations" are a common basis for the development of the majority of water quality models in use today. For a conservative substance (of concentration C) from (3.40) and (3.44) we obtain

$$\frac{\partial C}{\partial t} + \operatorname{div}(C\boldsymbol{v}) = \operatorname{div}(D \operatorname{grad} C). \tag{3.45}$$

The values of the diffusion coefficient D must be derived from field experience. Dependence on some empiricism cannot be avoided.

3.2.3. Advection–Diffusion Equations; Input–Output Models

Applying the methods described in section 3.1.5 (especially eqns. 3.7 and 3.14) to eqn. 3.40 or 3.45, we obtain two- and one-dimensional mass transport equations, called "advection–diffusion equations."

One-dimensional river flow models, which can be employed, for instance, to predict the biochemical oxygen demand (BOD) and/or the dissolved oxygen concentration (DO) in the river, are based on the differential equation 3.14, i.e.

$$\frac{\partial(AC)}{\partial t} + \frac{\partial}{\partial x}\left(QC - AE\frac{\partial C}{\partial x}\right) = AG. \tag{3.46}$$

The cross-sectional area A, the concentration C of the particular constituent $[\mathrm{M\,L^{-3}}]$, the net downstream flow Q $[\mathrm{L^3\,T^{-1}}]$, and the effective diffusion coefficient E may depend not only on time t but also on the downstream distance x. The source–sink term G is a function of C, x, and t. Equation 3.19 defines E. Table 3.2 gives an overview of mass balance equations of different types.

If we assume that a well mixed lake has a constant volume V_0, constant discharge Q_0 $[\mathrm{L^3\,T^{-1}}]$, constant input G_0 $[\mathrm{M\,T^{-1}}]$, and constant net decay coefficient K_0 $[\mathrm{T^{-1}}]$ of a nonconservative substance, we obtain the following input–output model of type (3.6):

$$\frac{\mathrm{d}C(t)}{\mathrm{d}t} = \frac{G_0}{V_0} - KC(t), \qquad K \equiv K_0 + \frac{Q_0}{V_0}. \tag{3.47}$$

Integration of (3.47) from the initial concentration $C(0)$ yields

$$C(t) = \frac{G_0}{KV_0} + \left(C(0) - \frac{G_0}{KV_0}\right)\exp(-Kt). \tag{3.48}$$

Models of this kind have been used, for instance by Vollenweider (1969), to predict the phosphorus concentrations in large lakes.

TABLE 3.2 Mass Balances and Advection–Diffusion Equations of Different Types.

Continuity equation for a single substance ($\rho_j = C$):	$\frac{\partial C}{\partial t} + \operatorname{div}(\boldsymbol{v}C + \boldsymbol{J}(C)) = G(C, \boldsymbol{r}, t)$, derived from	(3.40)
Use of $\boldsymbol{J}(C) = -D \operatorname{grad} C$:	$\frac{\partial C}{\partial t} + \operatorname{div}(\boldsymbol{v}C - D \operatorname{grad} C) = G(C, \boldsymbol{r}, t)$	(3.50)
One-dimensional case, $\boldsymbol{v} = \{\bar{u}(x, t), 0, 0\}$, etc.:	$\frac{\partial AC}{\partial t} + \frac{\partial}{\partial x}\left(QC - AE\frac{\partial C}{\partial x}\right) = AG(C, x, t)$	(3.46)
One-dimensional case without sources and sinks:	$\frac{\partial AC}{\partial t} + \frac{\partial}{\partial x}\left(QC - AE\frac{\partial C}{\partial x}\right) = 0$	
Integration of (3.50) over a finite time-independent volume V_0 and application of Green's theorem:	$\frac{\mathrm{d}}{\mathrm{d}t}\int_{V_0} C\,\mathrm{d}V + \oint_{A_V} (\boldsymbol{v}C - D \operatorname{grad} C)\mathrm{d}\boldsymbol{A} = \int_{V_0} G(C, \boldsymbol{r}, t)\mathrm{d}V$	
Introduction of the time-dependent mean values $\overline{C}(t), \overline{S}(t), \overline{G}(t)$:	$V_0 \frac{\mathrm{d}\overline{C}(t)}{\mathrm{d}t} = \underbrace{\overline{S}(t)}_{\text{input–output}} + \underbrace{V_0\overline{G}(t)}_{\text{internal production/destruction}}$	

In the special case of $\overline{G} = -K_0\overline{C}$ and $\overline{S} = G_0 - Q_0\overline{C}$, we obtain (3.47).

3.2.4. Biotic Components

The balance equations of the biotic components are of the type

$$\frac{\partial \rho_k}{\partial t} + \operatorname{div}(\rho_k \boldsymbol{v}) + \operatorname{div} \boldsymbol{J}(\rho_k) = Q_k - \rho_k R_k - \rho_k M_k, \tag{3.49}$$

$$\rho \frac{\mathrm{d}B_k}{\mathrm{d}t} + \operatorname{div} \boldsymbol{J}(B_k) = Q_k - \rho_k R_k - \rho_k M_k, \tag{3.50}$$

where, for $k = 1, 2, 3, \ldots,$

Q_k is the production and consumption term $[\mathrm{M\ L^{-3}\ T^{-1}}]$,
R_k is the rate of respiration of the kth component $[\mathrm{T^{-1}}]$,
M_k is the nonpredatory mortality rate of the kth biocomponent $[\mathrm{T^{-1}}]$.

Again, these equations are of the "advection–diffusion" type. The sources and sinks Q_k, R_k, M_k will be discussed in section 3.5.

3.2.5. Coupling of Chemical and Biotic Components

The chemical constituents and the biotic components are coupled in a very complicated manner by biochemical processes, represented by Y_j in (3.40), and by terms on the right-hand sides of (3.49) and (3.50). To illustrate these effects we use the balance of dissolved oxygen as an example. The local concentration of dissolved oxygen is influenced by chemical processes, such as oxidation and reduction, and by physical processes like advection, diffusion, and surface aeration. The coupling to biotic components is effected through photosynthetic oxygenation, respiration, and biochemical oxidation processes. The coupling of chemical and biotic components of the aquatic ecosystem is termed "biochemical" and is illustrated in Figure 3.4 (section 3.5).

3.3. CONSERVATION OF MOMENTUM; TYPES OF FLOW SYSTEMS

3.3.1. Navier–Stokes Equations

The hydrodynamic processes in streams, lakes, and reservoirs are integral components of the complex aquatic ecosystem. Water movements at different scales and of different types significantly influence not only the aggregation and distribution of microorganisms and plankton, but also the distribution of nutrients, dissolved gases, and temperature. Hence, circulation, convection, wave phenomena, and turbulent mixing are major influences on the distribution of biota and the productivity of natural water bodies.

A realistic hydrodynamic description of the behavior of the water body should be founded as far as possible on the basic hydrodynamic equations. Conservation of momentum is ensured by

$$\rho \frac{d\boldsymbol{v}}{dt} + \underbrace{2\rho\boldsymbol{\omega} \times \boldsymbol{v}}_{\text{Coriolis force}} = \underbrace{\sum_i \rho_i \boldsymbol{F}_i - \rho \operatorname{grad} \phi}_{\text{external forces}} - \underbrace{\operatorname{grad} p}_{\text{pressure}} + \underbrace{\nabla \cdot \mathbf{P}}_{\text{friction}}, \qquad \mathbf{(3.51)}$$

where

- $\boldsymbol{\omega}$ is the angular velocity of the Earth's rotation $[\mathrm{T}^{-1}]$,
- $\boldsymbol{F}_i$ represents the nonconservative external forces (per unit mass) acting on the component of density ρ_i,
- ϕ is the time-independent potential of external forces (e.g. gravity),
- p is the pressure $[\mathrm{M\,L^{-1}\,T^{-2}}]$,
- $\mathbf{P}$ is the dissipative part of the pressure tensor, i.e. the friction tensor including turbulent friction $[\mathrm{M\,L^{-1}\,T^{-2}}]$.

Viscous and turbulent stresses can be combined linearly and represented by

$$\nabla \cdot \mathbf{P} = \nabla \cdot \mathbf{A} \cdot \nabla \boldsymbol{v} \qquad \text{(Boussinesq/Prandtl)}, \qquad (3.52)$$

where $\mathbf{A}$ is the "eddy diffusivity tensor" (of rank 2). In isotropic fluid flows $\mathbf{A}$ reduces to a scalar quantity and (3.52) becomes

$$\nabla \cdot \mathbf{P} = \operatorname{div}(A \operatorname{grad} \boldsymbol{v}) = A\nabla \boldsymbol{v} + \operatorname{grad} A \cdot \nabla \boldsymbol{v}. \qquad (3.53)$$

The coefficient A or the tensor $\mathbf{A}$ must be determined, respectively, by measurements or by statistical investigations.

The Coriolis acceleration $2\boldsymbol{\omega} \times \boldsymbol{v}$ can be important in large embayments that are nearly circular and have diameters of 30 km or more. Using the transformation

$$\frac{d\boldsymbol{v}}{dt} = \frac{\partial \boldsymbol{v}}{\partial t} + \boldsymbol{v} \cdot \nabla \boldsymbol{v} = \frac{\partial \boldsymbol{v}}{\partial t} + \nabla(\tfrac{1}{2}v^2) - \boldsymbol{v} \times \operatorname{rot} \boldsymbol{v}, \qquad (3.54)$$

we obtain from (3.51):

$$\frac{\partial \boldsymbol{v}}{\partial t} - \boldsymbol{v} \times (2\boldsymbol{\omega} + \operatorname{rot} \boldsymbol{v}) = \sum_i \frac{\boldsymbol{F}_i \rho_i}{\rho} - \operatorname{grad}(\tfrac{1}{2}v^2 + \phi) - \frac{1}{\rho} \operatorname{grad} p + \nabla \cdot \mathbf{P}. \qquad \mathbf{(3.55)}$$

If nonconservative external forces are absent ($\boldsymbol{F}_i \equiv 0$) and if pressure and density are directly related, i.e. $\rho = \rho(p)$, we introduce

$$P_v(p) = \int_{p_0}^{p} \frac{dp'}{\rho(p')} \qquad (3.56)$$

and arrive at

$$\frac{\partial \boldsymbol{v}}{\partial t} - \boldsymbol{v} \times (2\boldsymbol{\omega} + \operatorname{rot} \boldsymbol{v}) = \nabla \cdot \mathbf{P} - \operatorname{grad}(\tfrac{1}{2}v^2 + \phi + P_v). \tag{3.57}$$

Modern mathematics offers a variety of tools for solving hydrodynamic problems (e.g. Lions, 1969). "Practical" solutions are normally computed using simplified basic equations. Some examples of those equations are summarized in the next section. Initial and boundary conditions must be added (cf. section 3.6.1).

3.3.2. Special Types of Fluid Flow

From (3.57) the following special cases can easily be derived.

(a) Stationary, rotational flow of a nonviscous fluid:

$$\tfrac{1}{2}v^2 + \phi + P_v = \text{constant along streamlines.} \tag{3.58}$$

(b) Irrotational motion of a nonviscous fluid (Coriolis acceleration omitted):

$$\operatorname{rot} \boldsymbol{v} = 0, \qquad \boldsymbol{v} = \operatorname{grad} \Psi(\boldsymbol{r}, t),$$

$$\frac{\partial \Psi}{\partial t} + \tfrac{1}{2}v^2 + \phi + P_v = f(t). \tag{3.59}$$

The kinetic energy per unit mass, $\tfrac{1}{2}v^2$, depends on Ψ. The quantities Ψ and P_v are unknown. A second equation results from the continuity equation (3.39):

$$\frac{\partial \rho}{\partial t} + \operatorname{grad} \Psi \cdot \operatorname{grad} \rho + \rho \Delta \Psi = 0. \tag{3.60}$$

Equations 3.56, 3.59, and 3.60 form a mathematically closed system.

(c) Stationary irrotational motion of a nonviscous incompressible fluid ($d\rho/dt = 0$) without Coriolis acceleration: from (3.60) and (3.59),

$$\Delta \Psi(\boldsymbol{r}) = 0, \tag{3.61}$$

where Δ is the Laplace operator, and

$$\tfrac{1}{2}v^2 + \frac{p}{\rho} + \phi = \text{constant (Bernoulli).} \tag{3.62}$$

The solution of (3.61) provides the velocity field, $\boldsymbol{v} = \operatorname{grad} \Psi$, from which, after substitution in (3.62), the pressure can be calculated.

(d) "Quasistatic approximation" (where vertical accelerations are neglected): the Coriolis force is approximated by its horizontal component,

$$2\boldsymbol{\omega} \times \boldsymbol{v} \approx \boldsymbol{e}_3 f \times \boldsymbol{v}_{\mathrm{h}} \qquad (f \equiv 2\omega \cos \Theta), \tag{3.63}$$

where $\boldsymbol{e}_3$ is the unit vector toward the zenith, f is the "Coriolis parameter," $\boldsymbol{\omega}$ is the angular velocity of the Earth's rotation, Θ is the geographical polar distance, and $\boldsymbol{v}_{\mathrm{h}} = \boldsymbol{e}_1 v_1 + \boldsymbol{e}_2 v_2$ is the horizontal component of the velocity. It follows that $\boldsymbol{v} = \boldsymbol{v}_{\mathrm{h}} + \boldsymbol{e}_3 v_3$. If we assume that $\boldsymbol{F}_i \equiv 0$ and that horizontal and vertical eddy viscosity coefficients A_{h} and A_3 differ from each other, then from (3.51) and (3.52):

$$\rho \frac{\mathrm{d}v_1}{\mathrm{d}t} - fv_2 + \frac{\partial p}{\partial x_1} = \frac{\partial}{\partial x_1}\left(A_{\mathrm{h}} \frac{\partial v_1}{\partial x_1}\right) + \frac{\partial}{\partial x_2}\left(A_{\mathrm{h}} \frac{\partial v_1}{\partial x_2}\right) + \frac{\partial}{\partial x_3}\left(A_3 \frac{\partial v_1}{\partial x_3}\right)$$

$$\rho \frac{\mathrm{d}v_2}{\mathrm{d}t} + fv_1 + \frac{\partial p}{\partial x_2} = \frac{\partial}{\partial x_1}\left(A_{\mathrm{h}} \frac{\partial v_2}{\partial x_1}\right) + \frac{\partial}{\partial x_2}\left(A_{\mathrm{h}} \frac{\partial v_2}{\partial x_2}\right) + \frac{\partial}{\partial x_3}\left(A_3 \frac{\partial v_2}{\partial x_3}\right)$$

$$\frac{\partial p}{\partial x_3} = \rho g. \tag{3.64}$$

In (3.64) we have used the gravitational acceleration g and the operator

$$\frac{\mathrm{d}}{\mathrm{d}t} = \frac{\partial}{\partial t} + \boldsymbol{v} \cdot \nabla = \frac{\partial}{\partial t} + \sum_{i=1}^{3} v_i \frac{\partial}{\partial x_i}.$$

Equations 3.64 are characteristic of "two-dimensional" circulation models of lakes (single-layer and multilayer models). Boundary conditions comprise surface wind stresses, bottom drag, and heat fluxes through the surface. Vertical velocities are computed from horizontal flow divergences if, additionally, the fluid is supposed to be incompressible:

$$\operatorname{div} \boldsymbol{v} = 0, \qquad \frac{\partial v_3}{\partial x_3} = -\frac{\partial v_1}{\partial x_1} - \frac{\partial v_2}{\partial x_2}. \tag{3.65}$$

(e) Equations describing the gross circulation patterns and the average mass transport in well mixed water bodies are obtained by integration over depth from the bottom of the basin to the free surface. However, from an ecological point of view even small parameter variations along the vertical axis may be very important. Biological cycles are related to solar influx at the air–water interface as well as to benthic processes and to vertical exchange in the water column.

(f) Ekman-type models of wind-generated currents in the ocean and in large lakes result from (3.64) and (3.65) by omission of the accelerations dv_i/dt. Vertical shearing stresses $A_3\, \partial^2 v_j/\partial x_3^2$ ($j = 1, 2$) are balanced by the Coriolis force (and by a pressure gradient, if a slope current exists). Table 3.3 provides simplified balance equations.

3.3.3. Waves, Currents, and Circulation; Three-Dimensional Models

Wind-generated surface waves (progressive waves) in shallow water can prevent aquatic plants from growing and may transport recently deposited material to deeper areas. Currents and circulation systems, but also internal waves and seiches in lakes and reservoirs, are of more or less importance in controlling the extent of heat intrusion into the hypolimnion. They may also influence nutrient return from the hypolimnion to the epilimnion, for instance in late summer and fall. For large water bodies with significant spatial gradients in nutrients and in biomass, three-dimensional hydrodynamic models are necessary, since some of the most important physical influences on the ecosystem originate from such relatively short-term phenomena as upwelling and from eddy diffusivity in localized areas. Turbulent eddies are able to circulate plankton and to influence sinking rates. Concentrated nutrient loadings may be redistributed and transported into nutrient-poor zones, etc.

Classical wave theory (Lamb, 1932) is based on (3.59) and (3.60). Modern theory involves additionally the concepts of random processes. Since the aim of this chapter is to summarize basic concepts and equations, we have chosen to eliminate details of the theories of waves, currents, circulation systems, upwelling, etc. from the discussion. The interested reader is referred to a number of excellent texts on these subjects, cited at the end of section 3.3.

3.3.4. Characteristic Magnitudes, Dimensionless Equations, and Dynamic Similarity

The characteristic magnitudes (maximum values) of the terms of the Navier–Stokes equation,

$$\frac{\partial \boldsymbol{v}}{\partial t} + \boldsymbol{v} \cdot \nabla \boldsymbol{v} + 2\boldsymbol{\omega} \times \boldsymbol{v} = -\boldsymbol{g} - \frac{1}{\rho}\,\mathrm{grad}\, p + \frac{1}{\rho} \nabla \cdot \mathbf{P}, \tag{3.66}$$

may be denoted by $|\partial \boldsymbol{v}/\partial t|_{\mathrm{m}}$, $|\boldsymbol{v} \cdot \nabla \boldsymbol{v}|_{\mathrm{m}}$, $|2\boldsymbol{\omega} \times \boldsymbol{v}|_{\mathrm{m}}$, etc. Here, $\boldsymbol{g}$ is the gravity vector. Introducing

$$\frac{\partial \boldsymbol{v}}{\partial t} = \left|\frac{\partial \boldsymbol{v}}{\partial t}\right|_{\mathrm{m}} \left(\frac{\partial \boldsymbol{v}}{\partial t}\right)', \qquad \boldsymbol{v} \cdot \nabla \boldsymbol{v} = |\boldsymbol{v} \cdot \nabla \boldsymbol{v}|_{\mathrm{m}} (\boldsymbol{v} \cdot \nabla \boldsymbol{v})',$$

TABLE 3.3 Simplified Versions of the Balance of Momentum.

Using eqn. 3.52 in eqn. 3.51 and assuming:	. . . we arrive at the following balance of momentum:	
(i) No nonconservative forces $\boldsymbol{F}_i$,	$\rho \dfrac{d\boldsymbol{v}}{dt} + 2\rho\boldsymbol{\omega} \times \boldsymbol{v} + \rho \operatorname{grad} \phi = -\operatorname{grad} p + \nabla \cdot \mathbf{A} \cdot \nabla \boldsymbol{v}$	
(ii) No vertical component of $\boldsymbol{v}$ or of the Coriolis force (cf. eqn. 3.63),	$\rho \dfrac{d\boldsymbol{v}_h}{dt} + \rho \boldsymbol{e}_3 f \times \boldsymbol{v}_h = -\operatorname{grad}_h p + \nabla \cdot \mathbf{A} \cdot \nabla \boldsymbol{v}_h$ $\rho \dfrac{\partial \phi}{\partial x_3} = -\dfrac{\partial p}{\partial x_3}$	(3.64)
(iii) No acceleration, $\mathbf{A}$ becomes a constant scalar A,	$\rho \boldsymbol{e}_3 f \times \boldsymbol{v}_h = -\operatorname{grad}_h p + A \Delta \boldsymbol{v}_h$	
(iv) $\boldsymbol{v}_h = \boldsymbol{v}_h(z)$, $p = p(z)$,	$\rho \boldsymbol{e}_3 f \times \boldsymbol{v}_h(z) = A \dfrac{d^2 \boldsymbol{v}_h(z)}{dz^2}$ Ekman	
(v) One-dimensional flow in the x direction: $\boldsymbol{v} = \{u(x, t), 0, 0\}$, without nonconservative forces and Coriolis acceleration,	$\rho \dfrac{\partial u}{\partial t} + \rho u \dfrac{\partial u}{\partial x} + \rho \dfrac{\partial \phi}{\partial x} = -\dfrac{\partial p}{\partial x} + \dfrac{\partial}{\partial x}\left(A \dfrac{\partial u}{\partial x}\right)$	

etc. into (3.66) and dividing, for instance, by $|\boldsymbol{v}\cdot\nabla\boldsymbol{v}|_{\mathrm{m}}$, we obtain the dimensionless form:

$$\frac{1}{\mathrm{St}}\left(\frac{\partial v}{\partial t}\right)' + (\boldsymbol{v}\cdot\nabla\boldsymbol{v})' + \frac{(2\boldsymbol{\omega}\times\boldsymbol{v})'}{\mathrm{Ro}} = -\frac{\boldsymbol{g}'}{\mathrm{Fr}} - \frac{[(1/\rho)\mathrm{grad}\, p]'}{\mathrm{Eu}} + \frac{1}{\mathrm{Re}}[(1/\rho)\nabla\cdot\mathbf{P}]', \qquad \mathbf{(3.67)}$$

where

$$\mathrm{Eu} = \frac{|\boldsymbol{v}\cdot\nabla\boldsymbol{v}|_{\mathrm{m}}}{|(1/\rho)\nabla p|_{\mathrm{m}}} \qquad \text{Euler or Ruark number}$$

$$\mathrm{Fr} = \frac{|\boldsymbol{v}\cdot\nabla\boldsymbol{v}|_{\mathrm{m}}}{|\boldsymbol{g}|_{\mathrm{m}}} \qquad \text{Froude number}$$

$$\mathrm{Re} = \frac{|\boldsymbol{v}\cdot\nabla\boldsymbol{v}|_{\mathrm{m}}}{|(1/\rho)\nabla\cdot\mathbf{P}|_{\mathrm{m}}} \qquad \text{Reynolds number} \qquad (3.68)$$

$$\mathrm{Ro} = \frac{|\boldsymbol{v}\cdot\nabla\boldsymbol{v}|_{\mathrm{m}}}{|2\boldsymbol{\omega}\times\boldsymbol{v}|_{\mathrm{m}}} \qquad \text{Rossby number}$$

$$\mathrm{St} = \frac{|\boldsymbol{v}\cdot\nabla\boldsymbol{v}|_{\mathrm{m}}}{|\partial\boldsymbol{v}/\partial t|_{\mathrm{m}}} \qquad \text{Strouhal number.}$$

If all these flow numbers are very small compared with unity, the nonlinearity $\boldsymbol{v}\cdot\nabla\boldsymbol{v}$ may be neglected in (3.66). It is possible to estimate these numbers with the help of the characteristic magnitudes of the variables $\boldsymbol{v}, \rho, p, \boldsymbol{r}, t$ or the "similarity constants" U, R, P, L, T. Two fluid motions are said to be dynamically similar if they are related by

$$\boldsymbol{v} = U\hat{\boldsymbol{v}}, \qquad \rho = R\hat{\rho}, \qquad p = P\hat{p}, \qquad \boldsymbol{r} = L\hat{\boldsymbol{r}}, \qquad t = (L/U)\hat{t}.$$

It can then be shown that, *inter alia*,

$$\mathrm{St} = \frac{UT}{L}, \qquad \mathrm{Ro} = \frac{U}{L\cdot 2\omega\cos\Theta}, \qquad \mathrm{Fr} = \frac{U^2}{g_{\mathrm{m}}L},$$
$$\mathrm{Eu} = \frac{U^2R}{P}, \qquad \mathrm{Re} = \frac{U^2R}{|\mathbf{P}|} = \frac{UL}{\nu_0}, \qquad (3.69)$$

where ν_0 is the characteristic magnitude of the kinematic viscosity.

The dimensionless forms of differential equations and the characteristic flow numbers are useful for:

- investigating the magnitudes of different terms in these equations (linearization and other simplifications of the mathematical model);
- solving these equations numerically;
- comparing laboratory experiments with field observations; etc.

The turbulent exchange coefficient can be expressed as a function of the Richardson number Ri. Strong vertical density gradients sometimes exist in a lake, e.g. between epilimnion and hypolimnion. As the difference between the flow velocities on both sides of this interface approaches some critical value, disturbances in the flow field grow steadily in amplitude and finally break into vortices. When the critical value is exceeded, a transition layer is generated across which there is a velocity gradient $\partial v/\partial z$ and a density gradient $\partial \rho/\partial z$. Turbulence is maintained or increased if

$$\mathrm{Ri} = \frac{g(\partial \rho/\partial z)}{\rho(\partial v/\partial z)^2} \leq 0.25. \tag{3.70}$$

The characteristic flow numbers allow the estimation of the type and stability of flow systems.

Out of the very large number of texts on hydrodynamics we mention: Lamb (1932), Proudman (1953), Lin (1955), Stoker (1957), Hinze (1959), Serrin (1959), Eckart (1960), Milne-Thomson (1960), Alder *et al.* (1964), Kinsman (1965), Monin and Jaglom (1965–67), Krauss (1966), Vasiliev *et al.* (1975), and Kamenkovich and Monin (1978).

3.4. ENERGY BALANCE; THERMAL ENERGY AND HEAT EXCHANGE

3.4.1. Energy Balance Equations

The following way of deducing the energy balance equations stems from basic principles. We start with the total energy of the system, which consists of internal, mechanical, and electromagnetic energy. The electromagnetic field is included because the propagation and absorption of electromagnetic radiation (as light energy) are often prominent in water quality and ecological models (e.g. eqns. 3.126 and 3.127).

In the basic equations we have to use field theory. Internal energy comprises the energy content of the chemical and biological constituents of the aquatic ecosystem. By subtracting the mechanical and electromagnetic energy (eqns. 3.72 and 3.73) from the total energy (eqn. 3.71), we shall derive the balance of the internal energy of the ecosystem (eqn. 3.75). Introduction of enthalpy and specific heat will yield the heat conduction equation 3.78. Use of (3.80) and of notation q for the right-hand side of (3.78), and specialization to one-dimensional problems will lead to (3.81). These steps will now be developed in more detail.

The balance of the *total energy* of the system is given by

$$\frac{\mathrm{d}}{\mathrm{d}t}\int_{V(t)} [u_e + \rho(u + \tfrac{1}{2}v^2 + \phi)]\mathrm{d}V = \int_{V(t)} \sum_i \rho_i \boldsymbol{v}_i \cdot \boldsymbol{F}_i \,\mathrm{d}V + \oint_{A(t)} [\boldsymbol{v}\cdot(\mathbf{P} - \mathbf{I}p) - \boldsymbol{W} - \boldsymbol{S}_e]\cdot \mathrm{d}\boldsymbol{A}, \tag{3.71}$$

which states that the time variation of the total energy results from the work of nonconservative external forces $\boldsymbol{F}_i$ (per unit mass) within the volume $V(t)$ occupied by the system, and from the actions of stress, pressure, heat flow $\boldsymbol{W}$, and flux $\boldsymbol{S}_e$ of electromagnetic energy on the surface $A(t)$ of $V(t)$. On the left-hand side of (3.71),

u is the internal energy density (per unit mass),
u_e is the electromagnetic energy density (per unit volume),
$\frac{1}{2}v^2 + \phi$ is the sum of kinetic and potential energy (per unit mass).

Multiplying (3.51) by $\boldsymbol{v}$, we obtain the balance of the *mechanical energy*:

$$\rho \frac{\mathrm{d}}{\mathrm{d}t}(\tfrac{1}{2}v^2 + \phi) = \underbrace{\sum_i \rho_i \boldsymbol{v} \cdot \boldsymbol{F}_i}_{\text{work of external forces}} - \underbrace{\boldsymbol{v} \cdot \nabla p}_{\text{pressure}} + \underbrace{(\nabla \cdot \mathbf{P}) \cdot \boldsymbol{v}}_{\text{friction}}. \tag{3.72}$$

The balance of the *electromagnetic energy* density u_e (cf. textbooks on electromagnetism, e.g. Stratton, 1941; Landau and Lifshitz, 1960; Phillips, 1962) is given by

$$\frac{\partial u_e}{\partial t} = -\mathrm{div}(\underbrace{\boldsymbol{S}_e}_{\text{flux of electromagnetic energy}} + \underbrace{u_e \boldsymbol{v})}_{\text{convection}} - \underbrace{\boldsymbol{j} \cdot \boldsymbol{E}}_{\text{Joule heat}} - \underbrace{i(I)}_{\text{absorption}}, \tag{3.73}$$

where the part $i(I)$ of radiation of intensity I taken up by biota ($k = 1, 2, 3, \ldots$) can be approximated by

$$i(I) = \sum_k \rho_k \int \kappa_k(\lambda) I(\boldsymbol{r}, t, \lambda) \mathrm{d}\lambda. \tag{3.74}$$

Subtracting (3.72) and (3.73) from (3.71), we derive the balance of the *internal energy* density u:

$$\rho \frac{\mathrm{d}u}{\mathrm{d}t} = \underbrace{\sum_i \boldsymbol{J}_i \cdot \boldsymbol{F}_i}_{\text{work of external forces}} - \underbrace{\mathrm{div}\, \boldsymbol{W} + \boldsymbol{j} \cdot \boldsymbol{E}}_{\text{heat supply}} - \underbrace{\rho p \frac{\mathrm{d}\alpha}{\mathrm{d}t}}_{\text{compression}} - \underbrace{\mathbf{P} \cdot\cdot \nabla \boldsymbol{v}}_{\text{dissipation}} + \underbrace{i(I)}_{\text{absorption}}. \tag{3.75}$$

The increase of internal energy results from the work of the nonconservative forces on diffusive motions $\boldsymbol{J}_i$, from external and internal heat supplies, from the transformation of convective energy by compression and friction (dissipation), and from the transformation $i(I)$ of radiative energy into biochemical energy, e.g. by photosynthesis. It is assumed that the specific absorption $\kappa_k(\lambda)$ of the species k, e.g. of primary producers, depends on the wavelength λ of the light, the intensity of which is I. In water quality modeling the Joule heat may be neglected. We introduce the density of enthalpy, $h = u + p\alpha$, where $\alpha = 1/\rho$, and use the continuity equation (3.39), so that the balance (3.75) becomes

$$\rho \frac{\mathrm{d}h}{\mathrm{d}t} - \frac{\mathrm{d}p}{\mathrm{d}t} + \operatorname{div} \boldsymbol{W} = \sum_i \boldsymbol{J}_i \cdot \boldsymbol{F}_i + \boldsymbol{j} \cdot \boldsymbol{E} + \mathbf{P} \cdot\cdot \nabla \boldsymbol{v} + i(I). \tag{3.76}$$

Assuming that $h = h(T, p, N_j, B_k)$ and taking into account that

$$\left.\frac{\partial h}{\partial T}\right|_{p, N_j, B_k} = c_p, \tag{3.77}$$

where c_p is the specific heat at constant pressure, we convert (3.76) into the "heat conduction equation" or "temperature equation":

$$\rho c_p \frac{\mathrm{d}T}{\mathrm{d}t} + \operatorname{div} \boldsymbol{W}^* = \mathbf{P} \cdot\cdot \nabla \boldsymbol{v} + \eta^*, \tag{3.78}$$

where (for $m \neq j$ and $n \neq k$)

$$\eta^* = \left(1 - \rho \left.\frac{\partial h}{\partial p}\right|_{T, N_j, B_k}\right) \frac{\mathrm{d}p}{\mathrm{d}t} + \sum_i \boldsymbol{J}_i \cdot \boldsymbol{F}_i + \boldsymbol{j} \cdot \boldsymbol{E} + i(I)$$
$$- \rho \sum_j \left.\frac{\partial h}{\partial N_j}\right|_{p, T, N_m, B_k} \frac{\mathrm{d}N_j}{\mathrm{d}t} - \rho \sum_k \left.\frac{\partial h}{\partial B_k}\right|_{p, T, N_j, B_n} \frac{\mathrm{d}B_k}{\mathrm{d}t}. \tag{3.79}$$

Very often in water quality modeling, η^* is omitted from (3.78), while

$$\boldsymbol{W}^* = \boldsymbol{W} + \boldsymbol{W}^{**} = -\kappa \operatorname{grad} T + \boldsymbol{W}^{**} \tag{3.80}$$

is taken to be the sum of heat conduction $\boldsymbol{W}$ and the non-Fourier heat supply $\boldsymbol{W}^{**}$, including radiation. Equations 3.41 and 3.49 may be introduced into (3.79) so that we can examine the role of chemical and biochemical processes in the heat balance (3.78). In general, the energy fluxes between the chemical and biological components of an aquatic ecosystem and the energy stored within these components are small compared with the hydrothermodynamic energy transformations.

3.4.2. Model Equations and Types of Solution

Since (3.75), (3.76), and (3.78) augmented by the continuity equation 3.39 may be transformed into balance equations of the type (3.2),

$$\rho \frac{\mathrm{d}u}{\mathrm{d}t} + \operatorname{div} \boldsymbol{W} = \frac{\partial}{\partial t} \rho u + \operatorname{div}(\rho u + \boldsymbol{W}) = \cdots,$$

the methods discussed in sections 3.1.4–3.1.7 apply to the treatment of these equations. Finite-difference and finite-element methods are well suited to numerical solution. By these means models with quasicontinuous temperature distributions, variable exchange coefficients, and time-dependent thermocline depths can be generated. Heat exchange with the atmosphere can be represented as a function of surface temperature. In simple cases (low inflow and outflow, small variation in thermocline depth, etc.) two- or even multi-layer models have been successfully developed. Ordinary differential equations like (3.6) can be used only for spatial averages within a finite volume V.

A very simple one-dimensional river flow model for the transport of heat is given by

$$\frac{\partial T(x,t)}{\partial t} + u\frac{\partial T(x,t)}{\partial x} = K\frac{\partial^2 T(x,t)}{\partial x^2} + \frac{q(x,t)}{\rho c_p}. \tag{3.81}$$

Similar equations are used also in reservoir temperature models, taking into account horizontal advection, as well as vertical eddy diffusivity. The flow velocity u and the coefficient K of turbulent heat transfer must be measured (or determined otherwise). Comparing (3.81) with (3.46), we note that mass diffusivity and heat diffusivity often differ even in the order of magnitude. Precise and concentrated measurements of temperature and other parameters (oxygen, phosphates, etc.) offer data for estimating mixing processes.

3.4.3. Energy Exchange

In (3.81) the heat supply factor $q(x, t)$ includes the absorption of shortwave solar radiation (after attenuation in the atmosphere and reflection at the water surface) and of longwave, indirect atmospheric radiation, as well as heat losses by the longwave radiation flux from the water, evaporation, etc. The flux of "sensible heat" (conduction across the interface and convection in the atmosphere) can be directed away from or toward the water surface. Heat exchange processes determine the boundary conditions of thermal models for lakes or reservoirs. The transmission and absorption of light within the water body are described approximately by the Lambert–Beer law:

$$I(z,t) = I_0(t)\exp(-\eta z), \tag{3.82}$$

where $I(z, t)$ is the light intensity or irradiance at depth z. The bulk extinction coefficient η depends on the wavelength and is a composite measure of light extinction by water and absorption by suspended particles and biotic components, as well as by dissolved compounds.

As a rule, absorption of solar radiation constitutes the main heat supply to lakes and reservoirs. However, indirect heating or cooling can become significant in lakes and reservoirs that have high-volume input from surface runoff or from groundwater sources. Energy input, temperature, and density stratifications are important regulators of nearly all physicochemical cycling processes and of biological productivity and metabolism. Therefore, the correct determination and simulation of energy exchange play dominant roles in water quality modeling.

The problems of this section have been treated by, for example, Meixner and Reik (1959), Serrin (1959), Truesdell and Toupin (1960), Jørgensen and Harleman (1978), Kamenkovich and Monin (1978), and Jørgensen (1979a).

3.5. CHEMICAL, BIOCHEMICAL, AND BIOLOGICAL PROCESSES

3.5.1. Basic Processes, Constituents, Interrelationships, and Equations

Potentially significant chemical processes in natural water bodies can be classified broadly into:

oxidation–reduction reactions
acid–base reactions
gas–solution processes and outgassing
coordination reactions of metal ions and ligands
precipitation and dissolution of solid phases
adsorption–desorption processes at interfaces.

We shall not go into the details of hydrochemistry, but shall confine the discussion to processes described by (3.41) and to the reaction equations in sections 3.5.2 and 3.5.3.

Aquatic organisms influence the concentrations of many substances by metabolic uptake, transformation, storage, and release. Particulate and dissolved organic matter serve as substrates for decomposer organisms, which, as a byproduct of their metabolism, generate inorganic substances. Primary production is the main process by which dissolved inorganic substances (CO_2, PO_4^{3-}, HPO_4^{2-}, NO_3^-, NO_2^-, etc.) are consumed. Excretion, egestion, and nonpredatory mortality produce dissolved and particulate organic substances; matter and energy are transported and stored in the food chain; and chemical elements are cycled through the ecosystem. To a certain extent, the element cycles are independent, e.g. luxury uptake of phosphorus by phytoplankton, and different rates of excretion of nitrogen and phosphorus by zooplankton and fish. Therefore, "stoichiometric models," coupling the intake and release of chemicals by the biotic components stoichiometrically to the growth and remineralization of biomass, are valid only within bounded changes of the amount and composition of the nutrient loadings to the aquatic ecosystem. Otherwise, "element cycle models" must be used, allowing for an independent element cycle description. Generally, for the improvement of the "causal" description of biological processes it is necessary to study certain details in the laboratory and others in the actual ecosystem in order to obtain the basic knowledge essential for a good working model.

To formulate the basic concepts of deterministic water quality models sufficiently generally, we consider an arbitrary number of *chemical constituents*:

dissolved inorganic matter DIM_j, $j = 1\,(1)\,n_1$;
dissolved organic matter DOM_j, $j = n_1 + 1\,(1)\,n_2$;
particulate organic matter POM_j, $j = n_2 + 1\,(1)\,n_3$.

We also consider *biotic components* BC_k:

several species of primary producers (phytoplankton and phytobenthos), $k = 1\,(1)\,L_1$;
several species of herbivorous zooplankton and zoobenthos, $k = L_1 + 1\,(1)\,L_2$;
several species of decomposers, $k = L_2 + 1\,(1)\,L_3$;
several species at every trophic level of the food chain.

The interrelationships of the chemical and biological components of the aquatic ecosystem are illustrated by Figure 3.4. *Chemical reactions* between the dissolved components DIM_j and DOM_j are described with the help of stoichiometric coefficients ν_{jr}, molar masses $\mathcal{M}_j$, and reaction rates w_r:

$$\rho \frac{dN_j}{dt} = \sum_r \nu_{jr} \mathcal{M}_j w_r + \cdots.$$

The mathematical description of *biochemical and biological processes* may be demonstrated by an example: the grazing of one species BC_k on several members BC_l that are lower in the food chain, and the effects of egestion and excretion by BC_k.

G_k is the grazing rate $[T^{-1}]$ of BC_k. It depends on "internal properties" of the predator population (physiological conditions, age structure, carrying capacity, etc.) and on physical and chemical state parameters of the environment (temperature, oxygen concentration, etc.);

$g_{kl} G_k$ is the partial feeding rate of the predator BC_k on the prey BC_l, where g_{kl} (dimensionless) describes the portions of different types of prey populations BC_l in the food of the predator BC_k and depends, *inter alia*, on the preference coefficient t_{kl} and on the relative densities of the populations (cf. eqn. 3.139);

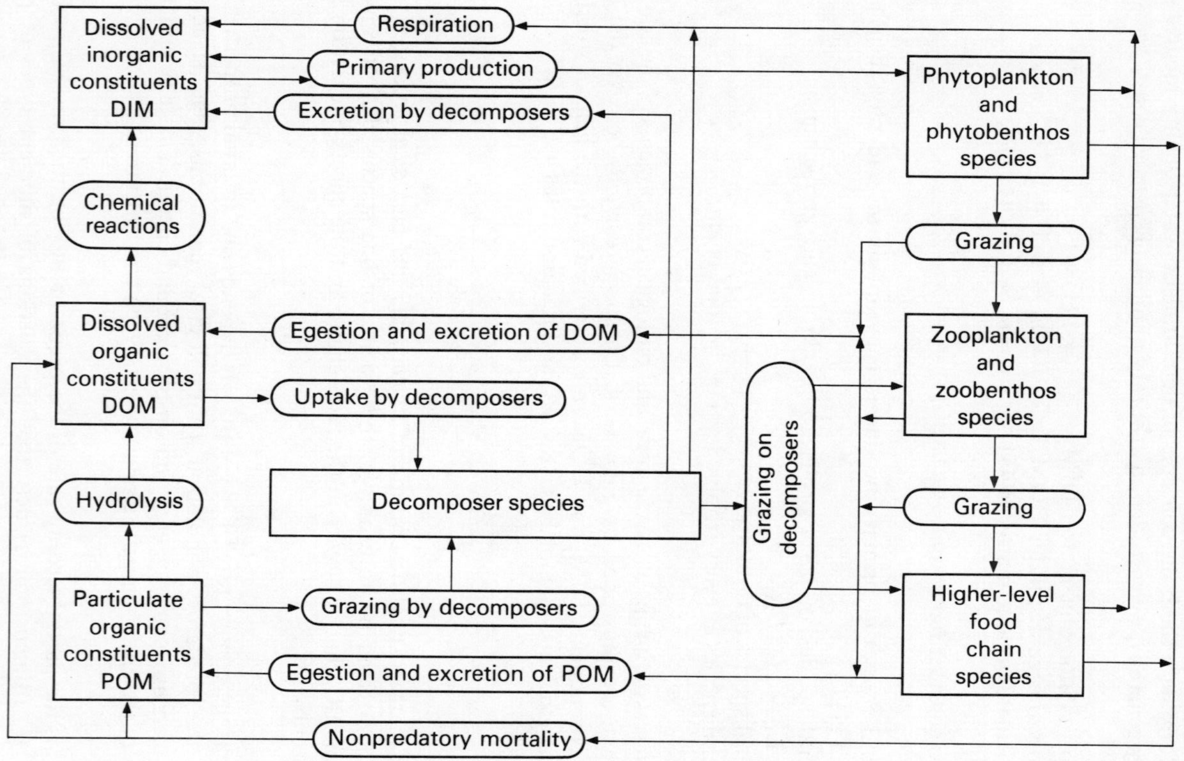

FIGURE 3.4 Interrelationships of the chemical and biological components of the aquatic ecosystem.

$$\frac{dB_l}{dt} = -g_{kl} B_k G_k$$

is the resulting decrease of one prey population BC_l;

$$\frac{dB_k}{dt} = \sum_l^* g_{kl} B_k G_k$$

is the increase of the predator population BC_k caused by grazing on several preys BC_l;

$p_{kl} g_{kl} B_k G_k$ is the partial rate $[T^{-1}]$ of egestion and excretion of POM by BC_k;
$s_{kl} g_{kl} B_k G_k$ is the partial rate $[T^{-1}]$ of egestion and excretion of DOM by BC_k (there is a more detailed description in section 3.5.7);

$$\frac{dB_k}{dt} = \sum_l^* (1 - p_{kl} - s_{kl}) g_{kl} B_k G_k$$

is the increase of the biomass of the predator population BC_k if egestion and excretion are taken into account.

Other processes in the aquatic ecosystem, e.g. uptake of DOM by decomposers, respiration, mortality, and hydrolysis of POM, are treated mathematically in a similar way. Figure 3.5 provides a general overview of the succeeding development of the necessary equations to describe these processes. We shall adopt the following summation symbols.

$\sum_i$ chemical and biotic components of the ecosystem including water ($i = 0$),
$\sum_j, \sum_m$ chemical components of the ecosystem,
$\sum_k, \sum_l$ biotic components of the ecosystem,
$\sum_k^{[1]}$ primary producers ($k = 1, 2, \ldots, L_1$),
$\sum_k^{[2]}$ herbivorous zooplankton and zoobenthos species ($k = L_1 + 1, L_1 + 2, \ldots, L_2$),
$\sum_k^{[3]}$ food chain members,
$\sum_k^{[4]}$ decomposers ($k = L_2 + 1, L_2 + 2, \ldots, L_3$),
$\sum_l^*$ members of the food chain before BC_k ($l < k$),
$\sum_l^{**}$ members of the food chain after BC_k ($l > k$), excluding decomposers,
$\sum_j^{[5]}$ DIM ($j = 1, 2, \ldots, n_1$),
$\sum_j^{[6]}$ DOM ($j = n_1 + 1, n_1 + 2, \ldots, n_2$),
$\sum_j^{[7]}$ POM ($j = n_2 + 1, n_2 + 2, \ldots, n_3$),
$\sum_r$ chemical reactions (excluding hydrolysis of POM).

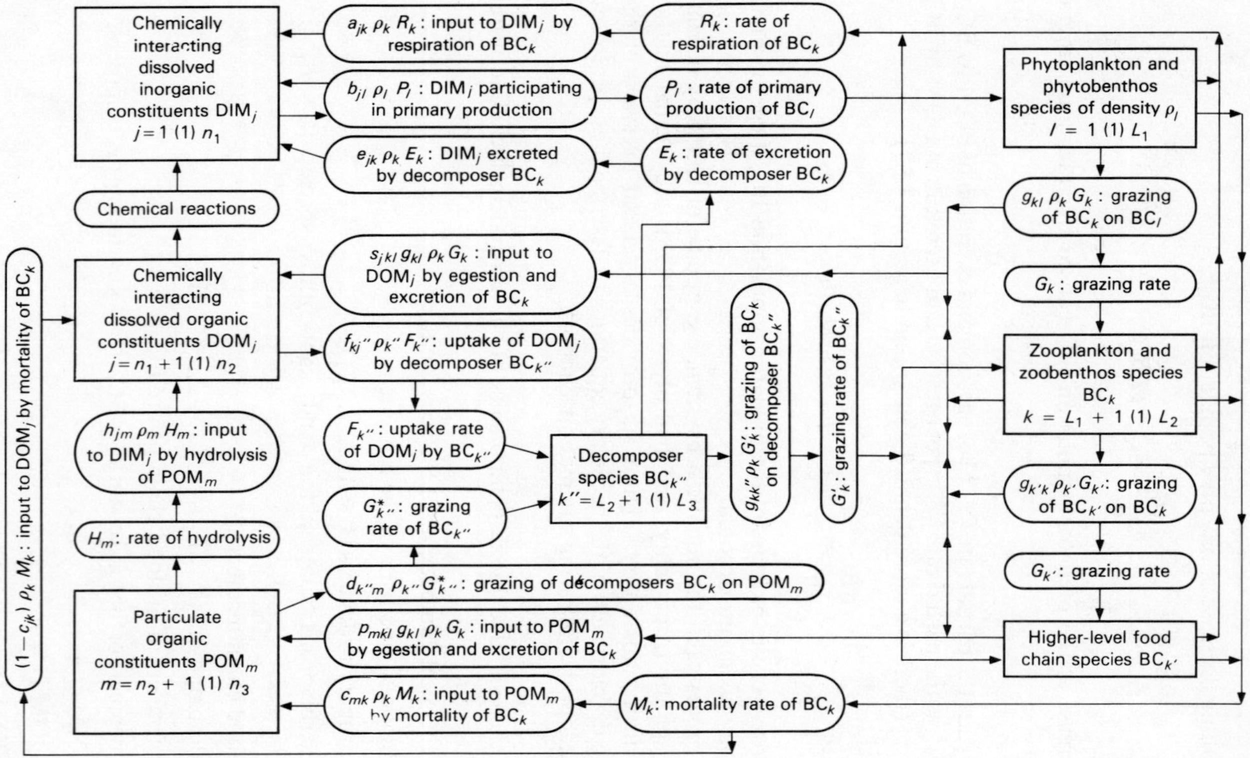

FIGURE 3.5 Illustration of the mathematical expressions describing the interrelationships between chemical and biological components of the aquatic ecosystem.

In the mass balances (3.41) and (3.50) of the chemical and biotic components of the ecosystem,

$$\rho \frac{dN_j}{dt} + \text{div}\, \boldsymbol{J}(N_j) = \sum_r \nu_{jr} \mathcal{M}_j w_r + Y_j, \qquad \textbf{(3.83)}$$

diffusion — chemical reactions — biochemical reactions

$$\rho \frac{dB_k}{dt} + \text{div}\, \boldsymbol{J}(B_k) = Q_k - \rho_k R_k - \rho_k M_k, \qquad \textbf{(3.84)}$$

respiration — mortality

the sources and sinks Y_j and Q_k are of the following types.

1. *The jth Dissolved Inorganic Constituent* DIM_j

$$Y_j = \sum_k a_{jk} \rho_k R_k + \sum_k^{[4]} e_{jk} \rho_k E_k - \sum_k^{[1]} b_{jk} \rho_k P_k, \qquad (3.85)$$

respiration — excretion by decomposers — primary production

where

R_k is the rate of respiration of BC_k $[T^{-1}]$,
E_k is the rate of excretion by BC_k $[T^{-1}]$,
P_k is the rate of primary production of BC_k by photosynthesis $[T^{-1}]$,
a_{jk} is the fraction of DIM_j produced by the respiration of BC_k,
$e_{jk} E_k$ is the fraction of the excreted DIM_j in E_k,
b_{jk} is the fraction of DIM_j used or generated by the primary production of BC_k ($b_{jk} < 0$ if DIM_j, such as oxygen, is produced).

In the aquatic ecosystem, DIM_j may be produced or consumed by biotic components through respiration and excretion, or by primary production.

2. *The jth Dissolved Organic Constituent* DOM_j

$$Y_j = \sum_k^{[3]} \sum_l{}^{*} s_{jkl} g_{kl} \rho_k G_k + \sum_k (1 - c_{jk}) \rho_k M_k$$

egestion/excretion — mortality

$$- \sum_k^{[4]} f_{kj} \rho_k F_k + \sum_m^{[7]} h_{jm} \rho_m H_m, \qquad (3.86)$$

uptake by decomposers — hydrolysis

where

G_k	is the grazing rate of the biocomponent BC_k $[T^{-1}]$,
g_{kl}	are coefficients describing trophic interactions,
$g_{kl}G_k$	is the fraction of the lth prey BC_l consumed by the kth predator BC_k $[T^{-1}]$,
s_{jkl}	is the input fraction to DOM_j by egestion and excretion,
M_k	is the rate of nonpredatory mortality of BC_k $[T^{-1}]$,
c_{jk}	is the fraction of POM_j generated by nonpredatory mortality of BC_k,
F_k	is the rate of uptake of DOM by the kth decomposer $[T^{-1}]$,
$f_{kj}F_k$	is the fraction of DOM_j in F_k $[T^{-1}]$,
H_m	is the rate of hydrolysis of POM_m $[T^{-1}]$,
$h_{jm}H_m$	is the input to DOM_j by the hydrolysis of POM_m.

DOM_j is produced by egestion, excretion (including defecation), and non-predatory mortality of the members of the food chain, and by hydrolysis of detritus. DOM_j is diminished through uptake by decomposers.

3. *The jth Particulate Organic Constituent* POM_j

$$Y_j = \underbrace{\sum_k^{[3]} \sum_l^{*} p_{jkl} g_{kl} \rho_k G_k}_{\text{egestion/excretion}} + \underbrace{\sum_k c_{jk} \rho_k M_k}_{\text{mortality}}$$

$$- \underbrace{\sum_k^{[4]} d_{kj} \rho_k G_k^*}_{\text{grazing by decomposers}} - \underbrace{\sum_m^{[6]} h_{mj} \rho_j H_j}_{\text{hydrolysis}}, \tag{3.87}$$

where

p_{jkl}	is the input fraction to POM_j by excretion and egestion,
G_k^*	is the rate of grazing on POM by decomposers,
d_{kj}	is the fraction of POM_j consumed by the kth decomposer.

Particulate organic components are generated by excretion (including defecation), egestion, and nonpredatory mortality, but are reduced by grazing and hydrolysis.

4. *The Primary Producers* $(k = 1, 2, 3, \ldots, L_1)$

$$Q_k = \rho_k P_k - \sum_l^{[2]} g_{lk} \rho_l G_l, \tag{3.88}$$

where P_k is the rate of primary production of BC_k by photosynthesis $[T^{-1}]$. In the mass balance, given by (3.83) and (3.88), for phytoplankton and/or phytobenthos, apart from primary production, respiration, and mortality, only the losses by zooplankton/zoobenthos are taken into account (BC_l is the predator; $g_{lk} G_l$ is the partial grazing rate).

5. *The Food Chain Groups* BC_k

$$Q_k = \underbrace{\sum_l^{*} (1 - p_{kl} - s_{kl}) g_{kl} \rho_k G_k}_{\text{grazing, egestion/excretion}} + \underbrace{\sum_l^{[4]} g_{kl} \rho_k G'_k}_{\text{grazing on decomposers}}$$

$$- \underbrace{\sum_l^{**} g_{lk} \rho_l G_l}_{\text{grazing by higher levels of the food chain}}, \tag{3.89}$$

where

$$p_{kl} = \sum_m^{[7]} p_{mkl} \quad \text{and} \quad s_{kl} = \sum_j^{[6]} s_{jkl} \tag{3.90}$$

are measures of the egestion and excretion of POM and DOM by BC_k (section 3.5.7). G'_k is the rate of grazing on decomposers by BC_k.

6. *Decomposers* ($k = L_2 + 1, L_2 + 2, \ldots, L_3$)

$$Q_k = \underbrace{\sum_m^{[7]} d_{km} \rho_k G_k^*}_{\text{grazing on POM}} + \underbrace{\sum_j^{[6]} f_{kl} \rho_k F_k}_{\text{uptake of DOM}}$$

$$- \underbrace{\sum_l^{[3]} g_{lk} \rho_l G'_l}_{\text{grazing by food chain members}} - \underbrace{\sum_j^{[5]} e_{jk} \rho_k E_k}_{\text{excretion by decomposers}}. \tag{3.91}$$

The uptake of POM and DOM is compensated by grazing and excretion.

The mathematical relationships used in (3.83–3.91) are shown by Figure 3.5.

3.5.2. Energy Balance for Chemical Reactions

In this section biotic components are neglected. The rth chemical reaction at constant pressure and temperature can be represented by

$$\underbrace{\nu_A A + \nu_B B + \cdots}_{\text{reactants}} \rightleftharpoons \underbrace{\nu_C C + \nu_D D + \cdots}_{\text{products}}$$

or

$$\sum_j \nu_{jr} \mathcal{M}_j = 0, \tag{3.92}$$

where the dimensionless stoichiometric coefficients are $\nu_{jr} > 0$ for products but $\nu_{jr} < 0$ for reactants. As a result of this reaction the Gibbs free enthalpy G_e is changed with respect to the extent ξ_r (mol) of the reaction:

$$d_r G_e = \sum_j \bar{\mu}_j \, dn_j = \sum_j \bar{\mu}_j \nu_{jr} \, d\xi_r = -A_r \, d\xi_r. \tag{3.93}$$

The summation is taken over all components and all phases;

$\bar{\mu}_j$ is the molar chemical potential;
n_j is the number of moles of the jth constituent.

$$A_r = -\sum_j \bar{\mu}_j \nu_{jr} \tag{3.94}$$

denotes the affinity, and $-A_r$ the free energy change of the rth reaction. The internal entropy production $d_i S$ of the total system is then given by

$$d_i S = -\frac{d_r G_e}{T} = \frac{A_r}{T} d\xi_r. \tag{3.95}$$

At equilibrium,

$$d_i S = 0, \qquad d_r G_e = 0, \qquad A_r = 0. \tag{3.96}$$

The Gibbs free enthalpy is a minimum. However, for a spontaneous reaction,

$$d_i S > 0, \qquad d_r G_e < 0, \qquad A_r > 0. \tag{3.97}$$

For the jth component in an aqueous solution, generally

$$\bar{\mu}_j = \bar{\mu}_j^0 + R_0 T \ln(\gamma_j x_j), \tag{3.98}$$

where

$\bar{\mu}_j^0$ is the standard state chemical potential,
R_0 is the gas constant,
γ_j are the activity coefficients,
x_j are the molar fractions.

As a convention, the ideal behavior of the solution is specified by $\gamma_j \to 1$ as $x_j \to 0$ for solutes (and as $x_j \to 1$ for the component(s) whose mole fraction(s) can approach unity in solution). For solutions of electrolytes, "mean activity coefficients," etc. must be introduced. The values of the standard state chemical potentials, concentrations, and activity coefficients depend on the scale used. The molal and molar concentration scales are commonly used for aqueous solutions. The mole fraction scale is appropriate to satisfy the Gibbs–Duhem relationship (3.170).

From (3.94) and (3.98),

$$A_r = -\sum_j \bar{\mu}_j^0 \nu_{jr} - R_0 T \ln Q_r = R_0 T \ln \frac{K_r}{Q_r}, \tag{3.99}$$

where

$$Q_r = \prod_j (\gamma_j x_j)^{\nu_{jr}} \tag{3.100}$$

is the reaction quotient, and

$$K_r(p, T) = \exp\left(-\frac{\sum_j \bar{\mu}_j^0 \nu_{jr}}{R_0 T}\right) \tag{3.101}$$

is the equilibrium constant. If $Q_r < K_r$, an irreversible reaction tends toward equilibrium, which is characterized by $A_r = 0$, $Q_r = K_r$. The quantities A_r, K_r, Q_r, γ_j, $\bar{\mu}_j$, etc. depend on temperature and pressure. In deep lakes and reservoirs the hydrostatic pressure may become large enough to require quantitative consideration of its effects on the chemical parameters and processes.

For the calculation of chemical equilibrium states and/or the direction of spontaneous reactions we need the following basic data (at standard pressure and temperature):

- $\bar{\mu}_j^0$ is the standard state chemical potential, which is equal to the partial molar free enthalpy $\bar{g}_j^0$,
- $\bar{h}_j^0$, $\bar{s}_j^0$ are the standard state enthalpy and entropy, since $\bar{g}_j^0 = \bar{h}_j^0 - T\bar{s}_j^0$,
- $\bar{V}_j$, $\bar{c}_{p_j}^0$ are the partial molar volume and heat capacity.

Thermodynamic data have been obtained by calorimetry, chemical equilibrium measurements, electromotive force measurements of galvanic cells, calculations of entropy based on the third law of thermodynamics, etc. Chemical potentials and other useful materials have been published by, for instance, Latimer (1952), Rossini *et al.* (1952), Kaye and Laby (1959), Lewis and Randall (1961), Klotz (1964), Wagman *et al.* (1968–69), Martell and Smith (1974–77), and Naumov *et al.* (1974).

3.5.3. Reaction Rates and Reaction Equations

In the kinetic approach, an elementary reversible chemical reaction at constant temperature and pressure is described by a mass balance (3.92) and by the relationship between the forward and backward reaction rates w_f, w_b:

$$w_f = \frac{K_r}{Q_r} w_b. \tag{3.102}$$

As a first approximation we assume that

$$w_f = k_f \prod_{\substack{i \\ \text{reactants}}} (\gamma_i x_i)^{|\nu_{ir}|}, \qquad w_b = k_b \prod_{\substack{k \\ \text{products}}} (\gamma_k x_k)^{\nu_{kr}}. \tag{3.103}$$

The rate coefficients k_f and k_b must be determined experimentally. The reaction quotient Q_r characterizes the direction of the net reaction:

$$Q_r < K_r \tag{3.104}$$

when the reaction (3.92) proceeds from left to right;

$$w_f = w_b, \qquad Q_r = K_r = \frac{k_f}{k_b} \tag{3.105}$$

at equilibrium. The resulting reaction rate $\bar{w}_r = w_f - w_b$ (mol s^{-1}) is, according to (3.99) and (3.102),

$$\bar{w}_r = w_f\left(1 - \frac{w_b}{w_f}\right) = w_f\left[1 - \exp\left(-\frac{A_r}{R_0 T}\right)\right]. \tag{3.106}$$

The conclusions (eqns. 3.104–3.106) remain valid even if there are rate-limiting steps and various pathways in a complex chemical reaction (for example, when the expressions (3.103) do not contain powers identical with the stoichiometric coefficients). Most realistic descriptions of the chemical reactions lie within the nonlinear range $A_r \geq R_0 T$. Thus, methods are required to study nonlinear thermodynamics of irreversible processes (section 3.6).

In simplified water quality modeling semiempirical expressions of the following types are widely used to describe chemical, biochemical, and biological processes:

(1)
$$\frac{dC}{dt} = KC^n. \tag{3.107}$$

The concentration C becomes irreversibly diminished by a reaction of order $n = 0$, by a first-order reaction ($n = 1$), or in a nonlinear manner ($n \neq 0, n \neq 1$, where n is determined empirically).

(2)
$$\frac{dC}{dt} = \frac{K_1 C}{K_2 + C} \quad \text{(Michaelis–Menten).} \tag{3.108}$$

For instance, the dependence of the uptake rate on the external nutrient concentration is approximated by this nonlinear expression.

(3)
$$\frac{dC_1}{dt} = KC_2^n \quad \text{or} \quad \frac{dC_1}{dt} = K(C_{2,0} - C_2)^n. \tag{3.109}$$

These simple laws of the nth order for the relationship between the concentrations C_1 and C_2 of two components (with or without saturation concentration $C_{2,0}$) are used, for instance, to describe reaeration or adsorption.

(4)
$$\frac{dC_1}{dt} = K_1 C_2^n - K_2 C_1, \qquad n \neq 0. \tag{3.110}$$

The Streeter–Phelps equation (eqn. 3.110 with $n = 1$) and its generalizations ($n \neq 1$) are used in theories of adsorption–desorption processes, of interrelationships between biotic and chemical constituents of the aquatic ecosystem, of the oxygen concentration in rivers, etc. For example, the source–sink terms for BOD and DO in (3.46) are:

$$G(\mathrm{BOD}, x, t) = -(K_1 + K_3)\mathrm{BOD} + Q, \tag{3.111}$$

$$G(\mathrm{DO}, x, t) = K_2(\mathrm{DO}_{\max} - \mathrm{DO}) - K_1 \cdot \mathrm{BOD} + P, \tag{3.112}$$

where

K_1 is the rate of deoxygenation $[\mathrm{T}^{-1}]$,
K_2 is the reaeration rate,
K_3 is the rate of sedimentation,
Q is the BOD addition due to runoff and local sources $[\mathrm{M\,L^{-3}\,T^{-1}}]$,
P is the net photosynthetic oxygen generated by primary producers.

By integration of these differential equations the location(s) of the critical point(s) of maximum oxygen deficit can be determined (oxygen sag curve).

(5)
$$\frac{\mathrm{d}C_1}{\mathrm{d}t} = K_1(C_{1,0} - C_1)C_2^n - K_2 C_1. \tag{3.113}$$

By (3.113) it is possible to describe adsorption bounded by the saturation value $C_{1,0}$ of the concentration of the adsorbed phase.

(6)
$$\frac{\mathrm{d}C_1}{\mathrm{d}t} = K_1 C_1 \frac{C_2}{K_2 + C_2} \qquad \text{(Monod).} \tag{3.114}$$

Equation 3.114 is used, for instance, for the change of the biomass C_1 of a primary producer if the growth rate is restricted by the concentration C_2 of a limiting nutrient. The half-saturation value K_2 cannot always be assumed to be a constant. In the case of nutrient inhibition, a Monod expression for competitive inhibition by a second nutrient C_3 is

$$\frac{\mathrm{d}C_2}{\mathrm{d}t} = K_1 C_1 \frac{C_2}{K_2 + C_2 + a_1 C_3}, \tag{3.115}$$

where a_1 is the inhibition constant. For allosteric inhibition we may assume that

$$\frac{\mathrm{d}C_2}{\mathrm{d}t} = K_1 C_1 \frac{C_2}{(K_2 + C_2)(1 + a_2 C_3)}. \tag{3.116}$$

Very rapid reactions can be described by

(7)
$$\frac{\mathrm{d}C_1}{\mathrm{d}t} = K_1 \frac{\mathrm{d}C_2}{\mathrm{d}t} - K_2 C_1 \qquad (K_2 \geq 0) \tag{3.117}$$

and by similar expressions. (References to the literature are on pp. 89–90.)

3.5.4. Chemical Reactions in Natural Water Bodies

In the hydrochemistry of natural water bodies we are seldom interested in only one chemical reaction. For a system of one or several phases ($\alpha = 1, 2, 3, \ldots$)

and several chemical components ($j = 1, 2, 3, \ldots$) the Gibbs potential is (neglecting biotic components)

$$G_p(p, T, n_j) = \sum_\alpha \sum_j n_j^\alpha \bar{\mu}_j^\alpha(p, T, x_j^\alpha), \tag{3.118}$$

where the summation is carried out for all species and phases of the system. Hence, the change of G_p is

$$\begin{aligned} dG_p &= \frac{\partial G_p}{\partial p}\bigg|_{T, n_j} dp + \frac{\partial G_p}{\partial T}\bigg|_{p, n_j} dT + \sum_j \frac{\partial G_p}{\partial n_j}\bigg|_{T, p, n_k} dn_j \qquad (k \neq j) \\ &= \sum_\alpha \left(V^\alpha \, dp^\alpha - S^\alpha \, dT^\alpha + \sum_j \bar{\mu}_j^\alpha \, dn_j^\alpha \right). \end{aligned} \tag{3.119}$$

At constant temperature and pressure we locally encounter

(a) equilibrium if

$$\sum_\alpha \sum_j \bar{\mu}_j^\alpha \, dn_j^\alpha = 0, \qquad G_p = \text{minimum}; \tag{3.120}$$

(b) spontaneous change if

$$\sum_\alpha \sum_j \bar{\mu}_j^\alpha \, dn_j^\alpha < 0. \tag{3.121}$$

If there are R chemical reactions or other physicochemical processes in an isobaric–isothermal system, the complete equilibrium state is defined by

$$d_r G_p = -A_r \, d\xi_r = 0 \qquad \text{for } r = 1, 2, 3, \ldots, R. \tag{3.122}$$

For a system of simultaneous reactions not at complete equilibrium the total internal entropy production is positive:

$$d_i S = -\frac{1}{T} \sum_r d_r G_p = \frac{1}{T} \sum_r A_r \, d\xi_r > 0. \tag{3.123}$$

If reactants and/or products are common to several reactions, so that the reactions cannot be treated independently, the direction of a particular reaction may not be deduced from its own affinity alone. In natural waters, for instance, hydrogen ions, electrons, hydroxide ions, oxygen, bicarbonate ions, and several metal ions are interlocking constituents of a large number of reactions.

The interpretation of multicomponent and multiphase chemical processes in aquatic ecosystems requires the solution of nonlinear systems of (differential) equations and, therefore, the utilization of modern numerical techniques and electronic computers. Most of the computerized models being used to calculate multicomponent–multiphase chemical equilibria can be classified either as an "equilibrium constant approach" or as a "Gibbs free energy approach" (Jenne *et al.*, 1979). In the former approach equilibrium constants are needed and the mass action expressions are substituted into the mass balance conditions, yielding a set of nonlinear equations that must be solved simultaneously or by an iterative procedure (Newton–Raphson method, linearized matrix inversion, etc.). This approach is often preferred, especially for the calculation of large complex systems.

In the Gibbs free energy approach, the total Gibbs potential is minimized for a given set of chemical species and their mole numbers subject to the mass balance requirement. Optimization techniques such as pattern search, linear programming, steepest descent, and gradient methods are appropriate to the solution of the mathematical problems. For relatively simple systems for which Gibbs free energy values are available and reliable, this approach is convenient. The method will probably find greater use as soon as accurate and consistent sets of thermodynamic data become available.

Computerized chemical models suited for calculating equilibria in aqueous systems have been described and compared by, e.g., Jenne *et al.* (1979), and van Zeggeren and Storey (1970).

Using one of the above-mentioned approaches, we assume that a unique solution exists. Natural aqueous systems (or aquatic ecosystems) may have several thermodynamically metastable states corresponding to local minima in the Gibbs function. Therefore, nonunique solutions of the equilibrium problem are possible, since the solution may occur not only at the most stable equilibrium point but also at a local minimum. For the application of the above-mentioned computerized models in aquatic chemistry, this problem of non-uniqueness is very important. For instance, the problem arises in the interpretation of solid precipitation and dissolution processes.

Furthermore, a number of processes in rivers, lakes, reservoirs, and other natural water systems cannot be treated by the "equilibrium theory" of chemical reactions. The flow of matter and energy, gradients of concentration and temperature, and exchanges of matter and energy with the surroundings characterize an open, continuous thermodynamic system that is poorly approximated by the equilibrium state of a closed homogeneous system. Knowledge of the temperature dependence of some of the standard quantities (section 3.5.2) is sometimes inadequate. Moreover, the characteristic time scales of river flows and reservoir management often do not allow for chemical equilibrium if the reactions are very slow. Local equilibrium conditions might be expected if the reaction rate is sufficiently high, even in a slowly changing environment (e.g. at the sediment–water interface). Local equilibrium within each small mass element can be realistically assumed to exist (section 3.6.3), but over larger regions such as the biologically active photic zone of a eutrophic lake, thermodynamic equilibrium is never realized. In such cases it is necessary to use the method of thermodynamics of irreversible processes (section 3.6).

3.5.5. Photosynthesis and Primary Production

Through the medium of photoactivated plant pigments (chlorophylls), light energy in the wavelength range 390–710 nm is transformed into chemical energy.

Carbohydrates, proteins, nucleic acids, and other metabolites are formed from CO_2, H_2O, and constituents containing N, P, S, Fe, etc. The photosynthetic production of algal protoplasm and the reverse destructive processes are summarized by

$$106\,CO_2 + 16\,NO_3^- + HPO_4^{2-} + 122\,H_2O + 18\,H^+ + \text{trace elements} + \text{energy} \rightleftharpoons C_{106}H_{263}O_{110}N_{16}P + 138\,O_2. \tag{3.124}$$

The net primary productivity is defined by gross photosynthesis (light and dark bottle) minus loss by respiration during the period of light minus losses due to respiration and mineralization over the daily cycle.

The primary production rate P_k of the kth biotic component BC_k of the ecosystem (phytoplankton or phytobenthos) is a complicated function of biomass, local light intensity, temperature, pH, vitality of BC_k, availability of intracellular and extracellular nutrient supply, and intracellular processes. In water quality modeling, P_k is approximated by expressions of the type

$$P_k = P_{k,0}\,f(I)f(T)f(\rho_1, \rho_2, \rho_3, \ldots), \tag{3.125}$$

where $f(\ldots)$ represents functions of different types. Different assumptions about $P_{k,0}$ and the functions of light intensity I, temperature T, and nutrient concentrations ρ_j are used. Some examples are given below.

(1) The maximum rate $P_{k,0}$ for optimum conditions of light, temperature, and nutrients is supposed to depend mainly on cell properties.

(2) For the description of the relationship between photosynthesis and irradiance, functions of the type

$$f(I) = \frac{I/I^*}{[1 + (I/I^*)^2]^{1/2}}, \qquad I(z, t) = I_0(t)\exp(-\eta z), \tag{3.126}$$

or

$$f(I) = \left(\frac{I}{I^*}\right)\exp\left(1 - \alpha\frac{I}{I^*}\right), \qquad \alpha > 0, \tag{3.127}$$

or a Michaelis–Menten expression, or other assumptions are used. $I_0(t)$ includes the annual and diurnal variations.

(3) The effects of temperature on biological rates are approximated, for instance, by

$$f(T) = a + b\frac{T}{T_0} \quad \text{or} \quad f(T) = K^{T-T_0}, \tag{3.128}$$

or

$$f(T) = \exp[-K(T - T_{\text{opt}})^2], \tag{3.129}$$

$$f(T) = \begin{cases} \exp\left(-\dfrac{K(T - T_{\text{opt}})^2}{(T_{\max} - T_{\text{opt}})^2}\right), & T > T_{\text{opt}} \\ \exp\left(-\dfrac{K(T_{\text{opt}} - T)^2}{(T_{\text{opt}} - T_{\min})^2}\right), & T \leq T_{\text{opt}}. \end{cases} \tag{3.130}$$

(4) The limiting expression $f(\rho_j)$ can be, for instance,

$$f(\rho_j) = \prod_{j'}^{[5]} \frac{\rho_{j'}}{K_{j'} + \rho_{j'}} \tag{3.131}$$

or

$$f(\rho_j) = \frac{\sum_{j'}^{[5]} \omega_{j'} \rho_{j'}/(K_{j'} + \rho_{j'})}{\sum_{j''}^{[5]} \omega_{j''}}, \tag{3.132}$$

where ω_j are weights, or

$$f(\rho_1, \rho_2, \rho_3, \ldots) = \frac{\rho_j^*}{K_j^* + \rho_j^*}, \tag{3.133}$$

where ρ_j^* is the mass density of the limiting nutrient.

For very low nutrient concentrations,

$$\frac{\rho_j}{K_j + \rho_j} \approx K\rho_j \qquad \left(K = \frac{1}{K_j}\right).$$

Sometimes, the Monod kinetics of nutrient uptake are replaced by other assumptions. An example is

$$f(\rho_1, \rho_2, \ldots) = 1 - \exp\left(-\frac{\sum_j c_{kj}(\rho_j - \rho_{j,\min})}{K_k \rho_k}\right). \tag{3.134}$$

The total photosynthesis per unit surface area of a lake can be determined by integrating over the depth z. If the extinction coefficient η is constant, then from (3.126):

$$\int_0^\infty f(I)\mathrm{d}z = \frac{1}{\eta} \ln\left\{\frac{I_0}{I^*} + \left[1 + \left(\frac{I_0}{I^*}\right)^2\right]^{1/2}\right\} \tag{3.135}$$

$$\approx \frac{1}{\eta} \ln\left(2\frac{I_0}{I^*}\right) \qquad \text{if } I_0 \gg I^*. \tag{3.136}$$

Logarithmic light expressions can be found in water quality modeling, but it should be recognized that the whole of (3.126) must be integrated over z and that, as a rule, η depends upon z.

Observations of algae kinetics show that intracellular concentrations ρ_{kj} of the jth inorganic nutrient in BC_k must be introduced; thus (3.85) is generalized to

$$Y_j = \sum_k^{[4]} a_{jk}\rho_k R_k + \sum_k e_{jk}\rho_k E_k - \sum_k^{[1]} \rho_k U_{kj}, \tag{3.137}$$

where U_{kj} is the rate of DIM_j uptake by the kth phytoplankton component. A simple assumption is that the uptake of nutrients is controlled by the vitality (or temperature) and by both the extracellular concentration ρ_j and intracellular concentration ρ_{kj} of the nutrient:

$$U_{kj} = U_{kj,0}\,\frac{\rho_{kj,\max} - \rho_{kj}}{\rho_{kj,\max} - \rho_{kj,\min}}\,\frac{\rho_j}{K_j + \rho_j}\,f(T). \tag{3.138}$$

If we suppose that photosynthesis and primary production are controlled by intracellular, but not by extracellular concentrations, the function $f(\rho_j)$ in (3.131)–(3.134) must be replaced by $f(\rho_{kj})$. Figure 3.6 shows this variant of primary production.

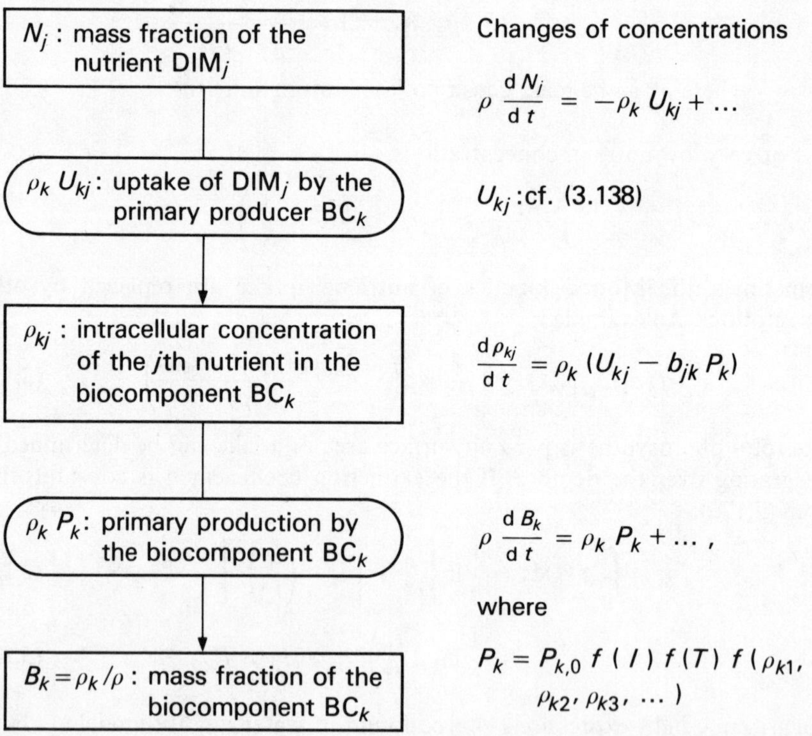

FIGURE 3.6 Intracellular storage of nutrients and primary production by the biocomponent BC_k.

3.5.6. Trophic Interactions (Grazing and Feeding Rates)

The feeding relationship forms an important nonlinear linkage between the biotic components of the ecosystem. At intermediate biomass densities the feeding rate depends on the biomass of both interacting populations. A preference term differentiates between various types of food. The feeding rate function exhibits a maximum at a specific temperature and is generally higher for juveniles than for adults. It decreases when the population reaches its maximum density.

Formulating the partial feeding rate $g_{kl} G_k$ of the predator (consumer) BC_k on the prey (food supply) BC_l, we use the Monod kinetic:

$$g_{kl} = \frac{t_{kl}(\rho_l - \rho_{l,\min})/\rho_k}{K_k(T) + \sum_m^* t_{km}(\rho_m - \rho_{m,\min})/\rho_k}, \tag{3.139}$$

$$G_k = c_k f(T) F\left(\frac{\rho_k}{\rho_{k,\max}}\right), \tag{3.140}$$

where

$$\rho_l \geq \rho_{l,\min}, \qquad \rho_k \leq \rho_{k,\max},$$

t_{kl} is the preference coefficient (for BC_l as food for BC_k),

$\rho_{l,\min}$ is the threshold concentration of the prey population below which the grazing on this population is zero; the assumption $\rho_{l,\min} \to 0$ is important for the simplification of the model,

$\rho_{k,\max}$ is the carrying capacity (maximum possible density) of the predator population,

$$F(\rho_k/\rho_{k,\max}) \to 0 \qquad \text{if } \rho_k \to \rho_{k,\max},$$

c_k is a correction factor (not specified here) that takes into account physiological effects, the age structure of the predator population, oxygen concentration, toxic impacts, etc.

Similar expressions are used for the rate of grazing, $g_{kl} G'_k$, on decomposers by the BC_k and for the partial rate of uptake, $f_{kj} F_k$, of DOM_j by the kth decomposer:

$$f_{kj} = \frac{t_{kj}(\rho_j - \rho_{j,\min})/\rho_k}{K_k(T) + \sum_m^{[6]}(\rho_m - \rho_{m,\min})/\rho_k}. \tag{3.141}$$

Equation 3.139 involves two asymptotic cases.

(a) For optimum feeding conditions the total feeding rate of BC_k is

$$\sum_l^* g_{kl} G_k \approx G_k. \tag{3.142}$$

(b) For very low food concentrations,

$$g_{kl} G_k \rho_k \approx \frac{G_k t_{kl}}{K_k}(\rho_l - \rho_{l,\min}). \tag{3.143}$$

Sometimes an exponential law is used instead of (3.139):

$$\sum_l^* g_{kl} G_k = G_k \left[1 - \exp\left(- \sum_l^* \frac{t_{kl}(\rho_l - \rho_{l,\min})}{\rho_k K_k} \right) \right], \tag{3.144}$$

leading also to the asymptotic expressions (3.142) and (3.143).

3.5.7. Egestion and Excretion

The fraction of food consumed but not assimilated by the kth component BC_k of the food chain (zooplankton, zoobenthos, and the higher trophic levels) consists of POM and DOM. The partial egestion rates are

$$\hat{p}_{jkl} g_{kl} G_k \text{ for POM} \quad \text{and} \quad \hat{s}_{jkl} g_{kl} G_k \text{ for } \mathrm{DOM}_j.$$

Hence, the partial egestion rate connected with the trophic interaction between the predator BC_k and the prey BC_l is

$$\left(\sum_j^{[7]} \hat{p}_{jkl} + \sum_j^{[6]} \hat{s}_{jkl} \right) g_{kl} G_k = (\hat{p}_{kl} + \hat{s}_{kl}) g_{kl} G_k. \tag{3.145}$$

The fraction of food assimilated and subsequently excreted (including defecation) by BC_k also consists of POM_j and DOM_j. The partial excretion rates are

$$\tilde{p}_{jk} \sum_l^* g_{kl} G_k \text{ for } \mathrm{POM}_j \quad \text{and} \quad \tilde{s}_{jk} \sum_l^* g_{kl} G_k \text{ for } \mathrm{DOM}_j.$$

Using the notations

$$p_{jkl} = \hat{p}_{jkl} + \tilde{p}_{jk}, \qquad s_{jkl} = \hat{s}_{jkl} + \tilde{s}_{jk}, \tag{3.146}$$

$$p_{kl} = \sum_j^{[7]} p_{jkl}, \qquad s_{kl} = \sum_j^{[6]} s_{jkl}, \tag{3.147}$$

we may combine the processes of egestion and excretion by BC_k:

$$E(BC_k) = \sum_l^* (p_{kl} + s_{kl}) g_{kl} G_k. \tag{3.148}$$

The relations (3.86), (3.87), and (3.89) are based on (3.148).

The rate of excretion of dissolved inorganic matter by the kth decomposer is approximated by

$$E_k = E_{k,0} \sum_j^{[6]} f_{kj} F_k \tag{3.149}$$

or by

$$E_k = E_{k,0} \sum_j^{[6]} (f_{kj} F_k - R_k) \tag{3.150}$$

if uptake $\sum_j^{[6]} f_{kj} F_k$ is greater than respiration. $E_{k,0}$ depends on physiological conditions, which are in turn related to the state of the environment.

3.5.8. Respiration and Nonpredatory Mortality

Respiration and mortality losses are considered proportional to biomass and to ingestion, respectively. Modifications result from the physiological responses to the structure and changes of environmental parameters (e.g. temperature, dissolved oxygen). Nongrazing mortality may be caused by overcrowding, if the population approaches the maximum that can occupy the ecosystem (the system carrying capacity).

One of the following types of expressions may be used for the respiration rate R_k of the BC_k:

$$R_k = a_k f(T), \tag{3.151}$$

$$R_k = b_k f(T) F\left(\frac{\rho_k}{\rho_{k,\max}}\right), \tag{3.152}$$

$$R_k = c_k f(T) \sum_l{}^* g_{kl} G_l, \tag{3.153}$$

$$R_k = d_k f(T) \sum_l{}^* (1 - \hat{p}_{kl} - \hat{s}_{kl}) g_{kl} G_k, \tag{3.154}$$

or a linear combination of (3.151) or (3.152) with (3.153) or (3.154).

For the rate of nonpredatory mortality M_k a number of different approximations are in use; for instance,

$$M_k = q_k f(T) F\left(\frac{\mathrm{d}T}{\mathrm{d}t}\right)\left(1 - \alpha_k \sum_l{}^{**} \frac{g_{lk} G_l \rho_l}{\rho_k}\right) \tag{3.155}$$

or laws of the types (3.151)–(3.154) as well as linear combinations of these. According to (3.155), in addition to the temperature effect $f(T)$, a second function F, depending on the time variation of the temperature, influences the mortality rate. Furthermore, the nonpredatory mortality is diminished by grazing losses ($0 < \alpha_k < 1$; $\alpha_k \approx 0.5$).

The reader interested in a more detailed exposition of hydrochemistry and hydrobiology is referred to the literature, e.g. Stumm and Morgan (1970), Rheinheimer (1971), Ciaccio (1971–73), Daubner (1972), Lehninger (1972), Romanenko and Kuznecov (1974), Golterman (1975), Salánki and Ponyi

(1975), Uhlmann (1975), Wetzel (1975), and Goldberg *et al.* (1977). Methods of mathematical modeling are dealt with in the following chapters and also by, e.g., Volterra (1932), Vinberg and Anisimov (1966), Menshutkin (1971), Biswas (1972), Romanenko and Kuznecov (1974), Bagotskiy and Bazykin (1975), Romanovskiy *et al.* (1975), Antomonov (1977), Mitropolskiy (1977), Jørgensen and Harleman (1978), and Jørgensen (1979a). Data are available in a handbook edited by Jørgensen (1979b).

3.6. FURTHER CONSTRAINTS ON WATER QUALITY MODELS

3.6.1 Initial and Boundary Conditions

To represent the water quality model by a well posed mathematical problem, it is necessary to supplement appropriate initial and boundary conditions of the system of differential equations. These equations describe the physical, chemical, and biological processes within the water body, while the boundary conditions reflect the constraints of the surrounding world acting upon the fluid motions and upon the components of the aquatic ecosystem.

The number and type of initial and boundary conditions depend strongly on the nature of the particular water body, on the specific problem of interest, and on the type of model that we intend to use. Therefore, the initial and boundary values in water quality modeling are of considerable variety. We confine ourselves to some general remarks, leaving the formulation of particular initial-boundary-value problems to later chapters that discuss special models.

An aquatic ecosystem exchanging energy and matter with the external world is called an "open system." In this case, at the surface separating the water body from its surroundings, we must prescribe the boundary values of

the fluxes of energy and matter across the surface, or
the temperature and the concentrations (e.g. densities ρ_i) of the constituents of the ecosystem, or
linear combinations of these quantities at the surface.

For instance, if thermal effects are included, a boundary condition on the temperature is required, specifying either the temperature or the heat flux, or a linear combination (boundary conditions of the first, second, or third type).

At the free surface of a water body or at an interface between two different water bodies (one of which we are investigating), a kinematic boundary condition must be fulfilled and the stress vector should be continuous. As a rule, this guarantees the continuity of the pressure and velocity distributions across the free surface or interface. The energy fluxes across this surface must be specified. There is no exchange of water through this type of surface.

At solid boundaries the fluid must adhere to the solid, at least at low tangential stresses and high pressures, if the viscosity is significant. If the viscosity is not large, however, we may safely assume that the fluid slips along the surface without appreciable drag. The energy fluxes across this boundary must be specified. If the solid boundary is moving a kinematic condition must be imposed.

Of very great importance in water quality modeling is the exchange of matter (nutrients, biota, detritus, etc.)

through free surfaces by precipitation, nitrogen fixation, outgassing, etc.,
across interfaces of water bodies by diffusion,
at solid boundaries by sedimentation, adsorption, desorption, chemical reactions, transport of interstitial water into the free water, etc.

Sometimes it is very difficult to prescribe the boundary values since they must be measured, but it should be carefully done, especially if the values are time-dependent. Changed environmental influences may lead to the activation of "possibilities" of the ecosystem not realized formerly.

3.6.2. Entropy

The concept of entropy was first introduced in physics with reference to thermodynamic phenomena. In recent years this idea has been so generalized that it has found fruitful application in fields far from thermal physics. For instance, in chemistry, cosmology, information theory, and biology, entropy has revealed itself as a powerful tool, as a comprehensive measure of irreversibility if we look at the system from outside, and as a measure of stability seen from inside the system.

The second law of thermodynamics postulates the existence of a state function S, the entropy, which is linked to other state functions by the fundamental Gibbs relation (eqn. 3.160). The change of entropy with time can be split into two parts, the entropy production inside the system and the exchange of entropy with the outside world (eqn. 3.186). The internal production, caused by dissipative processes, is never negative. The entropy flow across the boundary of the system is a necessary condition for its maintenance far from thermostatic equilibrium. Many of the physical and chemical processes in natural water bodies, as well as practically all biochemical processes in living systems, occur far from equilibrium. Therefore, the entropy principle offers the possibility of making deductions about the existence of stable, stationary states of an ecosystem.

Furthermore, without integrating the nonlinear model equations, we can derive from the entropy balance some conclusions concerning the evolution and further development of the aquatic ecosystem. Under unstable conditions, relatively small fluctuations (in space and/or time) of physical, chemical, and/or biological components or parameters of the water body (e.g. tides, seiches, day–night dependence of biological processes) can suddenly increase and give rise to a transition of the ecosystem to a "new" stationary state. This is especially likely when the environmental conditions or some internal variables/parameters of the ecosystem have undergone such variations that the inherent stability, characteristic of the stationary state of the ecosystem, has been lost. From this point of view, we may approximate the anthropogenically influenced history and further development of lakes, reservoirs, and other water resources as a succession of transitions from one stationary state to a new one caused by fluctuations of ecosystem parameters or components under unstable conditions.

The entropy principle also connects the phenomenological, stochastic, and cybernetic approaches to water quality modeling, as may be indicated by the following remarks and formulas.

(a) Entropy can be defined as "potential information" (Shannon):

$$S \approx -\sum_k p_k \lg_2 p_k, \qquad 0 \leq p_k \leq 1, \tag{3.156}$$

where p_k is the probability of each microstate. Entropy measures how much information we would gain if we knew not only the "macrostate" of the system but also the "microstate" within it (von Weizsäcker, 1972). In information theory and ecology the function (3.156) is called the Shannon index and is used as a measure of the information content of a code. It is a special case ($\delta \to 1$) of a more general class of "diversity indices." For a code of n kinds of symbols and proportions p_k of the kth kind ($k = 1, 2, 3, \ldots, n$), the function

$$H = \frac{1}{1-\delta} \lg_2 \sum_k p_k^{\delta} \tag{3.157}$$

is known as the entropy of order δ of the code or, equivalently, of the set (p_k) (Pielou, 1975).

As the number of microstates or "elementary arrangements" contained in a macrostate of the system increases, the entropy increases. If we take into account the binding energies and if the total energy of the system (or the temperature) is sufficiently small, macrostates of high order, like crystals, can have high entropy. Entropy is not the "general measure of disorder." Diffusion processes and heat conduction in gases, by which it is usually demonstrated that disorder increases together with entropy, are insufficient examples for the discussion of the meaning of increasing entropy, since gases generally are without structure.

(b) Microscale and macroscale fluctuations of physical, chemical, and/or biological components of the ecosystem are typical phenomena in water bodies. In spite of these fluctuations, some macroscale characteristics of the system are constant. Only after a fluctuation has occurred and been extended to a sufficiently large macroscopic range may it give rise to a change in the state of the ecosystem. Therefore, besides the stability of the stationary state we must specify the *a priori* probability of there being fluctuations and the probability that a fluctuation spreads and attains a large amplitude and range, allowing for changes in the state and structure of the ecosystem.

The determination of these probabilities belongs to the stochastic theory of aquatic ecosystems. The probability W of fluctuations around a stationary state is connected with the second variation $\delta^2 S$ of entropy by the Einstein formula:

$$W = W_0 \exp\left(\frac{\delta^2 S}{2k_B}\right), \tag{3.158}$$

where k_B is the Boltzmann constant. This equation offers a statistical foundation for the thermodynamic stability criteria.

(c) The processes in living systems are controlled by special regulating factors. Therefore, the theory of aquatic ecosystems should be based on a combination of thermodynamics and cybernetics. Cybernetic water quality models contribute valuable knowledge about

the feedback mechanisms in systems,
the isomorphic structures of complex systems,
the evolution of self-organizing systems, and
the hierarchy of systems and subsystems, etc.

The value of this knowledge for the investigation of water bodies has been pointed out, for instance, by Lyapunov (1972) and Straškraba (1977). On the other hand, the formation and stability of structures of aquatic ecosystems, the feedback mechanisms within them, etc. depend strongly on physical, chemical, and biological constraints. Therefore, only the combination of thermodynamic methods, taking into account the basic physical, chemical, and biological laws, with the stochastic and cybernetic investigations of the behavior of water bodies can ensure the further successful development of water quality modeling.

The progress of stochastic methods for the investigation of nonequilibrium systems has been extraordinary in recent years (cf. Nicolis and Prigogine, 1977.) Since this chapter is concerned primarily with deterministic water quality modeling, we confine ourselves to the thermodynamic aspects, especially to:

- formulation of the generalized Gibbs relation (section 3.6.3);
- derivation of the entropy balance equation for aquatic systems (section 3.6.5);
- stability criteria (section 3.6.7);
- ansatz equations for primary production (section 3.6.8).

3.6.3. The Generalized Gibbs Relation

An extension of the concept of entropy is needed for application to open systems and to situations far from thermostatic equilibrium since, from the thermodynamic point of view, the aquatic ecosystems of lakes, reservoirs, rivers, and other types of water resources are

- "open systems" in nature, exchanging energy and matter with their surroundings;
- far from thermostatic equilibrium because matter and energy flow through them;
- nonlinear systems because of the nonlinearity of the internal processes: the Reynolds stress terms, the temperature dependence of chemical reaction rates, etc.

Important exchange processes between the aquatic ecosystem and the external world include the transfer of energy, mass, and biomass across the water–water, air–water, water–plant, and water–soil interfaces, insolation, nitrogen fixation, input of chemicals by precipitation, and phosphate release from sediments. The exchange of energy and matter is an essential element of aquatic plant and animal populations.

The Gibbs relation may be generalized and applied to water bodies if

- the concentration gradients are limited, in that the mass densities of the constituents do not vary appreciably within distances of the order of the mean free path;
- chemical reactions are not too fast, in that the rate of elastic collisions is larger than the rate of reactive collisions, which in general is true also for all biological processes in dense media.

Interfaces inside the ecosystem call for special "jump conditions" (Glansdorff and Prigogine, 1971; Nicolis and Prigogine, 1977).

Thus, our basic postulate is that not only the physical and chemical processes but also the biological phenomena can be studied with reference to the laws of generalized thermodynamics, at least to a good approximation on the macroscopic scale. The system as a whole is out of thermostatic equilibrium. However, it is assumed that the macroscopic evaluation of internal energy, entropy, and entropy production, although very difficult for biological components, still remains possible, since "local equilibrium" exists within each mass element of the medium. Between the thermodynamic quantities u, s, T, etc. and the local macroscopic variables the same relations are supposed to be valid as between the state variables of the whole system at equilibrium. Thus, we assume that the internal energy density u (per unit mass) of the water body is determined by the local physical, chemical, and biological state parameters and one further, dimensionally independent scalar parameter, the entropy density s:

$$u = u(s, \alpha, N_j, B_k). \tag{3.159}$$

We also assume the Gibbs relation to be valid:

$$T\,\mathrm{d}s = \mathrm{d}u + p\,\mathrm{d}\alpha - \sum_j \mu_j\,\mathrm{d}N_j - \sum_k \beta_k\,\mathrm{d}B_k. \tag{3.160}$$

For the description of systems containing electrolytes the chemical potentials μ_j must be replaced by the electrochemical potentials. We introduce the potentials β_k of the biotic components. In a first approximation β_k is equal to the free enthalpy of BC_k. In this idealized description the trophic relationships are determined only by differences in the energy contents of BC_k and BC_l. The driving force is proportional to the difference, $\beta_k - \beta_l = g_k - g_l$. However, in reality the trophic relationship results also from regulatory processes that probably cannot be described adequately by chemical kinetics. We shall not go into detail here.

Differentiating (3.160) in the direction of center-of-mass motion produces

$$T\frac{\mathrm{d}s}{\mathrm{d}t} = \frac{\mathrm{d}u}{\mathrm{d}t} + p\frac{\mathrm{d}\alpha}{\mathrm{d}t} - \sum_j \frac{\mathrm{d}N_j}{\mathrm{d}t}\mu_j - \sum_k \frac{\mathrm{d}B_k}{\mathrm{d}t}\beta_k. \tag{3.161}$$

This is the generalized Gibbs relation, which may be applied to water quality modeling (Mauersberger, 1978, 1979).

3.6.4. Thermodynamic Relations

In this section we bring together some of the relations used in the thermodynamic theory of aquatic ecosystems. Adopting the "proper" independent variables of the thermodynamic quantities u, g, s, etc., we may easily derive from (3.159) and (3.160):

$$u(s, \alpha, N_j, B_k) = Ts - p\alpha + \sum_j \mu_j N_j + \sum_k \beta_k B_k, \tag{3.162}$$

$$\frac{\partial u}{\partial s} = T, \qquad \frac{\partial u}{\partial \alpha} = -p, \qquad \frac{\partial u}{\partial N_j} = \mu_j, \qquad \frac{\partial u}{\partial B_k} = \beta_k, \tag{3.163}$$

$$g(T, p, N_j, B_k) = u - Ts + p\alpha = \sum_j \mu_j N_j + \sum_k \beta_k B_k, \tag{3.164}$$

$$\frac{\partial g}{\partial T} = -s, \qquad \frac{\partial g}{\partial p} = \alpha, \qquad \frac{\partial g}{\partial N_j} = \mu_j, \qquad \frac{\partial g}{\partial B_k} = \beta_k. \tag{3.165}$$

Using further the independent variables T, p, N_j, B_k, we find from equation 3.165:

$$\left(\frac{\partial s}{\partial p}\right)_{T, N_j, B_k} = -\left(\frac{\partial \alpha}{\partial T}\right)_{p, N_j, B_k}, \tag{3.166}$$

$$\frac{\partial s}{\partial N_j} = -\frac{\partial \mu_j}{\partial T}, \qquad \frac{\partial s}{\partial B_k} = -\frac{\partial \beta_k}{\partial T}, \qquad \frac{\partial \mu_j}{\partial p} = \frac{\partial \alpha}{\partial N_j}, \tag{3.167}$$

$$\frac{\partial \beta_k}{\partial p} = \frac{\partial \alpha}{\partial B_k}, \qquad \frac{\partial \mu_i}{\partial N_j} = \frac{\partial \mu_j}{\partial N_i}, \qquad \frac{\partial \beta_k}{\partial N_j} = \frac{\partial \mu_j}{\partial B_k}. \tag{3.168}$$

The thermodynamic relations

$$\left(\frac{\partial \alpha}{\partial p}\right)_{T, N_j, B_k} = -\alpha\chi, \qquad \left(\frac{\partial s}{\partial T}\right)_{p, N_j, B_k} = \frac{c_p}{T} \tag{3.169}$$

and the Gibbs–Duhem relationship

$$-\alpha\,\mathrm{d}p + s\,\mathrm{d}T + \sum_j N_j\,\mathrm{d}\mu_j + \sum_k B_k\,\mathrm{d}\beta_k = 0 \tag{3.170}$$

are valid too (χ is the isothermal compressibility).

3.6.5. The Entropy Balance Equation

The entropy balance equation follows from the Gibbs relation and from the basic equations 3.75, 3.83, and 3.84:

$$\frac{\partial(\rho s)}{\partial t} + \operatorname{div} \boldsymbol{S} = \sigma. \tag{3.171}$$

The entropy flow density $\boldsymbol{S}$ is

$$\boldsymbol{S} = \rho s \boldsymbol{v} + \frac{\boldsymbol{W}}{T} - \sum_j \frac{\boldsymbol{J}(N_j)\mu_j}{T} - \sum_k \frac{\boldsymbol{J}(B_k)\beta_k}{T}. \tag{3.172}$$

For the dissipation $T\sigma$,

$$T\sigma = TW \operatorname{grad}\left(\frac{1}{T}\right) + \boldsymbol{j}\cdot\boldsymbol{E} + i(I) + \mathbf{P}\cdot\cdot\nabla\boldsymbol{v}$$

heat flow — Joule heat — absorption — friction

$$+ \sum_j \boldsymbol{J}(N_j)\cdot\left[\boldsymbol{F}_j - T\operatorname{grad}\left(\frac{\mu_j}{T}\right)\right]$$

diffusion of chemical components

$$+ \sum_k \boldsymbol{J}(B_k)\cdot\left[\boldsymbol{F}_k - T\operatorname{grad}\left(\frac{\beta_k}{T}\right)\right]$$

"diffusion" of biotic components

$$+ \sum_r w_r A_r + \sum_k^{[1]} \rho_k P_k A(P_k)$$

chemical reactions — primary production

$$+ \sum_k \rho_k (R_k A(R_k) + M_k A(M_k))$$

respiration — mortality

$$+ \sum_k^{[3]} \rho_k (G_k A(G_k) + G'_k A(G'_k))$$

grazing in the food chain and on decomposers

$$+ \sum_k^{[4]} \rho_k (G^*_k A(G^*_k) + F_k A(F_k) + E_k A(E_k))$$

grazing on POM, uptake of DOM, and excretion by decomposers

$$+ \sum_j^{[7]} \rho_j H_j A(H_j). \tag{3.173}$$

hydrolysis

Dissipation is produced by heat transfer, Joule heating, absorption of light (especially by primary producers), friction, diffusion of chemical and biological components, chemical reactions, primary production, respiration, mortality, grazing, excretion, and hydrolysis of particulate organic matter. The entropy production σ is a bilinear function of "flows" and "forces":

$$\sigma = \sum_n J_n X_n \geq 0. \tag{3.174}$$

Flows	$\boldsymbol{W}$	$\boldsymbol{j}$	w_r	$\rho_k P_k$	$\rho_k G_k$	...
Forces	$\operatorname{grad}(1/T)$	$\boldsymbol{E}$	A_r/T	$A(P_k)/T$	$A(G_k)/T$	...

The "affinities" are defined by

$$A_r = -\sum_j \nu_{jr}\mathscr{M}_j\mu_j \qquad \text{(chemical reactions)} \tag{3.175}$$

$$A(P_k) = \sum_j^{[5]} b_{jk}\mu_j - \beta_k \qquad \text{(primary production)} \tag{3.176}$$

$$A(R_k) = \beta_k - \sum_j^{[5]} a_{jk}\mu_j \qquad \text{(respiration)} \qquad ([illegible]$$

$$A(M_k) = \beta_k - \sum_j^{[6]} (1 - c_{jk})\mu_j - \sum_m^{[7]} c_{mk}\mu_m \qquad \text{(mortality)} \tag{3.178}$$

$$A(G_k) = {\sum_l}^{*} g_{kl}\left(\beta_l - \beta_k + \sum_j^{[6]} s_{jkl}(\beta_k - \mu_j) + \sum_m^{[7]} p_{mkl}(\beta_k - \mu_m)\right) \qquad \text{(grazing)} \tag{3.179}$$

$$A(G'_k) = \sum_l^{[4]} g_{kl}(\beta_l - \beta_k) \qquad \text{(grazing on decomposers)} \tag{3.180}$$

$$A(G^*) = \sum_m^{[7]} d_{km}(\mu_m - \beta_k) \qquad \text{(grazing of decomposers on POM)} \tag{3.181}$$

$$A(F_k) = \sum_j^{[6]} f_{kj}(\mu_j - \beta_k) \qquad \text{(uptake of DOM by decomposers)} \tag{3.182}$$

$$A(E_k) = \sum_j^{[5]} e_{jk}(\beta_k - \mu_j) \qquad \text{(excretion by decomposers)} \tag{3.183}$$

$$A(H_m) = \sum_j^{[6]} h_{jm}(\mu_m - \mu_j) \qquad \text{(hydrolysis).} \tag{3.184}$$

(The summation symbols were defined in section 3.5.1, p. 73.)

Integrating the entropy balance equation 3.171 over the whole ecosystem, we obtain the total entropy production P_s:

$$P = \int_V \sigma \, \mathrm{d}V = \int_V \sum_n J_n X_n \, \mathrm{d}V \geq 0 \tag{3.185}$$

and the balance equation

$$\frac{\partial S}{\partial t} = -\oint_A \boldsymbol{S} \cdot \mathrm{d}\boldsymbol{f} + P_s \equiv \frac{\partial_e S}{\partial t} + \frac{\partial_i S}{\partial t}, \tag{3.186}$$

arriving at the decomposition of the total change, $\partial S/\partial t$, of the entropy of the system into an external part $\partial_e S/\partial t$ related to the flux from or into the surroundings and an internal part $\partial_i S/\partial t$, which, by the second law of thermodynamics, can never be negative.

3.6.6. The Linear Region of Thermodynamic Theory

In the linear range of irreversible processes it is assumed that the Onsager relations are valid:

$$J_n = \sum_i L_{ni} X_i, \qquad L_{nn} > 0, \qquad L_{ni} = L_{in}. \tag{3.187}$$

Open linear systems subject to time-independent external constraints reach, after a sufficiently long time, a steady nonequilibrium state characterized by constant entropy production, $\sigma = \sigma_0$ or $P_s = P_0$. Stable steady states exist. To maintain such a state, a continuous negative flow of entropy into the system is necessary.

The stationary states and the evolution of linear open systems are characterized by the theorem of Prigogine:

$$P_s \geq 0, \qquad \frac{dP_s}{dt} < 0 \qquad \text{(evolution)} \tag{3.188}$$

$$P_s = P_0 = \text{minimum}, \qquad \frac{dP_s}{dt} = 0 \qquad \text{(steady state).} \tag{3.189}$$

The total entropy production P_s becomes a minimum compatible with the constant constraints applied to the system. According to Lyapunov's theorem, (3.188) guarantees the stability of the steady nonequilibrium state because the Eulerian derivative dP_s/dt is semidefinite of sign opposite to the Lyapunov functional P_s. Perturbations of this state (either internal fluctuations generated by the system itself or external excitations) give rise to an entropy production larger than σ_0 or P_0, respectively. According to (3.188) the system is driven back to the reference state σ_0 or P_0. In linear open systems entropy production plays the same role as thermodynamic potentials in closed systems.

The linear Onsager theory is valid, as far as transport phenomena are concerned, but in order to set up a satisfactory description of chemical and biochemical reactions or biological processes, it is necessary to extend the theory to the nonlinear range.

3.6.7. The Nonlinear Region of Thermodynamic Theory; Stability Criteria

In the nonlinear range of irreversible processes, neither eqns. 3.187 nor 3.188 and 3.189 are valid. Under the condition of fixed concentrations or zero fluxes at the boundaries, the evolution criterion of Glansdorff and Prigogine (1971),

$$\frac{\partial_x P_s}{\partial t} \equiv \int_V \sum_n J_n \frac{\partial X_n}{\partial t} \, dV \leq 0, \tag{3.190}$$

states that the change of the forces X_n always proceeds in such a way as to lower the entropy production P_s. However, $\partial_x P_s/\partial t = 0$ does not necessarily mean that P_s goes to a minimum. No state variable or potential is known that tends to a minimum or maximum if a nonlinear system is far from thermodynamic equilibrium. While no "global" equilibrium criterion exists, a "local" stability criterion may be derived. Starting from the Gibbs relation (3.160), we can calculate the second-order differential of the entropy $s(u, \alpha, N_j, B_k)$:

$$\delta^2 s = \delta\left(\frac{1}{T}\right)\delta u + \delta\left(\frac{p}{T}\right)\delta\alpha - \sum_j \delta\left(\frac{\mu_j}{T}\right)\delta N_j - \sum_k \delta\left(\frac{\beta_k}{T}\right)\delta B_k. \qquad (3.191)$$

Again, using (3.160) we arrive at

$$T\,\delta^2 s = -\delta T\,\delta s + \delta p\,\delta\alpha - \sum_j \delta\mu_j\,\delta N_j - \sum_k \delta\beta_k\,\delta B_k. \qquad (3.192)$$

If now p, T, N_j, B_k are supposed to be independent variables, from (3.192) and (3.166)–(3.170) it follows that

$$T\delta^2 s = -\left(\alpha\chi(\delta p)^2 + c_p(\delta T)^2 + \sum_j\sum_m \frac{\partial\mu_j}{\partial N_m}\,\delta N_m\,\delta N_j + \sum_k\sum_l \frac{\partial\beta_k}{\partial B_l}\,\delta B_l\,\delta B_k\right). \qquad (3.193)$$

Thus, the function

$$L(\delta u, \delta\alpha, \delta N_j, \delta B_k, \delta \boldsymbol{v}) = \delta^2 s - \frac{|\delta\boldsymbol{v}|^2}{T} \qquad (3.194)$$

may serve as a Lyapunov function if the necessary and sufficient conditions

$$\chi > 0, \qquad c_p > 0, \qquad \sum_j\sum_m \frac{\partial\mu_j}{\partial N_m}\,\delta N_m\,\delta N_j > 0, \qquad \sum_k\sum_l \frac{\partial\beta_k}{\partial B_l}\,\delta B_l\,\delta B_k > 0 \qquad (3.195)$$

and the sufficient condition

$$\frac{\mathrm{d}L}{\mathrm{d}t} \geq 0 \qquad (t \geq t_0) \qquad (3.196)$$

are fulfilled. Since L is a homogeneous function of second degree in the perturbations δN_j, δB_k, $\delta \boldsymbol{v}$, δp, δT (or δu, $\delta \alpha$), we obtain from (3.196) by means of Euler's theorem

$$\frac{1}{2}\frac{\mathrm{d}L}{\mathrm{d}t} = \delta\left(\frac{1}{T}\right)\frac{\mathrm{d}\,\delta u}{\mathrm{d}t} + \delta\left(\frac{1}{T}\right)\frac{\mathrm{d}\,\delta \alpha}{\mathrm{d}t} - \frac{1}{T}\delta \boldsymbol{v}\cdot\frac{\mathrm{d}\,\delta \boldsymbol{v}}{\mathrm{d}t} - \sum_k \delta\left(\frac{\beta_k}{T}\right)\frac{\mathrm{d}\,\delta B_k}{\mathrm{d}t} - \sum_j \delta\left(\frac{\mu_j}{T}\right)\frac{\mathrm{d}\,\delta N_j}{\mathrm{d}t} \geq 0. \tag{3.197}$$

The time derivatives on the right-hand side of (3.197) are given by the energy and mass balance equations of the perturbed state. The inequalities (3.195) and (3.197) represent a general criterion of the local stability of nonequilibrium states or processes (including convective as well as dissipative effects) and can be used to obtain explicit criteria for specific situations.

Let us consider a steady state solution that corresponds to thermodynamic equilibrium, e.g. to the minimum of the specific free enthalpy g_e for a closed system at given temperature and pressure. Far from equilibrium this steady state is stable if the excess entropy production $\delta_x P_s$ is positive definite:

$$\delta_x P_s = \int_V \sum_n \delta J_n \delta X_n \,\mathrm{d}V > 0 \qquad (t \geq t_0). \tag{3.198}$$

The vanishing of $\delta_x P_s$ determines the thermodynamic threshold (or "bifurcation point") that separates the stable regime ($\delta_x P_s > 0$) from the unstable regime ($\delta_x P_s < 0$). The relation

$$\delta_x P_s = 0 \tag{3.199}$$

enables us to calculate the constraints acting on the ecosystem, which are capable of inducing an instability of the state under consideration. Beyond this instability another solution of the nonlinear system of basic equations may exist, radically different from the former type of solution, describing a new type of structure of the ecological system. The new structure is maintained in conditions far from equilibrium by flows of energy and matter into, through, and out of the system. The variety of steady states accessible to an open system may become much larger in such conditions. Chemical and biological processes involve situations beyond the instability. Therefore, investigations of this type are of great importance in water quality modeling. Stability criteria derived from the entropy principle must play an important role in these investigations. (Of course, we have to differentiate exactly between the "biological stability" of the ecosystem, the stability of the mathematical model, and the stability of the numerical method for the solution of the model equations.)

3.6.8. Thermodynamics of Primary Production

Thermodynamic principles, acting as "constraints" in the theory of water quality modeling, may be used to improve the modeling of biological processes and, furthermore, their interrelationships with physical phenomena like diffusion. At least the types of mathematical functions suited for describing these processes can be inferred, for instance, from the local entropy production. Logistic growth laws are shown to govern the processes of net primary production and of grazing. As soon as accurate thermodynamic data are available, the method will probably find greater use in water quality modeling.

In this section, only the thermodynamics of primary production can be outlined. A more complete presentation is given by Mauersberger (1982).

Basic Assumptions

The thermodynamic theory of primary production is based on the following assumptions.

(a) Autotrophic species transform light energy into chemical energy via photoactivated pigments, mainly chlorophylls, and by means of a complex series of reactions. As a result, an amount ε_k of energy per unit mass of the kth species is stored, e.g. incorporated in adenosine triphosphate (ATP) molecules and other "internal components" ($f = 1, 2, 3, \ldots, f_k$). For every species the balance of this internally stored energy is formulated as

$$\frac{d\epsilon_k}{dt} = \sum_f \mu_{kf} \frac{dm_{kf}}{dt} \tag{3.200}$$

$$\frac{d\epsilon_k}{dt} = \Phi_k(T, I)K_k - \lambda_{P,k} P_k + \lambda_{R,k} R_k - \Psi_k(T) q_k, \tag{3.201}$$

where:

m_{kf}, μ_{kf} are the mass fraction and chemical potential of the fth storing component "inside" the kth species,

K_k is the specific absorption by the kth species [$L^2\ M^{-1}$],

I is the mean light intensity in the activating interval of wavelength [energy $L^{-2}\ T^{-1}$],

$\lambda_{P,k}$ is the energy demand for primary production [energy M^{-1}],

$\lambda_{R,k}$ is the increase in ϵ_k caused by respiration [energy M^{-1}],

q_k is the energy dissipated or used for reproduction [energy $M^{-1}\ T^{-1}$],

$\Phi_k(T, I)$, $\Psi_k(T)$ are the dimensionless functions describing the temperature dependence of biological processes in which enzymatically catalyzed reactions play an important role (cf. Figure 3.7).

(b) The internal storage of energy has to be taken into account in the generalized Gibbs relation by the addition of a term to the right-hand side of (3.161):

$$-\sum_k B_k \sum_f \mu_{kf} \frac{\mathrm{d}m_{kf}}{\mathrm{d}t}.$$

(c) The entropy flux connected with the "absorption" of solar radiation by the primary producers counteracts the local entropy production σ. Hence, the total production per unit volume is

$$\sigma_{\text{tot}} = \boldsymbol{W} \operatorname{grad}\left(\frac{1}{T}\right) + \cdots + \frac{1}{T} \sum_k^{[1]} \rho_k P_k A_{P,k}$$
$$+ \cdots - \frac{1}{T_0} \sum_k^{[1]} \rho_k K_k \Phi_k(T, I). \qquad (3.202)$$

The terms omitted from (3.202) can be found in (3.173). T_0 is the temperature of the light source, for instance the radiation temperature of the solar photosphere. $A_{P,k}$ denotes the affinity of primary production:

$$A_{P,k} = \sum_j^{[5]} b_{jk} \mu_j - \beta_k + \lambda_{P,k}. \qquad (3.203)$$

(d) The specific entropy s and specific free enthalpy g_e per unit mass of an aquatic ecosystem can be determined approximately by applying the theory of dilute solutions (cf. Planck, 1913; Stumm and Morgan, 1970; etc.) not only to the dissolved inorganic and organic chemical components and dead particulate organic materials in the water body, but also to the biota (Mauersberger, 1980a, 1981a). This is only a very rough approximation for higher trophic levels. Let us denote by $\mathscr{M}_0$ and $x_0 = n_0/n \approx 1$ the molar mass and molar fraction of pure water. The mean molar mass $\mathscr{M}$ of all the components of the ecosystem (water included) is defined by

$$\mathscr{M} = \frac{M}{n} = x_0 \mathscr{M}_0 + \sum_j x_j \mathscr{M}_j + \sum_k B_k \mathscr{M}. \qquad (3.204)$$

s and g_e are found to be

$$s = s_0 N_0 + \sum_j N_j \left(s_j(p, T) - \frac{R_0}{\mathscr{M}_j} \ln x_j \right)$$
$$+ \sum_k B_k \left[s_k(p, T, m_{k1}, \ldots, m_{kf_k}) - \frac{R_0}{\mathscr{M}_k} \ln\left(\frac{\mathscr{M}}{\mathscr{M}_k} B_k \right) \right] \qquad (3.205)$$

and

$$g_e = g_0 N_0 + \sum_j N_j \mu_j(p, T, N_j) + \sum_k B_k \beta_k(p, T, B_k), \qquad (3.206)$$

where x_j are the molar fractions of the chemical components and N_0, N_j, B_k are mass fractions, while R_0 denotes the universal gas constant. Furthermore,

$$\mu_j = g_j(p, T) + \frac{R_0 T}{\mathcal{M}_j} \ln x_j \tag{3.207}$$

$$\beta_k = g_k(p, T, m_{k1}, \ldots, m_{kf_k}) + \frac{R_0 T}{\mathcal{M}_k} \ln\left(\frac{\mathcal{M}}{\mathcal{M}_k} B_k\right) \tag{3.208}$$

$$g_k(p, T, m_{k1}, \ldots, m_{kf_k}) = g_k(p, T, 0, 0, \ldots, 0) + \sum_f \mu_{kf}(T) m_{kf} \tag{3.209}$$

$$s = -\frac{\partial g}{\partial T}, \qquad s_j = -\frac{\partial g_j}{\partial T}, \qquad s_k = -\frac{\partial g_k}{\partial T}. \tag{3.210}$$

From (3.205) it follows that the entropy increases with the number of species and with their specific entropies s_k. From (3.203), (3.207), and (3.208) it follows that

$$\exp\left(-\frac{\bar{A}_{P,k}}{R_0 T}\right) = \frac{\mathcal{M} B_k}{\mathcal{M}_k \prod_j^{[5]} x_j^{\bar{b}_{jk}}} \exp\left(-\frac{\bar{A}_{P,k}^{(0)}}{R_0 T}\right), \tag{3.211}$$

where

$$\bar{A}_{P,k}^{(0)} = \sum_j^{[5]} \bar{b}_{jk} \bar{g}_j(p, T) - \bar{g}_k(p, T, m_{k1}, \ldots, m_{kf_k}) + \bar{\lambda}_{P,k}. \tag{3.212}$$

The bar denotes molar quantities:

$$\begin{gathered} \bar{A}_{P,k} = \mathcal{M}_k A_{P,k}, \qquad \bar{g}_k = \mathcal{M}_k g_k, \qquad \bar{\lambda}_{P,k} = \mathcal{M}_k \lambda_{P,k}, \\ \bar{g}_j = \mathcal{M}_j g_j, \qquad \bar{b}_{jk} = \mathcal{M}_k b_{jk} / \mathcal{M}_j. \end{gathered} \tag{3.213}$$

From $A_{P,k} \geq 0$ it follows that

$$B_k \leq B_k^{\max} = \frac{\mathcal{M}_k}{\mathcal{M}} \left(\prod_j^{[5]} x_j^{\bar{b}_{jk}}\right) \exp\left(\frac{\bar{A}_{P,k}^{(0)}}{R_0 T}\right). \tag{3.214}$$

The biomass of every species in a given region is bounded by an upper limit $B_k^{\max}$, which depends on water temperature and pressure as well as on the concentrations of the dissolved chemicals that take part in primary production.

The Rate of Primary Production

From the local entropy production (3.202), by the rules of thermodynamics of irreversible processes we conclude that the rate of primary production, P_k, depends on the affinity $\bar{A}_{P,k}$, the light intensity I, the temperature T, and other quantities in the following manner:

$$P_k = P_k\left(\frac{\bar{A}_{P,k}}{T}, \Phi_k(T, I), \ldots\right). \tag{3.215}$$

Only for small values of $\bar{A}_{P,k}$ and $\Phi_k(T, I)$ is this relation linear (the Onsager relation). Furthermore, we know that P_k vanishes outside of the T and I intervals. In the nonlinear region of the theory we use an ansatz equation determined by an optimization principle (Mauersberger, 1982), from which the relations between rates and affinities of biological processes like primary production, respiration, grazing, etc. are inferred to be of exponential character:

$$P_k = P_k^{\max}\left[1 - \exp\left(-c_k \Phi_k(T, I) \frac{\bar{A}_{P,k}}{R_0 T}\right)\right], \tag{3.216}$$

where c_k is a coefficient. Introducing (3.211) and (3.214) into (3.216), we obtain the following expression for the primary production rate:

$$P_k = P_k^{\max}\left[1 - \left(\frac{B_k}{B_k^{\max}}\right)^{c_k \Phi_k(T, I)}\right]. \tag{3.217}$$

In a somewhat simplified version (3.217) can be written as

$$\frac{P_k}{P_k^{\max}} = 1 - \left(\frac{B_k}{C_{P,k}} \frac{[O_2]^{b_0}}{[NO_3]^{b_1}[PO_4]^{b_2}[\ldots]^{b_3}}\right)^{c_k \Phi_k(T, I)}. \tag{3.218}$$

The coefficient $C_{P,k}$ depends slightly on the temperature; $[NO_3]$ denotes the concentration of nitrate in water. By using (3.217) or (3.218) we can quantify the influences of light intensity, temperature, and pressure on P_k, and also the effects of the biomass and of the concentrations of nutrients, oxygen, CO_2, and so on. It should be stressed that these relations are inferred from the entropy balance equation. The type of the function $\Phi_k(T, I)$ is shown by Figures 3.7 and 3.8, while the dependences of P_k on the light intensity, temperature,

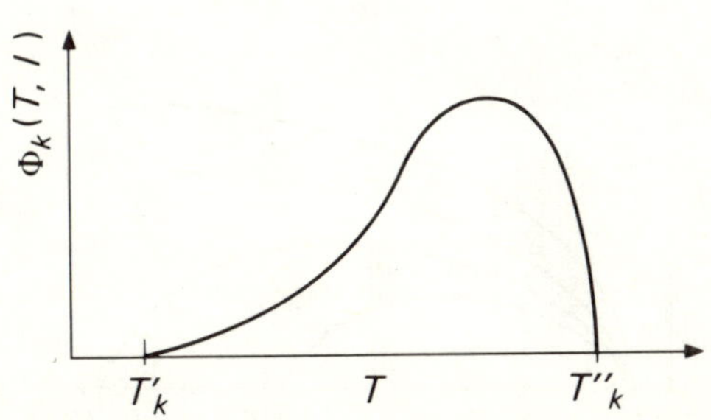

FIGURE 3.7 The function $\Phi_k(T, I)$ describes the temperature and light dependence of biological processes in which enzymatically catalyzed reactions play an important role. The function vanishes outside the interval from T_k' to T_k''. It is possible that this function has two peaks.

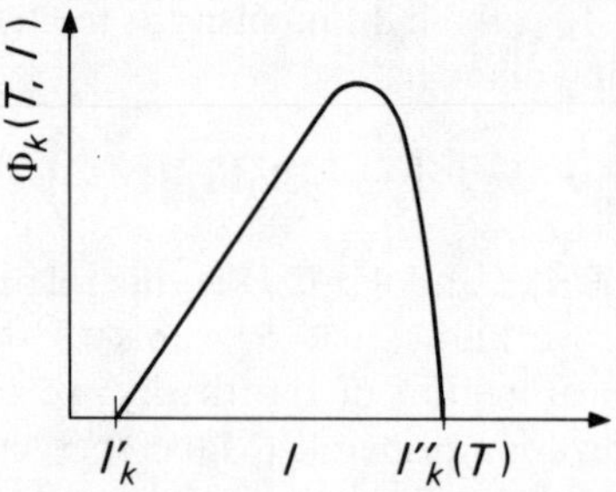

FIGURE 3.8 The type of dependence on light intensity I of the function $\Phi_k(T, I)$. While the lower boundary I_k' is approximately constant, the upper limit I_k'' depends on temperature T.

biomass, and concentrations of inorganic chemicals in the water body are shown qualitatively by Figures 3.9–3.14.

Growth Laws

Introducing (3.217) into (3.84), we can deduce the growth law for phytoplankton or the law governing gross primary production:

$$\frac{dB_k}{dt} = B_k P_k - \cdots = P_k^{\max} B_k \left[1 - \left(\frac{B_k}{B_k^{\max}}\right)^{c_k \Phi_k(T, I)}\right] - \cdots. \qquad (3.219)$$

Investigating respiration from the thermodynamic point of view, we find that the rate R_k is

$$R_k = R_k^{(0)} \left[\exp\left(c_k \psi_k(T) \frac{\bar{A}_{R,k}}{R_0 T}\right) - 1\right] \qquad (3.220)$$

$$= R_k^{(0)} \left[\left(\frac{B_k}{B_k^{\min}}\right)^{c_k \psi_k(T)} - 1\right] \approx R_k^{(1)} B_k^{c_k \psi_k(T)}. \qquad (3.221)$$

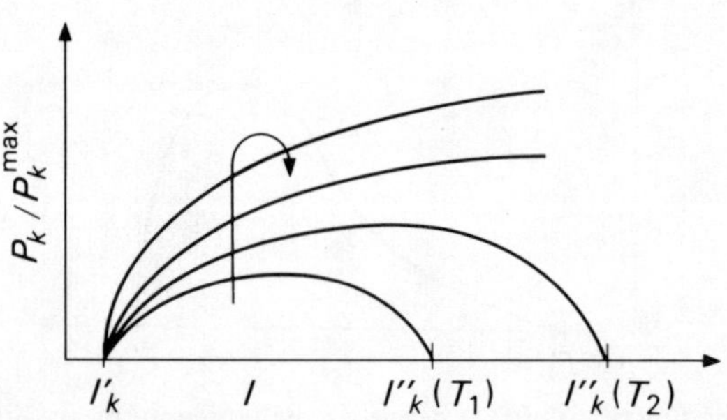

FIGURE 3.9 The rate of gross primary production P_k as a function of the light intensity I for different values of temperature. The arrow indicates the change of this function with increasing temperature.

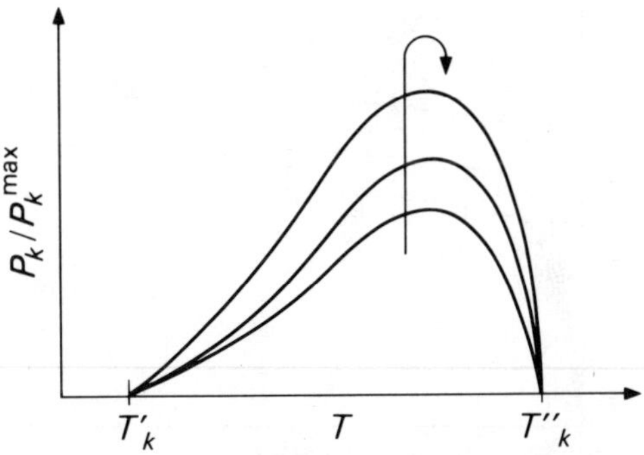

FIGURE 3.10 P_k as a function of temperature at constant values of nutrient concentrations and biomass. The arrow indicates the influence of increasing light intensity on this function.

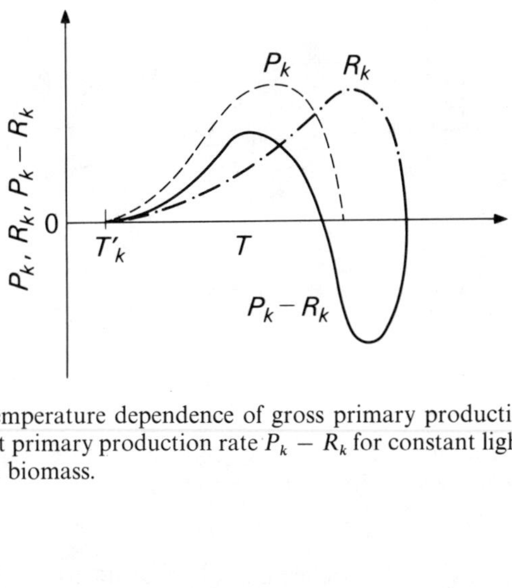

FIGURE 3.11 Temperature dependence of gross primary production rate P_k, respiration rate R_k, and net primary production rate $P_k - R_k$ for constant light intensity, nutrient concentrations, and biomass.

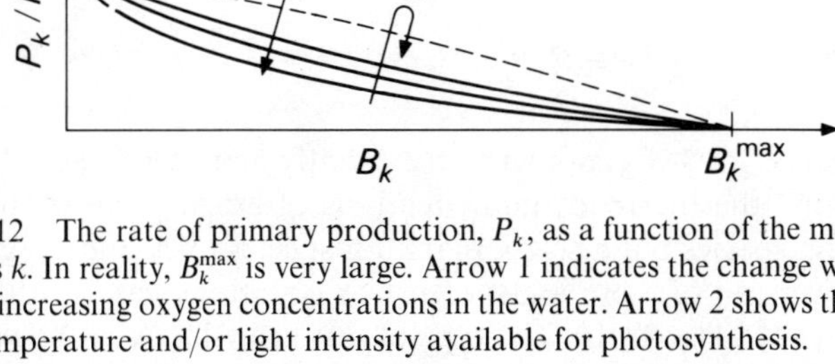

FIGURE 3.12 The rate of primary production, P_k, as a function of the mass fraction B_k of the species k. In reality, B_k^{max} is very large. Arrow 1 indicates the change with decreasing nutrient and increasing oxygen concentrations in the water. Arrow 2 shows the change with increasing temperature and/or light intensity available for photosynthesis.

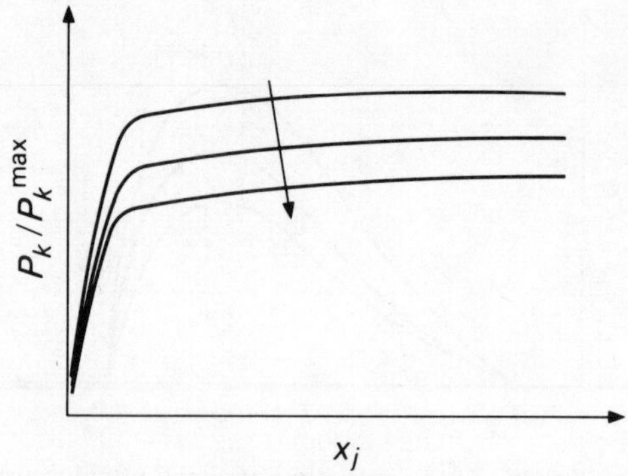

FIGURE 3.13 The rate of primary production as a function of the molar fraction x_j of one of the inorganic nutrients. The arrow indicates the change of this function with increasing biomass fraction B_k of the primary producer species k.

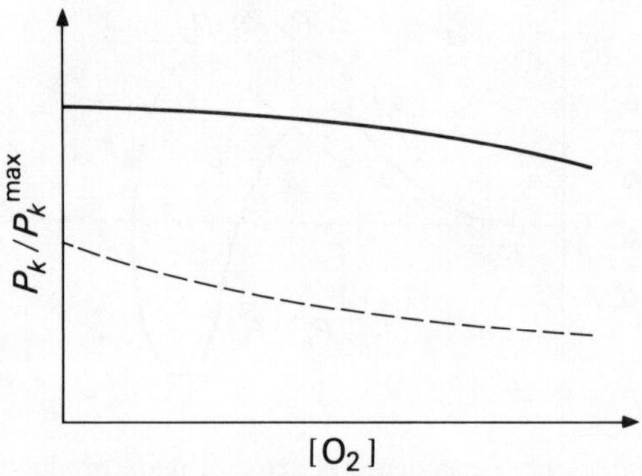

FIGURE 3.14 Dependence of P_k on the oxygen concentration $[O_2]$ in water. Positive curvature (broken line) is possible for small values of I and T.

The very small quantity $B_k^{\min}$ is determined by characteristics of the kth species and by environmental factors. Thus, the result is that the net production of the phytoplankton species k follows a "logistic" growth equation:

$$\frac{dB_k}{dt} = B_k(P_k - R_k) - \cdots = C_1 B_k - C_2 B_k^{1 + c_k \Phi_k} - C_3 B_k^{1 + c_k \psi_k} - \cdots. \quad (3.222)$$

The coefficients C_1, C_2, C_3, not written explicitly here, depend on T, I, and x_j.

Furthermore, the thermodynamic analysis of grazing shows that not only phytoplankton species but also zooplankton species obey logistic growth laws. A simplified version of these logistic growth equations has been discussed by Mauersberger (1980b, 1981b). The type of this growth law is shown by Figure

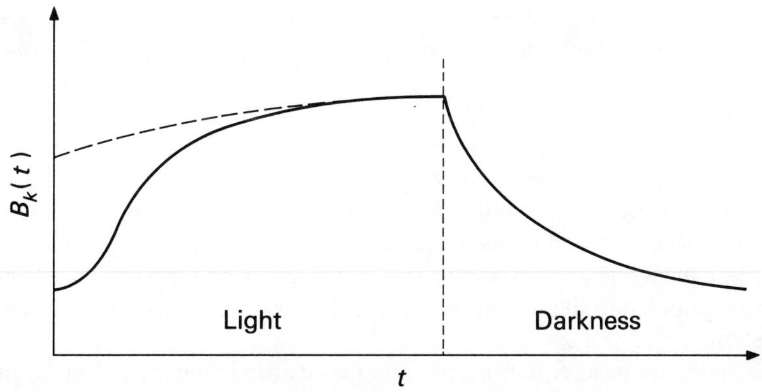

FIGURE 3.15 Characteristic growth of a phytoplankton population: mass fraction B_k as a function of time t. Broken line: very high initial value $B_k(0)$.

3.15. It agrees with the laws proposed and used by von Bertalanffy (1941), Straškraba (1978), and many others (cf. Majkowski and Uchmański, 1980).

Conclusion

The use of the entropy principle in ecosystems research is not only necessary (because the second law of thermodynamics cannot be ignored) but also of great advantage, supporting a powerful method for the analysis and synthesis of complex ecosystems, as has been demonstrated in this section.

3.7. PROBLEMS AND LIMITS OF DETERMINISTIC WATER QUALITY MODELING

Aquatic ecosystems are open, nonlinear systems far from thermostatic equilibrium, exchanging energy and matter with the surroundings and maintaining their more or less stationary state by fluxes of energy and matter through the systems. That is, they are able to use part of the energy and/or matter exchanged with the outside world to establish a macroscopic internal structure that is characterized by dissipative processes. There is a definite relation between the structure and the function of the ecosystem. These *dissipative structures* of a "self-organizing system" are not necessarily stable. The succession of structures and instabilities forms the anthropogenically influenced history of streams, lakes, reservoirs, and other water bodies.

The formation and maintenance of *self-organizing systems* are the result of nonequilibrium constraints, of appropriate nonlinear couplings, and of the *competition* between the entities constituting the ecosystem. This competition becomes significant whenever the resources necessary for the synthesis, growth, and/or survival of biological components are limited or become scarce. The result may be the elimination of some of the entities or a "dynamic equilibrium" of widely different, coexisting species.

Mathematically, the interactions between the dissipative system and the external world are specified either by boundary conditions on the separating surface or through constraints intervening explicitly within the differential equations. These interactions deeply influence the entropy-producing and entropy-reducing processes inside the water body and across its boundary, which are coupled to the structure, state, and further development of the aquatic ecosystem.

The nonlinear basic equations of water quality modeling (including boundary and initial conditions) may have more than one solution (*bifurcation*): more than one structure or state of the ecosystem may be possible. This fact should be taken into account in water quality modeling.

If the nonlinear basic equations describing the aquatic ecosystem allow for more than one solution without justifying preference for any one of them, internal *fluctuations* of components of the ecosystem (generated by the system itself or excited by the external world) have an important influence on the transition of the system to a new type of ordered configuration. This transition often occurs near a bifurcation point. Since fluctuations of physical, chemical, and/or biological parameters and/or components of aquatic ecosystems (e.g. day–night variations, seiches) are typical phenomena in water bodies, the investigation of fluctuations is a main task of water quality forecasting.

The responses of an ecosystem to radically changing conditions must be simulated by nonlinear relationships. The solution of the nonlinear initial-boundary-value problem representing the water quality model is not necessarily unique; stationary states are not necessarily stable. The *multiplicity of solutions* and the possibility of changing to another stationary state correspond to a gradual acquisition of autonomy from the environment. The existence of more than one solution to the model equations and the stability or instability of stationary states can depend upon:

the value of one or more parameters,
the numbers and types of variables/components taken into account, and
the types of connections between the components of the system.

If the stationary solution of the complex set of model equations is stable, this situation may change, for instance, if only one further component is taken into account. This corresponds to the change in behavior of an ecosystem caused by one new species, e.g. a virus.

The more complex the system of equations, the higher the degree of the characteristic equation determining the stability of the system and the greater the probability of an instability of a stationary state (May, 1973). This indicates a limit of the application of multicomponent deterministic water quality models. Of course, besides this "theoretical limit," which calls for the introduction of "new" variables allowing for the description of the ecosystem by a limited number of equations, there are practical demands for choosing the appropriate model complexity, e.g. to balance the modeling efforts with the data base available and/or to ensure that the model can be sustained economically for a sufficiently long time.

The scientific background, as well as other material, has been covered in works by La Salle and Lefshetz (1961), Morowitz (1970), Glansdorff and Prigogine (1971), Lyapunov (1972), Zotin (1972), Beserskiy and Popov (1975), Bautin and Leontovich (1976), Ebeling (1976), Rubin (1976), Jørgensen and Mejer (1977), Kogan (1977), Nicolis and Prigogine (1977), Dubois (1979), and Mejer and Jørgensen (1979).

REFERENCES

Alder, B., S. Fernbach, and M. Rotenberg (eds.) (1964). *Fundamental Methods in Hydrodynamics* (New York, NY: Academic Press) (Russian translation 1967, Moscow: Mir).

Antomonov, J. G. (1977). *Modeling of Biological Systems.* (Kiev: Naukova Dumka) (in Russian).

Bagotskiy, S. V., and A. D. Bazykin (1975). *Mathematical Ecology. A Bibliographic Index of Soviet Publications 1953–1974* (Moscow: Scientific Computing Centre, Academy of Sciences, USSR) (in Russian).

Bautin, N. N., and E. A. Leontovich (1976). *Methods and Procedures of the Qualitative Investigation of Dynamic Systems in Two Dimensions* (Moscow: Nauka) (in Russian).

von Bertalanffy, L. (1941). Stoffwechseltypen und Wachstumstypen. *Biologisches Zentralblatt* 61:510–532.

Beserskiy, V. A., and E. P. Popov (1975). *Theory of Systems of Automatic Control* 3rd edn. (Moscow: Nauka) (in Russian).

Biswas, A. K. (ed.) (1972). *Modelling of Water Resources Systems* (Montreal: Harvest House).

Ciaccio, L. C. (1971–73). *Water and Water Pollution Handbook* (New York, NY: Dekker).

Collatz, L. (1959). *The Numerical Treatment of Differential Equations*, 3rd edn. (Berlin: Springer).

Daubner, I. (1972). *Mikrobiologie des Wassers* (Berlin: Akademie-Verlag).

Dubois, D. M. (1979). Catastrophe theory applied to water quality regulation of rivers. *State of the Art in Ecological Modelling* vol. 7, ed. S. E. Jørgensen (Oxford: Pergamon), pp. 751–758.

Ebeling, W. (1976). *Strukturbildung bei Irreversibelen Prozessen* (Leipzig: Teubner).

Eckart, C. (1960). *Hydrodynamics of Oceans and Atmospheres* (Oxford: Pergamon).

Glansdorff, P., and I. Prigogine (1971). *Thermodynamic Theory of Structure, Stability and Fluctuations* (London: Wiley).

Goldberg, E. D., I. N. McCave, J. J. O'Brien, and J. H. Steele (eds.) (1977). *The Sea* vol. 6 Marine modeling (New York, NY: Wiley).

Golterman, H. L. (1975). *Physiological Limnology* (Amsterdam: Elsevier).

Hinze, J. O. (1959). *Turbulence* (New York, NY: McGraw-Hill) (Russian translation 1963, Moscow: Fizmatgiz).

Jenne, E. A., *et al.* (1979). Chemical modeling in aqueous systems. *Speciation, Sorption, Solubility, and Kinetics. Symposium Series of American Chemical Society.*

Jørgensen, S. E. (ed.) (1979a). Hydrophysical and ecological models of shallow lakes and reservoirs. Summary Report of an IIASA Workshop, April 11–14, 1978. *Collaborative Proceedings* CP-78-14 (Laxenburg, Austria: International Institute for Applied Systems Analysis).

Jørgensen, S. E. (ed.) (1979b). *Handbook of Environmental Data and Ecological Parameters* (Copenhagen: International Society for Environmental Modelling).

Jørgensen, S. E., and D. R. F. Harleman (eds.) (1978). Hydrophysical and Ecological modelling of deep lakes and reservoirs. Summary Report of a IIASA Workshop, December 12–15, 1977. *Collaborative Proceedings* CP-78-7 (Laxenburg, Austria: International Institute for Applied Systems Analysis).

Jørgensen, S. E., and H. Mejer (1979). A holistic approach to ecological modelling. *Ecological Modelling* 7: 169–189.

Kamenkovich, V. M., and A. S. Monin (eds.) (1978). *Oceanology and Physics of the Ocean* vol. 1 Hydrodynamics of the ocean; vol. 2 Hydrodynamics (Moscow: Nauka) (in Russian).

Kaye, G. W., and T. H. Laby (1959). *Tables of Physical and Chemical Constants*, 12th edn. (London: Longman) (Russian translation 1962, Moscow: Fizmatgiz).

Kinsman, B. (1965). *Wind Waves* (Englewood Cliffs, NJ: Prentice-Hall).

Klotz, I. M. (1964). *Chemical Thermodynamics: Basic Theory and Methods*, revised edn. (New York, NY: Benjamin).

Kogan, A. B. (1977). *Biological Cybernetics* (Moscow: Vishaya Shkola) (in Russian).

Krauss, W. (1966). Interne Wellen. *Methoden und Ergebnisse der Theoretischen Ozeanographie* vol. 2 (Berlin: Gebrüder Borntraeger) (Russian translation 1968, Leningrad: Gidrometizdat).

Lamb, H. (1932). *Hydrodynamics* (London: Cambridge University Press) (1945 New York, NY: Dover) (Russian translation 1947, Moscow: Gostekhizdat).

Landau, L., and E. Lifshitz (1960). *Electrodynamics of Continuous Media* (London: Pergamon) (Russian version 1959, Moscow: Fizmatgiz).

La Salle, J., and S. Lefshetz (1961). *Stability by Lyapunov's Direct Method* (New York, NY: Academic Press) (Russian translation 1964, Moscow: Mir).

Latimer, W. M. (1952). *Oxidation Potentials*, 2nd edn. (Englewood Cliffs, NJ: Prentice-Hall).

Lehninger, A. L. (1972). *Biochemistry* (New York, NY: Worth) (Russian translation 1976, Moscow: Mir).

Lewis, G. N., and M. Randall (1961). *Thermodynamics*, 2nd edn. (New York, NY: McGraw-Hill).

Lin, C. C. (1955). *The Theory of Hydrodynamic Stability* (London: Cambridge University Press) (Russian translation 1958, Moscow: Inostr. Lit.).

Lions, J. L. (1969). *Quelques Méthodes de Résolution des Problèmes aux Limites non Linéaires* (Paris: Dunod; Gauthier-Villars) (Russian translation 1972, Moscow: Mir).

Lyapunov, A. A. (1972). *Cybernetic Problems in Biology* (Moscow: Nauka) (in Russian).

Majkowski, J., and J. Uchmański (1980). Theoretical foundations of individual growth equations in animals. *Polish Ecological Studies* 6: 7–31.

Marchuk, G. I. (1975). *Methods of Numerical Mathematics* (New York, NY: Springer) (Russian version 1973, Novosibirsk: Nauka).

Martell, A. E., and R. M. Smith (1974–77). *Critical Stability Constants* vol. 1 Amino acids (1974); vol. 2 Amines (1975); vol. 3 Other organic ligands (1977); vol. 4 Inorganic complexes (1976) (New York, NY: Plenum).

Mauersberger, P. (1978). On the theoretical basis of modeling the quality of surface and subsurface waters. *Modeling the Water Quality of the Hydrological Cycle. Proceedings of Baden Symposium, September. International Association of Hydrological Sciences Publication* 125, pp. 14–23.

Mauersberger, P. (1979). On the role of entropy in water quality modelling. *Ecological Modelling* 7: 191–199.

Mauersberger, P. (1980a). Ergebnisse der Thermodynamik aquatischer Ökosystems. *Gerlands Beiträge zur Geophysik* 89: 339–341.

Mauersberger, P. (1980b). Growth equations of phyto- and zooplankton deduced from thermodynamic principles. *Proceedings of International Symposium on Simulation of Systems in Biology and Medicine: Si-Sy '80, Prague, 17–20 November*, vol. 3, pp. 55–65.

Mauersberger, P. (1981a). Entropie und freie Enthalpie im aquatischen Ökosystem. *Acta Hydrophysica* 26: 67–90.

Mauersberger, P. (1981b). Laws for population and community dynamics in aquatic ecosystems based on the balances of mass, energy, and entropy. *Proceedings of International Workshop on Ecosystem Dynamics in Freshwater Wetlands and Shallow Water Bodies, Minsk–Tskhaltoubo, USSR, 12–26, July.*

Mauersberger, P. (1982). Zur Bestimmung der nichtlinearen Beziehungen zwischen Raten und Affinitäten bei Produktions- und Abbauprozessen im aquatischen Ökosystem. *Acta Hydrophysica* 27: 125–130.

May, R. (1973). *Model Ecosystems* (Princeton, NJ: Princeton University Press).

Meixner, J., and H. R. Reik (1959). Thermodynamik der irreversibelen Prozesse. *Handbuch der Physik* vol. 3/2, ed. S. Flügge (Berlin: Springer), pp. 413–523.

Mejer, H., and S. E. Jørgensen (1979). Energy and ecological buffer capacity. *State of the Art in Ecological Modelling* vol. 7, ed. S. E. Jørgensen (Oxford: Pergamon), pp. 829–846.

Menshutkin, V. V. (1971). *Mathematical Modeling of Populations and Faunal Communities in Aquatic Ecosystems* (Leningrad: Nauka) (in Russian).

Milne-Thomson, L. M. (1960). *Theoretical Hydrodynamics*, 4th edn. (New York, NY: Macmillan) (Russian translation 1964, Moscow: Mir).

Mitchel, A. R., and R. Wait (1977). *The Finite Element Method in Partial Differential Equations* (New York, NY: Wiley).

Mitropolskiy, J. A. (ed.) (1977). *Mathematical Methods in Biology* (Kiev: Naukova Dumka) (in Russian).

Monin, A. S., and A. M. Jaglom (1965–67). *Statistical Hydrodynamics* vols. 1, 2 (Moscow: Nauka) (in Russian).

Morowitz, H. J. (1970). *Entropy for Biologists* (New York, NY: Academic Press).

Naumov, G. B., B. N. Ryzhenko, and I. L. Khodakovsky (1974). *Handbook of Thermodynamic Data.* US Geological Survey WRD-74-001. National Technical Information Service PB 226 722/AS.

Nicolis, G., and I. Prigogine (1977). *Self-Organization in Nonequilibrium Systems* (London: Wiley).

Phillips, M. (1962). Classical electrodynamics. *Handbuch der Physik* vol. 4, ed. S. Flügge (Berlin: Springer), pp. 1–108.

Pielou, E. C. (1975). *Ecological Diversity* (New York, NY: Wiley).

Pinder, G. F., and W. G. Gray (1977). *Finite Element Simulation in Surface and Subsurface Hydrology* (New York, NY: Academic Press).

Planck, M. (1913). *Vorlesungen über Thermodynamik*, 4th edn. (Berlin: Veit).

Proudman, J. (1953). *Dynamical Oceanography* (London: Academic Press) (Russian translation 1957, Moscow: Inostr. Lit.).

Rheinheimer, G. (1971). *Mikrobiologie der Gewässer* (Jena: VEB Gustav Fischer).

Richtmyer, E. D., and K. W. Morton (1967). *Difference Methods for Initial Value Problems*, 2nd edn. (New York, NY: Interscience).

Romanenko, V. I., and S. I. Kuznecov (1974). *Ecology of the Microorganisms of Freshwater Bodies* (Leningrad: Nauka) (in Russian).

Romanovskiy, J. M., N. V. Stepanova, and D. S. Chernavskiy (1975). *Mathematical Modeling in Biophysics* (Moscow: Nauka) (in Russian).

Rossini, F. D., D. D. Wagman, W. H. Evans, S. Levine, and I. Jaffe (1952). Selected values of chemical thermodynamic properties. *National Bureau of Standards, US Department of Commerce, Washington, DC Circular* 500.

Rubin, A. B. (1976). *Thermodynamics of Biological Processes* (Moscow: University Press) (in Russian).

Salánki, J., and J. E. Ponyi (eds.) (1975). *Limnology of Shallow Waters: Proceedings of Symposium, Tihany, Hungary, 1973* (Budapest: Akadémiai Kiadó).

Serrin, J. (1959). Mathematical principles of classical fluid flow. *Handbuch der Physik* vol. 8/1, ed. S. Flügge (Berlin: Springer), pp. 125–263.

Stoker, J. J. (1957). *Water Waves* (New York, NY: Interscience) (Russian translation 1959, Moscow: Inostr. Lit.).

Strang, G., and G. J. Fix (1973). *An Analysis of the Finite Element Method* (Englewood Cliffs, NJ: Prentice-Hall) (Russian translation 1977, Moscow: Mir).

Straškraba, M. (1977). Natural control mechanisms in models of aquatic ecosystems. *Proceedings of IFAC Symposium on Control Mechanisms in Bio- and Ecosystems, Leipzig, September* vol. 1, pp. 74–108.

Straškraba, M. (1978). Theoretical considerations on eutrophication. *Verhandlung Internationaler Vereinigung Limnologie* 20:2714–2720.

Stratton, J. A. (1941). *Electromagnetic Theory* (New York, NY: McGraw-Hill).

Stumm, W., and J. J. Morgan (1970). *Aquatic Chemistry* (New York, NY: Wiley).

Truesdell, C., and R. Toupin (1960). The classical field theories. *Handbuch der Physik* vol. 3/1, ed. S. Flügge (Berlin: Springer), p. 692.

Uhlmann, D. (1975). *Hydrobiologie* (Jena: VEB Gustav Fischer).

Vasiliev, O. F., V. I. Kvon, Ju. M. Litkin, and I. L. Rozovskiy (1975). Stratified flows. *Hydromechanika* 8:74–131 (in Russian).

Vinberg, G. G., and S. I. Anisimov (1966). Investigation of a mathematical model for an aquatic ecosystem. *Publications of All-Union Research Institute for Fishery Economics and Oceanography, Moscow* 67:49–75.

Vollenweider, R. A. (1969). Möglichkeiten und Grenzen elementarer Modelle der Stoffbilanz von Seen. *Archiv für Hydrobiologie* 66:1–36.

Volterra, V. (1932). *Leçons sur la Théorie Mathématique de la Lutte pour la Vie*. (Paris: Gauthier-Villars) (Russian translation 1976, Moscow: Nauka).

Wagman, D. D., W. H. Evans, V. B. Parker, I. Halow, S. M. Bailey, and R. H. Schumm (1968–69). Selected values of chemical thermodynamic properties. *National Bureau of Standards Technical Note* 270-3 (1968), 270-4 (1969).

von Weizsäcker, C. F. (1972). Evolution und Entropiewachstum. *Nova Acta Leopoldina, Neue Folge* 206:515–534.

Wetzel, R. G. (1975). *Limnology* (Philadelphia, PA: Saunders).

van Zeggeren, F., and S. H. Storey (1970). *The Computation of Chemical Equilibria* (London: Cambridge University Press).

Zienkiewicz, O. C., and Y. K. Cheung (1967). *The Finite Element Method in Structural and Continuum Mechanics* (London: McGraw-Hill).

Zotin, A. I. (1972). *Thermodynamic Aspects of Developmental Biology* (Basel: Karger).

CHAPTER 3: NOTATION

$a_{jk}, b_{jk}, c_{jk}, d_{km}, e_{jk}, f_{kj}, g_{kl}, h_{jm}, p_{jkl}, p_{jk}, s_{jkl}, s_{kl}$	fractions of ecosystem components produced or consumed by internal biochemical or biological processes
B_k	mass fraction of biota
BC_k	biotic component k of aquatic ecosystem
D, D^*	scalar diffusion coefficients
$\mathrm{DIM}(\mathrm{DIM}_j)$	dissolved inorganic matter (component j)

DOM(DOM_j)	dissolved organic matter (component j)
F_k	rate of uptake of DOM by kth decomposer
g_e	specific free enthalpy
G	sources, sinks
G_p	Gibbs potential
H_j	rate of hydrolysis of jth type of particulate organic matter (POM_j)
$\mathbf{j}$	electric current density
J_n	generalized flows
N_j	mass fraction of chemicals
P_s	total entropy function
P_v	pressure function
POM(POM_j)	particulate organic matter (component j)
Q_k	production and consumption term
Q_r	reaction quotient
s	entropy density
S	total entropy of system
$\mathbf{S}$	entropy flow density
t_{kl}	preference coefficient (for biocomponent l as food for k)
U_{kj}	rate of uptake of DIM_j by kth primary producer
X_n	generalized forces
Y_j	rates of biochemical reactions
α	specific volume
β_k	potentials of biocomponents
γ_j	activity coefficients
ϵ_k	coefficient of turbulent mixing (eddy diffusivity); internal energy per unit mass of species k
κ	thermal conductivity
λ	energy supply or demand
$\bar{\mu}_j$	$\mathscr{M}_j \mu_j$, molar chemical potential of jth chemical constituent
σ	entropy production rate
χ	isothermal compressibility.

4 Modeling the Ecological Processes

S. E. Jørgensen

4.1. BACKGROUND

4.1.1. The Ecosystem and Chemical–Biological Processes

The basic chemical, biochemical, and biological processes mentioned in section 3.5 have been quantitatively described by application of a great variety of equations, which will be surveyed and represented in this chapter in forms that can easily be translated into computer language (e.g. CSMP).

An ecosystem is defined by several biotic and abiotic variables, which will change in space and time owing to the above-mentioned processes. This is illustrated in Figure 4.1, which shows the flow of energy in various forms through a succession of trophic levels within an aquatic environment characterized by quality parameters such as the concentrations of dissolved oxygen, carbon dioxide, phosphorus, and nitrogen, and amounts of suspended matter, detritus, and bottom sediment. The mass and energy transformations are regulated by processes such as growth, respiration, mortality, and decomposition; these in turn are governed by environmental quality parameters such as temperature, toxicity, and nutrient concentrations. The system is highly coupled, as energy and mass balances for individual constituents are invariably linked to several others.

Ecological models, like other models, represent a compromise between the reality of the system and the simplicity of the mathematical description. Fitted regression equations might compare favorably in predictive capacity with causally based functions, but they provide little or no insight into the underlying processes and they are most often specific and cannot be used in several case studies without modification. Causal models require biological, chemical, and physical details in construction to enhance reliability over a broad environ-

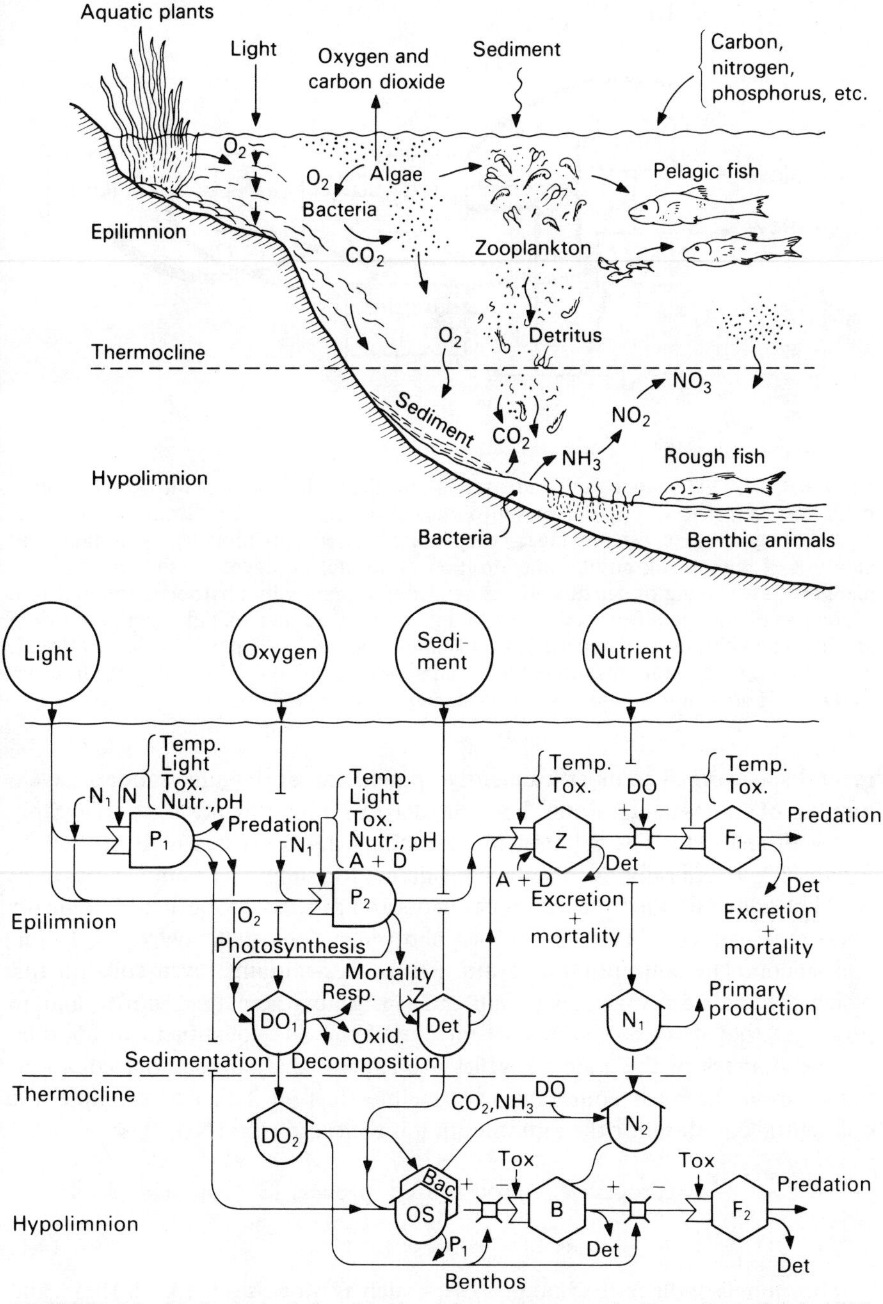

FIGURE 4.1 Ecological relationships in a lake environment (redrawn from Orlob, 1975). A, advection; B, benthic animals; Bac, bacteria; D, diffusion; Det, detritus; F_1, pelagic fish; F_2, rough fish; N_1, N_2, soluble nitrogen (NH_4^+ and NO_3^-); DO_1, DO_2, dissolved oxygen; OS, soluble carbonate; P_1, aquatic plants; P_2, algae; Tox, toxic substances; Z, zooplankton.

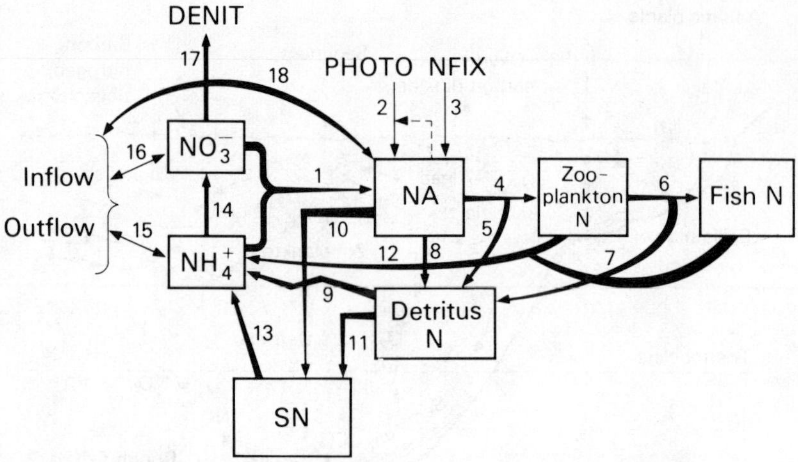

FIGURE 4.2 Lake model: only nitrogen is considered. 1, Uptake of nitrogen by phytoplankton; 2, growth of phytoplankton (photosynthesis); 3, nitrogen fixation; 4, grazing; 5, fecal (by grazing); 6, fish preying on zooplankton; 7, fecal (by process); 8, mortality and autolysis of phytoplankton; 9, mineralization of detrital nitrogen; 10, sinking of phytoplankton; 11, sinking of detritus; 12, excretion of ammonia by phytoplankton and fish; 13, release of ammonia from sediment; 14, nitrification; 15 and 16, inflow and outflow of ammonia and nitrate; 17, denitrification; 18, inflow and outflow of phytoplankton. DENIT, denitrification; NA, concentration of nitrogen in phytoplankton; NFIX, nitrogen fixation; PHOTO, photosynthetic rate; SN, nitrogen in upper layer of sediment.

mental spectrum. The most frequently applied causal dynamic models have a number of constituents: water, bacteria, detritus, phytoplankton, zooplankton, fish, sediment, etc. The relation to the underlying processes is illustrated by Figure 4.2, where only nitrogen is considered, for simplicity. Similar diagrams could be added to illustrate the constituents and processes for other biologically important elements such as carbon, phosphorus, oxygen, hydrogen, sulfur, and silicon. The conceptual diagram, Figure 4.2, contains seven constituents, which are the state variables (Chapter 2) in the mathematical model, and 18 processes that describe the flow of nitrogen from one constituent to another.

The changes of the state variables are expressed by differential equations based upon the mass conservation principle (section 3.2). For example, the differential equation for the ammonium ion concentration (NH_4^+) is:

$$\frac{dNH_4^+}{dt} = \text{process 9} + \text{process 13} + \text{process 12} \pm \text{process 15} - \text{process 14} - \text{process 1}. \tag{4.1}$$

The models of the individual processes, such as processes 9, 13, 12, 15, 14, and 1, and their related parameters, which can be termed submodels, will be surveyed in this chapter. The submodels are components of water quality models, which include ecological processes, such as the river model QUAL II (section 6.6) and the lake model CLEANER (section 9.2.1). The coupling of the submodels will be treated in relation to river and lake models in Chapters 6 and 9.

4.1.2. Brief Review of Development

Among the earliest ecological submodels were those that addressed the problem of predator–prey relationships, e.g. the set of equations referred to as the Lotka–Volterra predator–prey model (Lotka, 1924). Another field, which was approached at an early stage by the use of models, was the interaction between organisms and their feed or substrate. The so-called Monod kinetic has been widely used for many years:

$$\mu = \mu_{max} \frac{S}{K_S + S}, \tag{4.2}$$

where μ is the growth rate, μ_{max} the maximum rate, S the substrate concentration, and K_S the half-saturation constant. The equation is represented by Figure 4.3, which demonstrates that K_S is equal to S when $\mu = \frac{1}{2}\mu_{max}$. This equation was derived from the Michaelis–Menten formulation of biochemical enzyme kinetics.

The trend in ecological modeling over the past several years has been to include more biological realism, so that more and more detailed submodels attempt to take into consideration feedback mechanisms and couplings to other constituents and to environmental factors. The experience gained by working with river and lake ecological models has provoked new biological experiments devised to aid formulation of model construction by representing

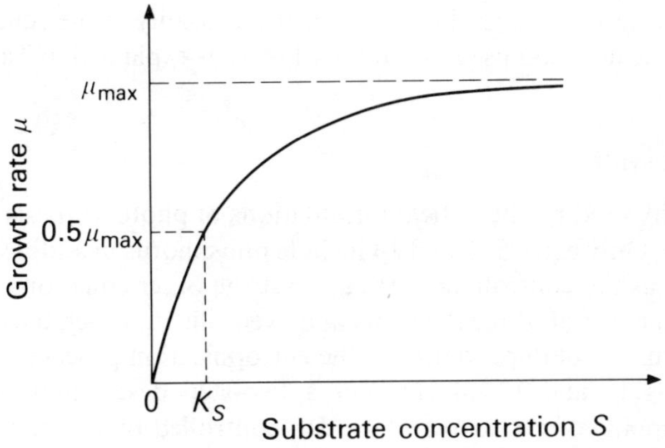

FIGURE 4.3 Michaelis–Menten kinetic expression for microbial substrate utilization and growth.

environmental responses for specific processes. As a result of this effort a wide range of mathematical formulations of the ecological processes in bodies of water is available today.

4.2. MODELING ECOLOGICAL PROCESSES IN AQUATIC ECOSYSTEMS

4.2.1. General Considerations

The plentiful sources of error in both our structural and measured knowledge of an ecological process make it hardly possible to select the only correct formulation for a specific process in an aquatic ecosystem. The reason for this is the difficulty in obtaining harmonized, regular, high-frequency spatial and temporal field data with reasonable accuracy. Such difficulties can partly be overcome by examination of the actual processes in the laboratory, but this is only, as indicated, partly a solution, since:

(i) the processes in the ecosystem can never be isolated from their environments, and
(ii) it is impossible to take account of all natural coupling mechanisms.

Consequently, each process has several alternative mathematical formulations that, at this stage of ecological modeling, are equally valid to a certain extent. As for the total model, the submodels should be selected in accordance with the problem, the data available, and the case study. The processes of interest for water quality modeling and their mathematical formulations will be surveyed in the following sections, and the realism and causality of the equations will be discussed. The notation used in Tables 4.1–4.12 is explained in Table 4.13.

4.2.2. Photosynthesis

The generally used mathematical formulations of photosynthesis are surveyed in Table 4.1. Only eqns. 5, 8, and 14 include phosphorus in addition to nitrogen and carbon as the controlling nutrients, but the other equations can easily be expanded to cover all three nutrients and even silicon, which must be included when diatoms are of importance for the eutrophication process.

Equations 10 and 12–14 consider a two-step description of the photosynthesis: uptake of nutrient, and growth controlled by the intracellular concentration of nutrient. If only the first process is indicated the growth is, or can be, formulated by use of the other growth equations. This two-step formulation

TABLE 4.1 Photosynthesis.

(1) Broqvist (1971)

$$\text{PHOTO} = \text{MY}(T) \cdot \text{PHYT} \cdot \min\left(\frac{I}{I(0)}, \frac{\text{PS}}{\text{PS}_0}, \frac{\text{NS}}{\text{NS}_0}\right)$$

(2) Chen (1970), Chen and Orlob (1975)

$$\text{PHOTO} = \text{MY}(T) \cdot \frac{I}{I + \text{IK}} \cdot \frac{\text{NS}}{\text{KN} + \text{NS}} \cdot \frac{\text{PS}}{\text{KP} + \text{PS}} \cdot \text{PHYT}$$

$$\text{MY}(T) = \text{MYMAX} \cdot f(T)$$

(3) Parker (1972)

$$\text{PHOTO} = \text{MY}(T) \cdot I \cdot \text{PS} \cdot \text{NS} \cdot \text{PHYT}$$

(4) Anderson (1973)

$$\text{PHOTO} = \text{MYMAX} \cdot \text{PHYT} \cdot (\text{PS} + \text{NS})$$

(5) Dahl-Madsen and Strange Nielsen (1974)

$$\text{PHOTO} = \text{MY}(T) \cdot f(I) \cdot \frac{\text{PS}}{\text{KP} + \text{PS}} \cdot \frac{\text{NS}}{\text{KN} + \text{NS}} \cdot \frac{\text{CS}}{\text{KC} + \text{CS}}$$

$$\text{MY}(T) = \text{MYMAX} \cdot g(T)$$

(6) Jansson (1972)

$$\text{PHOTO} = \text{MY} \cdot \text{PHYT} \cdot \text{PS} \cdot IT$$

(7) Lassen and Nielsen (1972)

$$\text{PHOTO} = \text{MY} \cdot \text{PHYT} \cdot f(\text{PS}) \cdot \text{FD} \cdot f(I)$$

(8) Patten *et al.* (1975)

$$\text{PHOTO} = \text{MY} \cdot \text{PHYT} \cdot f(I)(f(\text{NS}) + f(\text{PS}) + f(\text{CS}))/3$$

(9) Larsen *et al.* (1974)

$$\text{PHOTO} = \text{MY} \cdot \text{PHYT} \cdot \min(f(I), f(\text{PS}), f(\text{NS}))$$

$$f(I) = \frac{\bar{F}_{\max} R(t)}{\text{IK} + R(t)}$$

There are corresponding equations for $f(\text{PS})$ and $f(\text{NS})$.

(10) Bierman *et al.* (1974)

$$\text{UP} = \text{UPMAX}\left(\frac{1}{1 + \text{PKI} \cdot \text{PA}} - \frac{1}{1 + \text{PKI} \cdot \text{PS}}\right)$$

(11) Gargas (1976)

$$\text{MY} = \text{MYMAX} \cdot f(I) f(\text{PS}) f(\text{NS}) f(T) \cdot \text{FD} \cdot \text{FAC}$$

(*continued over*)

TABLE 4.1 (*continued*)

(12) Cloern (1978)

$$\text{PHOTO} = \text{MY}(T) \cdot \frac{I}{I_{\text{opt}}(T)} \exp\left(1 - \frac{I}{I_{\text{opt}}(T)}\right) f(\text{PA}) f(\text{NA})$$

$$\text{MY}(T) = 0.02 \exp(0.17\ T)$$

$$I_{\text{opt}}(T) = 0.06 \exp(0.22\ T)$$

$$f(\text{PA}) = \frac{\text{PAP}}{\text{PAP} - \text{PAMIN}}$$

$$f(\text{NA}) = \frac{\text{NAP}}{\text{NAP} - \text{NAMIN}}$$

$$\text{UP} = \text{UPMAX} \cdot \frac{\text{PS}}{\text{KP} + \text{PS}}$$

$$\text{UN} = \text{UNMAX} \cdot \frac{\text{NS}}{\text{KN} + \text{NS}}$$

(13) Nyholm (1978)

$$\text{PHOTO} = \text{MY} \cdot f(I) f(\text{NA, PA})$$

$$f(\text{NA, PA}) = \frac{2}{1/f(\text{PA}) + 1/f(\text{NA})}$$

(see also Bloomfield *et al.*, 1974)

$$f(\text{PA}) = \frac{\text{KPA} + \text{PAMAX} - \text{PAMIN}}{\text{PAMAX} - \text{PAMIN}} \cdot \frac{\text{PA/PHYT} - \text{PAMIN}}{\text{KPA} + \text{PA/PHYT} - \text{PAMIN}}$$

$$f(\text{PN}) = \frac{\text{NA/PHYT} - \text{NAMIN}}{\text{NAMAX} - \text{NAMIN}}$$

Excess supply of nutrients:

$$\frac{\text{dPS}}{\text{d}t} = \text{PAMAX} \frac{\text{dPHYT}}{\text{d}T}$$

$$\frac{\text{dNS}}{\text{d}t} = \text{NAMAX} \frac{\text{dPHYT}}{\text{d}T}$$

$$\text{UP} = \text{MY} \cdot \text{PAMAX}$$

$$\text{UN} = \text{MY} \cdot \text{NAMAX}$$

Limiting conditions: UP and UN are equal to supply.

TABLE 4.1 (*continued*)

(14) Jørgensen (1976)

$$\text{PHOTO} = \text{MYMAX} \cdot f(T) f(\text{PA}) f(\text{NA}) f(\text{CA})$$

$$f(\text{PA}) = \frac{\text{PA} - \text{PAMIN} \cdot \text{PHYT}}{\text{PA}}$$

$$f(\text{NA}) = \frac{\text{NA} - \text{NAMIN} \cdot \text{PHYT}}{\text{NA}}$$

$$f(\text{CA}) = \frac{\text{CA} - \text{CAMIN} \cdot \text{PHYT}}{\text{CA}}$$

$$\text{UC} = f(I) \cdot \text{UCMAX} \cdot \frac{\text{CAMAX} \cdot \text{PHYT} - \text{CA}}{\text{CAMAX} \cdot \text{PHYT} - \text{CAMIN} \cdot \text{PHYT}} \cdot \text{PHYT} \cdot \frac{\text{CS}}{\text{KC} + \text{CS}}$$

$$\text{UP} = \text{UPMAX} \cdot \frac{\text{PAMAX} \cdot \text{PHYT} - \text{PA}}{\text{PAMAX} \cdot \text{PHYT} - \text{PAMIN} \cdot \text{PHYT}} \cdot \text{PHYT} \cdot \frac{\text{PS}}{\text{KP} + \text{PS}}$$

$$\text{UN} = \text{UNMAX} \cdot \frac{\text{NAMAX} \cdot \text{PHYT} - \text{NA}}{\text{NAMAX} \cdot \text{PHYT} - \text{NAMIN} \cdot \text{PHYT}} \cdot \text{PHYT} \cdot \frac{\text{NS}}{\text{KN} + \text{NS}}$$

has the following advantages:

(i) It is in accordance with algal kinetic observations (Fuhs, 1969; Droop, 1974; Nyholm, 1975; Rhee, 1978).
(ii) Some case studies show better accordance between model and observations (Serruya and Berman, 1975; Jørgensen, 1976; Park *et al.*, 1979).

However, the disadvantage of a model consisting of independent element cycles is that more parameters are introduced into the model. This drawback can be partly eliminated, as the minimum and maximum concentrations of the elements in phytoplankton are well known. In lake studies where the ratios of element concentrations are kept fairly constant, the stoichiometric model might be used as a less complex alternative to description by independent element cycles (Jørgensen and Harleman, 1978).

The growth of phytoplankton is controlled by light and by nutrients. In Table 4.1 the light limitation is indicated by $f(I)$ and Table 4.2 gives some generally applied formulations for $f(I)$. Many of the equations include the extinction of light with increasing depth and use a Michaelis–Menten expression to describe the relationship between photosynthesis and irradiance. However, integration over depth is of course possible for all the formulations shown. Figures 4.4(a–i) present the relationships between photosynthesis and irradiance, excluding the influence of the depth Z on I.

Equation 3 in Table 4.2 includes the diurnal variation and eqns. 2 and 6 the length of day. Equations 2, 5, and 7 take the attenuation coefficients of water

TABLE 4.2 Light.

(1) Steele (1974)

$$f(I, Z) = \frac{I(0)\exp(-\epsilon Z)}{\text{IK}} \exp\left(1 - \frac{I(0)\exp(-\epsilon Z)}{\text{IK}}\right)$$

(2) Ahlgren (1973a, b)

$$f(I, Z) = \frac{2.72\ \text{FD}}{\epsilon Z}\left[\exp\left(\frac{\text{IDA}\cdot\exp(\epsilon Z)}{\text{IK}}\right) - \exp\left(\frac{\text{IDA}}{\text{IK}}\right)\right]$$

$$\text{FD} = 0.5 + 0.26\sin(t - 82)\ (\text{latitude } 60°)$$

$$\text{IDA} = I(0)/24\ \text{FD}$$

$$I(0) = 350 + 300\sin(t - 82) - 75[1 + \sin 2(t - 37)](\text{mWh cm}^{-2}\ \text{day}^{-1})$$

$$\epsilon = 0.5 + 14\ \text{PA}(1)$$

$$\epsilon = 0.5 + 29\ \text{PA}(2)$$

(3) Schofield and Krutchoff (1974)

$$I(0, t) = \text{IDA}\cdot K\left(\sin\frac{2\pi}{24}(t - 6)\right)^{1/2}$$

$$= 0 \qquad t < 6, t > 18$$

$$K = 2.6225$$

since

$$\int_6^{18} I(0, t)\mathrm{d}t = \text{IDA}$$

(4) Patten *et al.* (1975)

$$\text{REF} = 0.08 + 0.02\sin\left(\frac{2\pi}{52}\cdot\text{Week} - \frac{\pi}{2}\right)$$

(5) Gargas (1976)

$$f(I) = \begin{cases} \dfrac{I}{\text{IK}} & I \le \text{IK} \\ 1 & \text{IK} < I < \text{IH} \\ \max\begin{cases} 0 \\ 1 - \text{KLYS}(I - \text{IH}) \end{cases} & \text{IH} \le I \end{cases}$$

$$I = I(0)\exp[(-\alpha - \beta\text{PHYT} - \gamma\text{ZOO} - \delta\text{DET})Z]$$

(6) Nyholm (1978)

$$I(0) = \text{IDA}\cdot\text{CFGI/FD}$$

$$\text{IH} = 0.7I(0) + 3\cdot 10^5\ (\text{cal m}^{-2}\ \text{day}^{-1})$$

and as (5)

TABLE 4.2 (*continued*)

(7) Jørgensen (1976)

Growth rate of phytoplankton as

$$f(I) = \frac{A}{(\alpha + \beta \mathrm{PHYT})V} \ln \frac{I(0) + \mathrm{IK}}{I(0)\exp[(-\alpha - \beta \mathrm{PHYT})V/A] + \mathrm{IK}}$$

(8) Thomann *et al.* (1974)

Growth rate as

$$f(I) = \mathrm{MYMAX}\left[\frac{2.718}{\alpha + \beta \mathrm{PHYT}} \cdot \exp\left(-\frac{I(0)}{I_{\mathrm{opt}}}\exp[-(\alpha + \beta \mathrm{PHYT})Z]\right) - \exp\left(-\frac{I(0)}{I_{\mathrm{opt}}}\right)\right]$$

(9) Shelef *et al.* (1969)

$$f(I) = KI \qquad I \leq \mathrm{IK}$$

$$f(I) = \text{maximum} \qquad I \geq \mathrm{IK}$$

(10) Shelef *et al.* (1972)

$$f(I) = 1 - \exp(I/\mathrm{IK})$$

(11) Ikusima (1966)

$$f(I) = KI\left(1 - \frac{\mathrm{PHOTO}}{\mathrm{PHOTO}_{\max}}\right), \qquad K = \frac{1}{\mathrm{IK}}$$

(12) Talling (1957)

$$f(I) = \frac{I/\mathrm{IK}}{[1 + (I/\mathrm{IK})^2]^{1/2}}$$

IK corresponds here to I at PHOTO = 0.7 $\mathrm{PHOTO}_{\max}$.

(13) Takahashi *et al.* (1973)

$$f(I) = KI \exp(1 - KI/K_1 + K_2 T)$$

(14) Vollenweider (1965)

$$f(I) = \frac{I/\mathrm{IK}}{[1 + (I/\mathrm{IK})^2]^{1/2}} \cdot \frac{1}{[1 + (\alpha I)^2]^{K/2}}$$

IK corresponds here to I at PHOTO = 0.6 $\mathrm{PHOTO}_{\max}$.

(15) Steel (1973)

$$f(I) = 2\frac{I/2\mathrm{IK}}{1 + (I/2\mathrm{IK})^2}$$

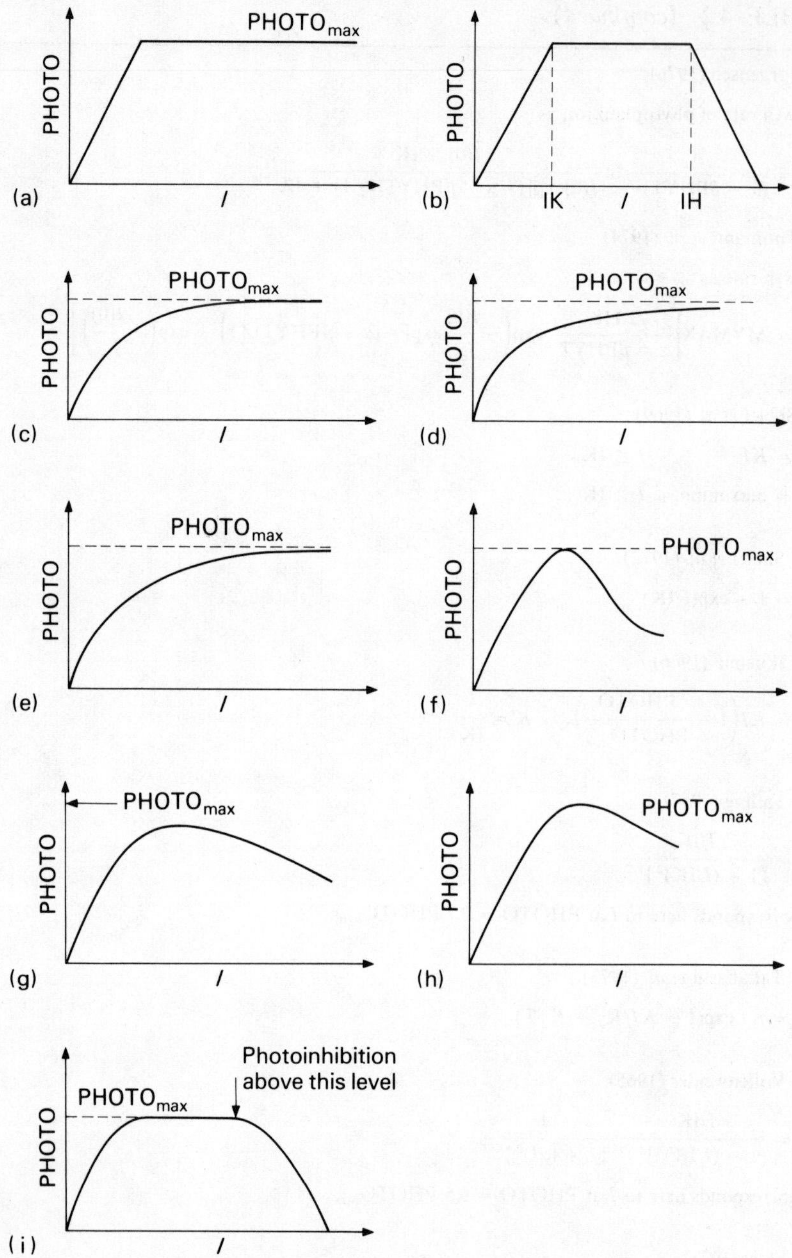

FIGURE 4.4 PHOTO = $f(I)$ according to different equations in Table 4.2: (a) eqns. 9; (b) eqns. 5; (c) eqn. 10; (d) eqn. 11; (e) eqn. 12; (f) eqn. 13; (g) eqn. 14; (h) eqn. 15; (i) eqn. 7, when the integration over depth is omitted. A Michaelis–Menten function is applied.

and of phytoplankton into consideration, and (5) even the attenuation by zooplankton and detritus. Equations 5, 7, 13, 14, and 15 also include the inhibition of irradiance above a certain level. However, for eutrophic lakes the inhibition will be significant only in the upper few millimeters of the body of water. As seen from Table 4.1, the existing mathematical models combine the effects of light, temperature, and nutrient concentration in a variety of ways. Some of the methods of combining the effects are described below (Park *et al.*, 1979).

(1) Multiplication of the effects (eqns. 5 of Table 4.1) has been used by Di Toro *et al.* (1971, 1975), Chen and Orlob (1975), Lehman *et al.* (1975), Youngberg (1977), and Kremer and Nixon (1978). Equations 14 of Table 4.1 use multiplication of temperature and nutrient-limiting effects, but the light-limiting effect only influences the uptake of carbon. This implies that the factors are independent of each other and are of equal importance.
(2) Some of the limitations are combined by using a mean expression and then multiplying it by the third limiting effect, often the temperature limitation (eqn. 8 of Tablc 4.1) (Bloomficld *et al.*, 1974; Park *et al.*, 1974; Patten *et al.*, 1975). This implies interaction between nutrients and light and results in a partial compensation for one of the other factors, while temperature is treated independently. Equation 8 uses an arithmetic average, while eqns. 13 use the harmonic mean.
(3) The minimum of two or more limitations is used (eqns. 1 and 9 of Table 4.1; Scavia and Park, 1976). This assumes that light and nutrients cannot be limiting at the same time.
(4) A complex interaction of the limitations, implying independence of the nutrient and light limitations, both being temperature-dependent (eqns. 12 of Table 4.1; Figure 9.11 on page 348 represents the complex combination used in MS CLEANER (Park *et al.*, 1979)).

4.2.3. Zooplankton Grazing

Zooplankton grazing plays an important role in lakes and reservoirs by limiting phytoplankton blooms. Table 4.3 summarizes the zooplankton grazing equations. Equation 5 describes classic logistic growth, which is modified in eqn. 6 by multiplication with a variable that is related to the availability of food and is dependent upon temperature and the concentrations of phytoplankton, oxygen, and toxic substances. Other authors use a Michaelis–Menten expression. Steele (1974) has modified this equation by using a threshold concentration below which the grazing ceases, while Frost (1975) uses a minimum feeding rate below the threshold concentration. This expression has the advantage of

TABLE 4.3 Zooplankton Grazing Rate.

(1) Dodson (1975)

$$\frac{\mathrm{dPHYT}}{\mathrm{d}t} = -K \cdot \mathrm{PHYT} \cdot \mathrm{ZOO}$$

(2) Steele (1974)

$$\mathrm{GRZ} = \mathrm{MYZ} \cdot \frac{\mathrm{PHYT} - \mathrm{KTR}}{\mathrm{KZ} + \mathrm{PHYT}} \cdot \mathrm{ZOO}$$

$$\mathrm{MYZ} = \mathrm{MYZMAX} \cdot f(T)$$

(3) Walsh and Dugdale (1971)

$$\mathrm{GRZ} = g(\mathrm{ZOO})(\mathrm{PHYT} - \mathrm{KTR})$$

For example,

$$g(\mathrm{ZOO}) = \mathrm{MYZ} \cdot \mathrm{ZOO}$$

(4) O'Brien and Wroblewski (1972)

$$\mathrm{GRZ} = \mathrm{ZOO} \cdot \mathrm{MYZ}\{1 - \exp[D_{\mathrm{p}}(\mathrm{PHYT} - \mathrm{KTR})]\}$$

$$\mathrm{MYZ} = \mathrm{MYZMAX} \cdot f(T)$$

(5) Lotka (1924)

$$\mathrm{GRZ} = \mathrm{MYZ} \cdot \left(1 - \frac{\mathrm{ZOO}}{\mathrm{CK}}\right) \cdot \mathrm{ZOO}$$

(6) Odum (1972)

$$\mathrm{GRZ} = \mathrm{MYZ} \cdot \mathrm{AV} \cdot \left(1 - \frac{\mathrm{ZOO}}{\mathrm{CK}}\right) \cdot \mathrm{ZOO}$$

$$\mathrm{AV} = f(\mathrm{PHYT}) f(\mathrm{OX}) f(T) f(\mathrm{TOX})$$

(7) Gargas (1976)

$$\mathrm{GRZ} = \begin{cases} \mathrm{MYZ} \cdot \dfrac{\mathrm{PHYT} - \mathrm{KTR}}{\mathrm{KZ} + \mathrm{PHYT}} \cdot \mathrm{ZOO} & \mathrm{PHYT} > \mathrm{KTR} \\ \mathrm{MYZ} \cdot \dfrac{\mathrm{PHYT}}{\mathrm{ZOO}} & \mathrm{PHYT} \leq \mathrm{KTR} \end{cases}$$

$$\mathrm{MYZ} = \mathrm{MYZMAX} \cdot f(T)$$

(8) Chen and Orlob (1975), Chrisholm *et al.* (1975), Canale *et al.* (1976)

$$\mathrm{GRZ} = \mathrm{MYZ} \cdot \frac{\mathrm{PHYT}}{\mathrm{KZ} + \mathrm{PHYT}} \cdot \mathrm{ZOO}$$

TABLE 4.3 (*continued*)

(9) Canale *et al.* (1976)

$$\mathrm{GRZ} = \mathrm{MYZ} \cdot \frac{\mathrm{KMFM} \cdot \mathrm{PHYT} + \mathrm{KFLM}}{\mathrm{PHYT} + \mathrm{KFLM}}$$

$$\mathrm{PREF} = \frac{\alpha_K^Z \,\mathrm{PHYT}}{\sum \alpha_K^Z \,\mathrm{PHYT}}$$

(10) Jost *et al.* (1973)

$$\mathrm{GRZ} = \mathrm{MYZ} \cdot \frac{\mathrm{PHYT}^2}{(\mathrm{KZ1} + \mathrm{PHYT})(\mathrm{KZ2} + \mathrm{PHYT})} \cdot \mathrm{ZOO}$$

(11) Gause (1934)

$$\mathrm{GRZ} = \frac{\mathrm{MYZ} \cdot \mathrm{PHYT}^{1/2} \cdot \mathrm{ZOO}}{\mathrm{KZ} + \mathrm{PHYT}^{1/2}}$$

(12) Dugdale (1975)

$$\mathrm{GRZ} = \mathrm{MYZ} \cdot \mathrm{ZOO}[1 - \exp(-\mathrm{KZ} \cdot \mathrm{PHYT} - \mathrm{KTR})]$$

(13) Richey (1977)

$$\mathrm{GRZ} = \mathrm{AK} \cdot (\mathrm{BL})^2 \cdot T$$

being in accordance with several observations, e.g. the grazing rate of filter feeders.

Canale *et al.* (1976) distinguish between raptors (eqn. 8), selective filter feeders (eqns. 9), and nonselective filters (eqn. 1). For the first two classes a constant assimilation efficiency is used, and for the latter Canale *et al.* use a simple but unconfirmed formulation:

$$\text{assimilation efficiency} = \text{maximum efficiency} \cdot \frac{\mathrm{HMFL}}{\mathrm{PHYT} + \mathrm{HMFL}},$$

where HMFL is the half-maximum-efficiency food level.

Jost *et al.* (1973) have developed a multiple kinetic model, which is in better agreement with their observations than was the Monod kinetic (eqn. 10). Gause (1934) used another model (eqn. 11), which predicts that the predator becomes more effective in capturing prey at a low prey concentration.

A persistent difficulty of the predicted zooplankton biomass is that it does not decrease sufficiently in the fall and winter. The model CLEANER contains a very complex construct for predator–prey interaction (section 9.2) (Park *et al.*, 1974; Scavia *et al.*, 1976), yet the model cannot predict very well the zooplankton biomass in the fall and winter. It is probably necessary to include starvation and the formation of resting stages, but unfortunately very few quantitative studies have been conducted on these processes.

4.2.4. Respiration and Nonpredatory Mortality of Phytoplankton and Zooplankton

Respiration and nonpredatory mortality are often described as first-order reactions (Table 4.4, eqns. 1) but a more causal expression is shown in eqn. 2 of Table 4.4, where the respiration is dependent upon the intracellular concentration of phytoplankton.

Some models do not include phytoplankton mortality as a separate factor, but combine it with settling, since each can be approximated as a first-order process.

The physiological state of algae is important in determining the point at which growth ceases and death (and settling) begins. When growth conditions become unfavorable (e.g. nutrient-deficient), algae become susceptible to decomposition. DePinto (1979) presents a comprehensive discussion of non-predatorial mortality.

In the model of Lehman *et al.* (1975) the death rate is a function of the number of days of algal growth in suboptimal conditions. Bierman (1976) uses a second-order algal death rate, as shown in (3) of Table 4.4. Thomann *et al.* (1975) and Canale *et al.* (1976) use a temperature-dependent mortality rate, as indicated in (4). In the model of Scavia and Park (1976) and in MS CLEANER (Park *et al.*, 1979) mortality rate is a function of temperature as well as of physiological state (Figure 9.11).

TABLE 4.4 Respiration and Mortality of Phytoplankton

(1) Gargas (1976)

$$\mathrm{RESP} = \mathrm{RESPK} \cdot f(T) \cdot \mathrm{PHYT}$$

$$\mathrm{MORT} = \mathrm{MORTK} \cdot \mathrm{PHYT}$$

(2) Lehman *et al.* (1975)

$$\mathrm{RESP} = \mathrm{RESPK} \cdot \left(\frac{\mathrm{CA}}{\mathrm{CAMAX}}\right) \cdot f(T) \cdot \mathrm{PHYT}$$

(as carbon)

(3) Bierman (1976)

$$\mathrm{MORT} = \mathrm{MORTK} \cdot \mathrm{PHYT}^2$$

(4) Thomann *et al.* (1975), Canale *et al.* (1976)

$$\mathrm{MORT} = \mathrm{MORTK} \cdot f(T)$$

(5) Straškraba (1976)

$$\mathrm{RESP} = 3.1\exp(0.09T)[\mathrm{mg\ O_2\ (mg\ chlorophyll)^{-1}\ h^{-1}}] \quad 0 \leq T \leq 25°\mathrm{C}$$

TABLE 4.5 Zooplankton Respiration, Excretion, and Mortality.

(1) Di Toro *et al.* (1975)

MORT = MORTK · ZOO

RESP = RESPK · $f(T)$ · ZOO

(2) Patten *et al.* (1975)

$\lg \text{RESP} = 0.0364T - 0.3418 \lg W + 0.6182$

A first-order kinetic is also often used for zooplankton mortality and respiration (Table 4.5). As discussed in section 4.2.3, it is, however, sometimes necessary to use more complex models, if a more realistic description of zooplankton dynamics is required.

4.2.5. Sinking, Mineralization, and Excretion

Sinking of phytoplankton is a function of shape, specific gravity, size, and the viscosity and turbulence of the surrounding water. Some algae are vacuolated and their sinking rate is controlled differently. Several factors controlling the sinking rate have been suggested:

- (i) physiological state: cells in the stationary phase sink two to four times faster than cells in the exponential phase of growth (Eppley *et al.*, 1967; Smayda, 1970);
- (ii) nutrient depletion (Smayda, 1974; Titman and Kilham, 1976);
- (iii) light (Eppley *et al.*, 1967; Steele, 1974);
- (iv) viscosity of water: is a function of temperature in fresh water (eqn. 2 of Table 4.6);
- (v) turbulence (Stefan *et al.*, 1976; Titman and Kilham, 1976).

Most submodels formulate sinking rate as a first-order reaction (eqn. 1 of Table 4.6), which is an oversimplification. Others include a reduction in sinking rate due to turbulence, applying the Reynolds number as a site variable. At present only the model of Scavia and Park (1976) considers the physiological state as a controlling factor.

The rate with which detritus, including dead algal cells, decomposes is important for understanding nutrient cycling. In most submodels the decomposition is formulated by a first-order reaction scheme as shown in Table 4.7. However, as discussed by Jørgensen (1979a), the detritus consists of a wide range of organic compounds with different decomposition rates. This implies that a more correct description should consider two or more fractions with different rate coefficients. It seems that especially allochthonous nitrogen compounds have a lower decomposition rate than other nitrogen compounds (discussed by Jørgensen, 1979; Jørgensen and Mejer, 1979).

TABLE 4.6 Settling.

(1) Chen and Orlob (1975)
$\text{SEV} = \dfrac{\text{SEV(P)}}{\text{SEV(O)}}$
(2) Jørgensen (1976), Jørgensen and Harleman (1978)
$\text{SETTL} = (\text{SEPH} \cdot \text{PHYT} + \text{SEDT} \cdot \text{DET}) \cdot \dfrac{T_a^{1/2}}{ZT_r^{1/2}}$
(3) Di Giarno *et al.* (1978)
$\text{SETTL} = \text{SEPH} \cdot \dfrac{\partial \text{PHYT}}{\partial Z}$
(4) Nyholm (1978)
$\text{SEPH} = \min \begin{cases} 0.20 \text{ m day}^{-1} \\ 0.05Z \text{ m day}^{-1} \end{cases}$
(5) Scavia and Park (1976)
$\text{SETTL} = \text{SEPH} \cdot \text{PHYT}(1 + \text{constant} \cdot T)$

TABLE 4.7 Detritus.

Jørgensen (1976)
$\text{DERV} = \text{MYDET} \cdot \text{DET}$
$\text{MYDET} = \text{MYDETMAX} \cdot f(T)$

Both phytoplankton and zooplankton (process 12 of Figure 4.2) excrete phosphorus, nitrogen, and organic carbon (discussed by Jørgensen, 1979). These processes are not included in most models, but phytoplankton excretion is accounted for in the two-step phytoplankton model (Table 4.1, eqns. 12–14) by consideration of the nutrient uptake as a net uptake. Zooplankton excretion is often omitted, as it is of minor importance for the nutrient budget. However, Patten *et al.* (1975) have used the formulation shown in Table 4.8.

TABLE 4.8 Excretion

Patten *et al.* (1975)	
$\text{EXCR} = \text{EXCRK} \cdot \text{ZOO-N}$	(ammonia excretion)
$\text{EXCR} = 1.35\text{ZOO}^{-0.65}$	(phosphorus excretion)

4.2.6. Nitrogen Fixation, Nitrification, and Denitrification

In many lakes, where nitrogen is a limiting factor, nitrogen-fixing algae may appear during the summer (Jørgensen, 1979b). In such case studies the nitrogen fixation must be accounted for to obtain a realistic nitrogen budget (Jørgensen and Mejer, 1979). As shown in Table 4.9, the rate of nitrogen fixation is related to the concentration of nitrogen-fixing algae. Equations 2 show how the ratio of the nitrogen-fixing species can be connected to the "lack of nitrogen" relative to phosphorus. Since the algae need approximately five times as much nitrogen as phosphorus, 5PS − NS expresses the surplus of phosphorus.

TABLE 4.9 Nitrogen Fixation and Nitrification.

(1) Gargas (1976)

$$\mathrm{NFIX} = \mathrm{KNFIX} \cdot \left(1 - \frac{\mathrm{NS}}{\mathrm{KN} + \mathrm{NS}}\right) \cdot \mathrm{PHYT(N)}$$

(2) Jørgensen and Harleman (1978)

$$\mathrm{NFIX} = \mathrm{KNFIX} \cdot \mathrm{PHYT(N)} \cdot (5\mathrm{PS} - \mathrm{NS})$$

or

$$\mathrm{NFIX} = \mathrm{KNFIX} \cdot \mathrm{PHYTN)}$$

$$\frac{\mathrm{PHYT(N)}}{\mathrm{PHYT}} = \frac{5\mathrm{PS} - \mathrm{NS}}{5\mathrm{PS}} \qquad 5\mathrm{PS} \geq \mathrm{NS}$$

$$\mathrm{PHYT(N)} = 0 \qquad \mathrm{NS} > 5\mathrm{PS}$$

(3) Jørgensen (1976)

$$\mathrm{NITRI} = \mathrm{NITRIK} \cdot \mathrm{NH4}$$

In most ecological water quality models, ammonia and nitrate are considered as one constituent: dissolved inorganic nitrogen. If the two compounds are separated, however, the nitrification rate will account for the change from ammonia to nitrate. The most generally applied mathematical formulation for this process is the first-order kinetic, eqn. 3.

In many eutrophic lakes the sediment becomes anaerobic during the summer, which implies that denitrification at the sediment–water interface may occur. Nitrogen balances of hypereutrophic lakes demonstrate that this process is of great significance for nitrogen balance. As much as 30% or more of the total nitrogen input may be removed by denitrification (case study by Jørgensen *et al.*, 1973).

Denitrification is a microbiological process and a more complete description must account for the dynamics of the denitrifying bacteria, as demonstrated by

TABLE 4.10 Denitrification.

(1) O'Connor *et al.* (1976)

$$\frac{dM}{dt} = \text{MYM} \cdot M$$

$$\text{MYM} = \frac{\text{MYMMAX} \cdot \text{NIT}}{\text{KNIT} + \text{NIT}}$$

$$M = M(0) + \text{STO} \cdot (\text{NIT}(0) - \text{NIT})$$

Substitution of the last two equations into the first gives upon integration:

$$\text{MYMMAX} \cdot t = \frac{\text{KNIT}}{M(0) + \text{STO} \cdot \text{NIT}(0)} \ln \frac{\text{NIT}(0)}{\text{NIT}(0) - 1/\text{STO}(M - M(0))} + \frac{\text{STO} \cdot \text{KNIT} + M(0) + \text{STO} \cdot \text{NIT}(0)}{M(0) + \text{STO} \cdot \text{NIT}(0)} \ln \frac{M}{M(0)}$$

(2) Jørgensen (1976)

$$\text{DENIT} = \text{DENITK} \cdot \text{NS}$$

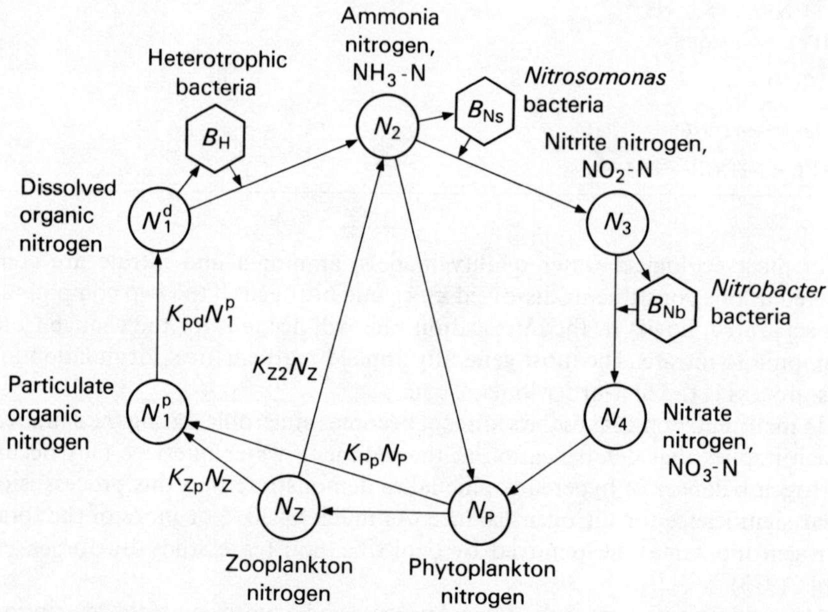

FIGURE 4.5 Aerobic nitrogen cycle.

O'Connor *et al.* (1976) (eqns. 1 of Table 4.10). However, a simple first-order formulation gives a satisfactory description in some cases (eqn. 2).

Najarian and Harleman (1975) and Harleman (1978) have suggested an even more detailed nitrogen cycle (Figure 4.5). Seven nitrogen constituents are considered: ammonia nitrogen, nitrite nitrogen, nitrate nitrogen, phytoplankton nitrogen, zooplankton nitrogen, particulate organic nitrogen, and dissolved organic nitrogen. In addition, three species of bacteria are considered: *Nitrosomonas*, *Nitrobacter*, and heterotrophic bacteria. However, normally in a lake the number of bacteria will generally be constant compared with the situation in a chemostat, where the bacterial concentration is low initially and increases exponentially in the first phase (Jørgensen and Harleman, 1978). Furthermore, the biomass of the bacteria is generally not known or, at best, is known with great uncertainty. Consequently, a simple quantitative model description of the nitrogen cycle is not only satisfactory, but must be recommended in most case studies, unless the data allow the calibration and validation of a more complex model.

4.2.7. Release of Nutrient from Sediment

A detailed description of release of nutrients from sediment is often significant for predictive models, as discussed by Jørgensen *et al.* (1973) and Jørgensen (1976). A completely different prognosis is produced, depending on whether a simple first-order kinetic or the more comprehensive submodel proposed by Kamp-Nielsen (1974, 1975) and Jørgensen *et al.* (1975) is applied. The principles of this submodel are demonstrated in Figure 4.6(a). The model distinguishes between exchangeable and nonexchangeable phosphorus, which is of great importance when the phosphorus loading is diminished. The phosphorus is released in a two-step process:

(i) decomposition of phosphorus compounds, raising the concentration of interstitial phosphorus, and
(ii) diffusion of interstitial phosphorus to the lake or river water.

Process (i) is different under aerobic and anaerobic conditions. The profile of the phosphorus indicates which part of the settled phosphorus is exchangeable and which part is not (Figure 4.6(b)). This model does not take all problems into consideration, e.g. groundwater infiltration.

Another complex submodel for bottom nutrient release has been suggested (Kozerski, 1977) and applied, with good results, to a lake by Schellenberger *et al.* (1978). The sediment model contains nine balance equations and represents the sedimentation and decay of organic material, the diffusion of phosphorus and iron from the sediment into water, the diffusion of oxygen into the sediment, the adsorption of phosphates on to iron compounds, the reduction and oxidation

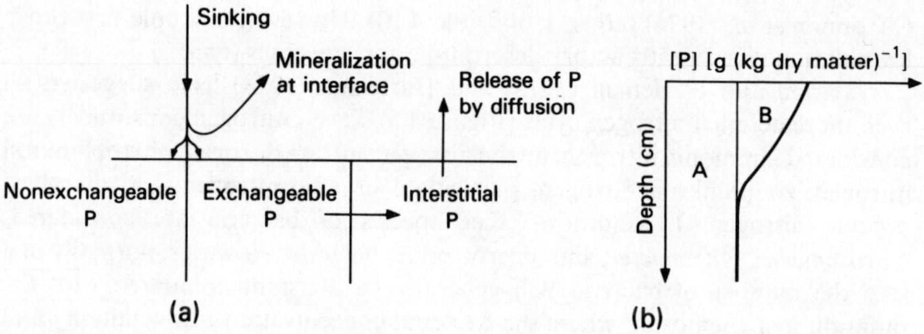

FIGURE 4.6 Submodel of the exchange of phosphorus between water and sediment. In (b), A represents the nonexchangeable phosphorus, and B the exchangeable phosphorus.

processes of iron and the compounds of calcium, and the transfer of material from the thin top sediment layer (5–10 mm) to the deeper, undisturbed layers. Langmuir's adsorption isotherm was used for the adsorption, and the order processes were simulated as first-order reactions. Most modelers, however, have used the first-order formulation (eqn. 1 of Table 4.11; Chen and Orlob, 1975). Since an essential part of the nutrient flow passes through the sediment, it is strongly recommended that a more detailed formulation of this submodel is considered, which implies that the process must be examined in the laboratory to obtain geochemical site constants.

TABLE 4.11 Release of Nutrient from Sediment.

(1) Nyholm (1978)

$$\text{REL} = \text{RELK} \cdot \text{SETTL}$$

(2) Jørgensen (1976)

$$\text{REL (N)} = \frac{4.0\ \text{SN} + 0.08}{1000} \cdot \frac{A}{V} f(T)$$

REL (P): submodel (Kamp-Nielsen, 1974, 1975; Jørgensen *et al.*, 1975)

4.2.8. Effect of Temperature on Process Rates

All chemical and biological process rates are strongly dependent upon temperature. The temperature response of phytoplankton and zooplankton should be modeled realistically, otherwise large discrepancies can arise between model and observations. Table 4.12 contains several formulations of the relationship between process rate and temperature; most of them consider an optimum temperature.

TABLE 4.12 Effects of Temperature upon Biological Rates.

(1) Chen and Orlob (1975)

$$K(T) = K_{20} \cdot \mathrm{KOT}^{T-20}$$

(2) Lassiter and Kearns (1974)

$$K(T) = K_{\mathrm{opt}} \exp[a(T - T_{\mathrm{opt}})]\left(\frac{T_{\max} - T}{T_{\max} - T_{\mathrm{opt}}}\right)^{a(T_{\max} - T_{\mathrm{opt}})}$$

(3) Lehman *et al.* (1975)

$$K(T) = K_{\mathrm{opt}} \exp[-2.3(T - T_{\mathrm{opt}})^2/(T_{\max} - T_{\mathrm{opt}})^2] \qquad T > T_{\mathrm{opt}}$$

$$K(T) = K_{\mathrm{opt}} \exp[-2.3(T_{\mathrm{opt}} - T)^2/(T_{\mathrm{opt}} - T_{\min})^2] \qquad T \leq T_{\mathrm{opt}}$$

(4) Jørgensen (1976)

$$K(T) = K_{\mathrm{opt}} \exp\left(-2.3\left|\frac{T - T_{\mathrm{opt}}}{15}\right|\right)$$

$$K(T) = K_{\mathrm{opt}} \exp(KT)$$

(5) Lamanna and Malette (1965)

$$K(T) = K_{\mathrm{opt}}\left(\frac{T}{T_{\mathrm{opt}}}\right)^n \exp\left[1 - \left(\frac{T}{T_{\mathrm{opt}}}\right)^n\right] \qquad 0 < T < T_{\mathrm{opt}}$$

$$K(T) = K_{\mathrm{opt}}\left[1 - \left(\frac{T - T_{\mathrm{opt}}}{T_{\max} - T_{\mathrm{opt}}}\right)^m\right] \qquad T_{\mathrm{opt}} < T < T_{\max}$$

(6) Park *et al.* (1979)

$$\mathrm{PHOTO}(T) = \exp[K(K_1 T^2 - K_2 T^{K_3} - 1)]$$

$$K = -\ln \mathrm{PHOTO}\ (0°\mathrm{C})$$

$$K_1 = K_2(T_{\max})^{K_3 - 2}$$

$$K_2 = \frac{1 + (\ln \mathrm{PHOTO}\ (T_{\mathrm{opt}}))/K}{(T_{\max})^{K_3 - 2}(T_{\mathrm{opt}})^{K_1} - (T_{\mathrm{opt}})^{K_3}}$$

$$\frac{2^{1/(K_3 - 2)}}{K_3} = \frac{T_{\mathrm{opt}}}{T_{\max}}$$

(7) Straškraba (1976)

$$T_{\mathrm{opt}} = T + 28 \exp(-0.115T)$$

$$\mathrm{PHOTO}(T) = \mathrm{PHOTO}(T_{\mathrm{opt}}) \cdot \exp[-K(T_{\mathrm{opt}} - T)^2]$$

The interactions between the responses of phytoplankton to temperature, nutrient, and light have been discussed in section 4.2.2. Seasonal adaptation to water temperature has been considered by Straškraba (1976) (eqns. 7 of Table 4.12).

4.2.9. Effect of Fish on Water Quality

A wide range of fish population models has been developed to optimize fishery, but the models are far too detailed to be included in water quality models and focus on a completely different problem. It might be tempting to model year classes of dominant species, as demonstrated by E. Ursin (1976, private communication), but the computational load is probably prohibitive. Fish is included as a constituent of a water quality model when there are trophic interactions with phytoplankton (herbivorous fish) and with zooplankton (carnivorous fish).

Chen and Orlob (1975) have suggested a compromise, whereby major growth stages are differentiated. This suggestion is also applied in MS CLEANER (Park *et al.*, 1979). The weakest point in this modeling approach is the population dynamics of the larval and immature stages; the model is highly sensitive to parameters used for these stages and it is hardly possible to obtain them by calibration. However, it seems satisfactory in most cases to simulate fish as one constituent using the same mathematical formulation as for zooplankton, or even to consider the fish population as a constant or as a driving variable based on fishery statistics.

MS CLEANER contains a function to determine the migration of fish (and of zooplankton). The function includes responses to dissolved oxygen levels, food availability, and temperature. It has also been suggested that the response to light intensity and swimming rates should be added.

4.2.10. Other Ecological Processes

Many toxic chemicals discharged into rivers and lakes may affect the water quality. The problem of modeling the distribution and effect of toxic substances is treated in Chapter 10. The relation between the concentrations of a toxic compound and of a biological constituent is often expressed as

$$\text{MORTALITY} = \text{MORT} + \beta C_x,$$

where MORT is the natural mortality, β the toxicity coefficient ($m^2\ g^{-1}\ day^{-1}$), and C_x the concentration of the toxic compound.

Toxic substances can also inhibit growth. There are two possibilities, competitive or allosteric inhibition, for which Michaelis–Menten expressions can be used:

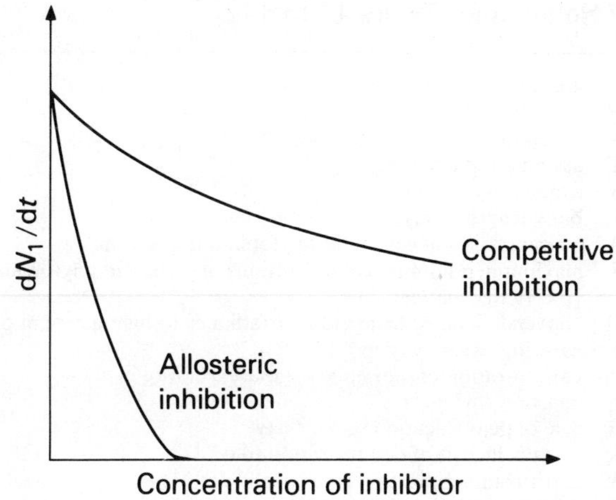

FIGURE 4.7 Process rate as a function of inhibitor concentration for both competitive and allosteric inhibition.

Competitive Inhibition

$$\frac{dN_1}{dt} = -a_{N_1B}\,\mu_{BN_1}^{\max}\frac{N_1}{K_{N_1} + N_1 + a_2N_2}B, \tag{4.3}$$

where N_1 is the concentration of inhibited nutrient, N_2 the concentration of inhibiting nutrient, a_2 a dimensionless inhibition constant for N_2, and B the concentration of bacteria.

Allosteric Inhibition

$$\frac{dN_1}{dt} = -a_{N_1B}\,\mu_{BN_1}^{\max}\frac{N_1}{(K_{N_1} + N_1)(1 + a_2N_2)}B. \tag{4.4}$$

It is assumed that the affinity of the bacteria for the inhibitor is as strong as for the inhibited nutrient. If the concentration N_2 is insignificant, then (4.3) and (4.4) are identical. In competitive inhibition the rate of N_1 degradation is less dependent on the concentration N_2 than in allosteric inhibition. An allosteric inhibitor will tend to block the degradation of a nutrient or constituent much faster than if it is only a competitive inhibitor. These distinctions and effects are illustrated in Figure 4.7.

4.2.11. Parameter Values and Stoichiometric Ratio

Table 4.14 summarizes some generally adopted values of parameters. Table 4.15 surveys the most used stoichiometric ratios, which might be employed to compute, for example, phytoplankton concentration from chlorophyll concentration, etc. These values should be considered only as means, since no

TABLE 4.13 Notation for Tables 4.1 to 4.12.

a	constant ($°C^{-1}$)
A	area (m^2)
AK	constant
ASSE	assimilation efficiency
AV	availability of food
BL	body length (mm)
CA	concentration of carbon in phytoplankton ($g\ C\ m^{-3}$)
CAMAX, CAMIN	maximum, minimum concentrations of carbon in phytoplankton [$g\ C\ (g\ dry\ matter)^{-1}$]
CFGI	conversion factor from global irradiance to light active in photosynthesis
CK	carrying capacity ($g\ m^{-3}$)
CS	concentration of inorganic soluble carbon ($g\ m^{-3}$)
D_p	constant ($m^3\ g^{-1}$)
DENIT	rate of denitrification ($g\ m^{-3}\ day^{-1}$)
DENITK	rate coefficient of denitrification (day^{-1})
DERV	detritus mineralization rate ($g\ m^{-3}\ day^{-1}$)
DET	detritus concentration ($g\ m^{-3}$)
EXCR	excretion rate ($g\ m^{-3}\ day^{-1}$)
EXCRK	excretion rate coefficient (day^{-1})
$f(x)$	function of x
$\bar{F}_{max}$	maximum fractional reduction in daily specific growth rate over eutrophic depth
FAC	correction factor for biochemical growth activities during dark periods
FD	relative day length
$g(y)$	function of y
GRZ	grazing rate ($g\ m^{-3}\ day^{-1}$)
I	irradiance†
$I(0)$	irradiance, surface intensity†
I_{opt}	optimum irradiance†
IDA	daily average irradiance†
IH	light saturation parameter†
IK	light saturation parameter†
K	unspecified constant
K_{20}	constant or coefficient at 20°C
$K(T)$	constant or coefficient at T
K_{opt}	constant at optimum temperature
KC	half-saturation constant of uptake of inorganic soluble carbon ($g\ m^{-3}$)
KFLM	food level, where multiplier is $\frac{1}{2}(1 + \text{KMFM})$
KLYS	constant for light inhibition
KMFM	minimum filtering rate multiplier
KN	half-saturation constant of uptake of soluble inorganic nitrogen ($g\ m^{-3}$)
KNFIX	nitrogen fixation coefficient (day^{-1})
KNIT	nitrogen fixation coefficient for denitrification ($g\ m^{-3}$)
KOT	temperature coefficient
KP	half-saturation constant of uptake of soluble inorganic phosphorus ($g\ m^{-3}$)
KPA	saturation constant for intracellular phosphorus
KTR	threshold concentration for grazing ($g\ m^{-3}$)
KZ	half-saturation concentration for grazing ($g\ m^{-3}$)
L_0	constant
m	constant (a number)
M	concentration of denitrification microorganism ($g\ m^{-3}$)
$M(0)$	concentration of denitrification microorganism at $t = 0$ ($g\ m^{-3}$)

† A number of different units are used.

TABLE 4.13 (*continued*)

MORT	mortality (g m^{-3} day^{-1})
MORTK	mortality coefficient (day^{-1})
MY	growth rate of phytoplankton (day^{-1})
MYDET	detritus mineralization rate coefficient (day^{-1})
MYDETMAX	maximum detritus mineralization rate coefficient(day^{-1})
MYM	growth rate coefficient of denitrification microorganism (day^{-1})
MYMAX	maximum growth rate of phytoplankton (day^{-1})
MYMMAX	maximum growth rate coefficient of denitrification microorganism (day^{-1})
MYZ	growth rate of zooplankton (day^{-1})
MYZMAX	maximum growth rate of zooplankton (day^{-1})
n	constant (a number)
NA	concentration of nitrogen in phytoplankton (g m^{-3})
NAMAX, NAMIN	maximum, minimum concentrations of nitrogen in phytoplankton [g N (g dry matter)$^{-1}$]
NAP	NA/PHYT
NFIX	nitrogen fixation (g m^{-3} day^{-1})
NH4	concentration of ammonia nitrogen (g m^{-3})
NH4L	NH4 beyond which nitrate nitrogen uptake is minimal
NIT	concentration of nitrate nitrogen (g m^{-3})
NIT(0)	concentration of nitrate nitrogen at $t = 0$ (g m^{-3})
NITRI	nitrification rate (g m^{-3} day^{-1})
NITRIK	nitrification rate coefficient (day^{-1})
NS	concentration of soluble inorganic nitrogen(g m^{-3})
NS_0	constant
OX	oxygen concentrations (g m^{-3})
PA	concentration of phosphorus in phytoplankton (g m^{-3})
PAMAX, PAMIN	maximum, minimum concentrations of phosphorus in phytoplankton [g P (g dry matter)$^{-1}$]
PAP	PA/PHYT
PHOTO	photosynthetic rate (g m^{-3} day^{-1})
PHYT	concentration of phytoplankton (g m^{-3})
PHYT(N)	concentration of phytoplankton with nitrogen fixation (g m^{-3})
PKI	equilibrium constant for reaction between phosphorus and carrier (l mol^{-1})
PREF	preference ratio
PS	concentration of soluble inorganic phosphorus (g m^{-3})
PS_0	constant
REF	reflected light (units as I)
$R(t)$	total daily radiation (units as I)
REL	rate of release of nutrients (g m^{-3} day^{-1})
RELK	release coefficient
RESP	respiration (g m^{-3} day^{-1})
RESPK	respiration coefficient (day^{-1})
SEDT	settling rate of detritus (m day^{-1})
SEPH	settling rate of phytoplankton (m day^{-1})
SETTL	rate of removal of nutrients by settling (g m^{-3} day^{-1})
SEV	settling rate (m day^{-1})
SEV(O)	surface overflow rate (m day^{-1})
SEV(P)	settling rate of particles (m day^{-1})
SN	nitrogen in upper layer of sediment (g l^{-1})
STO	stoichiometric constant [mg bacterial mass (mg substrate)$^{-1}$]
t	time (days)
T	temperature (°C): $T_j = T$ for j in element, T_i and T_o are T in inflow and outflow

(*continued over*)

TABLE 4.13 (*continued*)

T_a	absolute temperature (K)
T_{max}, T_{min}	maximum and minimum temperatures, at which rate coefficient is zero
T_{opt}	optimum temperature
T_r	reference temperature
$T_{Z,t}$, $T_{Z,0}$	temperature at depth z and time t or 0
TOX	concentration of toxic material (g m^{-3})
U, U_1, U_2	mass flow (kg day^{-1})
UC	uptake rate of inorganic carbon (g m^{-3} day^{-1})
UCMAX	maximum uptake rate of inorganic carbon (day^{-1})
UN	uptake rate of nitrogen (g m^{-3} day^{-1})
UNMAX	maximum uptake rate of nitrogen (day^{-1})
UNMIN(NIT)	minimum uptake rate of nitrate nitrogen (day^{-1})
UP	uptake rate of phosphorus (g m^{-3} day^{-1})
UPMAX	maximum uptake rate of phosphorus (day^{-1})
V	volume (m^3)
ΔV	change of volume (m^3)
W	weight (kg)
Z	depth (m)
ZOO	concentration of zooplankton (g m^{-3})
α	attenuation of water (m^{-1})
α_K^Z	preference coefficient
β	specific attenuation coefficient of phytoplankton (g^{-1} m^2)
γ	specific attenuation coefficient of zooplankton (g^{-1} m^2)
δ	specific attenuation coefficient of detritus (g^{-1} m^2)
ϵ	attenuation coefficient (m^{-1})
ρ	density (g m^{-3})

TABLE 4.14 Parameter Values.

ASSE	0.65[3] 0.60–0.65[5] 0.6[10] 0.6[11] 0.63[16]
CAMAX (g g^{-1})	0.6[16]
CAMIN (g g^{-1})	0.18[16]
CFGI	0.41[7]
DENITK (day^{-1})	0.002[16]
eddy diffusion coefficient (m^2 day^{-1})	0.1–0.5[9]
$\bar{F}_{max}$	0.27[6]
FAC	1.33[7]
IH	$20 \cdot 10^5$ (cal m^{-2} day^{-1})[1] 350 (ly day^{-1})[4] 0.03–0.2 (ly min^{-1})[13]
IK	0.006 (kcal m^{-2} s^{-1})[4] 300–350 (ly day^{-1})[5] 150 (ly day^{-1})[6] 300 (ly day^{-1})[10] 350 (ly day^{-1})[11] 300–350[12]
KC (g m^{-3})	0.5–0.6[4]
KFLM (g food C m^{-3})	0.2[17]
KMFM (g food C m^{-3})	0.1[17]
KN (g m^{-3})	0.05[1] 0.025[3] 0.3–0.4[4] 0.025[5] 0.014[6] 0.05[8] 0.025[10] 0.025[11] 0.005–0.008[12] 0.0014–0.01[13] 0.2[16]
KNFIX (g m^{-3} day^{-1})	0.02[1]
KOT	1.02[4] 1.14[7] 1.07[7] 1.066[11] 1.08[11] 1.02[16]

TABLE 4.14 (*continued*)

KP ($g\ m^{-3}$)	0.005[2] 0.01[3] 0.03–0.05[4] 0.001[6] 0.005–0.0[8] 0.01–0.1[9] 0.002[11] 0.005–0.015[12] 0.015–0.15[13] 0.03[16]
KTR ($g\ m^{-3}$)	0.5[12]
KZ ($g\ m^{-3}$)	0.5[4] 0.7[5] 4[7] 0.51[8] 1[11] 0.5[16]
MORTK (day^{-1}) (zooplankton)	0.015[1] 0.075[10] 0.001[12] 0.125[16]
MYDETMAX (day^{-1}) (organic N)	0.002T[3] 0.001[4] 0.035–0.14[5] 0.01[7] 0.007T[10] 0.1[10] 0.00175T[11] 0.1[16]
MYDETMAX (day^{-1}) (organic P)	0.02T[3] 0.001[4] 0.14–0.40[5] 0.1[7] 0.007T[10] 0.1[10] 0.007T[11]
MYMAX (day^{-1})	2.5[1] 0.1 + 0.06T[3] 1–2[4] 1.3–2.5[5] 2.4[6] 0.8–2.4[7] 1.8–3.9[10] 0.58[11] 1.1–1.6[12] 1–2[14] 2.0[16]
MYZMAX (day^{-1})	1.3[1] 0.012 + 0.021T[3] 0.15[4] 0.13–1.2[5] 0.1–0.6[7] 0.2–2.0[10] 0.06T[11] 0.6–0.85[12] 0.18[16]
NAMAX	0.1[7] 0.08[16]
NAMIN ($g\ g^{-1}$)	0.04[7] 0.02[16]
NITRIK (day^{-1})	0.002 + 0.0025T[3] 0.03[4] 0.04–0.2[5] 0.01T[10] 0.002T[11]
PAMAX	0.02[7] 0.03[16]
PAMIN	0.00146[7] 0.003[16]
RELK (day^{-1})	0.95–1.8[7]
RESPK (day^{-1}) (phytoplankton)	0.06[1] 0.01[12] 0.015[16]
RESPK (day^{-1}) (zooplankton)	0.0007($T - S$)[2,3] 0.01[4] 0.02–0.16[5] 0.001T[11]
SEDT ($m\ day^{-1}$)	0.05[1] 0.2[4] 0.01–0.6[7]
SEPH ($m\ day^{-1}$)	0.2[1] 0.05–0.2[4] 0.1[5] 0.05–0.3[7] 0.1–5[9] 0.1[11] 0.2[12] 0.1–0.4[14] 0.08[16]
T_{max} (°C) (blue-greens)	45[12]
T_{opt} (°C) (blue-greens)	33[12]
T_{max} (°C) (nanophytoplankton)	35[12]
T_{opt} (°C) (nanophytoplankton)	25[12]
T_{max} (°C) (net phytoplankton)	35[12]
T_{opt} (°C) (net phytoplankton)	20[12]
UCMAX (day^{-1})	0.45[16]
UNMAX (day^{-1})	0.0096[16]
UPMAX (day^{-1})	0.0035[16]
α (m^{-1})	0.16[6] 0.3[4] 0.27[16]
β ($m^2\ g^{-1}$)	0.2[4] 0.19[6] 0.1[12] 0.18[16]
γ ($m^2\ g^{-1}$)	0.2[1]
δ ($m^2\ g^{-1}$)	0.01[1]

[1] Gargas (1976)
[2] Dahl-Madsen and Strange Nielsen (1974)
[3] Di Toro *et al.* (1975)
[4] Chen and Orlob (1975)
[5] O'Connor *et al.* (1976)
[6] Larsen *et al.* (1974)
[7] Nyholm (1978)
[8] Ikeda and Adachi (1978)
[9] Lewis and Nir (1978)
[10] O'Connor *et al.* (1973)
[11] Thomann *et al.* (1975)
[12] Scavia and Park (1976)
[13] Lehman *et al.* (1975)
[14] Imboden (1974)
[15] Stumm and Morgan (1970)
[16] Jørgensen (1976)
[17] Canale *et al.* (1976)

TABLE 4.15 Stoichiometric Ratios.

C/chlorophyll	50[5] 50[3] 50[11]
N/chlorophyll	7–10[5] 7[3] 7.2[6] 10[10] 10[11]
P/chlorophyll	1[5] 1[3] 0.63[6] 1[10] 1[11]
O_2/detritus	2[4]
O_2/phytoplankton	1.6[4]
CO_2/BOD	0.2[4]
N/zooplankton	0.14[3]
P/zooplankton	0.02[3]
C/dry weight of phytoplankton	0.6[7] 0.33–0.43[10] 0.33[15]
N/dry weight of phytoplankton	0.044–0.084[10] 0.045[15]
P/dry weight of phytoplankton	0.011–0.029[10] 0.07[15]

[1] Gargas (1976)
[2] Dahl-Madsen and Strange Nielsen (1974)
[3] Di Toro *et al.* (1975)
[4] Chen and Orlob (1975)
[5] O'Connor *et al.* (1976)
[6] Larsen *et al.* (1974)
[7] Nyholm (1978)
[8] Ikeda and Adachi (1978)
[9] Lewis and Nir (1978)
[10] O'Connor *et al.* (1973)
[11] Thomann *et al.* (1975)
[12] Scavia and Park (1976)
[13] Lehman *et al.* (1975)
[14] Imboden (1974)
[15] Stumm and Morgan (1970)
[16] Jørgensen (1976)
[17] Canale *et al.* (1976)

biota have constant stoichiometry; the element cycles are independent although some realistic upper and lower limits can be set. A more comprehensive collection of parameters and stoichiometric ratios has been assembled by Jørgensen (1979b).

4.3. CONCLUDING REMARKS

As demonstrated in this chapter, the mathematical expressions describing ecological processes vary from model to model. It is generally possible in models or in the biological literature to find different mathematical descriptions of the same biological process. A more detailed examination usually reveals that the differences are caused by environmental factors being included or excluded. Theoretically, as many details as possible may be included in the description of biological processes in rivers and lakes. The question is whether these will provide additional advantages for the total model, in view of the objectives of the model. Generally, it must be recommended that the complexity of the submodels is selected in accordance with the objectives and the data available for the calibration and validation of the model. However, a *causal* submodel that has been used with good results in a number of case studies might be used where the data do not allow the calibration and validation of the submodel, but where previously experienced parameters can be applied. In most cases a causal submodel should be preferred to an empirical one, since it is easier to transfer from one study to another.

Very little has been published, up to the present, about the effect on ecosystem models of replacing one submodel with another. It seems absolutely necessary to intensify such studies so that modelers can gain more experience of where and how a certain submodel can be used. Such research should go hand in hand with laboratory or microscale investigation of important submodels.

It is a great advantage to know the process parameters within certain limits. The first attempt at a comprehensive collection of useful parameters was made by Jørgensen (1979b). If, however, such knowledge of limits to parameters is not available, it might be necessary to conduct experiments in the laboratory or *in situ* to obtain the basic data required for good parameter estimations (see also section 9.4 and Chapter 11). Generally, more complex submodels are used for processes directly influencing the state variables that are the model objectives. For instance, it is of more interest to have a good, detailed submodel for phytoplankton growth and nutrient uptake than it is for fish feeding on zooplankton in a eutrophication model. Consequently, rather complex phytoplankton growth models have been developed (Table 4.1) for inclusion in eutrophication models. The same is true for the sediment–water exchange of nutrients, which should be modeled accurately, as an essential part of the nutrients in an aquatic ecosystem goes through the sediment (discussed in section 4.2.7). However, more complex models contain more parameters to be calibrated, which again requires more data for calibration and validation. Nevertheless, if the state variable in focus (for eutrophication models it is the phytoplankton concentration) is sensitive to simplification, the more complex and causal model should be implemented and the parameters estimated by separate laboratory or *in situ* experiments (discussed in section 9.4), if the measurements are not sufficient to allow for a direct calibration.

REFERENCES

Ahlgren, I. (1973a). Limnologiska studier av Sjön Norrviken. III. Avlastningens effekter. *Scripta Limnologica Upsaliencia* No. 333.

Ahlgren, I. (1973b). Limnological studies of Lake Norrviken, a eutrophicated Swedish lake. Nutrients and phytoplankton before and after sewage diversion. *Scripta Limnologica Upsaliencia* No. 334.

Anderson, J. M. (1973). The eutrophication of lakes. *Towards Global Equilibrium*, ed. D. Meadows (Cambridge, MA: MIT Press), pp. 117–140.

Bierman, V. J. (1976). Mathematical model of the selective enhancement of blue-green algae by nutrient enrichment. *Modeling Biochemical Processes in Aquatic Ecosystems*, ed. R. P. Canale (Ann Arbor, MI: Ann Arbor Science), pp. 1–32.

Bierman, V. J., F. H. Verhoff, T. C. Poulson, and M. W. Tenney (1974). Multinutrient dynamic models of algal growth and species competition in eutrophic lakes. *Modeling the Eutrophication Process*, eds. E. Middlebrooks, D. H. Falkenberg, and T. E. Maloney (Ann Arbor, MI: Ann Arbor Science), pp. 89–109.

Bloomfield, J. A., R. A. Park, D. Scavia, and C. S. Zahorcak (1974). Aquatic modeling in the eastern deciduous forest biome, US International Biological Program. *Modeling the Eutrophication Process*, eds. E. Middlebrooks, D. H. Falkenberg, and T. E. Maloney (Ann Arbor, MI: Ann Arbor Science), pp. 139–158.

Broqvist, S., (1971). Matematisk modell för ëkosystemet i en sjö. Forskningsgruppen för Planeringsteori, Matematiske Institution, Tekniska Högskolan, Stockholm.

Canale, R. P., L. M. DePalma, and A. H. Vogel (1976). A plankton-based food web model for Lake Michigan. *Modeling Biochemical Processes in Aquatic Ecosystems*, ed. R. P. Canale (Ann Arbor, MI: Ann Arbor Science), pp. 33–74.

Chen, C. W. (1970). Concepts and utilities of ecological models. *Proceedings of the American Society of Civil Engineers, Journal of the Sanitary Engineering Division* 96(SA5): 1085–1097.

Chen, C. W., and G. T. Orlob (1975). Ecologic simulation of aquatic environments. *Systems Analysis and Simulation in Ecology* vol. 3, ed. B. C. Patten (New York, NY: Academic Press), pp. 476–588.

Chrisholm, S. W., R. G. Stross, and P. A. Nobbs (1975). Environmental and intrinsic control of filtering and feeding rates in Arctic *Daphnia*. *Journal of Fisheries Research Board of Canada* 32: 219–226.

Cloern, J. E. (1978). Simulation model of *Cryptomonas ovata* population dynamics in Southern Kootenay Lake, British Columbia. *Ecological Modelling* 4: 133–150.

Dahl-Madsen, K. I., and K. Strange Nielsen (1974). Eutrophication models for ponds. *Vand* 5: 24–31.

DePinto, J. V. (1979). Water column death and decomposition of phytoplankton: an experimental and modeling review. *Perspectives on Lake Ecosystem Modeling*, eds. D. Scavia and A. Robertson (Ann Arbor, MI: Ann Arbor Science), pp. 25–52.

Di Giarno, F. A., L. Lijklema, and G. van Straten (1978). Wind-induced dispersion and algal growth in shallow lakes. *Ecological Modelling* 4: 237–252.

Di Toro, D. M., D. J. O'Connor, and R. V. Thomann (1971). Nonequilibrium systems in natural water chemistry. *American Chemical Society, Advances in Chemistry Series* 106: 131–180.

Di Toro, D. M., D. J. O'Connor, R. V. Thomann, and J. L. Mancini (1975). Phytoplankton–zooplankton–nutrient interaction model for western Lake Erie. *Systems Analysis and Simulation Ecology* vol. 3, ed. B. C. Patten (New York, NY: Academic Press), pp. 423–474.

Dodson, I. S. (1975). Predation rates of zooplankton in arctic ponds. *Limnology and Oceanography* 20(3): 426–433.

Droop, M. R. (1974). The nutrient status of algal cells in continuous culture. *Memoranda of Marine Biological Association, UK* 54: 825–855.

Dugdale, R. C. (1975). Biological modeling. *Modeling of Marine Systems*, ed. J. C. Nihoul (Amsterdam: Elsevier), pp. 187–206.

Eppley, R. W., R. W. Holmes, and J. D. H. Strickland (1967). Sinking rates of marine phytoplankton measured with a fluorometer. *Journal of Experimental Marine Biology and Ecology* 1: 191–208.

Frost, B. W. (1975). A threshold feeding behavior in *Calanus pacificus*. *Limnology and Oceanography* 20: 263–266.

Fuhs, G. W. (1969). Phosphorus content and rate of growth in the diatoms *Cyclotella nana* and *Thalossiosira fluviatilis*. *Journal of Phycology* 5: 312.

Gargas, E. (1976). A three-box eutrophication model of a mesotropic Danish Lake. Water Quality Institute, Hørsholm, Denmark.

Gause, G. F. (1934). *The Struggle of Existence* (New York, NY: Hafner), p. 133.

Harleman, D. R. F. (1978). A comparison of water quality models for the aerobic nitrogen cycle. *Research Memorandum* RM-78-34 (Laxenburg, Austria: International Institute for Applied Systems Analysis).

Ikeda, S., and N. Adachi (1978). A dynamic water quality model of Lake Biwa. *Ecological Modelling* 4:151–172.

Ikusima, I. (1966). Ecological studies on the productivity of aquatic plant communities. II. *Botanical Magazine, Tokyo* 79:7–19.

Imboden, D. M. (1974). Phosphorus model for lake eutrophication. *Limnology and Oceanography* 19:297–304.

Jansson, B. O. (1972). Ecosystem approach to Baltic problem. *Swedish Natural Science Research Council Bulletins from Ecological Research Committee* No. 16.

Jørgensen, S. E. (1976). A eutrophication model for a lake. *Ecological Modelling* 2:147–165.

Jørgensen, S. E. (1979a). *Lake Management* (Oxford: Pergamon).

Jørgensen, S. E. (ed.) (1979b). *Handbook of Environmental Data and Ecological Parameters* (Oxford: Pergamon; Copenhagen: International Society for Ecological Modelling).

Jørgensen, S. E., and D. R. F. Harleman (eds.) (1978). Hydrophysical and ecological modelling of deep lakes and reservoirs. Summary Report of a IIASA workshop, December 12–15, 1977. *Collaborative Proceedings* CP-78-7 (Laxenburg, Austria: International Institute for Applied Systems Analysis).

Jørgensen, S. E., O. S. Jacobson, and I. Høi (1973). Prognosis for a lake. *Vatten* 29:383–404.

Jørgensen, S. E., L. Kamp-Nielsen, and O. S. Jacobsen (1975). A submodel for anaerobic mud–water exchange of phosphate. *Ecological Modelling* 1:133–146.

Jørgensen, S. E., and H. F. Mejer (1979). A holistic approach to ecological modelling. *Ecological Modelling* 7:169–190.

Jost, J. L., I. F. Drake, A. G. Frederickson, and H. M. Tsandriya (1973). Interactions of *Tetrahymena pyriformis, Escherichia coli, Az tobacter vinelandii* and glucose in minimal medium. *Journal of Bacteriology* 113:834–840.

Kamp-Nielsen, L. (1974). Mud–water exchange of phosphate and other ions in undisturbed sediment cores and factors affecting the exchange rate. *Archiv für Hydrobiologie* 2:218–237.

Kamp-Nielsen, L. (1975). A kinetic approach to the aerobic sediment–water exchange of phosphorus in Lake Esrom. *Ecological Modelling* 1:153–160.

Kozerski, H.-P. (1977). Ein einfaches mathematisches Modell für Phosphataustausch zwischen Sediment und Freiwasser. *Acta Hydrochimica et Hydrobiologia* 5:53–65.

Kremer, J. N., and S. W. Nixon (1978). *A Coastal Marine Ecosystem Simulation and Analysis* (Berlin: Springer), 217 pp.

Lamanna, C., and M. F. Malette (1965). *Basic Bacteriology* (Baltimore, MD: Williams and Wilkins).

Larsen, D. P., H. T. Mercier, and K. W. Malveg (1974). Modeling algal growth dynamics in Shagawa Lake, Minnesota. *Modeling the Eutrophication Process*, eds. E. J. Middlebrooks, D. H. Falkenberg, and T. E. Maloney (Ann Arbor, MI: Ann Arbor Science), pp. 15–33.

Lassen, H., and P. B. Nielsen (1972). Simple mathematical model for the primary production as a function of the phosphate concentration and the incoming solar energy applied to the North Sea. Danmarks Fiskeri- og Havundersøgelser, International Council for the Exploration of the Sea, Plankton Committee 1972.

Lassiter, R. R., and D. K. Kearns (1974). Phytoplankton population changes and nutrient fluctuations in a simple aquatic ecosystem model. *Modeling the Eutrophication Process*, eds. E. J. Middlebrooks, D. H. Falkenberg, and T. E. Maloney (Ann Arbor, MI: Ann Arbor Science), pp. 131–138.

Lehman, J. T., D. B. Botkin, and G. E. Likens (1975). The assumptions and rationales of a computer model of phytoplankton population dynamics. *Limnology and Oceanography* 3:343–364.

Lewis, S., and A. Nir (1978). A study of parameter estimation procedures of a model for lake phosphorus dynamics. *Ecological Modelling* 4:99–118.

Lotka, A. J. (1924). *Elements of Mathematical Biology* (New York, NY: Dover) (reprinted 1956).

Najarian, T. O., and D. R. F. Harleman (1975). A real-time model of nitrogen-cycle dynamics in an estuarine system. *MIT Department of Civil Engineering, R. M. Parsons Laboratory Report* 204.

Nyholm, N. (1975). Kinetics studies of phosphate-limited algae growth. *Thesis, Technical University of Copenhagen.*

Nyholm, N. (1978). A simulation model for phytoplankton growth and nutrient cycling in eutrophic, shallow lakes. *Ecological Modelling* 4:279–310.

O'Brien, J. J., and J. S. Wroblewski (1972). An ecological model of the lower marine trophic levels on the continental shelf of West Florida. *Geophysical Fluid Dynamics Institute, Florida State University, Tallahassee, FL Technical Report*, 170 pp.

O'Connor, D. J., R. V. Thomann, and D. M. Di Toro (1973). Dynamic water quality, forecasting and management. *US Environmental Protection Agency, Washington, DC Report* EPA 660/3-73-009.

O'Connor, D. I., R. V. Thomann, and D. M. Di Toro (1976). Ecologic models. *Systems Approach to Water Management*, ed. V. Biswas (New York, NY: McGraw-Hill), pp. 299–333.

Odum, H. T. (1972). An energy circuit language. *Systems Analysis and Simulation in Ecology* vol. 2, ed. B. C. Patten (New York, NY: Academic Press), pp. 139–211.

Orlob, G. T. (1975). Modeling of aquatic ecosystems. *Integrity of Water. US Environmental Protection Agency Report* (Washington, DC: US Government Printing Office), pp. 189–201.

Park, R. A., T. W. Groden, and C. J. Desormeau (1979). Modification to model CLEANER, requiring further research. *Perspectives on Lake Ecosystem Modeling*, eds. D. Scavia and A. Robertson (Ann Arbor, MI: Ann Arbor Science), pp. 87–108.

Park, R. A., *et al.* (1974). A generalized model for simulating lake ecosystems. *Simulation* 23:33–50.

Parker, R. A. (1972). Estimation of ecosystem parameters. *Verhandlung Internationaler Vereinigung Limnologie* 18:257–263.

Patten, B. C., D. A. Egloff, and T. H. Richardson (1975). Total ecosystem model for a cove in Lake Texoma. *Systems Analysis and Simulation in Ecology* vol. 3, ed. B. C. Patten (New York, NY: Academic Press), pp. 206–423.

Rhee, G. Yu. (1978). Effects of N:P atomic ratios and nitrate limitation on algal growth, cell composition, and nitrate uptake. *Limnology and Oceanography* 23:10–25.

Richey, J. E. (1977). An empirical and mathematical approach toward the development of a phosphorus model of Castle Lake. *Ecosystem Modeling in Theory and Practice*, eds. C. A. S. Hall and I. W. Day, Jr. (New York, NY: Wiley–Interscience), pp. 267–287.

Scavia, D., B. J. Eadie, and A. Robertson (1976). An ecological model for the Great Lakes. *Great Lakes Environmental Research Laboratory, National Oceanic and Atmospheric Association, Ann Arbor, MI Contribution* 64.

Scavia, D., and R. A. Park (1976). Documentation of selected constructs and parameter values in the aquatic model CLEANER. *Ecological Modelling* 2:33–53.

Schellenberger, G., H.-P. Kozerski, H. Behrendt, and S. Hoeg (1978). A mathematical ecosystem model applicable to shallow water bodies. *Modeling the Water Quality of the Hydrological Cycle. Proceedings of Baden Symposium, September. International Association of Hydrological Sciences Publication* 125, pp. 128–136.

Schofield, W. R., and R. G. Krutchoff (1974). Stochastic model for a dynamic ecosystem. *Virginia Water Resources Research Center Bulletin* VPI-WRRC 60.

Serruya, C., and T. Berman (1975). Phosphorus, nitrogen and the growth of algae in Lake Kinneret. *Journal of Phycology* 11(2):155–162.

Shelef, G., W. J. Oswald, and C. G. Golueke (1969). The continuous culture of algal biomass on wastes. *Proceedings of 4th Symposium on Continuous Cultivation of Microorganisms, Prague* (Oxford: Pergamon), pp. 601–629.

Shelef, G., M. Schwarz, and M. Schechter (1972). Prediction of photosynthetic production in biomass in accelerated algal–bacterial wastewater treatment systems. *Proceedings of 6th International Symposium on Water Pollution Research* (Oxford: Pergamon), Section 5, Paper a, pp. 1–10.

Smayda, T. J. (1970). The suspension and sinking of phytoplankton in the sea. *Oceanography and Marine Biology Annual Review* 8:353–414.

Smayda, T. J. (1974). Some experiments on the sinking rates of two freshwater diatoms. *Limnology and Oceanography* 19:628.

Steel, J. A. (1973). Reservoir algal productivity. The use of mathematical models in water pollution control. *University of Newcastle upon Tyne Symposium*, ed. A. James, pp. 107–135.

Steele, J. H. (1974). *The Structure of Marine Ecosystems* (Oxford: Blackwell), pp. 74–91.

Stefan, H., T. Skoglund, and R. Megard (1976). Wind control of algae growth in eutrophic lakes. *Proceedings of American Society of Civil Engineers, Journal of Environmental Engineering Division* 102(EE6):1201–1213.

Straškraba, M. (1976). Development of an analytical phytoplankton model with parameters empirically related to dominant controlling variables. *Umweltbiophysik*, eds. R. Glaser, K. Unger, and M. Koch (Berlin: Akademie-Verlag), pp. 33–65.

Stumm, W., and J. J. Morgan (1970). *Aquatic Chemistry* (New York, NY: Wiley), 432 pp.

Takahashi, M., K. Fujii, and T. R. Parsons (1973). Simulation study of phytoplankton photosynthesis and growth in the Fraser River Estuary. *Marine Biology* 19:102–116.

Talling, J. F. (1957). The phytoplankton populations as a compound photosynthetic system. *New Phytology* 56:133–149.

Thomann, R. V., D. M. Di Toro, and D. J. O'Connor (1974). Preliminary model of the Potomac Estuary phytoplankton. *Proceedings of American Society of Civil Engineers, Journal of Environmental Engineering Division* 100(EE3):699–715.

Thomann, R. V., D. M. Di Toro, R. P. Winfield, and D. J. O'Connor (1975). Mathematical modeling of phytoplankton in Lake Ontario. *US Environmental Protection Agency National Environmental Research Center, Corvallis, OR Report* EPA 660/3-75-005.

Titman, D., and P. Kilham (1976). Sinking in freshwater phytoplankton: some ecological implications of cell nutrient status and physical mixing processes. *Limnology and Oceanography* 21(3):409–417.

Vollenweider, R. A. (1965). Calculation models of photosynthesis–depth curves and some implications regarding day rate estimates in primary production. *Memorie dell Istituto Italiano di Idrobiologia* 18 Suppl.: 425–457.

Walsh, J. J., and R. C. Dugdale (1971). Simulation model of the nitrogen flow in the Peruvian upwelling system. *Investigacion Pesquera* 35:309–330.

Youngberg, B. A. (1977). Application of the aquatic model CLEANER to a stratified reservoir system. *Center for Ecological Modeling, Rensselaer Polytechnic Institute, Troy, NY Report* 1.

CHAPTER 4: NOTATION

a_{N_1B} constant
a_2 inhibition constant
B concentration of bacteria
C_x concentration of toxic compound
K_S half-saturation constant
N_1 concentration of inhibited nutrient
N_2 concentration of inhibiting nutrient
S concentration of substrate
β toxicity coefficient.

Table 4.13 lists the symbols used in Tables 4.1–4.12.

5 Simulation of the Thermal Regime of Rivers

J. Jacquet

5.1. PURPOSE OF MODELING

Water temperature has an important influence on the water quality, as well as the aquatic life, of a river. It varies according to a seasonal rhythm, upon which are superimposed random fluctuations that are closely linked to prevailing meteorological conditions. Knowledge of river temperature is more or less good, depending on the interest that investigators have attached to it; long time series of temperatures are available for several river sections, but for others temperature measurements are completely missing.

The increasing importance given to solving environmental problems has made it necessary to study the thermal regime of streams in considerable detail. The following points are important with regard to the impacts of technology.

(1) It is necessary to understand in detail the thermal regime of a river reach upon which a thermal power plant is to be located, in order to design the cooling system, which includes internal piping, pumping equipment, and the condenser.

(2) The design of the inlet and outlet structures of a thermal power plant must take into account the risks of recycling rejected hot water.

(3) One must understand that the thermal regime of a reach where a power plant is located is disturbed by all other installations upstream.

(4) The use of potable water from a stream requires maintenance of a minimum quality of water; specifically, the temperature should not exceed a certain threshold, beyond which problems might arise in water treatment and in delivery to the consumer.

From the environmental point of view, knowledge of the thermal regime of a stream is fundamental. A forecast of the impacts of any development on water quality and on aquatic life is often made by estimating the modification of this regime. Modification of water temperature can have direct, even lethal effects on the aquatic fauna or flora. It can also have an indirect effect through the disequilibrium imposed on the population by the competition between species reacting differently to the modification. Statistical methods, for example, permit one to indicate the domain of variations to which the distribution of populations is sensitive.

5.2. THERMAL REGIME OF A STREAM

The thermal regime of a stream depends essentially on the climatic conditions of the regions through which it passes, on its hydraulic characteristics, and on the temperatures of its successive water inputs. In practice, the water temperature becomes more closely linked to local climatic conditions as the distance from the source increases. Thus, the temperature of a river flowing slowly through a plain, where the climate is often uniform, depends almost entirely on local meteorological conditions, and it can be easily calculated from these. On the other hand, a mountain stream or a river with a steep slope goes through very different thermal regimes; the water temperature does not depend strongly on meteorological conditions, but reflects rather those of its tributaries.

Modification of the hydraulic characteristics of a stream or a change of inflows will disturb its thermal regime. One of the most common causes of modification is the development of a stream for navigation. By creation of successive flow levels, the flow is usually slowed and the depths are very often considerably increased. This type of development generally does not change the average temperature, but changes appreciably the thermal inertia and eliminates the daily fluctuations of temperature. In a region of natural cold waters an average temperature increase may be observed. Similarly, the construction of a reservoir on a stream or its tributaries can cause a serious change in the thermal regime downstream, depending on the scale of the development. Normally, the temperature will be lower in summer and higher in winter. A third cause is industrial or municipal discharge, which supplies significant amounts of water at a different temperature from that of the stream. This can modify the thermal regime of the receiving medium. Among such discharges are those from treatment plants, discharges from various industries (e.g. metallurgical industries), and, especially, discharges from thermal power plants, which are often the most important ones.

Other sources of modification of the thermal regime are the reduction of friction losses through the construction of hydroelectric power plants, resulting in cooling; heat input from navigation; and heat inputs or losses through biochemical reactions of self-purification in streams.

Each of these modifications to the thermal regime is of a quite different nature and each requires a different approach for it to be properly understood. The two most important causes of perturbation to the thermal energy balance of streams are (in order) heat discharge by industry, and the construction of large reservoirs. The study of the latter type of perturbation requires particular knowledge of stratified impoundments, to be discussed in Chapter 7. The problem posed for discussion in this chapter is that of the heat discharge to streams by large power plants.

5.3. AREAS FOR MODELING

The study of perturbations to the thermal regime of a stream, caused by the discharge of hot water or by modifications of the hydraulic characteristics, reveals two different problems. The "near field" problem is concerned with a very limited zone of a stream near the point of maximum perturbation. How does the thermal perturbation act locally? In other words, how, for example, do the hot waters mix with the receiving medium? What is the local modification in the daily thermal cycle due to hydraulic changes? This type of problem requires, most often, the use of physical hydraulic models or special mathematical models (Chapter 8, section 8.5: Mathematical models of cooling impoundments).

In "far field" problems, the main emphasis is placed on global effects. What is the effect of a perturbation on the thermal regime of a river a considerable distance downstream of the disturbance, after the initial mixing with hot water? Up to what distance will the average temperature increase of the stream remain significant? How might a series of power plants within the same river basin modify the thermal regime of the basin and its principal drainage courses? One considers the stream to be homogeneous in a given section and studies the perturbations in the average temperature, considered significant only at the basin level.

5.4. THERMODYNAMICS OF A RIVER

The thermal regime of a stream can be defined by a great number of parameters, depending on the area of interest:

- temperature time series corresponding to various sections of the stream;
- periodic characteristics, including the periods, phases, and amplitudes of daily and annual cycles;
- variations in temperature with space or time, including deviations observed simultaneously at a section of the river; deviations of temperature on the same day, deviations from one day to the next, from one year to the next, etc.;

extreme changes in temperatures of very short duration, or daily or monthly averages; and
stochastic structure of temperature, considered as a continuous process.

For example, to optimize the cooling circuit of a power plant, one would have to know the temperature time series; but to judge the development of a basin as a whole, one requires statistical parameters. The biologist who would like to estimate the possible impact of a development would be interested in still other parameters: extreme values, for example, or the thermal constraints applied to the aquatic life, estimated after the modification of average temperatures in a given critical period.

The thermal regime of rivers is governed generally by the following phenomena: exchange of energy with the atmosphere; exchange of energy with the banks; heat input by inflowing water, a tributary, or a waste discharge; heat exchange at a weir, waterfall, or rapids; artificial heat injection; evaporation and condensation; internal heat dissipation by friction; heat liberated by biochemical reaction; and convection and diffusion in the water mass. The study of the local modification of the temperature field does not usually require a completely reliable description of all these phenomena. Usually this is only necessary for a particular development, like the installation of a once-through cooled thermal power plant or a planned wastewater discharge. In practice, such a study is made either with a physical model or with a mathematical model that emphasizes the hydraulic mechanisms of heat transfer.

In contrast, problems related to water temperature changes over long distances or periods can only be treated with mathematical models. The more important processes considered in such models include energy exchange with the atmosphere or the ground.

5.5. SIMULATION MODELS

The use of either scaled physical or mathematical simulation models is essential for studying the effects of various developments on the thermal regime or local thermal structure of a stream. These models have to describe the evolution of the temperature regime (or of the field of temperatures in a given zone, in physical models) starting from limited data. The mechanisms governing this evolution must be represented with sufficient realism that the effects of a modification of the system can be studied. As suggested in section 5.3, two types of model have to be considered. There are those describing local phenomena, such as the field of temperatures in the proximity of a discharge, in which the phenomena are essentially governed by mechanisms internal to the water mass. Then there are models that describe the phenomena in an integrated fashion, where the exchanges of energy between water and the atmosphere play a most important role and hydraulic phenomena can be strongly schematized.

5.5.1. Deterministic Approach

Simulation of the evolution of the temperature regime along a river by considering the river in its entirety, or even as a hydrogeographic basin, is only done at the cost of certain simplifications. The first simplification is that the river is homogeneous in temperature in all cross sections. This hypothesis is not always verified, but it results in appreciable errors only in exceptional cases. In effect, heterogeneity of temperature in a cross section is only caused by local phenomena: zones of dead water, and mixture zones with tributary waters or with industrial discharges. These phenomena normally have only little influence on the thermal regime downstream. Kaisersŏt and Mitschel (1977) made a theoretical study of these extreme cases.

Hydraulic phenomena must also be schematized, sometimes even to an extreme. Practice shows, however, that these simplifications are justifiable. Turbulence and longitudinal dispersion, under special circumstances (sudden variations of stream discharge or sporadic industrial discharges), can sometimes cause appreciable errors if they are not taken into account.

With these simplifying assumptions, the evolution of water temperature in space and time can be described according to the conservation of energy principle by

$$\rho c H\left(\frac{\partial T}{\partial t} + u\frac{\partial T}{\partial x}\right) = R_s + R_a - R_e - E - C_v + P + \frac{\partial s}{\partial x}, \tag{5.1}$$

where

- ρ is the density of water (kg m^{-3}),
- c is the specific heat of water (J kg^{-1}),
- H is the average water depth (m),
- T is the water temperature (°C),
- x is the distance downstream (m),
- R_s is the solar radiation flux (ly day^{-1}),
- R_a is the atmospheric radiation flux (ly day^{-1}),
- R_e is the radiation flux emitted from the water surface (ly day^{-1}),
- E is the energy flux exchanged by evaporation, normalized per unit area of water surface (ly day^{-1}),
- C_v is the energy flux exchanged by convection at the air–water interface, normalized per unit area of water surface (ly day^{-1}),

P is the heat flux introduced artificially into a given section (ly day^{-1}),
$\partial s/\partial x$ is the heat flux exchanged with the ground and the banks of a stream, normalized per unit area of water surface (ly day^{-1}),
1 ly is one langley (1 calorie cm^{-2}).

Equation 5.1 can be written simply as

$$\rho c H\left(\frac{\partial T}{\partial t} + u\frac{\partial T}{\partial x}\right) = F(T, x, t). \tag{5.2}$$

5.5.2. Models for Steady State Conditions

There are various methods for solving (5.2). The first studies attempted merely to simplify the solution. Accordingly, we note that for a mass of water moving downstream,

$$\rho c H \frac{\partial T}{\partial t} = F(T, t). \tag{5.3}$$

The assumption of steady state conditions permits direct integration of this equation. Very often, for the equation in this form, the idea of an "equilibrium temperature" T_e is introduced. This may be taken to be the temperature of a mass of water without thermal inertia, i.e. for which $\partial T/\partial t$ would be zero, or

$$F(T_e, t) = 0.$$

Equation 5.3 can be written simply as

$$\begin{aligned} \rho c H \frac{dT}{dt} &= F(T, t) - F(T_e, t) \\ &= \frac{\partial F}{\partial T}(T - T_e) = K(T - T_e). \end{aligned} \tag{5.4}$$

5.5.3. Exponential Method

Under steady state conditions T_e is constant. If T is close to T_e, or varies only slightly along the stream, K can also be considered constant, which leads to

$$\frac{T - T_e}{T_0 - T_e} = \exp\left(\frac{-Kt}{\rho c H}\right), \tag{5.5}$$

where T is the temperature at time $t = 0$. Therefore, under steady state conditions the water temperature tends toward the equilibrium temperature exponentially with distance (or flow time) from the source. At this point confusion often arises between the natural water temperature T_n and the equilibrium temperature T_e, confusion that can lead to a mistake when the progressive disappearance of a temperature perturbation is studied. In fact, it would lead to writing (5.3) as

$$\rho c H \frac{\mathrm{d}(T - T_\mathrm{n})}{\mathrm{d}t} = F(T, t) = F(T_\mathrm{n}, t) = \frac{\partial F}{\partial t}(T - T_\mathrm{n}) = K'(T - T_\mathrm{n}). \quad (5.6)$$

K' has a different value from K in (5.5) because it corresponds to the derivative of the function F for another value of water temperature (Kahlig (1977) discusses this problem).

This type of approach permits simple assessment of questions of the evolution of a temperature perturbation generated, for example, by a discharge from a thermal power plant or by a release from a reservoir. Two essential weaknesses limit its usefulness.

(i) In practice, all stream parameters vary along the stream or in time. In particular, hydraulic parameters (e.g. average depth, stream discharge, roughness) vary considerably from upstream to downstream. Therefore, the simplified model can only be applied to reaches that are relatively short and along which the flow and geometric parameters vary only slightly.
(ii) The model, under steady state conditions, cannot be used to calculate the natural evolution of the water temperature along a stream, but only the evolution of the average difference in temperature with respect to either the equilibrium temperature or the natural temperature.

It is, therefore, impossible to compare calculated with measured values and to test the validity of the model. Nevertheless, provided that the coefficient K of energy exchange is correct, the calculations based on this method allow one to determine the approximate size of the residual temperature increase downstream of a section where heat was injected.

On the other hand, this method is completely inadequate for estimating the effect of a modification in flow on the thermal regime of a stream (e.g. the effect of channelization).

5.5.4. Differential Equation for Steady State Conditions

In order to carry out more realistic calculations it is possible to integrate (5.1) stepwise, still keeping the hydrometeorological conditions steady, but allowing geometric data to change from one step to the next. In this way the masses of water are followed downstream. This method was used by the Länderarbeitsgemeinschaft Wasser (1971) working group to study thermal energy changes in the Rhine, by the Hydraulic Institute of Karlsruhe for the Neckar River (Fleig and Flinsbach, 1973), and by the Hydrologisches Institut München (1976) for the Danube. The method allows one to simulate typical cases (winter, spring, summer, fall) by choosing fixed conditions for the water discharge of the stream, its natural temperature, and coefficients, representative of the season, for the exchanges of energy (average and extreme values).

5.5.5. Models for Nonsteady State Conditions

The approach outlined above can be modified still further by varying the set of parameters for the whole length of the stream. The calculation (for a given mass of water) is the same as given above, except that for each time step all parameters are allowed to vary, especially the hydrometeorological parameters. Various approaches have been proposed to determine values of these parameters.

Some case studies use real data (chosen more or less arbitrarily to represent the periods or conditions given, such as seasons, or periods of low water level). An example of this approach is a model developed by Raphael (1962) and applied by Motor-Columbus to the Rhine. Each season was represented by a period of ten days. The model introduces a longitudinal diffusion term by the following artifice. For each section the calculation is made for two masses of water flowing at different speeds; the residence times are different and the temperatures of the masses evolve differently; at each end of a reach the masses of water are mixed, and the average temperature calculated corresponds to the temperature of the mixture. Then the model is calibrated according to the evaporation coefficient. This calibration consists of the determination of the value of the coefficient that gives the best agreement between the calculated and measured temperatures for the four ten-day periods.

5.5.6. Simulation Models for Nonsteady State Conditions

Instead of considering some typical periods for this type of simulation, one can carry out the calculation over very long periods. Smitz (1975) proposed this solution for the Meuse River, for which he simulated the thermal regime for the whole year, with 24-hour time steps. He took into consideration two phenomena that have been neglected in the models previously described: heat exchange with the river bed; and longitudinal dispersion.

Smitz showed that in the calculation of the natural temperature of a stream, the first of these terms plays an important role, whereas longitudinal dispersion can almost always be neglected. This does not apply, however, to the calculation of the temperature of water influenced by a time-varying heat discharge. Heat exchanges with the ground were calculated by a far-field method based on simple conduction in the soil (Smitz, 1975; Jobson, 1977).

During the one-year application of a nonsteady temperature model to the Meuse, Smitz simulated a great number of hydrometeorological conditions. Thus, it was possible to make a statistical estimation of temperature increases.

Smitz did not proceed further with calibration of his model, which gave a good agreement between calculations and measurements for the year studied.

A similar approach was used by Gras and Jacquet (1971), Gras and Martin (1976), and a working group of the Commission Internationale pour la Protection du Rhin Contre la Pollution (1973–74). The calculation technique is very similar to that of the Smitz model but has the following differences.

(i) It permits simulation over a great number of years (between ten and twenty).

(ii) It is possible to adjust the exchange coefficients through comparison of calculated and measured temperatures for one year. The adjustment criteria are based on the annual average deviation and the typical deviation of the daily differences between calculated and measured values. Then, this adjustment is verified for another year.

(iii) A "loop" term is introduced to take into account globally the phenomena that cannot be estimated separately:

exchange of heat with the ground;

water inputs at different temperatures from that of the stream (from underground waters, small tributaries, etc.);

heat inputs from turbulence, navigation, internal chemical reactions, and small industrial or urban discharges; and

meteorological data that are poorly representative of the larger data set, especially the more or less systematic (seasonal) deviations between the air temperature and humidity over the water surface and the corresponding observations at the meteorological station.

The loop term is also estimated by adjustment between the calculated and measured temperatures.

(iv) Since the hydrometeorological sample is very important, the results obtained can be examined statistically.

(v) Since the number of cases studied is especially large, it is possible to introduce other considerations into the simulation. For example, one can take into account the random character of heat discharges from power plants.

On the other hand, this approach does not deal with longitudinal dispersion. Neglecting this phenomenon can introduce appreciable errors in the calculation of instantaneous values or of temperature increases in sections close to the heat injection point. In practice, however, it does not influence results that are treated statistically.

Gras (1970) used this method to generate time series of natural temperatures in those sections of the stream that were sufficiently far from the source. It is sufficient for this purpose to integrate (5.3) stepwise, considering the water mass to be stationary. He demonstrated that, independently of the initial temperature taken, T converges rapidly toward the same temperature $T(t)$, which is the natural temperature of a water mass of average depth H influenced by local meteorological conditions.

This method is very convenient for completing an insufficient set of data. For example, it may be used to calculate the temperatures of tributaries or discharges for which no measurements are available or to construct the temperature time series that serves as an input to a model.

5.6. DATA REQUIRED FOR IMPLEMENTATION

The collection of necessary data for the application of a model constitutes an essential step in its implementation. Very often difficulties arise that require large amounts of data to resolve them, especially for simulation models of nonsteady state conditions.

5.6.1. Simplified Models for Steady State Conditions

These models are used most often in specific case studies where the choice of data is more or less arbitrary. The stream to be studied is divided into reaches delimited by special sections, such as the point of heat injection, the confluence with another stream, or the point where the temperature increase or the water temperature itself is stipulated (e.g. a potable water intake).

The equation of temperature evolution can be written in either of two forms:

$$\frac{T - T_e}{T_0 - T_e} = \exp\left(\frac{-Kt}{\rho c H}\right), \qquad \frac{T - T_n}{T_0 - T_n} = \exp\left(\frac{-K't}{\rho c H}\right) \tag{5.7}$$

or

$$\frac{T - T_e}{T_0 - T_e} = \exp\left(\frac{-KA}{\rho c Q}\right), \qquad \frac{T - T_n}{T_0 - T_n} = \exp\left(\frac{-K'A}{\rho c Q}\right). \tag{5.8}$$

In practice, the second expression, using the water surface area A (defined reach by reach) and the flow Q, is much more convenient. The exchange coefficients K and K' can be taken from the literature. Gras and Jacquet (1971) have provided a convenient diagram for estimation. Sweers (1976) formulated K as a function of T. The equilibrium temperature T_e or the water temperature T_n, as well as the stream discharge Q, is arbitrarily chosen depending on the case studied.

5.6.2. Simulation Models for Nonsteady State Conditions

Whichever specific model is chosen, nonsteady state simulation requires the collection of a considerable amount of data. Calculations are invariably made with very small time and space steps. The geometry of the stream as a function of the stream discharge has to be known with good precision. In addition to the length of the reach, which is specified before constructing the model, at least two of the following parameters have to be known:

average depth	$H = H(Q)$
average velocity	$v = v(Q)$
average width of reach	$l = l(Q)$.

The stream discharge is given by

$$Q = vlH.$$

In practice, there are no great difficulties in collecting data for these parameters, but results are not particularly sensitive to errors in the parameters. If necessary, missing data can be supplied by interpolation, correlation, or regression techniques.

To simulate the thermal regime it is necessary to define time series of water temperatures, stream discharges, and representative meteorological data.

Temperatures of the stream at the head of the first reach studied and the temperatures of effluents, especially the main ones, must be specified. Often these time series are not completely available. Gras (1970) proposed a method for generating the series from only meteorological data, by considering only the natural temperatures of the stream in sections that were relatively distant from the source.

Stream discharges must include discharges for each of the reaches and the principal tributaries.

Representative meteorological data include, as a minimum, wind speed, air temperature, humidity, atmospheric pressure, and solar and atmospheric radiation. The time series collected through the established meteorological network are generally sufficient. In practice, it is advisable to limit the number of meteorological stations used, even for very long reaches. For example, for a study of a 700 km reach of the Rhine, initially seven meteorological stations were taken into consideration; but it was found that this number could be reduced to three without affecting appreciably the quality of the results. Therefore, it is essential to verify whether the meteorological stations selected are indeed representative.

For models that do not proceed to a calibration stage, the choice of meteorological stations has to be made very carefully (Smitz, 1975). The quality of the results depends completely on selection of the most appropriate stations.

For models that are to be calibrated, the adjustment of evaporation coefficients can compensate for differences between meteorological parameters, especially between the wind velocity at the water level and the wind velocity at the meteorological station, where these values are sufficiently well correlated.

If u_r and u_m are the average wind velocities at the water level and as measured at the meteorological station, respectively, and $e(u_r)$ is the evaporation coefficient as a function of the wind velocity, then the necessary adjustment equations are

$$u_r = \alpha u_m + \epsilon \tag{5.9}$$

and

$$e(u_r) = a + bu_r,$$

where a, b, and ϵ are constants to be adjusted. It follows that

$$e(u_r) = a + b\alpha u_m + b\epsilon, \tag{5.10}$$

where a, b, α, and ϵ are constants to be adjusted. The adjustments of a, b, and α will take into account explicitly the relationship between u_m and u_r.

Direct measurements of solar and atmospheric radiation are very often deficient or lacking altogether. Calculating methods allow estimation of these fluxes from other meteorological data, measured or estimated at the ground station. These usually include dry and wet bulb air temperatures, humidity, cloud cover or duration of radiation, wind velocity, atmospheric pressure, and sometimes pan evaporation rates. This subject has been discussed by Gras (1970), Wunderlich and Hsieh (1971), and Klein and Momal (1979). The appendix summarizes a set of heat flux equations of this type developed by the Tennessee Valley Authority (1972).

At this point, attention is focused on the quality of the data assembled, which is often questionable. The data must be analyzed before they are introduced into a model. Experience shows, for example, that it is rare for temperature time series not to be affected by important errors. Certain errors can be detected easily, such as those originating from poor measurement techniques or scale errors that occur in carrying forward the values read. Other errors can be found only by careful study of the time series, especially systematic errors caused by a change of instruments or use of a nonrepresentative measurement station.

The direct comparison of temperature time series can help to detect these errors, but often the model itself has to be used to complete this data analysis. On these occasions it is often possible to reveal the influence of particular phenomena. An example might be a temperature measurement station under the influence of a heated water discharge, even though it is a significant distance downstream from the injection point. Another might be a measurement station upstream of a thermal power plant where hot water is recycled sporadically. Measurement stations under the influence of a single tributary or heat source may reveal systematic differences that can be discovered only by careful probing of the time series.

5.6.3. Choosing a Model

The choice of a model depends, of course, on the problem to be solved. Local problems, e.g the temperature field in the locale of a water discharge or the risk of recirculating water between intake and discharge, will be handled by models adapted to mixing zones. When problems of the thermal regime of a stream have to be studied, one-dimensional models, which consider the water to be homogeneous in temperature throughout the cross section, have to be used.

It appears that, of the methods available, nonsteady state modeling permits solution of the greatest variety of problems. Nevertheless, steady state solutions are often useful for preliminary studies and for checking more costly nonsteady state model results.

5.7. CALCULATION TECHNIQUES

Two general methods of computation, describing the thermal regime in a river, are indicated. The exponential method gives a broad, generalized treatment of the heat balance; and the simulation method provides considerable detail in space and time, even to the point of facilitating statistical treatment of results.

Implementation of models on the macrolevel with the exponential method does not present any particular difficulty. It can be applied with a simple pocket calculator. The simulation method, on the other hand, requires access to a computer. Calculations are generally independent of the model used, although in simulations using several years of data the results of calculation may be so voluminous as to require storage on tape or disk for later statistical analysis.

When the simulation technique is employed for large and complex river systems, like the Meuse, it is necessary to continually update the time series of temperatures at intervals corresponding to the selected time step. The technique for this calculation will now be described briefly.

The temperature profile along a stream is $T = f(x)$ at time t, as shown in Figure 5.1. Each section of water moves downstream at a velocity $u_1(x)$ and, warming in the process, creates a new temperature profile at $t + \Delta t$. Two time series, for t and $t + \Delta t$, are thus defined at intervals of $u_1(x)\Delta t$ along the x axis. Intermediate values of the temperature at particular stations (x_s) are interpolated between adjacent members of the same series with respect to either distance or time, depending on which suits better the requirements of the problem.

If the flow changes from Q_1 to Q_2 in the interval from $t + \Delta t$ to $t + 2\Delta t$, one can interpolate on the temperature profile for $t + \Delta t$ at intervals of $u_2(x)\Delta t$ to obtain a new time series as shown in Figure 5.2. This can, in turn, be compared with the series representing the state of the system at $t + 2\Delta t$, and so on, following the procedure outlined in Figure 5.1. The method allows calculation of

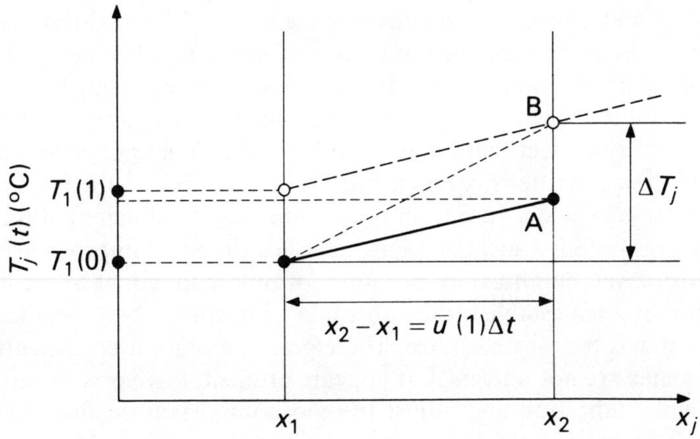

FIGURE 5.1 Temperature profile change along stream, under steady state conditions. A and B are the profiles at $t = 0$ and Δt, respectively. ΔT_j is the temperature increase during flow $Q(1)$ from x_1 to x_2 in time Δt. $\bar{u}(t) = Q(t)/\frac{1}{2}(A_1(0) + A_2(0))$.

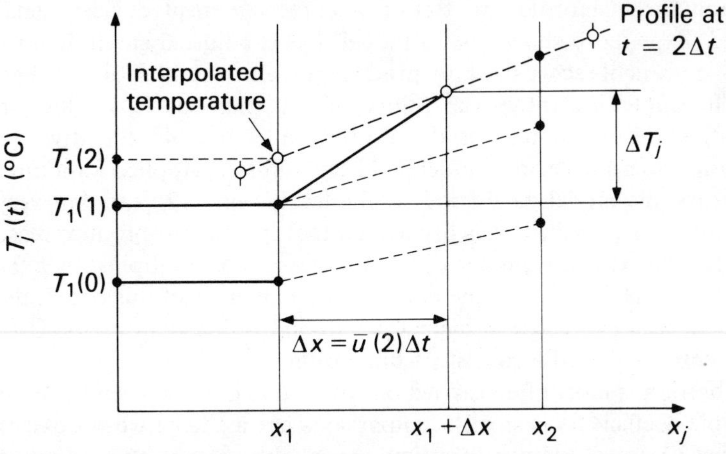

FIGURE 5.2 Adjusted temperature profile after change in rate of flow. ΔT_j is the temperature increase during flow $Q(2)$ from x_1 to $x_1 + \Delta x$.

temperature time series for the entire system with a constant time interval Δt (taken as 3 h in the Meuse study) between values.

5.8. MODEL ADJUSTMENTS AND SENSITIVITY ANALYSIS

After critical analysis of the water temperature and hydrometeorological data, it is the usual practice to adjust the empirical coefficients used, i.e. to make a calibration. For macroscale models, this adjustment is most often made to the evaporation and convection coefficients, e.g. in the Motor-Columbus model. Also, it may be performed on the loop coefficient, which takes globally into account the simplifications made in the model, such as neglecting heat exchanges with the ground or heat inputs from minor sources. The loop coefficient may also account for the fact that the meteorological data are not accurately representative of the conditions prevailing at the stream, that is, the meteorological station may be some distance from the stream and at a different altitude.

The justification for such an adjustment lies simply in improved simulation results. Moreover, empirical evaporation formulas, abundant in the literature, have generally been established by an analog adjustment between the observation station and the water surface. Therefore, evaporation coefficients used in these formulas are not universal. It appears prudent, therefore, to carry out an independent calibration and adjust the more important coefficients for each new case (or model) studied.

The problem that remains is one of uniqueness. Will an adjustment made for a given year remain valid for another? The answer can be found by testing a year other than the one initially studied, as long as the system remains the same and the meteorological station and measurement devices remain unchanged. Experience shows that a model that is adjusted carefully for one or two given years of records usually produces excellent results for all other years.

An anecdote illustrates the utility of the method. In a study by the International Commission for the Protection of the Moselle River Against Pollution, a thermal regime model had been adjusted. Applied for a time series of 13 years, the model gave good results for the first 12 years but very poor results for the last year. It was discovered that in order to produce reasonable results for that year the wind speed would have to be multiplied by a factor of about 2. A careful review of the data revealed an error of this order: the wind speed had been expressed in meters per second rather than knots. Use of the proper units produced a consistent simulation.

Calibration is most often carried out by trial and error when the number of adjustable coefficients is small. Comparisons are made between observed and simulated values of temperature time series with the objective of minimizing the absolute value of the annual average difference between measured and calculated temperatures, and of minimizing the sum of squared errors between measured and calculated temperatures.

Chapter 11 provides additional discussion of the topics of sensitivity testing, calibration, and validation.

5.9. MODEL APPLICATION

Implementation of a temperature simulation model will be illustrated by an application to the Moselle River by Gras and Martin (1976). Their objective was to evaluate the impact of construction of a nuclear power plant a few kilometers upstream of the France–Luxemburg border, i.e. to determine whether the resultant water temperature and temperature increase would conform with the European standards for international streams. An additional consideration was the plan of Luxemburg to construct a second power plant just downstream of the border. The investigation was conducted in five phases.

Phase 1. Selection of Model

European standards for international streams include limiting the fraction of the time for which the temperature standards may be exceeded. On the other hand, the operator of the power plant delivering rejected heat to the stream can limit the load of the plant for short periods. In order to make his operation decisions, he must know accurately the risk he takes. The risk must, therefore, be expressed by the mathematical probability of load reductions. The model chosen for simulation of temperature in the stream must be sufficiently disaggregated to permit analysis of extreme cases and to evaluate the probabilities of their occurrence while at the same time calculating relatively modest changes in temperature. Since one of the criteria for decision making is the value of the extreme temperature, the model must not be biased toward the highest temperatures.

Having considered all of these factors, the investigators studying the Moselle chose a simulation model for nonsteady state conditions, which could be calibrated for the river and would provide statistical results.

Phase 2. Data Collection and Analysis

Two stations of the meteorological network near the Moselle, at Metz and at Nancy, could have been used for the study. Calculation of the natural temperature using data from both stations revealed that the data from Nancy gave somewhat better results. Two distinct periods seem to be evident from the data of the Metz station. This anomaly was attributed to a change of meteorological equipment at the station. Consequently, data from only the Nancy station were retained for the study.

Daily records of flow were available for several locations along the reach studied. Various time series of water temperature data, unfortunately quite dissimilar in the locations and periods covered, were available from stations at Millery (1973–74), Metz (1970–73), La Maxe (1972–73), Palzem (1959–62, 1971–75), Trier (1959–64), Cochem (1960–75), Müden (1971–75), and Güls (1972–74). Since 1964 two relatively large power plants with once-through cooling have been established on the Moselle (KKW1 and KKW2, shown in Figure 5.3). Their discharges disturb the water temperatures downstream, so it was necessary to take into account these disturbances in order to study the coherence of the temperature time series. It had to be established whether there existed a systematic difference that could be attributed either to a real

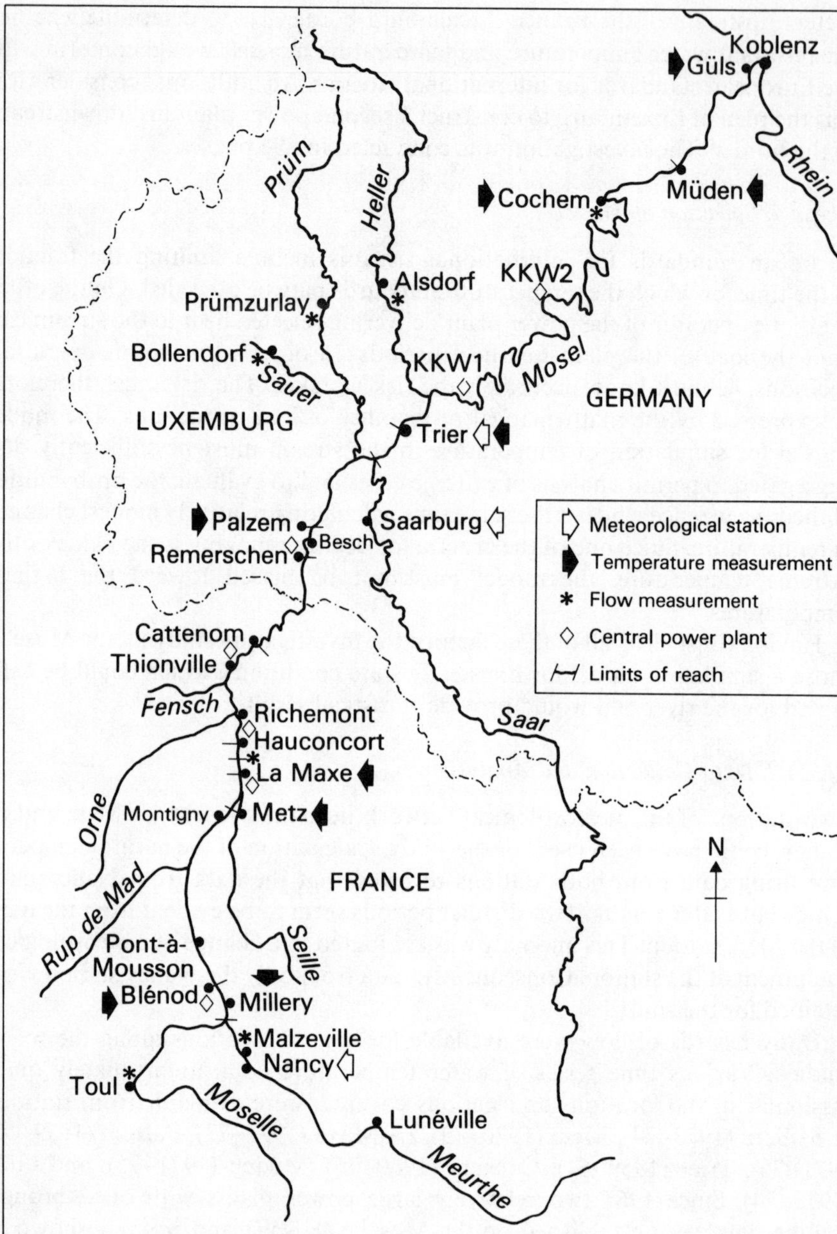

FIGURE 5.3 Sketch of the Moselle River basin, showing the division of the river into reaches.

phenomenon, e.g. climatological modifications along the stream, the effect of a tributary, etc., or to systematically incorrect measurements, and whether the precision of measurements was acceptable in view of the natural dispersion of temperature observations.

To simplify the calculations the Palzem station was chosen as a reference by virtue of its favorable location near the border and about midway along the reach. Observations at the other stations were correlated to those of Palzem over their common periods. Analysis of these correlations revealed that the differences observed between the stations could, in large part, be explained by the effects of the existing power plants. The temperature time series were considered as coherent.

Phase 3. Development of the Model

The Moselle was divided into reaches by taking as endpoints for the calculation sections: sites of power plants; confluences with important tributaries (e.g. the Saar); stream discharge gauging stations; and temperature measurement stations. Figure 5.3 shows how the division was made. For each reach, $H = H(Q)$ and $v = v(Q)$, the average depth and average velocity as functions of discharge, were determined.

Phase 4. Model Calibration

Minimum residuals between calculated and observed values of temperature were obtained by successive adjustment of the evaporation coefficient e and the loop coefficient a. The evaporation coefficient was defined *a priori* to be of the form

$$e(u) = a(1 + u)$$

for the whole set of reaches to which meteorological data from the Nancy station were applied. Figure 5.4 shows the evaluation of these two parameters considered in the adjustment. The minimum of $\sigma/x = y$ corresponds closely to the value of a for which ΔT_m is zero, independently of the measurement station considered.

Experience shows that the loop coefficient can be represented by a function of the air temperature. In other words, it is a term of seasonal character that makes adequate recognition of phenomena like heat exchange with the ground or temperature inputs from small tributaries that are not otherwise explicitly included in the model. For ease of calculation, the authors combined this coefficient with that of atmospheric radiation, which can also be expressed as a function of the air temperature.

Figures 5.5 and 5.6 show comparisons between the calculated and measured temperatures of the Moselle at the France–Luxemburg border for 1972 and 1973, respectively.

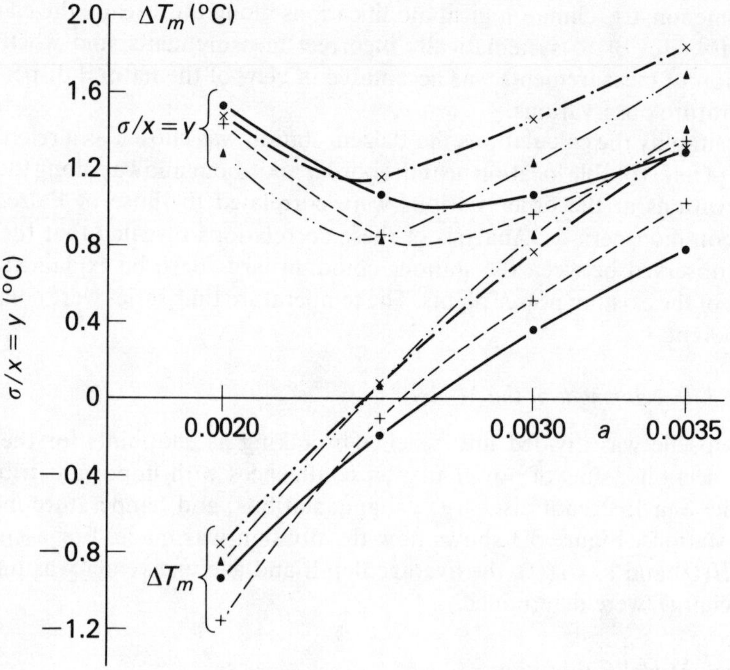

FIGURE 5.4 Determination of the optimum of the evaporation coefficient $e(u) = a(1 + u)$. The data are from stations at Millery (●), Metz (+), La Maxe (×), and Palzem (▲).

Phase 5. Results of Simulation

Simulations of the Moselle to study the influence of power plants were performed using data for the period 1961–73. The primary results of this investigation comprised:

(i) time series of natural stream temperatures at different key stations;
(ii) time series of perturbed stream temperatures downstream of power plants after mixing of cooling water discharges;
(iii) time series of temperatures of effluent discharges from power plants situated along the Moselle;
(iv) time series of temperature increases above the natural state downstream of power plants and at the France–Luxemburg border; and
(v) distribution curves of these parameters, indicating the frequency of occurrence of stipulated levels of temperature and temperature increase.

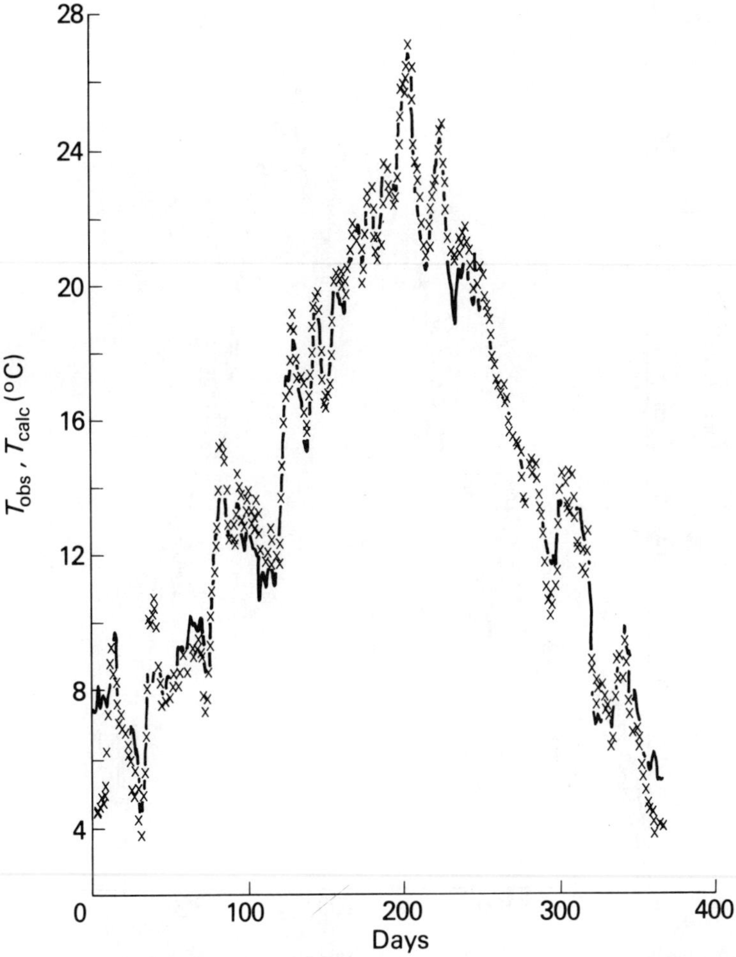

FIGURE 5.5 Observed (broken line) and calculated (×) temperatures of the Moselle River at the France–Luxemburg border in 1972.

It was possible to use this information to evaluate the effects of different regulation hypotheses based on maximum temperatures and allowable increases above the natural state. In the specific case of the power plant at Cattenom (Figure 5.3) it was possible to make an important change in the cooling system to relieve the thermal load on the Moselle. This plant will now be equipped with cooling towers, the blow-down from which will be diverted through a cooling pond of several hundred hectares before discharge to the river. In calculation of the operation of the cooling towers, the meteorological conditions have been simulated, as well as the thermal behavior of the cooling pond.

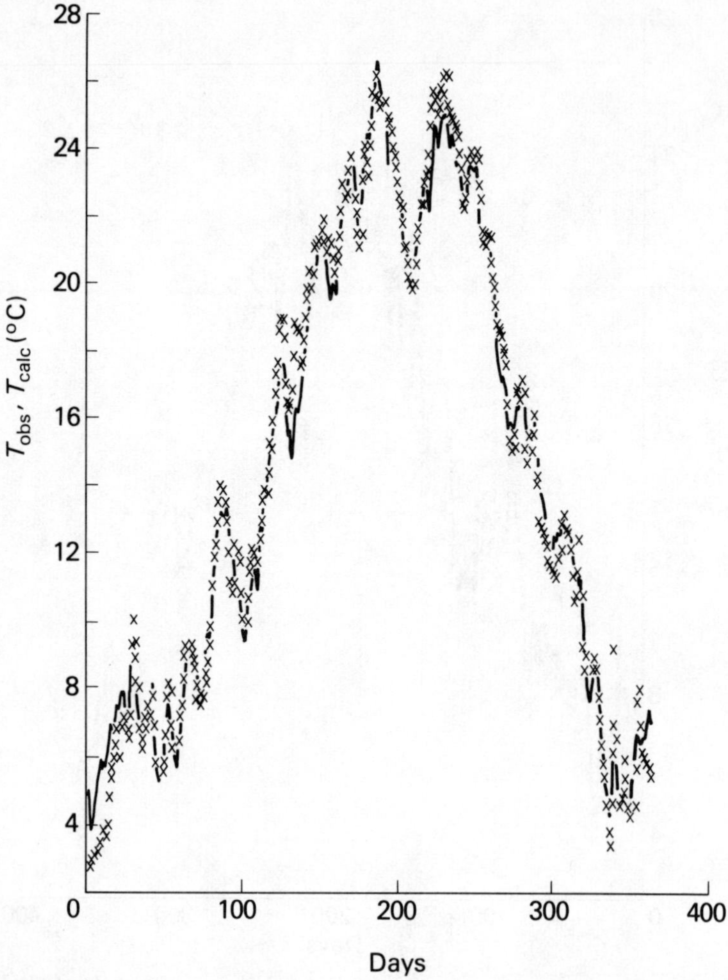

FIGURE 5.6 Observed (broken line) and calculated (×) temperatures of the Moselle River at the France–Luxemburg border in 1973.

APPENDIX. SURFACE HEAT EXCHANGE

Heat exchanges through the air–water interface depend upon both the internal hydromechanical behavior of the water body and the physics of its interaction with the overlying air mass. Meteorological factors such as solar radiation, wind, humidity, pressure, and cloudiness figure prominently in the many physical processes involved. The aggregate effect of the most important of these processes is represented in the general heat flux equation:

$$q_n = q_{sn} + q_{at} + q_{ws} + q_e + q_h, \tag{A5.1}$$

where

q_n is the net energy flux across the air–water interface [J L^{-2} T^{-1}],
q_{sn} is the net solar radiation flux,
q_{at} is the net atmospheric radiation flux,
q_{ws} is the water surface radiation flux,
q_e is the evaporation heat flux,
q_h is the sensible heat flux.

Among the early investigations to formulate explicitly the independent energy flux terms, one of the most comprehensive was undertaken in the late 1960s by the Tennessee Valley Authority Engineering Laboratory. This led to publication of Report 14 (TVA, 1972), from which the following formulations have been derived.

Net Solar Radiation Flux q_{sn}

The net solar radiation flux is the residual flux through the water surface of the solar radiation that is received at the top of the atmosphere and reduced through attenuation by the atmospheric column, interception by clouds, and reflection at the water surface:

$$q_{sn} = q_0 f(A)(1 - R_t)(1 - 0.65C^2), \tag{A5.2}$$

where

q_0 is the solar radiation flux on a horizontal plane at the top of the atmosphere [J L^{-2} T^{-1}],
$f(A)$ is the atmospheric attenuation factor, a function of the optical properties of the air column, dust content, moisture, and terrestrial surface reflectance (Klein, 1948),
R_t is the reflectivity of the water surface ($R_t \approx 0.03$),
C is the degree of cloudiness ($0 < C < 1.0$).

The solar radiation flux at the top of the atmosphere at a particular location and time is given by

$$q_0 = (I_0/r^2)\sin\alpha, \tag{A5.3}$$

where

I_0 is the solar constant, equal to the solar radiation incident normal to the top of the atmosphere when the earth is at its mean distance from the sun [J L^{-2} T^{-1}] ($I_0 \approx 2$ ly min^{-1}),

r is the radius vector, i.e. the ratio of the actual earth–sun distance to the mean distance ($0.9833 < r < 1.0167$ and $0.9669 < r^2 < 1.0337$),

α is the solar altitude (radians), a function of latitude, and of the declination and local hour angle of the sun ($0 < \alpha < \pi/2$).

Net Atmospheric Radiation Flux q_{at}

The atmospheric radiation flux is the net longwave flux from the atmospheric air mass, including clouds, after reflection from the water surface. It is a function of absolute air temperature, cloudiness, and water surface reflectivity:

$$q_{\text{at}} = C_{\text{at}}\, \sigma\, T_2^6\, (1 + 0.17\, C^2)(1 - R_t), \qquad \text{(A5.4)}$$

where

q_{at} is the longwave atmospheric radiation flux from a cloudy sky [$\text{J L}^{-2}\,\text{T}^{-1}$],

C_{at} is an empirical coefficient: $0.906 \cdot 10^{-5} < C_{\text{at}} < 0.999 \cdot 10^{-5}$; Swinbank's coefficient is $0.938 \cdot 10^{-5}$,

σ is the Stefan–Boltzmann constant, $2.041 \cdot 10^{-3}\ \text{J m}^{-2}\,\text{h}^{-1}\,\text{K}^{-4}$,

T_2 $= 273.2 + \theta_{a2}$ is in kelvins; θ_{a2} is the dry bulb air temperature 2 m above the ground (°C).

Water Surface Radiation Flux, q_{ws}

The water surface radiation flux is the longwave flux [$\text{J L}^{-2}\,\text{T}^{-1}$] from the water mass emitting as a black body at a specific surface temperature. It is a function of water surface emissivity and surface temperature:

$$q_{\text{ws}} = -\,\epsilon_{\text{w}}\, \sigma\, T_{\text{a}}^4, \qquad \text{(A5.5)}$$

where

ϵ_{w} is the emissivity of the water surface, i.e. the ratio of the radiation emission of the water surface to that of a perfect black body; $\epsilon_{\text{w}} \approx 0.96$ (dimensionless),

T_a $= 273.2 + \theta_{\text{a}}$ is in kelvins; θ_{a} is the average water surface temperature (°C).

Evaporation Heat Flux q_e

Heat is lost from the water surface in the form of the latent heat of evaporation:

$$q_{\text{e}} = -\,\rho_{\text{w}}\, L\, E, \qquad \text{(A5.6)}$$

where

ρ_{w} is the density of water [$\text{force} \cdot \text{T}^2\,\text{L}^{-4}$ or M L^{-3}]: $\rho_{\text{w}} = 998.2\ \text{kg m}^{-3}$ at 20°C,

L is the latent heat of evaporation [cal M^{-1}]: $L = 597.1 - 0.57\,\theta_0$ (cal g^{-1}),

E is the evaporation rate [L T^{-1}].

The TVA Engineering Laboratory study provides an extensive review of various evaporation formulas, concluding that the Marciano–Harbeck formula, derived from detailed investigation of evaporation on Lake Hefner in Oklahoma (Marciano and Harbeck, 1954), probably gives the most consistent practical results from standard weather observations. This formula (converted to metric units) is

$$E = 4.33 \cdot 10^{-5}\, u(e_0 - e_a), \tag{A5.7}$$

where

E is the evaporation rate (mm day^{-1}),
u is the wind velocity (km h^{-1}),
e_0 is the saturation vapor pressure at the water surface temperature (mm Hg),
e_a is the vapor pressure of air (mm Hg).

Adjusting for observations made at specific elevations above the water surface yields

$$E = 7.44 \cdot 10^{-5}\, u_4(e_0 - e_2), \tag{A5.8}$$

where the subscripts 2 and 4 refer to observations made at 2 and 4 m above the surface.

An alternative form of the Lake Hefner expression that takes account of stability of the overlying air mass is

$$E = N(P/P_0)u(C_0 - C_a)f(\mathrm{Ri}), \tag{A5.9}$$

where

E is the evaporation rate (m h^{-1}),
N is the evaporation coefficient (dimensionless),
P is the atmospheric pressure (mbar),
P_0 is the standard atmospheric pressure at sea level, 1013 mbar; P/P_0 may be approximated by $P/P_0 = (1 - 0.0226z)^{5.256}$ with z in kilometers above mean sea level,
u is the wind speed (km h^{-1}),
C_0 $= 0.622\, e_0/P$ is the water vapor concentration of the saturated air at the water surface (kg vapor/kg air),
e_0 is the vapor pressure of saturated air (mbar),
e_a is the vapor pressure of air (saturation vapor pressure at dewpoint) (mbar),
$f(\mathrm{Ri})$ is a function of the Richardson number:

$$f(\mathrm{Ri}) = \begin{cases} (1 - 22\mathrm{Ri})^{0.80} & \text{for } 0 \geq \mathrm{Ri} \geq -1 \\ (1 + 34\mathrm{Ri})^{-0.80} & \text{for } 0 \leq \mathrm{Ri} \leq 2 \\ 1 & \text{for } \mathrm{Ri} = 0 \text{ (neutral case)} \end{cases}$$

$$\mathrm{Ri} = -\frac{g(\rho_a - \rho_0)z}{\rho_a u^2},$$

where

g	is the acceleration due to gravity [$L\,T^{-2}$],
ρ_a	is the density of air (usually at $z = 2$ m) [$M\,L^{-3}$],
ρ_0	is the density of saturated air at the water surface temperature [$M\,L^{-3}$],
u	is the wind speed (usually at $z = 2$ m) [$L\,T^{-1}$],
z	is the height above the water surface [L].

Values of N in (A5.9) depend on particular conditions of meteorological observations:

for airport data,

$$N = 1.17 \cdot 10^{-6}, \qquad z = 5.5\,\text{m};$$

for land/lake data,

$$N = 1.40 \cdot 10^{-6}, \qquad z = 2\,\text{m};$$

P and e_a observed at an upwind land station are unaffected by lake evaporation;

for lake data,

$$N = 1.54 \cdot 10^{-6}, \qquad z = 2\,\text{m}.$$

Sensible Heat Flux q_h

Sensible heat is transferred between air and water by conduction and transported from or toward the air–water interface by convection associated with the moving air mass:

$$q_h = -f(\text{Ri})\,\rho_w\,c_p\,N(P/P_0)u(\theta_0 - \theta_a), \qquad \text{(A5.10)}$$

where

q_n	is the forced convective sensible heat flux [$J\,L^{-2}\,T^{-1}$],
θ_0	is the water surface temperature (°C),
θ_a	is the dry bulb air temperature (°C),
c_p	is the specific heat of air at constant pressure [$J\,M^{-1}\,°C^{-1}$].

Other Heat Losses

In addition to the five principal heat fluxes cited in (A5.1), heat may be transferred to the water by condensation from an overlying supersaturated air mass or into or out of the system by advective transfers associated with precipitation, evaporation, and/or condensation. The condensation flux is of the same general form as the combination resulting from (A5.6) and (A5.9) for evaporation, except that C_a and C_0 are interchanged and C_a is required to be less than the saturation vapor concentration of the air. The heat gain by condensation is usually relatively small, so it is often neglected in energy calculations or considered as a part of the net evaporative heat loss. Heat exchanges associated with advective transport of water through the interface are usually neglected except in the most detailed studies.

REFERENCES

Commission Internationale pour la Protection du Rhin Contre la Pollution (1973–74). Plan de charge thermique du Rhin. *Rapports Internes du Sous-groupe: Eaux de Refroidissement du Groupe de Travail T.*

Fleig, H., and D. Flinsbach (1973). Wärmelastplan Neckar. Ministerium für Ernährung, Landwirtschaft, und Umwelt, Baden-Württemberg.

Gras, R. (1970). Estimation des éléments du bilan radiatif pour le calcul des températures naturelles d'un secteur d'un cours d'eau. *Electricité de France Direction des Etudes et Recherches, Centre de Chatou Rapport* HF 41-70, No. 26.

Gras, R., and J. Jacquet (1971). Simulation du comportement d'un réseau hydrographique équipé d'un ensemble de centrales thermiques, application au bassin de la Seine. *XIes Journées de l'Hydraulique Paris, 1970, Hydrotechnical Society of France, Question IV Rapport* 11, p. 10.

Gras, R., and P. Martin (1976). Modèle de simulation du comportement thermique de la Moselle. *Electricité de France Direction des Etudes et Recherches, Centre de Chatou Rapport* E 31-76, No. 18.

Hydrologisches Institut München (1976). Wärmelastplan Donau. Bayrisches Landesamt für Wasserversorgung und Gewässerschutz.

Jobson, H. (1977). Bed conduction computation for thermal models. *Proceedings of American Society of Civil Engineers, Journal of Hydraulics Division* 103(HY10): 1213–1237.

Kahlig, P. (1977). Zur Exponentialmethode der Flußtemperaturberechnung. *Archiv für Meteorologie, Geophysik und Biologie, Serie* A 25: 357–365.

Kaisersöt, K., and H. Mitschel (1977). Ein Beitrag für die Berechnung des Wärmeaustauschkoeffizienten bei der Abkühlung eines künstlich erwärmten Flußes. *Brennstoff-Wärme-Kraft* 29(6): 250–255.

Klein, P., and P. Momal (1979). Présentation et mise en oeuvre numérique d'une méthode d'estimation du rayonnement solaire. *Electricité de France Direction des Etudes et Recherches, Centre de Chatou Rapport* E 31-79, No. 15.

Klein, W. H. (1948). Calculation of solar radiation and the solar heat. *Journal of Meteorology* 5, No. 4.

Länderarbeitsgemeinschaft Wasser (1971). *Grundlage für die Beurteilung der Wärmebelastung von Gewässern* (Wiesbaden: Kochler und Hermann).

Marciano, T. T., and G. E. Harbeck, Jr. (1954). Mass transfer studies. *Water Loss Investigation, Lake Hefner Studies, Technical Report, United States Geological Survey Professional Paper* 269.

Raphael, J. M. (1962). Prediction of temperature in rivers and reservoirs. *Proceedings of American Society of Civil Engineers, Journal of Power Division* 88(PO2): 157–181.

Smitz, J. S. (1975). Modèle mathématique de la température naturelle des rivières. *Bulletin of Royal Society for Science, Belgium.*

Sweers, H. E. (1976). A nomogram to estimate the heat exchange coefficient at the air–water interface as a function of wind speed and temperature, a critical survey of some literature. *Journal of Hydrology* 30: 375–401.

Tennessee Valley Authority (1972). Heat and mass transfer between a water surface and the atmosphere. *TVA Water Resources Research Engineering Laboratory, Norris, TN Report* 14.

Wunderlich, W. O., and L. W. Hsieh (1971). The computation of daily mean stream temperatures. *TVA Water Resources Research Engineering Laboratory, Norris, TN Report* 25.

6 Stream Quality Modeling

M. J. Gromiec, D. P. Loucks, and G. T. Orlob

6.1. INTRODUCTION

The achievement of regional water quality goals, especially in the more developed areas of the world, often involves substantial capital investments and changes in public attitudes concerning resource management. Economic impacts may include not only the cost of facilities designed to reduce the discharge of contaminants into natural waters or to improve the quality of waters receiving waste, but also any limitations on economic development in a particular region or river basin. Those responsible for the formulation and approval of water quality plans or management policies must have a means of estimating and evaluating the temporal and spatial economic, environmental, or ecological impacts of these plans and policies. This need has stimulated the development and application of a wide range of mathematical modeling techniques for predicting various impacts of alternative pollution control plans and policies.

There are many different types of stream quality model. The appropriate model and the required data depend on the purpose of the specific study. Long-range regional water quality planning does not require the detail that is appropriate, for example, when evaluating a single proposed industrial waste outfall or discharge site. There is no best single water quality model for all streams and for all planning situations. An important decision that must be made early in the planning process is the selection of the modeling method or methods appropriate for planning and capable of development, calibration, validation, and execution within the limits of available time and money.

Most stream quality predictive models in use today apply to streams receiving wastewater from point sources. As the quantities of wastes discharged from point sources are reduced, nonpoint or distributed sources of wastewater from agricultural and urban runoff become increasingly important. Models are needed to help predict nonpoint source waste inputs to surface waters. The outputs of these nonpoint source wastewater generation models provide the inputs to water quality models of the receiving stream. For example, urban stormwater management models are often integrated from various models for runoff, sewer routing, and prediction of the quality of the waste-receiving water.

This chapter reviews a number of the more typical stream quality predictive models developed for and applied to waste-receiving streams. The models range from the simple to the complex, yet each has proven effective in certain planning situations. The inclusion of various water quality management alternatives, and their costs, within these predictive models will be reviewed in Chapter 12.

6.2. TYPES OF STREAM QUALITY MODEL

It is useful to distinguish between certain types of model and to discuss their characteristics. This will provide an opportunity to define a few of the terms and to illustrate some of the concepts used by those who develop and apply water quality models. Many of the models in use are extensions of two simple equations proposed by Streeter and Phelps in 1925 for predicting the biochemical oxygen demand (BOD) of various biodegradable constituents, and the resulting dissolved oxygen concentration (DO) in rivers (Thomann, 1972). Often used with these BOD–DO models are other fairly simple first-order exponential decay, dilution, and sedimentation models for additional nonconservative and conservative substances.

More complex multiconstituent water quality models have also been proposed and have been applied to predict the physical, chemical, and biological interactions of many constituents and organisms found in natural water bodies. These multiconstituent simulation models generally require more data and computer time, but they also can provide much more detailed and comprehensive information on the quantity and quality of water resulting from various water and land management policies.

Water quality models can be used to evaluate steady state conditions, for which the values of the water quality and quantity variables do not change with time. They can also be used to evaluate dynamic or time-varying conditions. The latter type of model permits an evaluation of transient phenomena such as nonpoint stormwater runoff and spills of pollutants. Steady state models are usually simpler and require less computational effort than dynamic or transient models, and are more relevant to long-term planning than to short-term management and control.

Assumptions pertaining to the mixing of pollutants in water bodies dictate the spatial dimensions of the model. Sufficient accuracy may be obtained in many river systems by modeling only one or two dimensions. One-dimensional models of river systems assume complete vertical and lateral mixing. One-dimensional lake models usually assume complete mixing in all but the vertical direction. Two-dimensional models may assume either lateral mixing, as in stratified estuaries or lakes, or vertical mixing, as in relatively shallow and wide rivers.

Undoubtedly, the most data-demanding model type is the stochastic or probabilistic model as compared with its deterministic counterpart. Most deterministic models yield estimates of mean values of various quality constituents, whereas probabilistic models explicitly take into account the randomness or uncertainty of various physical, biological, or chemical processes. Validation of stochastic models is especially difficult because of the quantity of data on the water body that is necessary for comparing probability distributions of variables rather than just their expected or mean values. This introduction will be confined to a review of one-dimensional deterministic models of steady and unsteady state flow. These relatively simple models are used for long- and short-term planning, management, and control of water quality during periods when complete mixing exists in the other two dimensions and flows are constant.

6.3. DISSOLVED OXYGEN AND BIOCHEMICAL OXYGEN DEMAND

In connection with a study of the Ohio River, Streeter and Phelps (1925) developed the first important water quality model describing the BOD–DO relationship in a stream. In their pioneering work the simplest system was considered, in which biodegradable waste is discharged to the stream and consumes oxygen, atmospheric reaeration being the only source of oxygen. The rate of change in the dissolved oxygen deficit, $\mathrm{d}D/\mathrm{d}t$, was assumed to be directly proportional to the unsatisfied oxygen demand and to the oxygen deficit in the stream. Therefore, the differential equation for the process can be written as

$$\frac{\mathrm{d}D}{\mathrm{d}t} = K_{\mathrm{d}} L - K_{\mathrm{a}} D, \tag{6.1}$$

where

- D is the oxygen saturation deficit $[\mathrm{M\ L^{-3}}]$ or the difference between the DO saturation concentration and the concentration at time t [T],
- L is the carbonaceous biochemical oxygen demand $\mathrm{BOD^c}$ $[\mathrm{M\ L^{-3}}]$,
- K_{d} is the deoxygenation rate coefficient $[\mathrm{T^{-1}}]$,
- K_{a} is the reaeration rate coefficient $[\mathrm{T^{-1}}]$.

The solution to (6.1) is

$$D_t = \frac{K_d L_0}{K_a - K_d}[\exp(-K_d t) - \exp(-K_a t)] + D_0 \exp(-K_a t) \tag{6.2}$$

or

$$D_t = \frac{k_d L_0}{k_a - k_d}(10^{-k_d t} - 10^{-k_a t}) + D_0 10^{-k_a t}, \tag{6.3}$$

where

D_t is the dissolved oxygen saturation deficit after time t,
L_0 is the oxygen demand at an initial reference time ($t = 0$),
D_0 is the dissolved oxygen deficit at $t = 0$,

$$k_d = 0.434K_d, \qquad k_a = 0.434K_a.$$

Equation 6.2 or 6.3 is the Streeter–Phelps "oxygen sag formula," and a profile of DO along the stream is referred to as a "dissolved oxygen sag curve" (Figure 6.1). With D_0, L_0, K_d, K_a known the deficit D_t may be computed at any time t, and when D_0, D_t, K_d, L_0 are given (6.2) or (6.3) may be solved for K_a.

In many cases, the main interest is directed to the critical point, which is determined by the critical deficit D_c and critical time t_c. At this point there is no

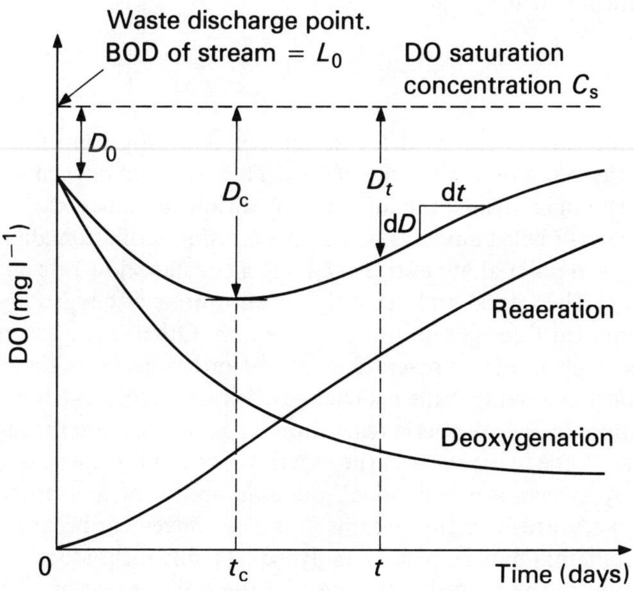

FIGURE 6.1 Dissolved oxygen sag curve.

change in the deficit (reaeration balances deoxygenation):

$$\frac{dD}{dt} = K_d L - K_a D = 0. \tag{6.4}$$

Therefore, the critical deficit is

$$D_c = \frac{K_d}{K_a} L = \frac{K_d}{K_a} L_0 \exp(-K_d t_c) \tag{6.5}$$

or

$$D_c = \frac{k_d}{k_a} L_0 10^{-k_d t_c}, \tag{6.6}$$

and the critical time can be expressed as

$$t_c = \frac{1}{k_a - k_d} \lg\left[\frac{k_a}{k_d}\left(1 - \frac{D_0(k_a - k_d)}{L_0 k_d}\right)\right]. \tag{6.7}$$

Fair (1939) introduced a new constant $f = k_a/k_d$, called the self-purification factor, and defined the critical time by

$$t_c = \frac{1}{k_d(f-1)} \lg\left[f\left(1 - (f-1)\frac{D_0}{L_0}\right)\right] \tag{6.8}$$

and the critical deficit by

$$D_c = \frac{L_0}{f} 10^{-k_d t_c}. \tag{6.9}$$

These relationships were used for studies of the self-purification of natural streams. t was taken to be the time of flow. The dissolved oxygen sag equations are also useful in analyzing the influence of various parameters.

The Streeter–Phelps model is based on a constant pollution load discharged at a single given point along a stream having a constant flow rate and a uniform cross section. The lateral and vertical concentrations of oxygen and BOD are assumed uniform throughout any cross section. Other basic assumptions are that the deoxygenation and reaeration are first-order reactions, that the reaction rates are constant, and that the net change of oxygen deficit is a function only of deoxygenation and atmospheric reaeration by gas absorption through a liquid–air interface. While in many cases the kinetics of the BOD reaction can be better described by a second-order model, the assumption of first-order kinetics is reasonably accurate for many streams. It is also convenient because it describes a linear system that may be readily analyzed by many techniques.

In addition to the bacterial oxidation of the organic matter and the atmospheric reaeration considered in the Streeter–Phelps model, numerous processes are taking place in natural streams. Dobbins (1964) summarized some of these processes:

(1) the reduction of BOD by sedimentation or adsorption;
(2) the increase of BOD from scoured bottom deposits or from the diffusion of partly decomposed organic products from the benthic layer into the water above;
(3) the increase of BOD along the stream from runoff;
(4) the removal of oxygen from the water by diffusion into the benthic layer to satisfy the oxygen demand in the aerobic zone of this layer;
(5) the removal of oxygen from the water by the purging action of gases rising from the benthic layer;
(6) the addition of oxygen by the photosynthetic action of plankton and fixed plants;
(7) the removal of oxygen by respiration of plankton and fixed plants;
(8) the continuous redistribution of both BOD and DO by longitudinal dispersion.

Frankel and Hansen (1968) stated that the following factors should also be considered:

(9) the variation of K_d with time, particularly at the onset of nitrification, which precludes assuming K_d to be constant for larger values of t (time of travel);
(10) changes in channel configuration that alter the characteristics of surface turbulence and consequently the rate of transfer of oxygen from the atmosphere;
(11) the effects of suspended and dissolved substances upon the rate of diffusion of oxygen from the surface into the main body of the stream;
(12) diurnal variation in oxygen content, BOD, temperature, and flow rate of influent discharges.

Various modifications of the Streeter–Phelps model have been proposed to take into account some of these processes. Such models, used to describe the BOD profiles along a river reach, have been based on various assumptions. A brief summary of some of these models is presented below.

Thomas (1948) pointed out that part of the BOD can be removed by sedimentation without consumption of the dissolved oxygen, and that the removal rate is directly proportional to the remaining BOD. Therefore,

$$\frac{\mathrm{d}L}{\mathrm{d}t} = -(K_d + K_s)L, \tag{6.10}$$

where K_s is the rate constant for BOD removal by sedimentation [T^{-1}]. Then

$$L_t = L_0 \exp[-(K_d + K_s)t] \tag{6.11}$$

and

$$D_t = \frac{K_d L_0}{K_a - (K_d + K_s)} \{\exp[-(K_d + K_s)t] - \exp(-K_a t)\} + D_0 \exp(-K_a t). \tag{6.12}$$

In this case, the amount of bottom sediment is assumed to be small and the reaction rate so low that no measurable oxygen is drawn from the sediment; photosynthesis is not considered.

Camp (1963) proposed that the BOD and DO profiles can be described by the following differential equations:

BOD profile,

$$\frac{\mathrm{d}L}{\mathrm{d}t} = -(K_d + K_s)L + B; \tag{6.13}$$

DO profile,

$$\frac{\mathrm{d}D}{\mathrm{d}t} = -K_a D + K_d L - P, \tag{6.14}$$

where

- B is the rate of addition of BOD to the overlying water from the bottom deposits [$M\,L^{-3}\,T^{-1}$],
- P is the rate of oxygen production in the euphotic zone by photosynthesis [$M\,L^{-3}\,T^{-1}$].

The solutions of these equations are

$$L_t = \left(L_0 - \frac{B}{K_d + K_s}\right)\exp[-(K_d + K_s)t] + \frac{B}{K_d + K_s} \tag{6.15}$$

and

$$D_t = \frac{K_d}{K_a - (K_d + K_s)}\left(L_0 - \frac{B}{K_d + K_s}\right)\{\exp[-(K_d + K_s)t] - \exp(-K_a t)\}$$
$$+ \frac{K_d}{K_a}\left(\frac{B}{K_d + K_s} - \frac{P}{K_d}\right)[1 - \exp(-K_a t)] + D_0 \exp(-K_a t). \tag{6.16}$$

If the BOD is added to the overlying water from the bottom sediments, and the reduction of BOD by settling is negligible, these equations can be presented as

$$L_t = \left(L_0 - \frac{B}{K_d}\right)\exp(-K_d t) + \frac{B}{K_d} \tag{6.17}$$

and

$$D_t = \frac{K_d}{K_a - K_d}\left(L_0 - \frac{B}{K_d}\right)[(\exp(-K_d t) - \exp(-K_a t)]$$

$$+ \frac{B - P}{K_a}[1 - \exp(-K_a t)] + D_0 \exp(-K_a t). \tag{6.18}$$

If BOD is reduced by settling, but benthic demand has no measurable effect on the oxygen deficit of the stream, then (6.16) becomes

$$D_t = \frac{K_d L_0}{K_a - (K_d + K_s)} \{\exp[-(K_d + K_s)t] - \exp(-K_a t)\}$$

$$- \frac{P}{K_a}[1 - \exp(-K_a t)]. \tag{6.19}$$

If K_s, B, and P are zero, (6.18) becomes the original Streeter–Phelps oxygen sag equation.

In addition, Camp (1963) considered a stream with significant longitudinal mixing and in which BOD and DO do not vary with the temporal mean depth. An element of unit width, length, and depth was considered, and the equations for the BOD and DO profiles in the x direction at steady state were written as:

BOD profile,

$$E\frac{d^2L}{dx^2} - U\frac{dL}{dx} - (K_d + K_s)L + B = 0; \tag{6.20}$$

DO profile,

$$E\frac{d^2C}{dx^2} - U\frac{dC}{dx} + K_a(C_s - C) - K_d L + P = 0, \tag{6.21}$$

where

E is the turbulent transport (longitudinal mixing) coefficient $[L^2 T^{-1}]$,
U is the average stream velocity $[L T^{-1}]$,
C is DO $[M L^{-3}]$,
C_s is DO at saturation $[M L^{-3}]$.

O'Connor (1962) assumed that in natural streams mass transport by turbulent diffusion (i.e. longitudinal mixing) is insignificant. Assuming only deoxygenation by organic matter oxidation and atmospheric reaeration, he defined the nonsteady state distribution of dissolved oxygen by

$$\frac{\partial C}{\partial t} = U\frac{\partial C}{\partial x} - K_d L - K_a(C_s - C). \tag{6.22}$$

Under the steady state condition $\partial C/\partial t = 0$, (6.22) reduces to

$$U\frac{dC}{dx} - K_d L - K_a(C_s - C) = 0. \tag{6.23}$$

Dividing through by U gives

$$\frac{dC}{dx} - \frac{K_d}{U}L + \frac{K_a}{U}(C_s - C) = 0. \tag{6.24}$$

If the variable L is expressed as a function of the downstream distance x, where $L = L_0 \exp(-K_r x/U)$, and K_r is the coefficient of BOD removal in the stream, which may be different from that of oxidation, K_d, (6.24) may be integrated to

$$C = C_s - \frac{j_d L_0}{j_a - j_r}[\exp(-j_r x) - \exp(-j_a x)] - (C_s - C_0)\exp(-j_a x). \tag{6.25}$$

The boundary condition for (6.24) is $C = C_0$ at $x = 0$, and the coefficients are defined as $j_a = K_a/U$, $j_d = K_d/U$, $j_r = K_r/U$. The units of the coefficients are in terms of distance, i.e. L^{-1}.

The following assumptions were made by Dobbins (1964):

(1) The stream flow is steady and uniform.
(2) The process for the stretch is in a steady state; the conditions at every cross section do not change with time.
(3) BOD removal by bacterial oxidation and by sedimentation and/or adsorption are first-order reactions; the rates of removal at any cross section are proportional to the BOD present.
(4) The removal of oxygen by benthic processes and by plant respiration, the addition of oxygen by photosynthesis, and the increase of BOD from the benthic layer or local runoff are all uniform along the stretch.
(5) The BOD, L, and DO, C, are uniformly distributed over each cross section, thus permitting the equations to be written in the usual one-dimensional form.

Using these assumptions, Dobbins derived a set of two differential equations describing the BOD and DO profiles as functions of distance downstream:

BOD profile,

$$E\frac{d^2L}{dx^2} - U\frac{dL}{dx} - (K_d + K_s)L + L_{add} = 0; \tag{6.26}$$

DO profile,

$$E\frac{d^2C}{dx^2} - U\frac{dC}{dx} + K_a(C_s - C) - K_d L - D_b = 0, \tag{6.27}$$

where

L_{add} is the rate of increase of BOD along the stretch [$M\ L^{-3}\ T^{-1}$],
D_b is the rate of removal of oxygen caused by benthic demand and plants [$M\ L^{-3}\ T^{-1}$].

The Camp–Dobbins models take no account of the nitrification stage, since it was assumed that all deoxygenation occurs as first-stage deoxygenation.

Hansen and Frankel (1965) suggested an important modification of Camp–Dobbins models. They assumed that the classical pattern of diurnal DO profiles in a stream can be represented by a periodic function. Therefore, if factors involving BOD are neglected, (6.14) may be written as

$$\frac{dD}{dt} = -K_a D + P_m \cos(\omega t + \phi), \tag{6.28}$$

in which P_m is the maximum rate of oxygen production/consumption by photosynthesis/respiration [$M\ L^{-3}\ T^{-1}$]. Equation 6.28 can be integrated to

$$D_t = [D_0 - a(K_a \cos\phi + \omega \sin\phi)]\exp(-K_a t)$$
$$+ a[K_a \cos(\omega t + \phi) + \sin(\omega t + \phi)], \tag{6.29}$$

in which $a = P_m/(\omega^2 + K_a^2)$, ω is the frequency, and ϕ the phase. The maximum or minimum deficit (depending on the sign of the function) is given by $D_m = (P_m/\omega)\sin\alpha = (P_m/K_a)\cos\alpha$, where $\alpha = \omega t + \phi = \tan^{-1}(\omega/K_a)$.

It was proposed that the maximum rate [$M\ L^{-2}$] of oxygen production/consumption should be related to the solar energy input, the mean depth of the stream, and turbidity. The rate of addition of BOD to overlying water from the benthic layer, B [$M\ L^{-3}\ T^{-1}$], was assumed to represent the rate of addition through only physical mechanisms; therefore,

$$\frac{1}{H}\frac{dL_d}{dt} = -K_b L_d - B, \tag{6.30}$$

where

L_d is the total areal BOD of the benthos [$M\ L^{-2}$],
H is the depth of the stream [L],
K_b is the areal demand rate constant [T^{-1}].

Integration of (6.30) results in

$$\frac{L_{d_t}}{H} = \left(\frac{L_{d_0}}{H} + \frac{B}{K_b}\right)\exp(-K_b t) - \frac{B}{K_b}. \tag{6.31}$$

The rate of change of BOD from suspended and dissolved matter, L_s, was given by Frankel and Hansen (1968) as

$$\frac{dL_s}{dt} = -KL_s + B, \tag{6.32}$$

which can be integrated to

$$L_{s_t} = \left(L_{so} - \frac{B}{K}\right)\exp(-Kt) + \frac{B}{K}, \tag{6.33}$$

where

$K \;= K_d + K_r + K_s\,[\mathrm{T}^{-1}]$,
K_d is the laboratory rate of BOD increase by deoxygenation,
K_r is the river rate of BOD increase by deoxygenation due to attached aquatic growths and slimes,
K_s is the rate of BOD reduction by sedimentation and/or adsorption.

Also, when sedimentation occurs rather than scour, the fractional change in suspended BOD that is due to sedimentation was expressed by the ratio K_s/K, and

$$\frac{1}{H}\frac{dL_d}{dt} = -\frac{K_s}{K}\frac{dL_s}{dt}. \tag{6.34}$$

The total oxygen deficit resulting from biological oxidation of the suspended and dissolved BOD load, oxygen consumption by anaerobic decomposition products of the benthos, "decay" of the initial oxygen deficit, and production/consumption of oxygen by photosynthesis/respiration is expressed as

$$\begin{aligned} D_t = {} & \frac{(K_d + K_r)(1 + K_s/K)}{K_a - K}\left(L_{so} - \frac{B}{K}\right)[\exp(-Kt) - \exp(-K_a t)] \\ & + \frac{K_b}{K_a - K_b}\left(\frac{L_{do}}{H} + \frac{B}{K_b}\right)[\exp(-K_b t) - \exp(-K_a t)] \\ & + [D_0 - a(K_a \cos\phi + \omega \sin\phi)]\exp(-K_a t) \\ & + a[K_a \cos(\omega t + \phi) + \omega \sin(\omega t + \phi)]. \end{aligned} \tag{6.35}$$

O'Connor and Di Toro (1968, 1970) assumed that the temporal form of the photosynthetic oxygen source in streams may be represented by a half-cycle sine wave. The daily rate of photosynthetic oxygen production as a function of time, $P(t)$, was expressed as

$$\begin{aligned} P(t) &= P_m \sin[(\pi/p)(t - t_s)] \quad && \text{when } t_s \le t \le t_s + p \\ P(t) &= 0 && \text{when } t_s + p \le t \le t_s + 1, \end{aligned} \tag{6.36}$$

where

t_s is the time, expressed as a fraction of the day, at which the source becomes active,

p is the fraction of the day during which the source is active (period of sunlight).

A periodic extension of the temporal and spatial variations of photosynthetic oxygen production was expressed as a Fourier series:

$$P(x, t) = P_m\left(\frac{2p}{\pi} + \sum_{n=1}^{\infty} b_n \cos[2\pi n(t - t_s - p/2)]\right), \tag{6.37}$$

in which

$$b_n = \frac{4\pi/p}{(\pi/p)^2 - (2\pi n)^2} \cos n\pi p.$$

The following equation for the DO deficit, $D(x, t) = C_s - C(x, t)$, was set forth:

$$\frac{\partial D(x, t)}{\partial t} = -\frac{Q}{A}\frac{\partial D(x, t)}{\partial x} - K_a D(x, t) + K_d L(x) + K_n N(x) + S(x) + R(x) - P(x, t), \tag{6.38}$$

where

A is the cross-sectional area $[L^2]$,

Q is the flow rate $[L^3\,T^{-1}]$,

U $= Q/A$ is the stream velocity $[L\,T^{-1}]$,

$L(x)$ is the concentration of carbonaceous BOD at x $[M\,L^{-3}]$,

$N(x)$ is the concentration of nitrogenous BOD at x $[M\,L^{-3}]$,

K_d is the deoxygenation rate coefficient of carbonaceous BOD $[T^{-1}]$,

K_n is the deoxygenation rate coefficient of nitrogenous BOD, which reflects both the removal of ammonia and the utilization of oxygen $[T^{-1}]$,

K_a is the reaeration rate coefficient $[T^{-1}]$,

$S(x)$ is the benthic respiration rate (sink) $[M\,L^{-3}\,T^{-1}]$,

$R(x)$ is the algal respiration rate (sink) $[M\,L^{-3}\,T^{-1}]$,

$P(x, t)$ is the algal photosynthetic oxygen production rate (source) $[M\,L^{-3}\,T^{-1}]$.

The respiration sinks in (6.38) were assumed to be constant. The solution of (6.38) may be expressed as

$$\begin{aligned} D(x, t) = {} & D_0(t - x/U)\exp(-j_a x) + F_{d,r}[\exp(-j_r x) - \exp(-j_a x)] \\ & + F_{n,r}[\exp(-j_n x) - \exp(-j_a x)] \\ & + (S/K)[1 - \exp(-j_a x)] + (R/K)[1 - \exp(-j_a x)] \\ & - P_m\{(2p/\pi K_a)[1 - \exp(-j_a x)] + f(x, t)\}, \end{aligned} \tag{6.39}$$

where

$$j_a = \frac{K_a}{U}, \qquad j_r = \frac{K_r}{U}, \qquad j_n = \frac{K_n}{U}$$

$$F_{d,r} = \frac{K_d L_0(t - x/U)}{K_a - K_r}, \qquad F_{n,r} = \frac{K_n N_0(t - x/U)}{K_a - K_r}$$

$$f(x, t) = \sum_{n=1}^{\infty} \frac{b_n}{[K_a^2 + (2\pi n)^2]^{1/2}} \cos\left[2\pi n\left(t - t_s - \frac{p}{2}\right) - \tan^{-1}\frac{2\pi n}{K_a}\right]$$

$$- \exp(-j_a x) \sum_{n=1}^{\infty} \frac{b_n}{[K_a^2 + (2\pi n)^2]^{1/2}}$$

$$\cdot \cos\left[2\pi n\left(t - t_s - \frac{p}{2} - \frac{x}{U}\right) - \tan^{-1}\frac{2\pi n}{K_a}\right].$$

Equation 6.39 was considered to be applicable to streams in which the primary cause of the diurnal variation in DO is the algal oxygen production. However, it was recognized that larger, rooted plants may also be important in some cases.

Appendix 6.1 presents several rate coefficients for the BOD decay processes that are frequently of major importance in river water quality modeling. Many excellent reviews of stream reaeration have been written (Bennett and Rathbun, 1972; Lau, 1972a, b; Wilson and MacLeod, 1974; Rathbun, 1977). A detailed survey of formulas for the reaeration coefficient is summarized in Appendix 6.2.

6.4. DOSAG I AND QUAL I

DOSAG I is one of two stream quality models developed by the Texas Water Development Board (1970a) to simulate the spatial and temporal variations in BOD and DO under various conditions of flow and temperature. The other model, QUAL I, was designed to simulate the spatial and temporal variations in water temperature and conservative mineral concentration in addition to BOD and DO. These programs may be used as complements to each other. DOSAG I analyzes a multiple-reach, branching river system.

A junction is the confluence between two streams within the river basin being modeled.
A stretch is the length of river between two junctions.
A headwater stretch is the length of river from a headwater to its first junction with another stream.
A reach is the subunit of length within any stretch.

When a significant change occurs in the hydraulic, biological, or physical characteristics of the stream (including the addition of a waste load or withdrawal of water from the stream), then a new reach is added to any point in the stretch.

The purpose of DOSAG I is to predict the BOD and the minimum DO in a stream, as well as to estimate the required flow augmentation to bring the DO up to the target level. The removal of carbonaceous biochemical oxygen demand BOD^c and nitrogenous biochemical oxygen demand BOD^n is expressed by first-order equations:

$$\frac{dBOD^c}{dt} = -K^c \cdot BOD^c \tag{6.40}$$

and

$$\frac{dBOD^n}{dt} = -K^n \cdot BOD^n. \tag{6.41}$$

The bio-oxidation rates for BOD^c and BOD^n, K^c and K^n, are considered to be constant for a given reach. The standard equation for atmospheric reaeration is used in the form

$$\frac{dDO}{dt} = K_a(DO^s - DO) \tag{6.42}$$

and the DO equation is given as

$$\begin{aligned} DO_t = DO^s &+ \frac{K^c}{K_a + K^c}[\exp(-K_a t) - \exp(-K^c t)]BOD_0^c \\ &+ \frac{K^n}{K_a + K^n}[\exp(-K_a t) - \exp(-K^n t)]BOD_0^n \\ &- (DO^s - DO_0)\exp(-K_a t). \end{aligned} \tag{6.43}$$

A Lagrangian solution technique, in which a coordinate system moves with a particle of water downstream, was applied to solve the DO equation. At each change in the reach and at every junction a mass balance is computed to determine the BOD and DO in the next reach. The model has provisions for waste discharges and for withdrawal of water at any location within the stream system. As an example, this model was applied to the upper portion of the San Antonio River basin (Figure 6.2).

QUAL I was developed during 1969–70 by F. D. Masch and Associates in collaboration with the Texas Water Development Board (1970b). This integrated system of interrelated mathematical models of water quality is capable of routing through a one-dimensional completely mixed branching stream system: (a) conservative minerals, (b) BOD/DO, and (c) temperature. QUAL I

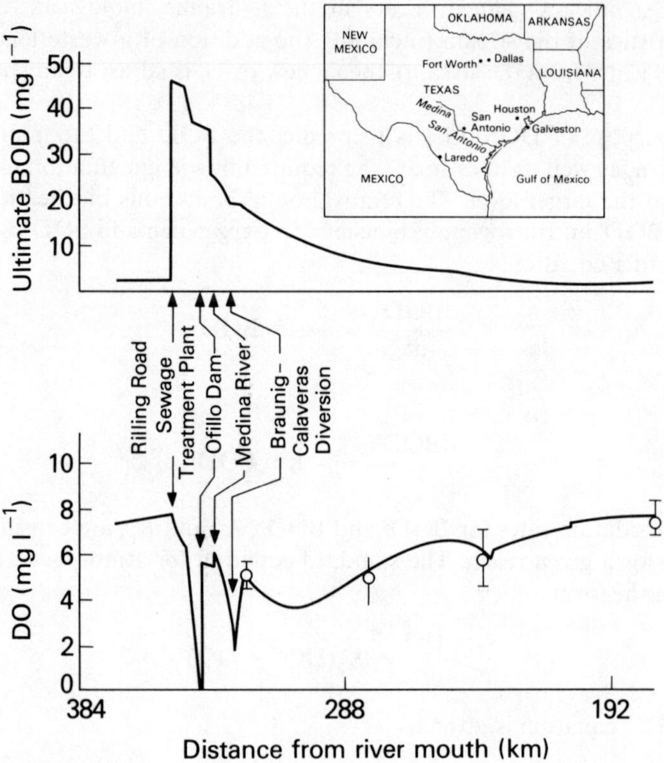

FIGURE 6.2 DOSAG I model applied to San Antonio River basin (Texas Water Development Board, 1970b).

possesses the following capabilities and characteristics (Texas Water Development Board, 1971):

(1) The stream may be discretized into elements of suitable length and variable cross section to obtain any degree of resolution that is warranted.
(2) It can account for heat exchange across the air–water interface and is capable of handling waste inputs and withdrawals at selected points along the channel axis.
(3) It allows for transport by advective and dispersive mechanisms along the principal axis of flow (longitudinal axis).
(4) Solutions provide for a temporal and spatial description of conservative material, BOD and DO, and temperature variation throughout a stream system.
(5) It provides for determining flow augmentation requirements based on selected minimum allowable concentrations of dissolved oxygen.

(6) It has the capability of an integrated system so that the results of any submodel can be used as the input to another model provided that such feedback is required; each submodel also has a stand-alone capability.
(7) It is so structured as to be completely general, and can be applied to any stream system by choosing the appropriate parameters and providing the necessary data that relate to a specific case.

QUAL I is a series of mass balances of a given water quality constituent over a time interval Δt on a stream segment of cross-sectional area A and of length Δx along the principal axis of flow. In steady state nonuniform flow, conservative and nonconservative constituents in a stream may be described by

$$\frac{\partial C}{\partial t} A = \frac{\partial(AE\,\partial C/\partial x)}{\partial x} - \frac{\partial(AUC)}{\partial x} \pm AS, \tag{6.44}$$

where

C is the concentration of the constituent $[\mathrm{M\,L^{-3}}]$,
E is the longitudinal dispersion coefficient $[\mathrm{L^2\,T^{-1}}]$,
S represents sources or sinks of a nonconservative constituent $[\mathrm{M\,L^{-3}\,T^{-1}}]$.

There is a similar equation to (6.44) in which C represents temperature and S represents sources of heat. When (6.44) is written for a control volume V_i of element i in the stream system (as illustrated in Figure 6.3 for steady state nonuniform hydraulics), the resulting expression is

$$\frac{\partial C_i}{\partial t} = \frac{(AE\,\partial C/\partial x)_{i+1/2} - (AE\,\partial C/\partial x)_{i-1/2}}{V_i} - \frac{Q_{i-1/2}C_{i-1} - Q_{i+1/2}C_i \pm Q_{x_i}C_{x_i}}{V_i} \pm S_i, \tag{6.45}$$

where

V_i is the volume of a control element $[\mathrm{L^3}]$, $V_i = \bar{A}_i \Delta x$,
$\bar{A}_i$ is the mean cross-sectional area of the control element $[\mathrm{L^2}]$, $\bar{A}_i = \frac{1}{2}(A_{i-1/2} + A_{i+1/2})$,
$(AE\,\partial C/\partial x)_{i-1/2}$ is the total longitudinal dispersion of the constituent $[\mathrm{L^3\,T^{-1}/M\,L^{-3}}]$ or the temperature $[\mathrm{L^3\,T^{-1}/^\circ C}]$ on the inflow side of the control volume,
$(AE\,\partial C/\partial x)_{i+1/2}$ is the total longitudinal dispersion of the constituent $[\mathrm{L^3\,T^{-1}/M\,L^{-3}}]$ or the temperature $[\mathrm{L^3\,T^{-1}/^\circ C}]$ on the outflow side of the control volume,
$Q_{i-1/2}$ is the rate of flow into the control volume $[\mathrm{L^3\,T^{-1}}]$,
$Q_{i+1/2}$ is the rate of flow out of the control volume $[\mathrm{L^3\,T^{-1}}]$,
Q_{x_i} represents local inflows or withdrawals $[\mathrm{L^3\,T^{-1}}]$,
C_{i-1} is the concentration of the constituent $[\mathrm{M\,L^{-3}}]$, or temperature (°C), in the inflowing water,

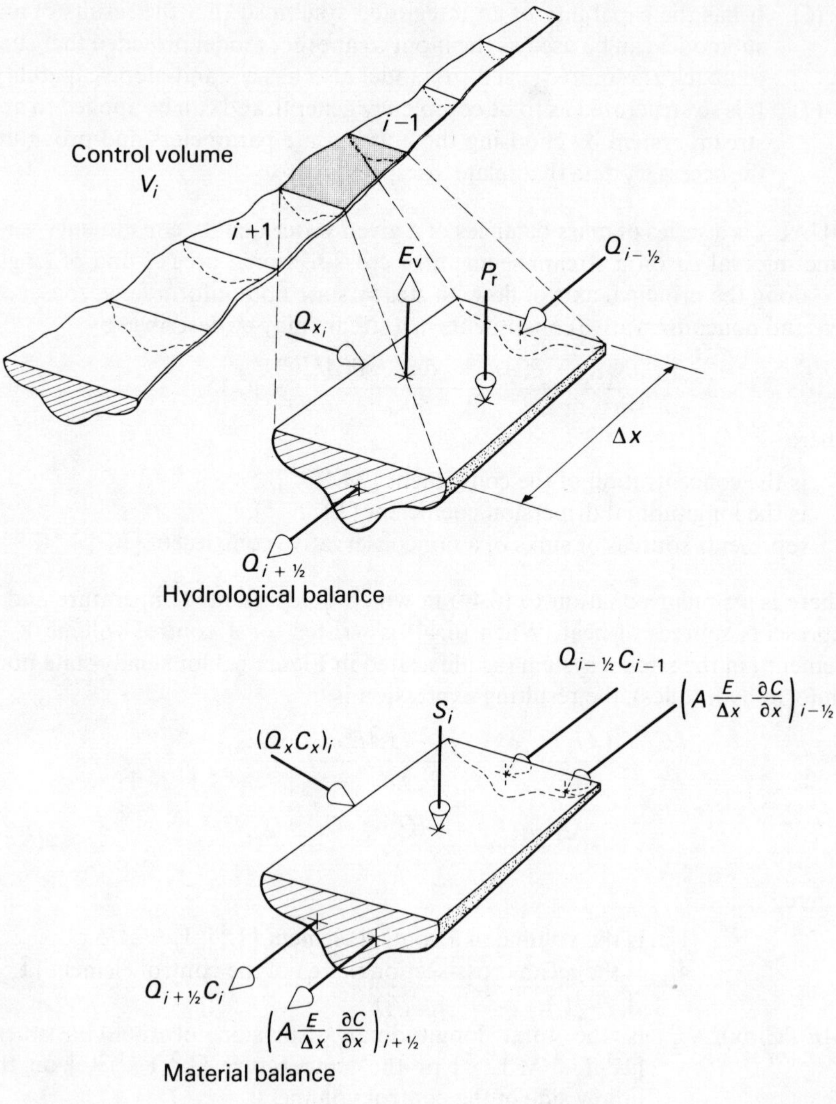

FIGURE 6.3 Discretized stream system (Water Resources Engineers, Inc., 1967).

C_i is the concentration of the constituent [M L^{-3}], or temperature (°C), in the control volume,

C_{x_i} is the concentration of the constituent [M L^{-3}], or temperature (°C), in Q_{x_i},

S_i represents sources or sinks of a nonconservative constituent [M L^{-3}], or temperature (°C).

Under the assumption of steady state conditions a hydraulic balance for the control volume has been represented by the following continuity equation:

$$Q_{i-1/2} - Q_{i+1/2} \pm Q_{x_i} + P_r - E_v = 0, \tag{6.46}$$

where

P_r is the precipitation rate,
E_v is the evaporation rate.

In many cases E_v and P_r are neglected. Equation 6.44 without a source or sink term S may be used to describe the behavior of a conservative mineral within a stream:

$$\frac{\partial M}{\partial t} A = \frac{\partial(AE\,\partial M/\partial x)}{\partial x} - \frac{\partial(AUM)}{\partial x}, \tag{6.47}$$

where M is the concentration of the conservative mineral [M L^{-3}]. Also, if there is complete mixing, (6.44) written with a source term for temperature is

$$\frac{\partial T}{\partial t} A = \frac{\partial(AE\,\partial T/\partial x)}{\partial x} - \frac{\partial(AUT)}{\partial x} \pm \frac{A}{\gamma c_p} S_T, \tag{6.48}$$

where

γ is the density of water [M L^{-3}],
c_p is the specific heat of water [cal M^{-1} °C^{-1}],
S_T is the heat source term (the net rate of heat gain or loss per unit volume of water) [cal L^{-3} T^{-1}].

For a stream of length Δx and mean surface area W the total rate of heat transfer across the air–water interface is $T_N W$, where T_N is the net energy flux passing through the interface. This heat is distributed uniformly throughout the underlying volume $\bar{A}\Delta x$. The heat term can be defined as

$$S_T = \frac{T_N W}{\bar{A}\Delta x} = T_N H^{-1}, \tag{6.49}$$

where H is the mean hydraulic depth of the stream. Therefore, the basic one-dimensional heat transport equation is

$$\frac{\partial T}{\partial t} A = \frac{\partial(AE\,\partial T/\partial x)}{\partial x} - \frac{\partial(AUT)}{\partial x} \pm \frac{AT_N}{\gamma c_p H}. \tag{6.50}$$

For practical applications the parameters γ and c_p are considered constant.

The rate of oxygen utilization due to BOD is expressed by the first-order reaction (the ultimate BOD is considered), and the reaeration process is expressed in the same way as in DOSAG I. If the transverse section of the stream is completely mixed, (6.44) can be written for DO and for BOD, respectively, as

$$\frac{\partial \mathrm{DO}}{\partial t} A = \frac{\partial(AE\,\partial \mathrm{DO}/\partial x)}{\partial x} - \frac{\partial(AU\mathrm{DO})}{\partial x} \pm AS_{\mathrm{DO}} \tag{6.51}$$

and

$$\frac{\partial \mathrm{BOD}}{\partial t} A = \frac{\partial(AE\,\partial \mathrm{BOD}/\partial x)}{\partial x} - \frac{\partial(AU\mathrm{BOD})}{\partial x} \pm AS_L, \tag{6.52}$$

where

$S_{\mathrm{DO}} = K_a(\mathrm{DO}^s - \mathrm{DO}) - (K_d + K_s)\mathrm{BOD}$,

$S_L = -(K_d + K_s)\mathrm{BOD}$,

K_d is the deoxygenation rate constant $[\mathrm{T}^{-1}]$,

K_s is the rate constant for BOD removal by sedimentation $[\mathrm{T}^{-1}]$.

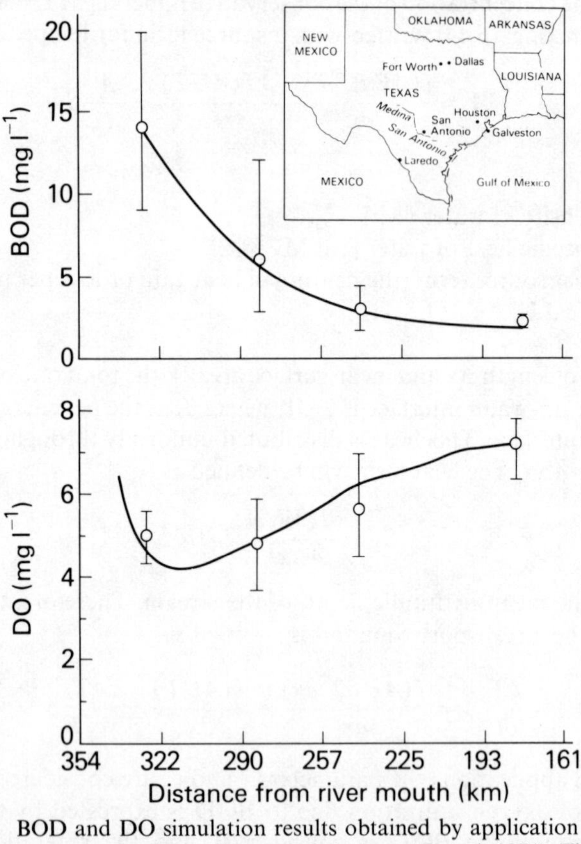

FIGURE 6.4 BOD and DO simulation results obtained by application of QUAL I to the San Antonio River basin (Texas Water Development Board, 1971). The graphs show the observed points and the computed curves. K_d was estimated from field data.

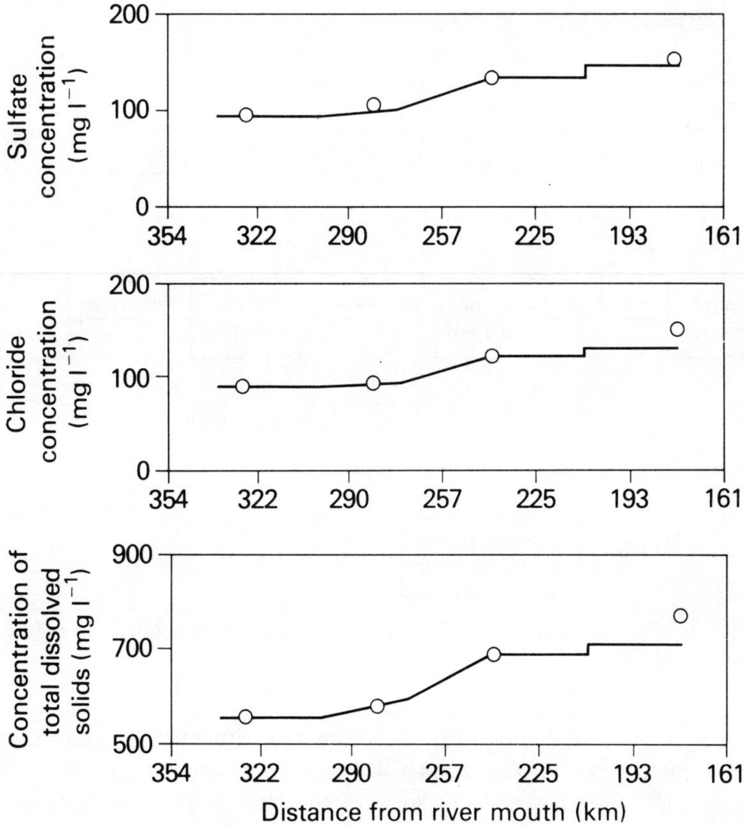

FIGURE 6.5 Results of simulations for conservative minerals obtained by application of QUAL I to the San Antonio River basin (Texas Water Development Board, 1971). The graphs show the observed points and the computed curves.

A finite-difference method is used to obtain an analytical solution. QUAL I uses a longitudinal dispersion coefficient E [$L^2\ T^{-1}$], defined as $E = CU^*H$, where U^* is the bed shear velocity [$L\ T^{-1}$]. Since for steady state flow the shear velocity was described as $U^* = R^{2/3}nS_e^{0.5}$, where R is the hydraulic radius, S_e is the slope of the energy gradient, and n is the Manning coefficient for the reach, then the resulting dispersion equation was expressed as $E = C'nUH^{0.833}$. The mean velocity U and the mean depth of flow in a reach, H, are calculated in a similar way as in DOSAG I, by equations of the form $U = \alpha Q^\beta$ and $H = \gamma Q^\zeta$, where constants α, β, γ, ζ are usually determined from stage-discharging rating curves and Q is the mean discharge in a reach. QUAL I was applied to part of the San Antonio River basin. The results from the BOD and DO simulations are shown in Figure 6.4. The simulations for conservative minerals are shown in Figure 6.5.

6.5. NITROGEN CYCLE

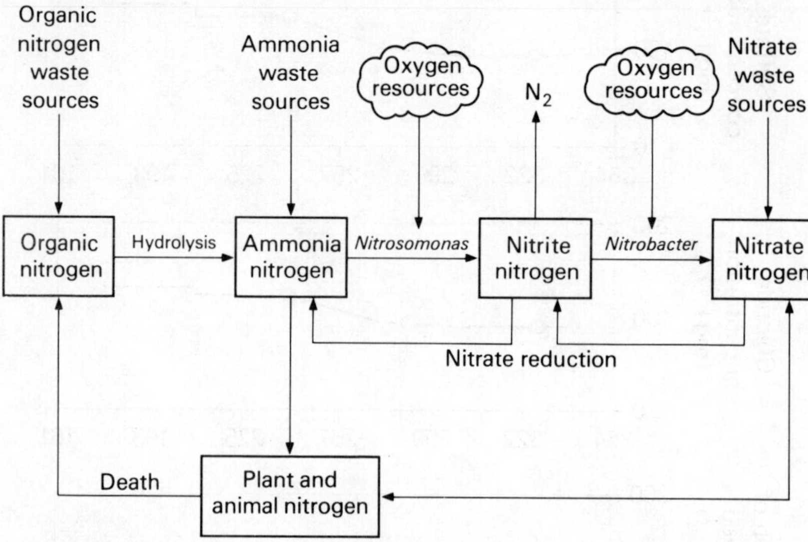

FIGURE 6.6(a) Major features of the nitrogen cycle, showing nitrification (oxidation) and nitrate reduction.

In river systems where nitrogen is a major constituent, the assumption of a constant reaction rate K^n (as in eqn. 6.41) may not be appropriate. The rate may vary along the system simply because of the changing concentrations of the various forms of nitrogen, each form having its own rate constant. A modification of (6.44), together with data on the input of each form of nitrogen and its corresponding reaction rate, will permit the prediction of the separate components of the nitrification process as illustrated in Figures 6.6(a) and (b).

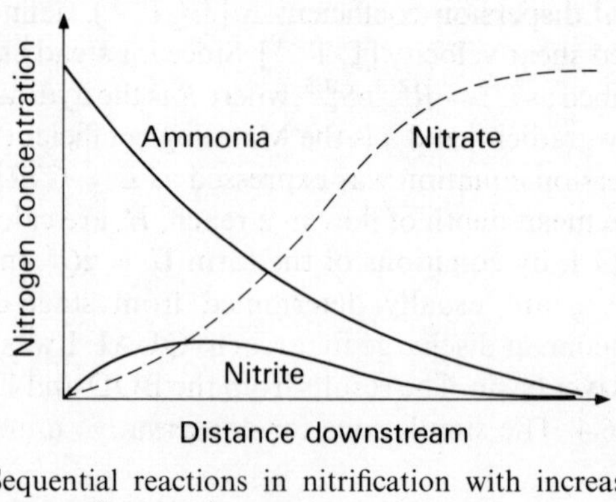

FIGURE 6.6(b) Sequential reactions in nitrification with increasing distance downstream from point source of ammonia waste.

O'Connor and Di Toro (1970) proposed some sequential reaction models to account for each nitrogen component.

Let us consider the following concentration components [M L^{-3}]: N_1, organic nitrogen; N_2, ammonia nitrogen (NH_3-N); N_3, nitrite nitrogen (NO_2-N); and N_4, nitrate nitrogen (NO_3-N). We shall let K_i^n now represent the first-order decay (including settling) of nitrogen form i [T^{-1}], $K_{i,i+1}^n$ the forward reaction coefficient [T^{-1}], and $W_i(X)$ the discharge of form i at site X [M L^{-3} T^{-1}]. The solution of the following four equations will permit a prediction of the concentration of each form of nitrogen in steady state river systems:

$$0 = E\frac{d^2N_1}{dX^2} - U\frac{dN_1}{dX} - K_1^n N_1 + W_1(X)$$

$$0 = E\frac{d^2N_i}{dX^2} - U\frac{dN_i}{dX} - K_i^n N_i + K_{i-1,i}^n N_{i-1} + W_i(X) \qquad i = 2, 3, 4, \qquad (6.53)$$

where

E is the dispersion coefficient (applicable to estuaries) [L^2 T^{-1}],
U is the net downstream velocity [L T^{-1}].

The decrease in the dissolved oxygen concentration in such river systems due to the individual nitrogen constituents is caused by oxidation of ammonia, $3.43K_{23}^n N_2$, and of nitrite, $1.14K_{34}^n N_3$. The dissolved oxygen deficit concentration D^n resulting from the oxidation of these two nitrogen forms can be predicted by

$$0 = E\frac{d^2D^n}{dX^2} - U\frac{dD^n}{dX} - K_a D^n + 3.43\,K_{23}^n N_2(X) + 1.14\,K_{34}^n N_3(X), \qquad (6.54)$$

in which K_a is the reaeration rate constant [T^{-1}] and $N_2(X)$ and $N_3(X)$ are obtained from the solution of eqns. 6.53.

This model can be further expanded to include denitrification and algal utilization. Denitrification may involve the reduction of NO_3 to NO_2 and the conversion of NO_2 to N_2 gas. These reactions occur in low concentrations of dissolved oxygen. Assimilatory nitrate reduction can convert NO_3 to NH_3. The utilization of ammonia nitrogen and nitrate nitrogen by phytoplankton may also take place, and hence produce organic nitrogen, thereby completing a very simplified version of the nitrogen cycle. This process is generally defined by reaction rate constants K_{ij}^n $(i, j = 1, 2, 3, 4)$. Feed-forward reactions involve reaction rates K_{ij}^n $(i < j)$ where j is usually $i + 1$, and feedback reactions involve reaction rates K_{ij}^n $(i > j)$ where j is usually $i - 1$. All these reactions $(K_{ij}^n N_i)$ are then included in (6.53):

$$0 = E\frac{d^2N_i}{dX^2} - U\frac{dN_i}{dX} - K_i^n N_i + \sum_{j \neq i} (K_{ij}^n N_j) + W_i(X) \qquad i = 1, 2, 3, 4. \qquad (6.55)$$

In general, each $K_i^n \geq \sum_j K_{ij}^n, j \neq i$.

By finite-difference approximations four algebraic equations, one for each form of nitrogen, can be written for each finite section k of the river system as appropriate.

$$0 = \sum_j [-Q_{kj}(\alpha_{kj} N_{ik} + \beta_{kj} N_{ij}) + E'_{kj}(N_{ij} - N_{ik})] - V_k(K^{n}_{ik} N_{ik} - \sum_h K^{n}_{hik} N_{hk}) + W_{ik} \qquad i = 1, 2, 3, 4, \forall k, \qquad (6.56)$$

where

Q_{kj} is the net flow from an adjacent section k to section j (positive if from k to j, negative if from j to k) $[L^3 T^{-1}]$,
V_k is the volume of segment k $[L^3]$,
α_{kj} is a finite-difference weight (max $(\frac{1}{2}, 1 - E'_{kj}/Q)$),
β_{kj} is $1 - \alpha_{kj}$,
E'_{kj} is the exchange (bulk dispersion) coefficient $[L^3 T^{-1}]$, defined by

$$E'_{kj} = \frac{2E_{kj} A_{kj}}{L_k + L_j}, \qquad (6.57)$$

N_{ik} is the concentration $[M\, L^{-3}]$ of nitrogen form i in section k,
K^{n}_{hik} are the reaction rate constants $[T^{-1}]$ for the conversion of nitrogen form h to form i in section k,
W_{ik} is the direct discharge $[M\, T^{-1}]$ of nitrogen form i into section k.

In equation 6.57,

E'_{kj} is the dispersion coefficient between sections k and j,
A_{kj} is the interfacial cross-sectional area,
L_k, L_j are the lengths of sections k and j.

The spatial distribution of the dissolved oxygen deficit concentration D^{n}_k caused by ammonia oxidation (from NH_3 to NO_2, reaction rate $3.43K^{n}_{23}$) and by nitrite oxidation (from NO_2 to NO_3, reaction rate $1.14K^{n}_{34}$) can be predicted from the simultaneous solution of similar finite-difference equations:

$$0 = \sum_j [-Q_{kj}(\alpha_{kj} D^{n}_k + \beta_{kj} D^{n}_j) + E'_{kj}(D^{n}_j - D^{n}_k)] - K_{ak} D^{n}_k V_k + 3.43 K^{n}_{23} N_{2k} V_k + 1.14 K^{n}_{34} N_{3k} V_k, \qquad (6.58)$$

in which N_{2k} and N_{3k} represent, respectively, ammonia and nitrite nitrogen concentrations $[M\, L^{-3}]$ in section k and K_{ak} is the reaeration rate constant $[T^{-1}]$ for section k.

Equations 6.56 and 6.58 represent five equations for each section of the river. They can be combined and put into a convenient matrix form for solution using any of a number of computer programs available for solving large sets of linear equations. Since some of the feedback reaction coefficients are functions of the dissolved oxygen deficit, parameter adjustments will be necessary to ensure an adequate prediction of the concentrations of dissolved oxygen and of each form of nitrogen. Just how these adjustments are made is indeed part of the art of water quality modeling. The results of these nitrogen models, such as those applied to the Delaware Estuary, are illustrated in Figure 6.7 (O'Connor *et al.*, 1976).

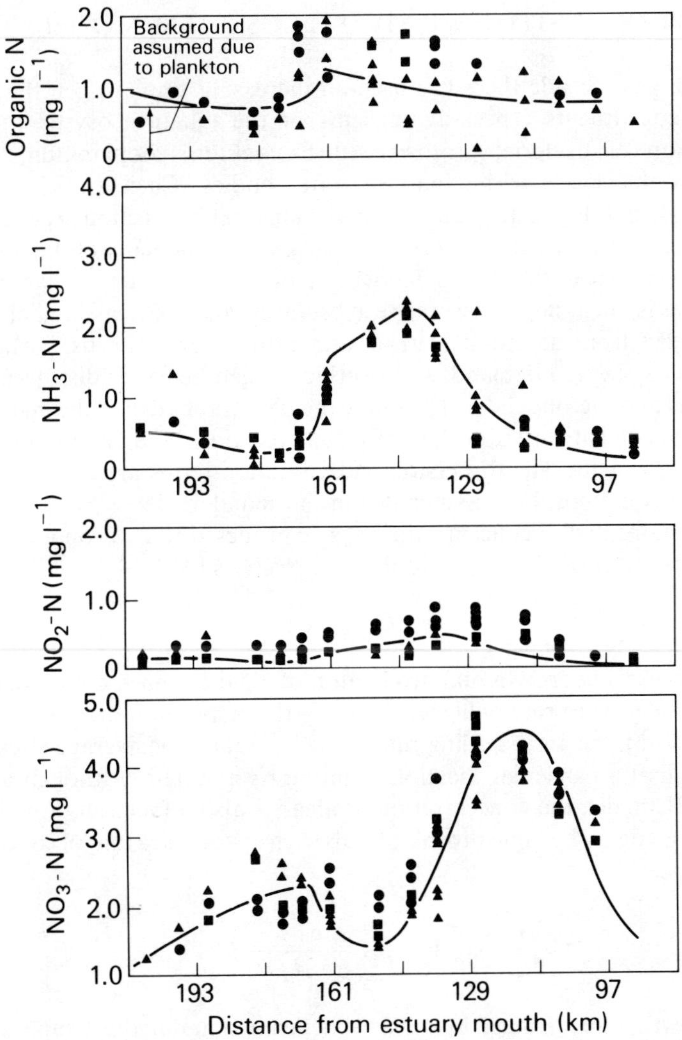

FIGURE 6.7 Observed (points) and computed (curves) nitrogen profiles for the Delaware Estuary (O'Connor *et al.*, 1976). Observations were made in 1964 on 30 July (▲), 10 August (■), and 31 August (●).

The models discussed above are all first-order kinetic steady state models. While they vary in complexity and in their requirements for data, they remain relatively simple. In the models, flows and temperatures do not vary with time and the complex nonlinear kinetic interactions between the microorganisms and the constituents are approximated by linear or first-order reactions.

6.6. MULTICONSTITUENT RIVER ECOSYSTEM MODELS

Over the past decade there has been an increasing emphasis on the effects of various constituents, especially nutrients, on the aquatic ecosystem, i.e. on the production of bacteria, protozoa, phytoplankton, zooplankton, fish, and other trophic levels within natural water bodies (Chen and Orlob, 1975). QUAL II, developed for the US Environmental Protection Agency (Water Resources Engineers, Inc., 1973), is a good example of many of the operational aquatic ecosystem simulation models. It predicts a variety of water quality constituents, including conservative substances; algal biomass and chlorophyll *a*; nitrogen from ammonia, nitrite, and nitrate; phosphorus, carbonaceous biochemical oxygen demand, and benthic oxygen demand; dissolved oxygen; coliforms and radionuclides. The following paragraphs define the mathematical relationships that describe the individual reactions and interactions among these constituents. The dispersion and advection terms are not shown in the following equations, but are included in the actual model.

The chlorophyll *a* concentration C_a is considered to be proportional to the concentration A of phytoplanktonic algal biomass $[\mathrm{M\,L^{-3}}]$:

$$C_a = \alpha_0 A. \tag{6.59}$$

The time-varying growth and production of algal biomass, $\mathrm{d}A/\mathrm{d}t$, depends on the specific growth rate of algae, $\mu_A\,[\mathrm{T^{-1}}]$, the respiration rate (or specific loss rate) $\rho\,[\mathrm{T^{-1}}]$, the algal settling rate $\sigma_1\,[\mathrm{L\,T^{-1}}]$, and the average stream depth $H_{\mathrm{a}}[\mathrm{L}]$, all at a particular location X in the river system. Although not in the QUAL II model, the concentration of algae is also a function of mortality M and of grazing G_Z by higher trophic levels such as zooplankton (of concentration Z). Hence,

$$\frac{\mathrm{d}A}{\mathrm{d}t} = \mu_A A - \rho A - \frac{\sigma_1 A}{H_{\mathrm{a}}} - MA - G_Z. \tag{6.60}$$

The growth rate of algal biomass is dependent upon the temperature and availability of nutrients (nitrogen, carbon, phosphorus) and upon light. The standard Michaelis–Menten formulation, illustrated in Figure 6.8, defines the specific growth rate μ at a given site in a river system.

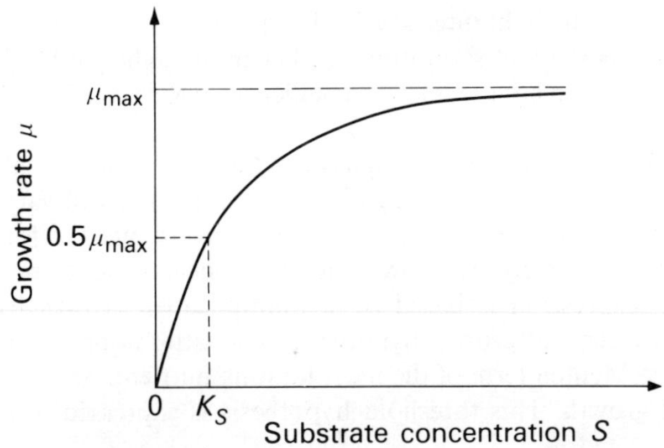

FIGURE 6.8 Michaelis–Menten kinetic expression for microbial substrate utilization and growth. Growth rate $\mu = \mu_{max}[S/(K_S + S)]$, where K_S is the half-saturation concentration.

This Michaelis–Menten formulation assumes that μ is a parabolic function of the essential substrate concentration S:

$$\mu = \mu_{max} \frac{S}{K_S + S}. \tag{6.61}$$

The parameter K_S has the value of S when μ is half of the maximum growth rate μ_{max}, and is called the half-saturation concentration. At low substrate concentrations, μ is proportional to S. For high values of S, μ approaches the limiting saturation value μ_{max}. In QUAL II the algal growth rate μ_A may be limited by nitrate nitrogen NO_3, phosphorus P, carbon C, light intensity, and temperature. Hence,

$$\mu_A = (\mu_{A,20}^{max}\theta^{T-20})\left(\frac{NO_3}{NO_3 + K_{NO_3}}\right)\left(\frac{P}{P + K_P}\right) \cdot \left(\frac{C}{C + K_C}\right)\frac{1}{\lambda H_a}\ln\left(\frac{K_I + I}{K_I + I\exp(-\lambda H_a)}\right), \tag{6.62}$$

in which

- $\mu_{A,20}^{max}$ is the maximum growth rate of algal biomass at 20°C [T^{-1}],
- θ is a temperature coefficient ranging from 1.02 to 1.06,
- T is the water temperature (°C),
- NO_3, P, C are the nitrate nitrogen, orthophosphate phosphorus, and carbon concentrations [$M\,L^{-3}$],
- K_{NO_3}, K_P, K_C are the temperature-dependent half-saturation concentrations for algal growth [$M\,L^{-3}$],

I is the light intensity [ly T^{-1}],
K_I is the half-saturation coefficient for light [ly T^{-1}],
λ is the light extinction coefficient in the river [L^{-1}].

Equation 6.62 couples algal biomass production to the available supply of nutrients, light, and heat; hence algae and chlorophyll *a* will vary in time and space in response to the concentrations of elements needed for growth. By eqn. 6.62, if any of the critical growth element concentrations are zero, then μ_A is zero. This expression is based on a multiplicative growth hypothesis, in contrast to a threshold growth hypothesis. The latter approach includes only the Michaelis–Menten term of the most limiting nutrient, or of light, that constrains algal growth. This threshold hypothesis is expressed in the following equations:

$$\begin{aligned}
\mu_A &= (\mu_{A,20}^{\max}\,\theta^{T-20})\left(\frac{\mathrm{NO_3}}{\mathrm{NO_3}+K_{\mathrm{NO_3}}}\right) && \text{if } \mathrm{NO_3} \text{ is limiting}\\
\mu_A &= (\mu_{A,20}^{\max}\,\theta^{T-20})\left(\frac{\mathrm{P}}{\mathrm{P}+K_{\mathrm{P}}}\right) && \text{if P is limiting}\\
\mu_A &= (\mu_{A,20}^{\max}\,\theta^{T-20})\left(\frac{\mathrm{C}}{\mathrm{C}+K_{\mathrm{C}}}\right) && \text{if C is limiting}\\
\mu_A &= (\mu_{A,20}^{\max}\,\theta^{T-20})\left[\frac{1}{\lambda H_{\mathrm{a}}}\ln\left(\frac{K_I+I}{K_I+I\exp(-\lambda H_{\mathrm{a}})}\right)\right] && \text{if } I \text{ is limiting.}
\end{aligned} \tag{6.63}$$

Respiration is also temperature-dependent:

$$\rho_T = \rho_{20}\,\theta^{T-20}, \tag{6.64}$$

where ρ_{20} is the respiration rate [T^{-1}] at 20°C. All temperature-dependent rate constants are defined by equations having the form of eqn. 6.64, the constant θ being defined for each rate constant.

The mortality rate M [T^{-1}] of algal biomass can be expressed as a linear function of the natural mortality rate M_{N} and the mortality caused by the toxicity of the water, βC_{x}:

$$M = M_{\mathrm{N}} + \beta C_{\mathrm{x}}, \tag{6.65}$$

where

β is a toxicity coefficient (rate of mortality per unit concentration) [$T^{-1}\,M^{-1}\,L^3$],
C_{x} is the concentration of the toxic constituents [$M\,L^{-3}$].

The growth rate μ_Z of zooplankton can be defined by

$$\mu_Z = (\mu_{Z,20}^{\max}\,\theta^{T-20})\left(\frac{A}{K_A+A}\right), \tag{6.66}$$

where K_A is the temperature-dependent half-saturation concentration for algal biomass [M L^{-3}]. The product of zooplankton growth rate μ_Z [T^{-1}], concentration Z [M L^{-3}], and a conversion coefficient $F_{A,Z}$ [M M^{-1}], which indicates the mass of algal biomass required per unit mass of zooplankton, is an estimate of the loss in algal biomass due to zooplankton grazing G_Z. The reliability of current models for predicting trophic levels higher than phytoplankton (such as zooplankton and fish) is relatively poor (Russell, 1975), hence the omission of these higher trophic levels from operational models such as QUAL II.

The nitrogen cycle in QUAL II is described by differential equations governing the transformation of nitrogen from one form to another. For the ammonia nitrogen concentration NH_3 [M L^{-3}],

$$\frac{dNH_3}{dt} = \alpha_1 \rho A - \beta_1 NH_3 + \frac{\sigma_3}{A_X}, \tag{6.67}$$

where

α_1 is the fraction of respired algal biomass resolubilized as ammonia nitrogen by bacteria,
β_1 is the temperature-dependent rate of biological oxidation of NH_3 [T^{-1}],
σ_3 is the benthos source rate for NH_3 [M T^{-1} L^{-1}],
A_X is the average stream cross-sectional area at location X [L^2].

For the nitrite nitrogen concentration NO_2 [M L^{-3}],

$$\frac{dNO_2}{dt} = \beta_1 NH_3 - \beta_2 NO_2, \tag{6.68}$$

in which

β_1 is the rate of oxidation of NH_3 to NO_2 [T^{-1}],
β_2 is the rate of oxidation of NO_2 to NO_3 [T^{-1}].

For the nitrate nitrogen concentration NO_3 [M L^{-3}],

$$\frac{dNO_3}{dt} = \beta_2 NO_2 - \alpha_1 \mu_A A. \tag{6.69}$$

Equations 6.67–6.69, together with eqn. 6.64, close the loop of the nitrogen cycle. Equations 6.67 and 6.69 demonstrate that the fraction of respired algal biomass resolubilized as NH_3-N is assumed to equal the fraction of biomass that is NO_3-N.

The phosphorus cycle is modeled in less detail than the nitrogen cycle. Only the interaction of phosphorus and algae and a sink term are considered. Thus the differential equation describing the time-varying concentration P [M L^{-3}] of orthophosphate phosphorus is written as

$$\frac{dP}{dt} = \alpha_2 A(\rho - \mu_A) + \frac{\sigma_2}{A_X}, \tag{6.70}$$

in which

α_2 is the fraction of algal biomass that is phosphorus,
σ_2 is the benthos source rate for phosphorus $[\mathrm{M\,T^{-1}\,L^{-1}}]$.

Carbonaceous biochemical oxygen demand $\mathrm{BOD^c}$ $[\mathrm{M\,L^{-3}}]$ is formulated as a first-order reaction:

$$\frac{\mathrm{dBOD^c}}{\mathrm{d}t} = -K^{\mathrm{c}}\mathrm{BOD^c} - K_{\mathrm{s}}\mathrm{BOD^c}, \tag{6.71}$$

where

K^{c} is the temperature-dependent deoxygenation or decay rate of carbonaceous BOD $[\mathrm{T^{-1}}]$,
K_{s} is the rate constant for loss of carbonaceous BOD by settling $[\mathrm{T^{-1}}]$.

The benthic oxygen demand $\mathrm{BOD^b}$ $[\mathrm{M\,L^{-3}}]$ is assumed to be fixed, dependent on the cross-sectional area A_X at location X:

$$\frac{\mathrm{dBOD^b}}{\mathrm{d}t} = \frac{K_{\mathrm{b}}}{A_X}, \tag{6.72}$$

where K_{b} is a constant benthic source rate $[\mathrm{M\,T^{-1}\,L^{-1}}]$.

The differential equation that describes the rate of change in dissolved oxygen is

$$\frac{\mathrm{dDO}_X}{\mathrm{d}t} = K_{\mathrm{a}}(\mathrm{DO}_X^{\mathrm{s}} - \mathrm{DO}_X) + (\alpha_3\mu_A - \alpha_4\rho)A - K^{\mathrm{c}}\mathrm{BOD^c} - \frac{K_{\mathrm{b}}}{A_X} - \alpha_5\beta_1\mathrm{NH_3} - \alpha_6\beta_2\mathrm{NO_2}, \tag{6.73}$$

where

$\mathrm{DO}_X^{\mathrm{s}}$ is the temperature-dependent dissolved oxygen saturation at location X $[\mathrm{M\,L^{-3}}]$,
DO_X is the dissolved oxygen concentration at X $[\mathrm{M\,L^{-3}}]$,
K_{a} is the temperature-dependent reaeration rate constant $[\mathrm{T^{-1}}]$,
α_3 is the rate of oxygen production through photosynthesis per unit of algal biomass $[\mathrm{M\,M^{-1}}]$,
α_4 is the rate of oxygen uptake from respiration per unit of algal biomass $[\mathrm{M\,M^{-1}}]$,
α_5 is the rate of oxygen uptake per unit of oxidation of ammonia nitrogen $[\mathrm{M\,M^{-1}}]$,
α_6 is the rate of oxygen uptake per unit of nitrite nitrogen oxidation $[\mathrm{M\,M^{-1}}]$.

The complete set of equations that is numerically solved by QUAL II includes the effects of dispersion, advection, constituent reactions, and interactions up to the phytoplankton trophic level. The set also includes a source term S [M T^{-1}] that is assumed uniform over the length ΔX of the river section at location X.

6.6.1. Bacteria and Protozoa in Pollutant Degradation

The QUAL II model structure, a part of which has just been outlined, is typical of many multiconstituent water quality models. In their application some assumptions have been changed and modifications made to suit particular river conditions. However, an alternative modeling approach that explicitly includes the role of bacteria and protozoa in the degradation of biodegradable pollutants and in the reduction of dissolved oxygen in aerobic river systems has been developed by Stehfest (1977). Michaelis–Menten kinetics for bacterial and protozoan growth are assumed, including the possibility of differing nutrient uptake rates (caused by the presence of more than one type of enzyme) or inhibition, i.e. the reduction in the rate of degradation of one nutrient due to the presence of another nutrient, toxic substance, or degradation byproduct.

The overall model structure can be introduced by first considering a simple self-purification process consisting of a single nutrient concentration N, such as that of a particular form of nitrogen, a bacterial biomass concentration B, and a dissolved oxygen concentration DO, in an aquatic environment suitable for bacterial growth. In this situation the rate of change $\mathrm{d}N/\mathrm{d}t$ in concentration will be a function of the consumption of N by bacteria:

$$\frac{\mathrm{d}N}{\mathrm{d}t} = -a_{NB}\,\mu_{BN}^{\max}\left(\frac{N}{K_N + N}\right)B, \tag{6.74}$$

where

a_{NB} is the mass of nutrient N consumed per unit increase of bacterial biomass [M M^{-1}],

$\mu_{BN}^{\max}$ is the maximum growth (increase) of bacterial biomass resulting from a unit mass of nutrient N per unit time [M M^{-1} T^{-1}],

K_N is the nutrient half-saturation concentration for bacterial growth [M L^{-3}].

The rate of change $\mathrm{d}B/\mathrm{d}t$ in the bacterial biomass concentration will depend on bacterial growth and endogenous respiration:

$$\frac{\mathrm{d}B}{\mathrm{d}t} = \mu_{BN}^{\max}\left(\frac{N}{K_N + N}\right)B - \rho_B B, \tag{6.75}$$

where ρ_B is the endogenous respiration rate of bacteria, i.e. the mass decrease of a unit mass of bacteria per unit time [$M\,M^{-1}\,T^{-1}$].

The rate of change in the concentration of dissolved oxygen, DO, will be governed by physical reaeration and by the consumption of oxygen from nutrient degradation and endogenous respiration:

$$\frac{\mathrm{dDO}}{\mathrm{d}t} = K_a(\mathrm{DO}^s - \mathrm{DO}) - a_{OB}^N \mu_{BN}^{max} \left(\frac{N}{K_N + N}\right) B - a_{OB}\, \rho_B B, \quad (6.76)$$

where

a_{OB}^N is the mass of oxygen required per unit increase in bacterial biomass resulting from nutrient N [$M\,M^{-1}$],
a_{OB} is the mass of oxygen required per unit decrease in bacterial biomass due to endogenous respiration [$M\,M^{-1}$].

The Michaelis–Menten expression on the right-hand side of (6.74) represents a two-parameter approximation of the kinetics of nutrient degradation. For low nutrient concentrations the rate of change of N will be approximately proportional to the product of the nutrient and bacterial biomass concentrations, NB. For high nutrient concentrations $\mathrm{d}N/\mathrm{d}t$ will be independent of N; it will depend only on B.

The first term on the right-hand side of (6.75) is identical to $-a_{BN}\,\mathrm{d}N/\mathrm{d}t$, where a_{BN} is the mass of bacteria increase resulting from a unit mass of nutrient ($1/a_{NB}$). Hence the change in bacterial biomass per unit change in nutrient concentration, $\mathrm{d}B/\mathrm{d}N$, is assumed constant and equal to $-a_{BN}$. The second term on the right-hand side assumes that the decrease of bacterial biomass through endogenous respiration is proportional to the biomass concentration.

In (6.76) oxygen consumption in nutrient degradation is assumed proportional to $\mathrm{d}N/\mathrm{d}t$. Oxygen consumption in endogenous respiration is assumed constant per unit of bacterial biomass.

If a second nutrient is added, it could be one that is degraded independently of the first constituent. The resulting model would include an equation similar to (6.74) for the second nutrient, and appropriate terms in (6.75) and (6.76). The procedure can be extended to include any number of independently degraded constituents.

Nutrient growth can be hindered by either competitive or allosteric inhibition. A Michaelis–Menten expression for competitive inhibition is

$$\frac{\mathrm{d}N_1}{\mathrm{d}t} = -a_{N_1 B}\mu_{BN_1}^{max}\left(\frac{N_1}{K_{N_1} + N_1 + a_2 N_2}\right)B, \quad (6.77)$$

where

N_1 is the concentration of the inhibited (or more slowly utilized) nutrient,
N_2 is the concentration of the inhibiting nutrient,
a_2 is the inhibition constant for N_2, a dimensionless parameter.

An expression for allosteric inhibition is

$$\frac{dN_1}{dt} = -a_{N_1B}\,\mu_{BN_1}^{\max}\left(\frac{N_1}{(K_{N_1} + N_1)(1 + a_2 N_2)}\right)B \tag{6.78}$$

as long as the affinity of the bacteria to the inhibitor is as strong as it is to the inhibited nutrient.

If N_2 is insignificant, (6.77) and (6.78) are identical. In competitive inhibition the rate of N_1 degradation is less dependent on the concentration of N_2 than it is in allosteric inhibition. An allosteric inhibitor will tend to block the degradation of a nutrient or constituent much faster than will a competitive inhibitor. These distinctions and effects are illustrated in Figure 6.9.

The consideration of protozoa feeding on bacteria requires another term in (6.75) that reflects the consequent decrease in bacterial biomass, a separate equation to reflect the rate of change in protozoa mass P, and finally terms in (6.76) defining the oxygen requirements for protozoa feeding and respiration. Hence, in addition to (6.74), which defines dN/dt, the single-nutrient model will include the following three equations:

$$\frac{dB}{dt} = \mu_{BN}^{\max}\left(\frac{N}{K_N + N}\right)B - a_{BP}\,\mu_P^{\max}\left(\frac{B}{K_B + B}\right)P - \rho_B B \tag{6.79}$$

$$\frac{dP}{dt} = \mu_P^{\max}\left(\frac{B}{K_B + B}\right)P - \rho_P P \tag{6.80}$$

$$\frac{d\mathrm{DO}}{dt} = K_a(\mathrm{DO}^s - \mathrm{DO}) - a_{\mathrm{O}B}^N\,\mu_{BN}^{\max}\left(\frac{N}{K_N + N}\right)B - a_{\mathrm{O}P}^B\,\mu_P^{\max}\left(\frac{B}{K_B + B}\right)P - a_{\mathrm{O}B}\,\rho_B B - a_{\mathrm{O}P}\,\rho_P P, \tag{6.81}$$

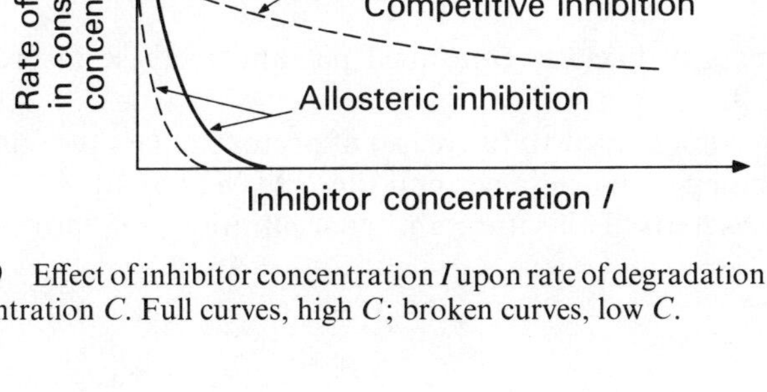

FIGURE 6.9 Effect of inhibitor concentration I upon rate of degradation of a constituent having concentration C. Full curves, high C; broken curves, low C.

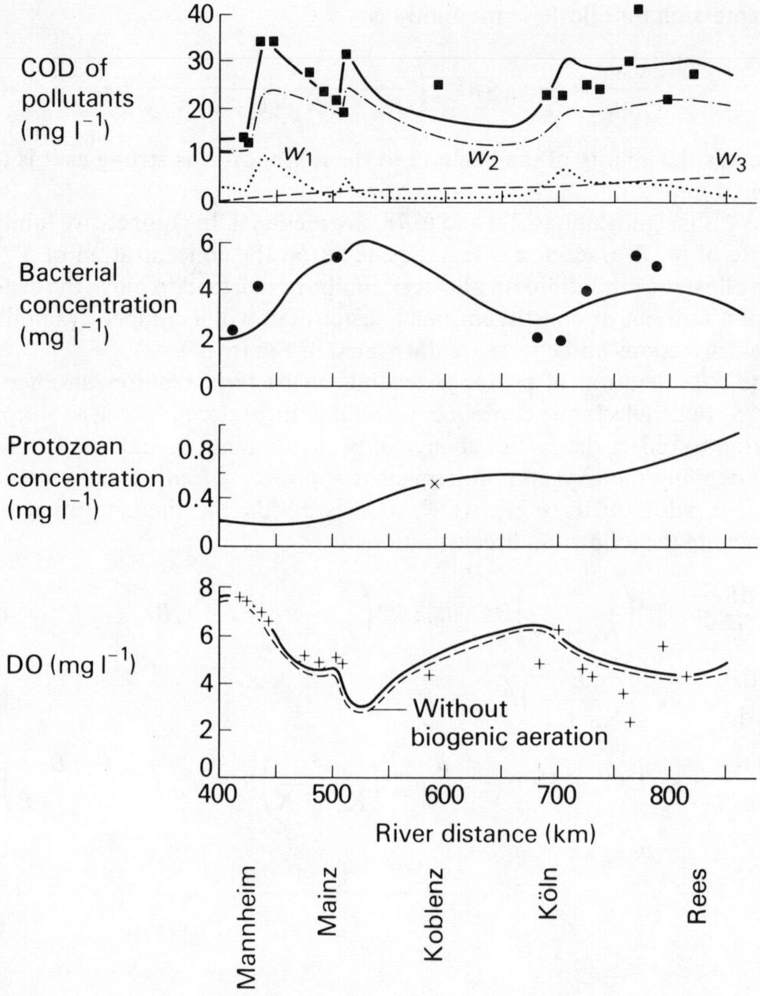

FIGURE 6.10 Solution (full curve) of model of the River Rhine, and some measured values (Stehfest, 1977). w_1, w_2, and w_3 are the concentrations of easily degradable, slowly degradable, and nondegradable pollutants, respectively.

where

- a_{BP} is the mass of bacteria consumed per unit increase in protozoa mass $[\mathrm{M\,M^{-1}}]$,
- $\mu_P^{\max}$ is the maximum growth (increase) of protozoa mass resulting from unit mass feeding on bacteria per unit time $[\mathrm{M\,M^{-1}\,T^{-1}}]$,
- K_B is the bacteria half-saturation concentration for protozoa growth $[\mathrm{M\,L^{-3}}]$,

ρ_P is the endogenous respiration rate of protozoa mass [M M^{-1} T^{-1}],
a_{OP}^B is the mass of oxygen required per unit increase in protozoa mass resulting from feeding on bacteria [M M^{-1}],
a_{OP} is the mass of oxygen required per unit decrease in protozoa mass due to endogenous respiration [M M^{-1}].

The modeling approach just defined has been applied to the prediction of organic pollution, and dissolved oxygen, in the Rhine in Germany, a relatively benthos-free, fast-flowing, and deep river (Stehfest, 1977). In this application the parameter values were identified on the basis of measurements of the concentrations of chemical oxygen demand COD, bacterial biomass B, protozoa mass P, and dissolved oxygen DO. COD was divided into easily and slowly degradable nutrients, N_1 and N_2, respectively (COD $= N_1 + N_2$). Equation 6.74 describes the degradation kinetics of N_1, whereas (6.77) or (6.78) was considered appropriate for the degradation kinetics of N_2, in which it was assumed that N_1 acted as an inhibitor to reduce the degradation rate of N_2. Equation 6.77 defining competitive inhibition was considered more appropriate.

For a constant mass discharge rate of COD (where αCOD $= N_1$ and $(1 - \alpha)$COD $= N_2$ [M T^{-1}]) into a river reach having a flow rate Q [L^3 T^{-1}] and velocity U [L T^{-1}], the predictive model for the river reach, excluding advection and dispersion terms, is defined in Tables 6.1 and 6.2. The result, applied to the Rhine, is illustrated in Figure 6.10.

6.6.2. Michaelis–Menten Models of Aerobic Nitrogen Cycle

Figure 6.11 summarizes an extension of this Michaelis–Menten kinetic modeling approach using the aerobic nitrogen cycle as an example (Harleman, 1978). The components of the aerobic nitrogen cycle include nitrogen in the chemical forms of ammonia, nitrite, nitrate, and the organic nitrogen content of phytoplankton and zooplankton. The cycle also includes particulate and dissolved organic nitrogen. Not included are free nitrogen and the exchange of nitrogen between the atmosphere and bottom sediments. The transformation of phytoplankton directly to dissolved organic nitrogen is assumed to be negligible relative to other nitrogen transformation processes.

The model assumes a bacterially mediated transformation from dissolved organic nitrogen to ammonia, to nitrite, and finally to nitrate nitrogen. Michaelis–Menten kinetics are assumed for the uptake of nitrate and ammonia by phytoplankton and the grazing of zooplankton on phytoplankton.

The concentration of heterotrophic bacteria affects the rate of conversion of dissolved organic nitrogen to ammonia nitrogen. The concentration of *Nitrosomonas* bacteria affects the rate of conversion of ammonia nitrogen to nitrite nitrogen and the *Nitrobacter* bacteria concentration affects the conversion rate of nitrite to nitrate nitrogen.

TABLE 6.1 Water Quality Interactions for a Benthos-Free River (Stehfest, 1977).

Easily degradable inhibiting nutrient concentration (N_1)	$\frac{dN_1}{dt} = -a_{N_1B}\mu_{BN_1}^{\max}\left(\frac{N_1}{K_{N_1}+N_1}\right)B + \frac{U\alpha \text{COD}}{Q}$
Slowly degradable inhibited nutrient concentration (N_2)	$\frac{dN_2}{dt} = -a_{N_2B}\mu_{BN_2}^{\max}\left(\frac{N_2}{K_{N_2}+N_2+a_1N_1}\right)B + \frac{U(1-\alpha)\text{COD}}{Q}$
Bacterial biomass concentration (B)	$\frac{dB}{dt} = \mu_{BN_1}^{\max}\left(\frac{N_1}{K_{N_1}+N_1}\right)B + \mu_{BN_2}^{\max}\left(\frac{N_2}{K_{N_2}+N_2+a_1N_1}\right)B - a_{BP}\mu_P^{\max}\left(\frac{B}{K_B+B}\right)P - \rho_B B$
Concentration of protozoa mass (P)	$\frac{dP}{dt} = \mu_P^{\max}\left(\frac{B}{K_B+B}\right)P - \rho_P P$
Dissolved oxygen concentration (DO)	$\frac{d\text{DO}}{dt} = K_a(\text{DO}^s - \text{DO}) - a_{\text{OB}}^{N_1}\mu_{BN_1}^{\max}\left(\frac{N_1}{K_{N_1}+N_1}\right)B - a_{\text{OB}}^{N_2}\mu_{BN_2}^{\max}\left(\frac{N_2}{K_{N_2}+N_2+a_1N_1}\right)B - a_{\text{OP}}^{B}\mu_P^{\max}\left(\frac{B}{K_B+B}\right)P - a_{\text{OB}}\rho_B B - a_{\text{OP}}\rho_P P$

TABLE 6.2 Parameter Values for Rhine River Model (Table 6.1) (20°C, 1.25 · Mean Flow) (Stehfest, 1977).

Parameter	Definition	Value
a_{N_1B}	Mass of N_1 consumed per unit increase of bacterial biomass	2.6 mgN_1/mgB
a_{N_2B}	Mass of N_2 consumed per unit increase of bacterial biomass	3.4 mgN_2/mgB
a_{BP}	Mass of bacterial biomass consumed per unit increase in protozoa mass	3.0 mgB/mgP
$a_{OB}^{N_1}$	Mass of oxygen required per unit increase in bacterial biomass from nutrient N_1	1.6 mgO/mgB
$a_{OB}^{N_2}$	Mass of oxygen required per unit increase in bacterial biomass from nutrient N_2	2.4 mgO/mgB
a_{OP}^{B}	Mass of oxygen required per unit increase in protozoa mass from feeding on bacteria	2.0 mgO/mgP
a_{OB}	Mass of oxygen required per unit decrease in bacterial biomass from endogenous respiration	1.0 mgO/mgB
a_{OP}	Mass of oxygen required per unit decrease in protozoa mass from endogenous respiration	1.0 mgO/mgP
a_1	Inhibition constant for nutrient N_1	3.0
K_a	Reaeration rate constant	varies, near 0.25 h^{-1}
K_{N_1}	Nutrient N_1 half-saturation concentration for bacterial growth	20.0 mg l^{-1}
K_{N_2}	Nutrient N_2 half-saturation concentration for bacterial growth	20.0 mg l^{-1}
K_B	Bacterial half-saturation concentration for protozoa growth	12.0 mg l^{-1}
Q	Volume flow in river reach $[L^3 T^{-1}]$	varies
U	Flow velocity in river reach $[L T^{-1}]$	varies
α	Proportion of COD that is nutrient N_1	varies
$\mu_{BN_1}^{max}$	Maximum growth rate of bacterial biomass resulting from nutrient N_1	0.48 h^{-1}
$\mu_{BN_2}^{max}$	Maximum growth rate of bacterial biomass resulting from nutrient N_2	0.10 h^{-1}
μ_P^{max}	Maximum growth rate of protozoa mass resulting from feeding on bacteria	0.36 h^{-1}
ρ_B	Endogenous respiration rate for bacterial biomass	0.06 h^{-1}
ρ_P	Endogenous respiration rate for protozoa mass	0.07 h^{-1}

In addition to the endogenous respiration rates, the first-order coefficients include K_{Z2}, the fraction of zooplankton nitrogen N_Z converted to ammonia nitrogen N_2 per unit time; K_{Pp}, the fraction of phytoplankton nitrogen N_P converted to particulate organic nitrogen N_1^p per unit time; and K_{pd}, the fraction of particulate organic nitrogen N_1^p that is transformed to dissolved organic nitrogen N_1^d per unit time. Model constants also include the maximum

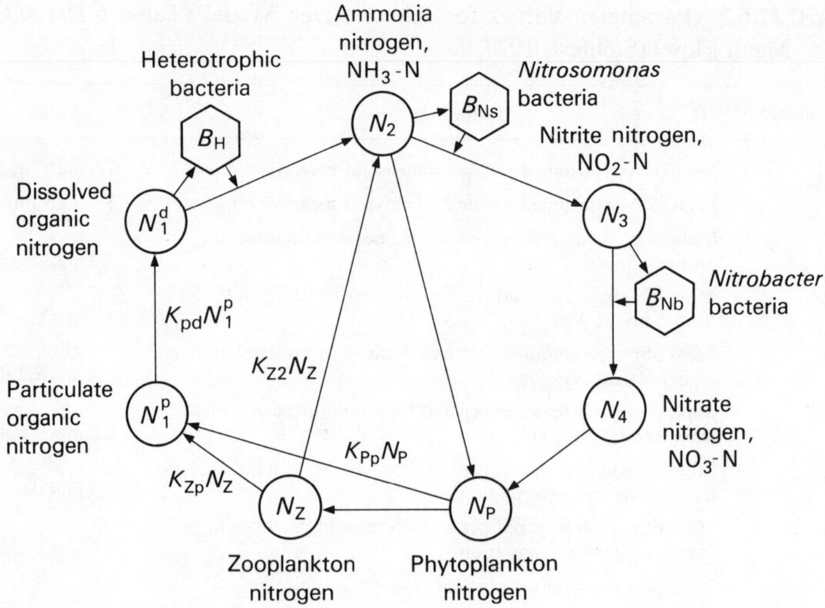

FIGURE 6.11 Aerobic nitrogen cycle.

phytoplankton nitrogen uptake rates and half-saturation constants for ammonia nitrogen ($\mu_{N_P N_2}$, $K^{P}_{N_2}$) and nitrate nitrogen ($\mu_{N_P N_4}$, $K^{P}_{N_4}$), and the zooplankton grazing rates ($\mu_{N_Z N_P}$, $K^{Z}_{N_P}$). Inhibition by ammonia of nitrate uptake by phytoplankton has been observed and could be included in the model (eqns. 6.77 and 6.78), as could ammonia regeneration by phytoplankton and the temperature and light dependence of phytoplankton uptake rates and zooplankton grazing rates. The constants a^{N}_{ij} [M M^{-1}] are the previously defined coefficients indicating, for this model, the mass of nitrogen of form i required per unit increase in the mass of material j resulting from the uptake of nutrient N (if other than i).

The model, as illustrated at the end of this section, has 22 constants and requires ten initial concentrations [M L^{-3}] for the seven nitrogen compounds (N^{p}_{1}, N^{d}_{1}, N_2, N_3, N_4, N_P, N_Z) and the three species of bacteria (B_{Ns}, B_{Nb}, B_H). All first-order constants have dimension T^{-1}, as have all maximum growth rates μ. Hence each term in each equation has dimensions M L^{-3} T^{-1}. The values of most of these constants are available in the literature.

This combination of first-order Michaelis–Menten models has been compared with the strictly first-order kinetic models (section 6.5) and has been found to fit observed data more accurately. Whether or not the increased accuracy, and expense, is warranted will depend on the management problem being solved as well as on the availability of data (Harleman, 1978).

1. *Ammonia nitrogen, N_2*

$$\frac{dN_2}{dt} = -a_{N_2}B_{Ns}\mu_{B_{Ns}}\left(\frac{N_2}{K_{N_2} + N_2}\right)B_{Ns}$$

decrease in N_2 from growth of *Nitrosomonas*, B_{Ns}

$$+ a_{N_2B_H}^{N_1^d}\mu_{B_H}\left(\frac{N_1^d}{K_{N_1^d} + N_1^d}\right)B_H + K_{Z2}N_Z$$

increase in N_2 from growth of heterotrophic bacteria, B_H

increase in N_2 from regeneration by zooplankton, N_Z

$$- \mu_{N_PN_2}\left(\frac{N_2}{K_{N_2}^P + N_2}\right)N_P$$

decrease in N_2 from growth of phytoplankton, N_P

2. *Nitrite nitrogen, N_3*

$$\frac{dN_3}{dt} = a_{N_2B_{Ns}}\mu_{B_{Ns}}\left(\frac{N_2}{K_{N_2} + N_2}\right)B_{Ns} - a_{N_3B_{Nb}}\mu_{B_{Nb}}\left(\frac{N_3}{K_{N_3} + N_3}\right)B_{Nb}$$

increase in N_3 from growth of *Nitrosomonas*, B_{Ns}

decrease in N_3 from growth of *Nitrobacter*, B_{Nb}

3. *Nitrate nitrogen, N_4*

$$\frac{dN_4}{dt} = a_{N_3B_{Nb}}\mu_{B_{Nb}}\left(\frac{N_3}{K_{N_3} + N_3}\right)B_{Nb} - \mu_{N_PN_4}\left(\frac{N_4}{K_{N_4}^P + N_4}\right)N_P$$

increase in N_4 from growth of *Nitrobacter*, B_{Nb}

decrease in N_4 due to growth of phytoplankton, N_P

4. *Phytoplankton nitrogen, N_P*

$$\frac{dN_P}{dt} = \mu_{N_PN_2}\left(\frac{N_2}{K_{N_2}^P + N_2}\right)N_P + \mu_{N_PN_4}\left(\frac{N_4}{K_{N_4}^P + N_4}\right)N_P$$

increase in N_P from phytoplankton uptake of ammonia nitrogen, N_2

increase in N_P from phytoplankton uptake of nitrate nitrogen, N_4

$$- \mu_{N_ZN_P}\left(\frac{N_P}{K_{N_P}^Z + N_P}\right)N_Z - K_{P_P}N_P$$

decrease in N_P from zooplankton grazing

decrease in N_P from endogenous respiration

5. *Zooplankton nitrogen, N_Z*

$$\frac{dN_Z}{dt} = \mu_{N_Z N_P}\left(\frac{N_P}{K_{N_P}^{Z} + N_P}\right)N_Z \quad - \quad K_{Z2}N_Z \quad - \quad K_{Zp}N_Z$$

increase in N_Z from zooplankton grazing of phytoplankton, N_P — decrease in N_Z due to regeneration of ammonia nitrogen, N_2 — decrease in N_Z from endogenous respiration

6. *Particulate organic nitrogen, N_1^p*

$$\frac{dN_1^p}{dt} = K_{Pp}N_P + Z_{Zp}N_Z \quad - \quad K_{pd}N_1^p$$

increase in N_1^p from endogenous respiration of phytoplankton, N_P, and death and defecation of zooplankton, N_Z — concentration of N_1^p transformed to dissolved organic nitrogen, N_1^d

7. *Dissolved organic nitrogen, N_1^d*

$$\frac{dN_1^d}{dt} = K_{pd}N_1^p \quad - a_{N_2 B_H}^{N_1^d}\mu_{B_H}\left(\frac{N_1^d}{K_{N_1^d} + N_1^d}\right)B_H$$

increase in N_1^d from solution of particulate organic nitrogen, N_1^p — decrease in N_1^d from growth of heterotrophic bacteria, B_H, and production of ammonia nitrogen, N_2

8. *Nitrosomonas, B_{Ns}*

$$\frac{dB_{Ns}}{dt} = \mu_{B_{Ns}}\left(\frac{N_2}{K_{N_2} + N_2}\right)B_{Ns} \quad - \quad \rho_{Ns}B_{Ns}$$

growth of B_{Ns} from uptake of ammonia nitrogen, N_2 — reduction of B_{Ns} from endogenous respiration

9. *Nitrobacter, B_{Nb}*

$$\frac{dB_{Nb}}{dt} = \mu_{B_{Nb}}\left(\frac{N_3}{K_{N_3} + N_3}\right)B_{Nb} \quad - \quad \rho_{Nb}B_{Nb}$$

growth of B_{Nb} from uptake of nitrite nitrogen, N_3 — reduction of B_{Nb} from endogenous respiration

10. *Heterotrophic bacteria, B_H*

$$\frac{dB_H}{dt} = \mu_{B_H}\left(\frac{N_1^d}{K_{N_1^d} + N_1^d}\right)B_H \quad - \quad \rho_H B_H$$

growth of B_H from uptake of dissolved organic nitrogen, N_1 — reduction of B_H from endogenous respiration

6.7. QUALITY ROUTING IN UNSTEADY FLOW

All of the models described in this chapter are based on an assumption of hydraulic steady state, i.e. $dQ/dt = 0$. Whenever the flow changes rapidly with time, as may happen when a hydraulic transient is admitted to the stream, it becomes necessary to give greater attention to temporal description of the flow regime and its effect on the transport and mixing of quality constituents identified with the flow. Such cases of interest include the routing of load patterns associated with diurnal fluctuations in wastewater discharges, "pollutographs" of storm runoff to stream systems, accidental spills of pollutants, and periodic flow oscillations in the tidal reaches of stream systems.

6.7.1. Hydrodynamic Equations

To provide the necessary spatial and temporal description of the quality routing problem it is required that we include the equations for unsteady open-channel flow (following the one-dimensional approach of Vasiliev (1976)):

Equation of continuity

$$\frac{\partial A}{\partial t} + \frac{\partial Q}{\partial x} = q, \tag{6.82}$$

where

A is the area of flow cross section $[L^2]$,
Q is the discharge $[L^3\,T^{-1}]$, $Q = UA$,
U is the mean velocity $[L\,T^{-1}]$,
q is the lateral discharge along the axis of flow (sources or sinks) $[L^2\,T^{-1}]$.

Momentum equation

$$\frac{\partial}{\partial t}(AU) + \frac{\partial}{\partial x}(QU) = -gA\left(\frac{\partial z}{\partial x} + \frac{U|U|}{C^2R} + \frac{H_c}{\rho}\frac{\partial \rho}{\partial x}\right), \tag{6.83}$$

where

g is the acceleration due to gravity $[L\,T^{-2}]$,
z is the water surface elevation $[L]$,
C is the Chezy resistance coefficient $[L^{1/2}\,T^{-1}]$,
R is the hydraulic radius $[L]$,
ρ is the density $[M\,L^{-3}]$,
H_c is the mean channel depth $[L]$.

6.7.2. Convection–Diffusion Equation

Following the traditional approach in water quality modeling, we may couple the hydrodynamic equations to water quality through the convection–diffusion equation, which for conservation of temperature may be written as

$$\frac{\partial}{\partial t}(A\Theta) + \frac{\partial}{\partial x}(Q\Theta) = \frac{\partial}{\partial x}\left(AD\frac{\partial \Theta}{\partial x}\right) - K_T A(\Theta - \Theta_E) + q\Theta_1, \quad (6.84)$$

where

Θ is the temperature (°C),
D is the diffusion coefficient $[L^2 T^{-1}]$,
K_T is the heat transfer coefficient $[T^{-1}]$,
Θ_E is the equilibrium temperature (°C),
Θ_1 is the temperature of lateral inflow (sources or sinks) (°C).

For a conservative substance carried with the flow the mass balance equation is

$$\frac{\partial}{\partial t}(AS) + \frac{\partial}{\partial x}(QS) = \frac{\partial}{\partial x}\left(AD\frac{\partial S}{\partial x}\right) + qS_1, \quad (6.85)$$

where S is the concentration of a conservative substance $[M L^{-3}]$. These equations must be supplemented with an equation of state relating water density ρ to temperature and salinity:

$$\rho = f(\Theta, s). \quad (6.86)$$

To describe the hydrochemical and hydrobiological processes in the stream, (6.85) may be modified by adding terms representing the appropriate chemical–biological linkages and interactions. This step follows closely the structure of QUAL II.

6.7.3. Solutions for Unsteady Flow

Numerical solutions of (6.82)–(6.86) have been obtained by Vasiliev and Voyevodin (1975) and Vasiliev *et al.* (1976) for unsteady flows in open-channel networks.

Depending on the particular case examined, it may prove practical to simplify the equations before seeking a solution. For example, in stream systems or well mixed estuaries it may be reasonable to neglect the interdependence of the hydrodynamic and advection–diffusion equations through the equation of state (6.86), thus uncoupling the equation set and allowing independent solution of the hydrodynamic problem. This approach has been successful in the modeling of channel networks and shallow estuaries with braided channels (Orlob, 1972).

Additionally, longitudinal dispersion may prove troublesome because it is empirical in nature and some numerical techniques introduce "mixing" effects (Bella and Dobbins, 1968). These difficulties have sometimes been successfully overcome by considering that longitudinal dispersion is negligible near the downstream boundary (Daily and Harleman, 1972; Vasiliev, 1976). Problems with numerical mixing, especially troublesome in explicit solution techniques, have sometimes been partially resolved by relying on a combination of the numerical and physical processes to describe what is actually observed in the water body. This approach has not been wholly successful since it does not address the fundamental question of a rigorous description of the internal mixing processes.

6.8. PERFORMANCE OF RECEIVING-WATER QUALITY SIMULATION MODELS

The stream quality prediction models outlined in this chapter provide only an introduction to the variety of types of steady state model used to simulate or predict water quality in water bodies. The relative reliabilities of various constituent concentration predictions provided by current water quality models are listed in Table 6.3. The table also summarizes the major impacts of various constituents in natural river systems.

TABLE 6.3 Quality Impacts and Current State of Modeling.

Component	Quality Impacts	Model Reliability	
		Streams	Estuaries
Transport, steady state		Good	Fair
Transport, dynamic		Fair	Poor
Conservative substances	Water supply	Fair	Fair
Suspended solids	Water supply Recreation	Poor	Poor
Bacteria, protozoa	Water supply Recreation	Fair	Poor
BOD, DO	Aquatic ecosystem	Good	Good
Simple chemicals and metals	Water supply Ecosystem	Fair	Fair
Synthetic chemicals and complex metals	Water supply Ecosystem	Poor	Poor
Nutrients	Aquatic ecosystem Recreation	Fair	Fair
Eutrophication (algae)	Recreation	Fair	Poor
Zooplankton and fish	Recreation	Unsatisfactory	Unsatisfactory
Temperature	Aquatic ecosystem	Fair	Fair
Virus	Water supply Recreation	Unsatisfactory	Unsatisfactory
Floating substances	Recreation	Unsatisfactory	Unsatisfactory
Color and turbidity	Recreation Water supply	Unsatisfactory	Unsatisfactory

6.9. SUMMARY

This chapter has been an introduction to water quality simulation modeling of rivers. The modeling procedures or approaches have ranged from fairly simple one-dimensional steady state approximations to procedures that are considerably more involved.

These simulation models, or their more complex extensions, are relatively crude approximations of the interactions among various constituents that occur in water bodies. Yet in spite of their current limitations, they are the only reasonable means available for predicting surface water quality. The state of the art in water quality modeling and an understanding of the physical, chemical, and biological processes that affect water quality are improving rapidly. Readers interested in pursuing this area of modeling activity are encouraged to study in greater detail alternative water quality modeling approaches and solution procedures, many of which are cited in the reference list.

APPENDIX 6.1. RATE COEFFICIENTS OF PRINCIPAL BOD DECAY PROCESSES.

Processes	Equations	Coefficients: Definitions	Coefficients: Comments	Illustrations	Temperature Coefficients
Biochemical oxidation of organic matter in laboratory	Oxidation of carbonaceous organic matter: $\frac{dL}{dt} = -KL$ $L_t = L_0 \exp(-Kt)$ $Y = L_0(1 - 10^{-kt})$ Y is the carbonaceous BOD exerted in time t [M L^{-3}]. $k_T = k_{(20°C)}\theta^{T-20}$, T is the temperature (°C).	K, carbonaceous BOD decay coefficient	Depends on type and concentration of carbonaceous organic matter, time, temperature, pH, inorganic and organic nutrients, and toxic materials. Second-order and higher-order models are also used.	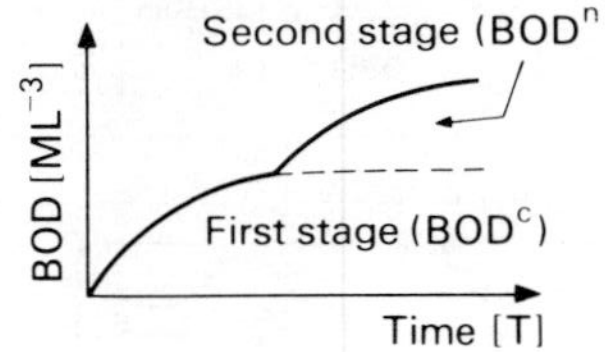	1.047–1.135
Biochemical oxidation of organic matter in streams	Oxidation of carbonaceous organic matter, L: $L = L_0 \exp(-K_d x/U)$	K_d, deoxygenation rate coefficient (for soluble carbonaceous organic material) [T^{-1}]	Depends mainly on type and concentration of organic matter remaining, and temperature. May be obtained by determining the BOD of filtered samples.	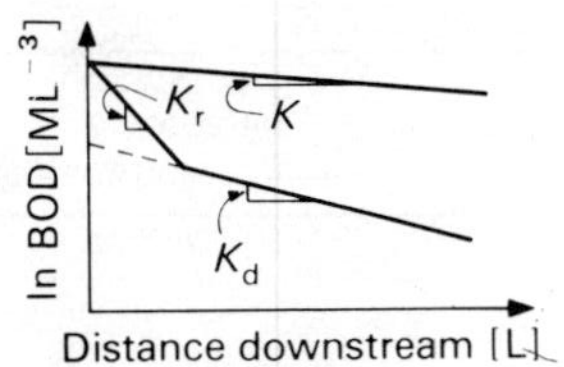	1.047–1.075
	$L = L_0 \exp(-K_r x/U)$	K_r, total removal rate coefficient for carbonaceous organic matter [T^{-1}]	Depends on concentration of total organic matter and temperature, and is affected by sedimentation, scour, and volatilization of certain organic compounds.		1.00–1.075
	$K_s = K_r - K_d$	K_s, coefficient for rate of removal by mechanisms other than oxidation [T^{-1}]	Depends on concentration of organic matter that can settle, and stream velocity. Has a small positive value in a rapidly moving stream ($U > 0.3$ m s^{-1}), and a large value in a more sluggish stream ($U < 0.18$ m s^{-1}).	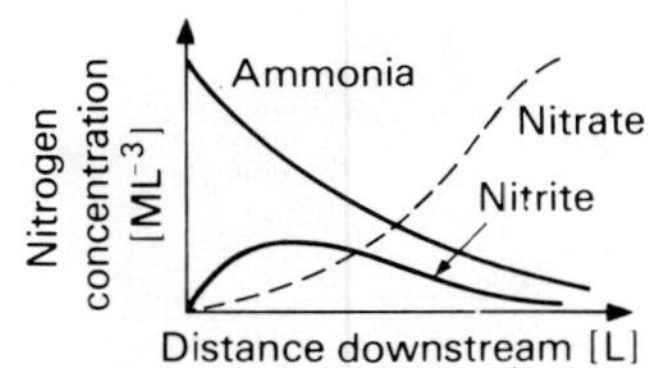	1.106 for *Nitrosomonas* 1.072 for *Nitrobacter*

(*continued over*)

APPENDIX 6.1. (*continued*)

	Oxidation of nitrogenous organic matter, N: $N = N_0 \exp(-K_n x/U)$ or $N = N_0\{1 + \exp[K_n N_0(t_{1/2} - t)]\}$ $t_{1/2}$ is the time for half completion of reaction.	K_n, rate coefficient for nitrification $[T^{-1}]$	Depends on type and concentration of nitrogenous organic matter, temperature, dissolved oxygen level, pH.		
Benthic decomposition	$-\frac{dL_b}{dt} = \frac{K_b}{1 + r_c} L_b$ L_b is the areal BOD of the bottom deposits $[M\ L^{-2}\ T^{-1}]$, r_c is the coefficient of retardation.	K_b, benthic demand rate coefficient $[T^{-1}]$	Depends on the diffusion of soluble oxidizable products of anaerobic decomposition into a stream, aerobic stabilization of the top layers of the benthos, the transport of degradable particles back into the flowing body of water, etc.	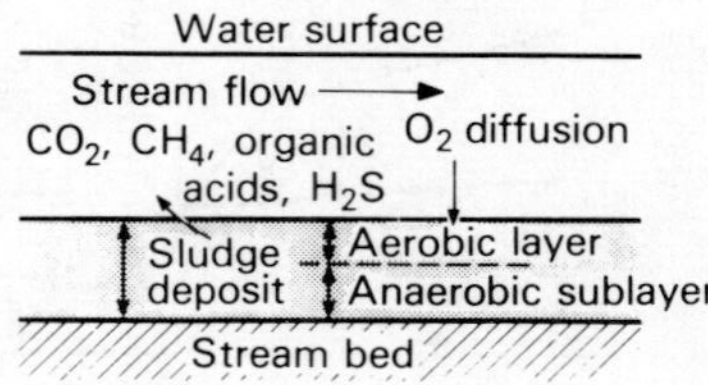	
Photosynthetic oxygen production and respiration (see also Chapter 4).	$P_{(x,t)} = P_m\left(\frac{2p}{\pi} + \sum_{n=1}^{\infty} b_n \cos 2\pi nt\right)$ $t^* = t - t_s - \frac{p}{2}$ $b_n = \frac{4\pi/p}{(\pi/p)^2 - (2\pi n)^2} \cos n\pi p$	P_m, maximum rate of photosynthetic oxygen production $[M\ L^{-3}\ T^{-1}]$	Depends mainly on solar energy; therefore it is affected by the availability and intensity of sunlight, the availability of a carbon source and nutrients, and temperature. The respiration of plants is a reverse reaction, with oxygen being consumed in the oxidation of organic matter.	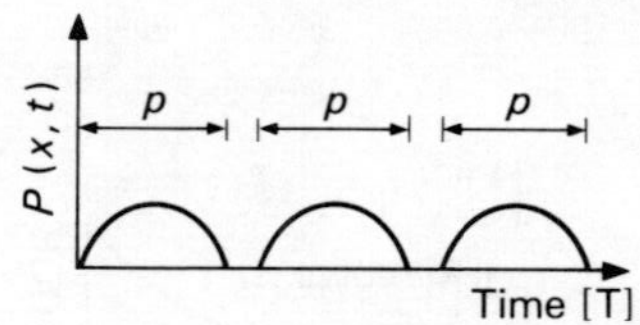	
Reaeration	$\frac{dC}{dt} = \frac{K_L A}{V}(C_s - C)$ $\frac{dD}{dt} = -\frac{K_L}{H} D = -K_a D$	K_L, mass transfer coefficient $[L\ T^{-1}]$ K_a, reaeration coefficient $[T^{-1}]$	Depends mainly on turbulence and temperature of air and water, and the presence of surface-active agents.	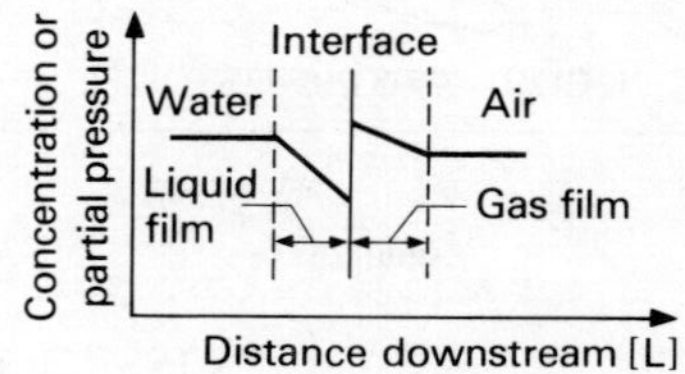	1.0241

APPENDIX 6.2. PREDICTIVE MODELS FOR REAERATION COEFFICIENT.

Investigators	Formulas for k_a (base 10) (day^{-1})	Comments and Units of Variables
Streeter and Phelps (1925)	$k_a = cU^n H_0^{-2}$	U is the mean velocity of flow ($m\ s^{-1}$) H_0 is the mean depth of water above extreme low water (m) c, n are empirical constants n ranged from 0.57 to 5.40 c varied between 23.96 and 13.06.
O'Connor and Dobbins (1956)	Anisotropic turbulence $k_a = 356.7 D_M^{0.5} S^{0.25} H^{-1.25}$ $k_{a(20°C)} = 4.8 S^{0.25} H^{-1.25}$ Isotropic turbulence $k_a = 127 D_M^{0.5} U^{0.5} H^{-1.5}$ $k_{a(20°C)} = 1.71 U^{0.5} H^{-1.5}$	D_M is the coefficient of molecular diffusivity ($m^2\ d^{-1}$) $D_{M(20°C)} = 0.000181\ m^2\ d^{-1}$ S is the slope of the river channel ($m\ m^{-1}$) H is the mean depth of flow (m) U is the mean velocity of flow ($m\ s^{-1}$).
Krenkel and Orlob (1962)	$k_{a(20°C)} = 2.4 \cdot 10^{-2} E_1^{1.321} H^{-2.32}$ $k_{a(20°C)} = 1.98 \cdot 10^{-1} E_2^{0.408} H^{-0.66}$	E_1 is the longitudinal dispersion coefficient ($m^2\ min^{-1}$) E_2 is the energy dissipation per unit mass of flow ($m^2\ min^{-3}$) H is in meters. A multiple correlation analysis was applied to the data from experiments in a laboratory flume.
Krenkel and Orlob (1963)	$k_a = 2.6 E_y^{1.237} H^{-2.087}$	E_y is the average vertical eddy diffusivity ($m^2\ s^{-1}$) H is in meters. A regression analysis was applied to data from a laboratory flume.
Churchill *et al.* (1962)	$k_{a(20°C)} = 2.18 U^{0.969} H^{-1.673}$ $k_{a(20°C)} = 2.26 U H^{-1.67}$	H is in meters, U in $m\ s^{-1}$. A dimensional analysis and multiple regression techniques were applied to the data from several tributary streams in the upper Tennessee Basin.
Owens *et al.* (1964)	$k_{a(20°C)} = 2.316 U^{0.67} H^{-1.85}$ $k_{a(20°C)} = 3.0 U^{0.73} H^{-1.75}$	H is in meters, U in $m\ s^{-1}$. Regression analysis was performed on the data from Gameson *et al.* (1955), Churchill *et al.* (1962), and Owens *et al.* (1964). Regression analysis was performed on the restricted data set from the Water Pollution Research Laboratory.

(*continued over*)

APPENDIX 6.2. *(continued)*

Investigators	Formulas for k_a (base 10) (day^{-1})	Comments and Units of Variables
Langbein and Durum (1967)	$k_{a(20°C)} = 2.23UH^{-1.33}$	H is in meters, U in m s^{-1}. Regression analysis was performed on the field data of Churchill *et al.* (1962), together with the data O'Connor and Dobbins (1956), the laboratory data of Streeter *et al.* (1936), and the data of Krenkel and Orlob (1962).
Isaacs and Gaudy (1968)		H is in meters, U in m s^{-1}.
	$k_{a(20°C)} = 2.22UH^{-1.5}$	A least-squares analysis was applied to the field data of Churchill *et al.* (1962).
	$k_{a(20°C)} = 2.064UH^{-1.5}$	Obtained from 29 of the 30 data points of Churchill *et al.* (1962).
	$k_{a(20°C)} = 1.347UH^{-1.5}$	A least-squares analysis was used on the data of Krenkel (1960) and Krenkel and Orlob (1962).
	$k_{a(20°C)} = 1.685UH^{-1.5}$	Obtained from the laboratory data of Isaacs and Gaudy (1968).
Isaacs and Maag (1969)	$k_a = 1.645\alpha_1\alpha_2UH^{-1.5}$ $k_a = 2.057UH^{-1.5}$	α_1 is a nondimensional variable that changes with the channel geometry. α_2 is a nondimensional variable that is a measure of surface velocity. H is in meters, U in m s^{-1}. For the data of Churchill *et al.* (1962) the average value of α_1 was 1.078, and the average of α_2 was 1.16.
Thackston and Krenkel (1969)	$k_a = 10.8(1 + \mathrm{Fr}^{0.5})U_*H^{-1}$	H is in meters, U in m s^{-1}. $\mathrm{Fr} = U(gH)^{-0.5}$ is the Froude number U_* is the shear velocity (m s^{-1}) g is the acceleration of gravity (m^2 s^{-1}).
Cadwallader and McDonnell (1969)	$k_a = 25.7E^{0.5}H^{-1}$	E_D is the energy dissipation per unit mass of flow (m^2 s^{-3}). H is in meters A multivariate analysis was performed using a restricted data set, and the data of Churchill *et al.* (1962), Owens *et al.* (1964), and the Water Pollution Research Laboratory channel data.

Negulescu and Rojanski (1969)	$k_a = 4.74U^{0.85}H^{-0.85}$	Developed from a flume with depth less than 0.5 m and velocity not greater than 1.2 m s^{-1}.
Padden and Gloyna (1971)	$k_{a(20°C)} = 1.963U^{0.703}H^{-1.0545}$	H is in meters, U in m s^{-1}. During this experimental study performed in a research flume, the velocities ranged from 0.03 to 0.14 m s^{-1} and depths from 0.035 to 0.19 m.
Bennett and Rathbun (1972)	$k_a = 2.33U^{0.674}H^{-1.865}$	H is in meters, U in m s^{-1}. The equation was developed using the data of Churchill *et al.* (1962) and Owens *et al.* (1964).
Tsivoglou (1967) Tsivoglou and Wallace (1972)	$k_{a(20°C)} = 1.63\Delta Ht^{-1}$	ΔH is the water surface elevation change (m) t is time of flow (days) Corrected to 20°C using $\theta_r = 1.0241$. A gas tracer technique was applied to nontidal streams.
Lau (1972a,b)	$k_{a(20°C)} = 1088.64U_*^3U^{-2}H^{-1}$	H is in meters, U in m s^{-1}. U_* is the friction velocity (m s^{-1}) A dimensional analysis was performed using the data of Krenkel (1960), Churchill *et al.* (1962), and Thackston and Krenkel (1969).
Foree (1976)	$k_{a(20°C)} = 0.116 + 2147.8S^{1.2}$	S is the channel slope (m m^{-1}) Corrected to 20°C using $\theta_r = 1.0241$. A gas tracer technique was used on streams in Kentucky.

REFERENCES

Bella, D. A., and W. E. Dobbins (1968). Difference modeling of stream pollution. *Proceedings of American Society of Civil Engineers, Journal of Sanitary Engineering Division* 94(SA5):995–1016.

Bennett, J. P., and E. R. Rathbun (1972). Reaeration in open-channel flow. *US Geological Survey Professional Paper* 737.

Cadwallader, T. E., and A. J. McDonnell (1969). A multivariate analysis of reaeration data. *Water Research* 3:731–742.

Camp, T. R. (1963). *Water and Its Impurities* (London: Chapman and Hall).

Chen, C. W., and G. T. Orlob (1975). Ecological simulation of aquatic environments. *Systems Analysis and Simulation in Ecology* vol. 3, ed. B. C. Patten (New York, NY: Academic Press), ch. 12.

Churchill, M. A., H. L. Elmore, and R. A. Buckingham (1962). The prediction of stream reaeration rates. *Tennessee Valley Authority, Chattanooga, TN Report.*

Daily, J. E., and D. R. F. Harleman (1972). Numerical model for the prediction of transient water quality in estuary networks. *MIT Department of Civil Engineering, R. M. Parsons Laboratory Report* 158, p. 226.

Dobbins, W. E. (1964). BOD and oxygen relationships in streams. *Proceedings of American Society of Civil Engineers, Journal of Sanitary Engineering Division* 90(SA3):53–78.

Fair, G. M. (1939). The dissolved oxygen sag—An analysis. *Sewage Works Journal* 11(3):445.

Foree, E. G. (1976). Reaeration and velocity prediction for small streams. *Proceedings of American Society of Civil Engineers, Journal of Environmental Engineering Division* 102(EE5):937–952.

Frankel, R. J., and W. W. Hansen (1968). Biological and physical responses in a freshwater dissolved oxygen model. *Advances in Water Quality Improvement*, eds. E. F. Gloyna and W. W. Eckenfelder (Austin, TX and London: University of Texas Press), pp. 126–140.

Gameson, A. L. H., G. A. Truesdale, and A. L. Downing (1955). Reaeration studies in lakeland beck. *Journal of Institution of Water Engineers* 9:571–594.

Hansen, W. W., and R. J. Frankel (1965). Economic evaluation of water quality, a mathematical model of dissolved oxygen concentration in freshwater streams. *Sanitary Engineering Laboratory, University of California, Berkeley Report* 65-11.

Harleman, D. R. F. (1978). A comparison of water quality models of the aerobic nitrogen cycle. *Research Memorandum* RM-78-34 (Laxenburg, Austria: International Institute for Applied Systems Analysis).

Isaacs, W. P., and A. F. Gaudy (1968). Atmospheric oxidation in a simulated stream. *Proceedings of American Society of Civil Engineers, Journal of Sanitary Engineering Division* 94 (SA2): 319–344.

Isaacs, W. P., and J. A. Maag (1969). Investigation of the effects of channel geometry and surface velocity on the reaeration rate coefficient. *Purdue University, Lafayette, IN Engineering Bulletin* vol. 53.

Krenkel, P. A. (1960). Turbulent diffusion and the kinetics of oxygen adsorption. *Ph.D. Dissertation, University of California, Berkeley.*

Krenkel, P. A., and G. T. Orlob (1962). Turbulent diffusion and the reaeration coefficient. *Proceedings of American Society of Civil Engineers, Journal of Sanitary Engineering Division* 88(SA2): 53–83.

Krenkel, P. A., and G. T. Orlob (1963). Discussion of turbulent diffusion and the reaeration coefficient. *Transactions of American Society of Civil Engineers* 128(3):293–334.

Langbein, W. B., and W. H. Durum (1967). The reaeration capacity of streams. *US Geological Survey Circular* 542.

Lau, L. Y. (1972a). A review of conceptual models and prediction equations for reaeration in open-channel flow. *Department of Environment, Ottawa Technical Bulletin* 61.

Lau, L. Y. (1972b). Prediction equation for reaeration in open-channel flow. *Proceedings of American Society of Civil Engineers, Journal of Sanitary Engineering Division* 96(SA6): 1063–1068.

Negulescu, M., and V. Rojanski (1969). Recent research to determine reaeration coefficient. *Water Research* 3:189–202.

O'Connor, D. J. (1962). The effect of stream flow on waste assimilation capacity. *Proceedings of 17th Industrial Waste Conference, Purdue University Lafayette, IN, 1–3 May. Engineering Extension Series* No. 112, pp. 608–629.

O'Connor, D. J., and D. M. Di Toro (1968). The distribution of dissolved oxygen in a stream with time-varying velocity. *Water Resources Research* 4(3):639–646.

O'Connor, D. J., and D. M. Di Toro (1970). Photosynthesis and oxygen balance in streams. *Proceedings of American Society of Civil Engineers, Journal of Sanitary Engineering Division* 96(SA2): 547–571.

O'Connor, D. J., and W. E. Dobbins (1956). The mechanism of reaeration in natural streams. *Proceedings of American Society of Civil Engineers, Journal of Sanitary Engineering Division* 82: 1–30.

O'Connor, D. J., R. V. Thomann, and D. M. Di Toro (1976). Ecologic models. *Systems Approach to Water Management*, ed. A. K. Biswas (New York, NY: McGraw-Hill), ch. 8.

Orlob, G. T. (1972). Mathematical models of estuarial systems. *Proceedings of International Conference on Water Resource Systems, Ottawa, May* vol. 1, pp. 78–128.

Owens, M., R. W. Edwards, and J. W. Gibbs (1964). Some reaeration studies in streams. *International Journal of Air and Water Pollution* 8:469–486.

Padden, T. J., and E. F. Gloyna (1971). Simulation of stream process in a model river. *Center for Research in Water Resources, University of Texas at Austin Report* EHE-70-23, CRWR-72.

Rathbun, R. E. (1977). Reaeration coefficient of streams—State of the art. *Proceedings of American Society of Civil Engineers, Journal of Hydraulics Division* 103(HY4):409–424.

Russell, C. S. (ed.) (1975). *Ecological Modeling in a Resource Management Framework. Resources for the Future Working Paper* QE-1 (Baltimore, MD: Johns Hopkins University Press).

Stehfest, H. (1977). Mathematical modeling of self-purification of rivers. *Professional Paper* PP-77-11 (Laxenburg, Austria: International Institute for Applied Systems Analysis).

Streeter, H. W., and E. B. Phelps (1925). A study of the pollution and natural purification of the Ohio River. *US Public Health Service, Washington, DC Bulletin* 146.

Streeter, H. W., C. T. Wright, and R. W. Kehr (1936). Measures of natural oxidation in polluted streams, Part III, An experimental study of atmospheric reaeration under stream flow conditions. *Sewage Works Journal* 8(2):282–316.

Texas Water Development Board (1970a). DOSAG I simulation of water quality in streams and canals. *Program Documentation and User's Manual.*

Texas Water Development Board (1970b). QUAL I simulation of water quality in streams and canals. *Program Documentation and User's Manual.*

Texas Water Development Board (1971). Simulation of water quality in streams and canals. *Report* 128, PB 202-975.

Thackston, E. L., and P. A. Krenkel (1969). Reaeration prediction in natural streams. *Proceedings of American Society of Civil Engineers, Journal of Sanitary Engineering Division* 95(SA1):65–94.

Thomann, R. V. (1972). *Systems Analysis and Water Quality Management* (New York, NY: Environmental Research and Applications, Inc.).

Thomas, H. A. (1948). Pollution load capacity of streams. *Water and Sewage Works* 95:409.

Tsivoglou, E. C. (1967). Tracer measurement in stream reaeration. *Federal Water Pollution Control Administration, US Department of Interior, Washington, DC Report.*

Tsivoglou, E. C., and J. R. Wallace (1972). Characterization of stream reaeration capacity. *Office of Research and Monitoring, US Environmental Protection Agency, Washington, DC Report* EPA R3-72-012.

Vasiliev, O. F. (1976). Mathematical modeling of water quality in rivers and water bodies. *Proceedings of 4th All-Union Hydrological Congress* vol. 9 (Leningrad: Gidrometizdat), pp. 161–168.

Vasiliev, O. F., and A. F. Voyevodin (1975). Mathematical modeling of water quality in systems of open channels. *Dinamica Spoloshnoi Sredy, Institut Gidrodinamiki SO AN USSR, Novosibirsk* No. 22, pp. 73–88.

Vasiliev, O. F., A. F. Voyevodin, and A. A. Atavin (1976). Numerical methods for the calculation of unsteady flow in systems of open channels and canals. *International Symposium on Unsteady Flow in Open Channels, IAHR, Newcastle upon Tyne, April*, Paper E2, pp. 15–22.
Water Resources Engineers, Inc. (1967). Prediction of thermal energy distribution in streams and reservoirs. *Report to California Department of Fish and Game, June* (*revised August 1968*), 90 pp.
Water Resources Engineers, Inc. (1973). Computer program documentation for the stream quality model QUAL II. *Report to US Environmental Protection Agency, Washington, DC.*
Wilson, G. T., and N. MacLeod (1974). A critical appraisal of empirical equations and models for the prediction of the coefficient of reaeration of deoxygenated water. *Water Research* 8: 341–366.

CHAPTER 6: NOTATION

B	rate of addition of BOD to overlying water from bottom deposits
D	dissolved oxygen saturation deficit
D_b	rate of removal of oxygen caused by benthic demand and plants
E	turbulent transport coefficient; longitudinal dispersion coefficient
E_v	evaporation rate
L	carbonaceous BOD
L_d	total areal BOD of benthos
n	Manning coefficient
N_1, N_2	concentration of inhibited nutrient, of inhibiting nutrient
p	period of sunlight
P	rate of oxygen production; protozoa mass
P_r	precipitation rate
q	lateral discharge along axis of flow
R	algal respiration rate; hydraulic radius
S	benthic respiration rate
S_e	slope of energy gradient
U, U^*	average stream velocity, bed shear velocity
w	concentration of pollutant
z	water surface elevation
α_1	fraction of respired algal biomass resolubilized as ammonia nitrogen
α_2	fraction of algal biomass that is phosphorus
α_3 to α_6	rates of oxygen uptake
β	toxicity coefficient
β_1, β_2	rates of oxidation
γ	density of water
Θ	temperature
λ	light extinction coefficient
ρ	respiration rate
σ_1	algal settling rate
σ_2, σ_3	benthos source rate for phosphorus, for ammonia.

7 One-dimensional Models for Simulation of Water Quality in Lakes and Reservoirs

G. T. Orlob

7.1. NEED FOR MATHEMATICAL MODELS

Mathematical modeling of surface water impoundments received its greatest impetus in the early 1960s with an awakened interest worldwide in environmental conservation and pollution control. Among the more obvious problems in need of solution were those identified with accelerating nutrient enrichment of both natural and artificial impoundments. Such widely publicized examples as Lake Erie and the Zurich See, which were often characterized as "dead"—beyond hope of recovery—drew the attention of environmentalists, pollution control regulatory authorities, and scientists to this special class of environmental problems. "Eutrophication" became synonymous with extremes in quality degradation, rather than merely a change in state of the biodynamic cycle. Efforts were intensified to find practical solutions for eliminating or significantly reducing nutrient accretions to lakes and reservoirs, often without an understanding of the consequences to the ecological and water quality balances. Even though technology then existed for removal of phosphorus from wastewaters, denitrification, and regulation of flows, there was uncertainty concerning the effectiveness of such measures in achieving desired responses in the water body. This lack of general understanding of the fundamentals of hydrodynamic, water quality, and ecological behavior of natural water bodies, particularly lakes, and of their responses to external stimuli encouraged additional field work. This work produced data from water bodies and stimulated development of analytical techniques suited to prediction of changes resulting from alternative strategies for management of these important water resources. One such technique is the one-dimensional mathematical model for prediction of thermal energy changes in a stratified impoundment, which, as will be shown later, is an essential first stage leading toward development of a general capability to describe the responses of water quality and ecological systems of natural and artificial impoundments.

7.2. ONE-DIMENSIONAL APPROXIMATION

The earliest limnological studies of the thermal structure of lakes (Hutchinson, 1957) recognized that during the period of greatest stratification, especially with the formation of a distinct thermocline, there was comparatively little variation in temperature over a horizontal plane parallel to the water surface. Although in large lakes longitudinal and lateral gradients were easily identified with transient phenomena, e.g. extreme hydrological episodes and wind disturbances, these were often rapidly dissipated by gravitational forces and the lake was restored to a condition in which the dominant variations in temperature over the greater part of the annual cycle were in the vertical direction. (Strictly speaking, we should refer to density rather than temperature although observations were seldom made of density.) For lakes of small to medium size, say less than 50 km along the major axis, the representation of the temperature structure as one-dimensional has been found by experience to be reasonable. Exceptions are lakes that are relatively narrow and deep and of relatively small volume compared with the peak rate of inflow.

A criterion that gives some guidance to the applicability of the one-dimensional approximation was suggested by Water Resources Engineers, Inc. (1969). The lake is characterized by a densimetric Froude number that compares the inertial force, represented by an average flow-through velocity, with the gravitational force tending to maintain densimetric stability:

$$\mathrm{Fr} = \frac{U}{[(\Delta\rho/\rho_0)\,gd]^{1/2}} < \frac{1}{\pi}, \tag{7.1}$$

where

U $= Q/bd$ is the average flow-through velocity [L T^{-1}],
Q is the volumetric discharge [L^3 T^{-1}],
d is the average depth [L],
$\Delta\rho$ is the density difference over depth d [M L^{-3}],
ρ_0 is the reference density [M L^{-3}],
g is the acceleration due to gravity [L T^{-2}].

If the length and volume of the impoundment are introduced as characteristic parameters, (7.1) becomes

$$\mathrm{Fr} = \frac{l}{d}\frac{Q}{V}\left(\frac{\rho_0}{g\beta}\right)^{1/2}, \tag{7.2}$$

where

l is the length [L],
V $= lbd$ is the volume [L^3],
β $= \Delta\rho/d$ is the density gradient [M L^{-4}].

TABLE 7.1 Stratification Characteristics of Selected Impoundments.

Impoundment	Location	Length l (km)	Mean Depth d (m)	Discharge: volume Q/V (s^{-1})	Densimetric Froude Number Fr	Comments
Lake Roosevelt	Washington	200	70	$5.0 \cdot 10^{-7}$	0.46	Weakly stratified
Hungry Horse Reservoir	Montana	47	70	$1.2 \cdot 10^{-8}$	0.0026	Strongly stratified
Wells Reservoir	Washington	46	26	$6.7 \cdot 10^{-6}$	3.8	Fully mixed
Cayuga Lake	New York	60	55	$4.9 \cdot 10^{-9}$	0.015	Strongly stratified
Lake Päijänne	Finland	120	17	$3.3 \cdot 10^{-10}$	0.0004	Strongly stratified
Ross Lake	Washington	32	122	$1.2 \cdot 10^{-7}$	0.0042	Strongly stratified

Deep, well stratified impoundments, for which one-dimensional models are best suited, are those for which $\mathrm{Fr} \ll 1/\pi$. Weakly stratified impoundments, for which a two-dimensional representation is sure to be necessary to describe the temperature (density)–velocity relationship, are generally those for which $0.1 < \mathrm{Fr} < 1.0$. Fully mixed systems are defined by $\mathrm{Fr} > 1.0$.

Illustrations of this classification system are summarized in Table 7.1. Typically, Hungry Horse Reservoir in Montana ($l = 47$ km, $d = 70$ m, $Q/V = 1.2 \cdot 10^{-8}\ \mathrm{s}^{-1}$) is a strongly stratified system, i.e. $\mathrm{Fr} = 0.0026 \ll 1/\pi$. In contrast, Wells Reservoir on the Columbia River ($l = 46$ km, $d = 26$ m, $Q/V = 6.7 \cdot 10^{-6}\ \mathrm{s}^{-1}$), studied by Raphael (1962a) in one of the early attempts to model the temperature regime in reservoirs, is classified as fully mixed: $\mathrm{Fr} = 3.8$. Between these extremes, Lake Roosevelt behind Grand Coulee Dam ($l = 200$ km, $d = 70$ m, $Q/V = 5.0 \cdot 10^{-7}\ \mathrm{s}^{-1}$) is a weakly stratified reservoir: $\mathrm{Fr} = 0.46$. Indeed, observations of this impoundment reveal isothermal "planes" inclined downstream toward the outlet, a physical circumstance that clearly indicates strong coupling between hydrodynamic behavior and density changes brought about by heat influx through the air–water interface along the major axis of the reservoir (Water Resources Engineers, Inc., 1969).

This chapter will deal exclusively with strongly stratified impoundments ($\mathrm{Fr} \ll 1/\pi$), for which the assumption of horizontal isothermal planes through the thermally stratified water body is reasonably consistent with reality. This case will be regarded, therefore, as one-dimensional in the mathematical sense and it will be implicit that mixing of heat introduced in the horizontal plane is instantaneous and complete. The only gradients treated will be those along the vertical axis. We begin with consideration only of thermal energy changes induced by advected flows, diffusional transport, and heat exchange across the air–water interface. The models that have evolved from these assumptions, restrictive though they may seem at first, are an important class, covering many situations of practical interest and providing basic foundations for further advances in water quality and ecological modeling.

7.3. BRIEF REVIEW OF DEVELOPMENT

7.3.1. Early Attempts to Simulate Lake Temperatures

The first important attempts to describe mathematically the annual thermal cycle in lakes appear to have been those of the pioneering limnologists, who, having observed the development of thermal stratification in small ponds and lakes, sought to quantify the vertical transfer of heat. McEwen (1929) was among the first to estimate heat transfer coefficients using assumed temperature profiles typical of thermally stratified lakes when a thermocline was evident. Ertel (1954) designed a diffusional model of the thermocline formation, assuming a constant coefficient of diffusion, independent of time or depth. Hutchinson (1957), in his monumental work, *A Treatise on Limnology*, reported these efforts and his own using data derived from field studies to explain quantitatively the formation of the thermocline, that feature of the annual thermal cycle that is so pronounced in its influence on physical, chemical, and biological responses of the lake. However, these efforts were largely unsuccessful in producing credible representations of the entire annual cycle of thermal energy changes that had been so well documented in the annals of limnological research. The most apparent deficiency at this point in the development of a quantitative description of the thermal stratification process was an adequate account of heat exchange through the air–water interface.

In the period after the Second World War, the booming construction of large dams and impoundments worldwide focused some (if not sufficient) attention on the impacts of such developments on the quality of waters released downstream. Particular concerns developed where anadromous fish migrations were likely to be affected by temperature changes and flow regulation, as in the unusually intense development of the Columbia River in the northwestern United States, a major salmon-producing river with an enormous hydroelectric power potential. Under pressure of fisheries interests, power companies sought to predict the modification that could be expected in the downstream temperature as a result of storage and flow regulation. Raphael (1961, 1962a) was among the first to devise a method for quantifying the thermal energy budget of an operating reservoir. His technique, which was carried out by tedious manual calculation, gave specific attention to heat energy gains and losses by advection, insolation, evaporation, and conduction. Results of his method, applied first to the well mixed reservoirs of the Middle Columbia River ($Fr > 1.0$), compared favorably with field observations. Subsequently, he adapted the method to predict temperatures of releases from deep reservoirs like Oroville Reservoir in California (Raphael, 1962b). The method, while giving reasonable estimates of downstream temperatures for specific projects, was not capable of describing the distribution of heat energy within the impoundment. Moreover, it was not developed in model form or computerized for general use. Similar methods were developed and applied by Burt (1958, 1960, 1963) to several reservoirs in the Pacific Northwest, but like the Raphael procedures, these were essentially manual and were not adopted by others.

7.3.2. Quantification of Heat Exchange

Among the first concerted efforts to quantify the heat exchange process, the landmark work in the late 1960s of the Tennessee Valley Authority (TVA) Engineering Laboratory under the direction of Rex A. Elder stands out as unique. Elder's co-workers, W. O. Wunderlich and R. Gras, are credited with the thorough and careful research leading to the authoritative report, *Heat and*

Mass Transfer Between a Water Surface and the Atmosphere (Tennessee Valley Authority, 1972). This document served as an important building block for parallel development of the first working mathematical models of the thermal stratification process in deep reservoirs. It is the primary source for the quantitative description of the heat exchange process to be described in this chapter (see also the appendix to Chapter 5).

7.3.3. First Attempts at One-Dimensional Temperature Modeling

Formal mathematical modeling of temperature changes in deep, stratified impoundments appears to have been stimulated by the efforts of both the TVA and the California Department of Fish and Game, the latter agency being concerned with the effects of large impoundments, e.g. Oroville Reservoir, on salmon migration. In 1965 the Department contracted with Water Resources Engineers, Inc. to develop a predictive model (Orlob, 1965). In the following year the TVA and WRE collaborated in developing the model. The TVA quantified heat exchange phenomena and conducted field studies on several of its reservoirs to provide data for model calibration and validation (Elder and Wunderlich, 1968). This combined effort culminated in a working model that was first applied to Fontana Reservoir in the TVA system; it was later revised (WRE, Inc., 1968) as a result of experience with several reservoirs in the northwestern United States (WRE, Inc., 1969). The characteristics of the model and preliminary test results were first reported by Orlob and Selna (1967, 1970). Subsequently, it was thoroughly documented for the Environmental Protection Agency (Gaume and Duke, 1975). It is currently being used in various forms by many United States governmental agencies.

In a parallel research and development effort, spanning the same period in the late 1960s and also in collaboration with the TVA Engineering Laboratory, D. R. F. Harleman and his co-workers at the Massachusetts Institute of Technology developed a comparable one-dimensional temperature simulation model (Huber *et al.*, 1972). The development effort at MIT focused more strongly at first on fundamental heat transfer mechanisms, utilizing laboratory models as prototypes for mathematical development (Dake and Harleman, 1966). Subsequently, however, the MIT model was extended to simulation of actual reservoirs, e.g. Fontana Reservoir.

The MIT model, which has also been well documented and tested, is presently used by the TVA and other United States governmental agencies. Apart from some refinements in treating inflow and withdrawal processes, which will be discussed later, the MIT model is substantially equivalent in performance to the WRE model.

A one-dimensional temperature model designed for deep, stratified lakes was developed at the Cornell Aeronautical Laboratory (Sundaram *et al.*, 1969) and applied to Cayuga Lake in upper New York State in a study of power plant cooling water discharges. The model is based on the one-dimensional diffusion equation and is limited by assumptions of a constant cross-sectional area (horizontal plane), absorption of all incoming heat energy in the surface layer, and neglect of heat advected either laterally or vertically in the water column, except that associated with power plant withdrawals or discharges. Wind effects are included. Apparently, because of its case-specific nature, the model has been applied only to Cayuga Lake.

7.3.4. Extensions to Segmented Impoundments

Other significant developments in mathematical modeling of temperature in deep, stratified impoundments include several attempts to extend the one-dimensional concept to segmented, weakly stratified reservoirs. Water Resources Engineers, Inc. (1968) modeled Lake Roosevelt as a six-segment system and Baca *et al.* (1974), using a modification of the WRE model, simulated American Falls Reservoir with a three-segment system. In each instance, model results compared favorably with observations of the impoundment, but difficulties were experienced in interfacing of segments according to the densimetric criteria used for introducing advective flows into the water column.

7.3.5. Dual-Purpose One-Dimensional Models

Several one-dimensional models, capable of simulating temperature in addition to other quality constituents, are worthy of note at this point for reason of completeness, although they will be discussed more fully in section 7.9, which deals with ecological and water quality models.

The model LAKECO, also developed by WRE, includes the temperature simulation routines of the earlier Deep Reservoir Model in addition to the capability for simulation of some 20–25 abiotic and biotic state variables. This model, in turn, is embedded in the package WQRRS (Water Quality River–Reservoir Simulation) developed for the Hydrologic Engineering Center of the United States Army Corps of Engineers by WRE and others (HEC, 1974). The solution technique used in the original temperature model (WRE, Inc., 1968, 1969) has become an integral part of each of these packages.

A one-dimensional water-quality–ecological model developed by Baca and Arnett (1976) utilizes a finite-element technique for solution of the governing equations, including those for thermal energy balance. The solution technique is apparently superior in some respects to those of the WRE and MIT models, coping more reliably in the one case with steep gradients and in the other with avoiding instabilities. The technique will be presented in some detail later in the chapter.

A unique model based on principles of energy conservation, including kinetic energy induced by wind shear, has been developed by Imberger *et al.*

(1978). The model, which is considered suitable for small to medium-sized reservoirs (less than 10 km long), simulates temperature and salinity variations in one-dimensional systems that are stratified in density. The model will be described more fully in section 7.13.2. A somewhat similar approach has been applied by Stefan and Ford (1975) to simulate temperature dynamics of small lakes in the north-central United States.

7.3.6. One-Dimensional Destratification Simulation

Henderson-Sellers (1978) described use of a one-dimensional model to simulate the result of jet-induced destratification. Details of the model were not provided; however, it appears that inflows and outflows (associated with destratification) are treated in much the same manner as with the WRE and MIT models, by observing the requirements for neutral buoyancy of reinjected flows and densimetric stability in the water column. Similar applications of the WRE temperature and LAKECO models have been made in studies of the effects of recirculation in pumped storage schemes (Chen and Orlob, 1972).

7.4. CONCEPTUAL REPRESENTATION OF A ONE-DIMENSIONAL LAKE OR RESERVOIR

In most of the models described above (an exception being that of Imberger *et al.*, 1978), the one-dimensional impoundment is conceptualized from an Eulerian viewpoint as a continuum of horizontal slices, usually of equal thickness, as illustrated in Figure 7.1. The slices, or volume elements, are of fixed volume and constant thickness, except for that at the surface, which is allowed to vary in some models in accordance with changes in impoundment volume. (In the WRE model, for example, slices are added or subtracted as the water surface fluctuates.) Each volume element is capable of receiving laterally advected flows, discharging from the impoundment, and transferring advected flows along the vertical axis while preserving mass continuity. Heat or mass passes through the bounding horizontal planes by advection and diffusion, and heat energy by direct insolation depending on the location of the slice. Of course, heat may be transferred into or out of a slice laterally by advection. Inflows occur in accordance with densimetric criteria, that is at elevations where there is a correspondence between inflow density and the density of the water within the slice. Outflows occur at specified withdrawal points, including the surface in the case of natural lakes and losses due to evaporation. The general notation applied to successive slices is illustrated in Figure 7.2.

The notation adopted in Figure 7.1 corresponds to that of the WRE model, but is virtually the same as for the MIT model. It will be used in development of the general formulation for the one-dimensional temperature model as follows.

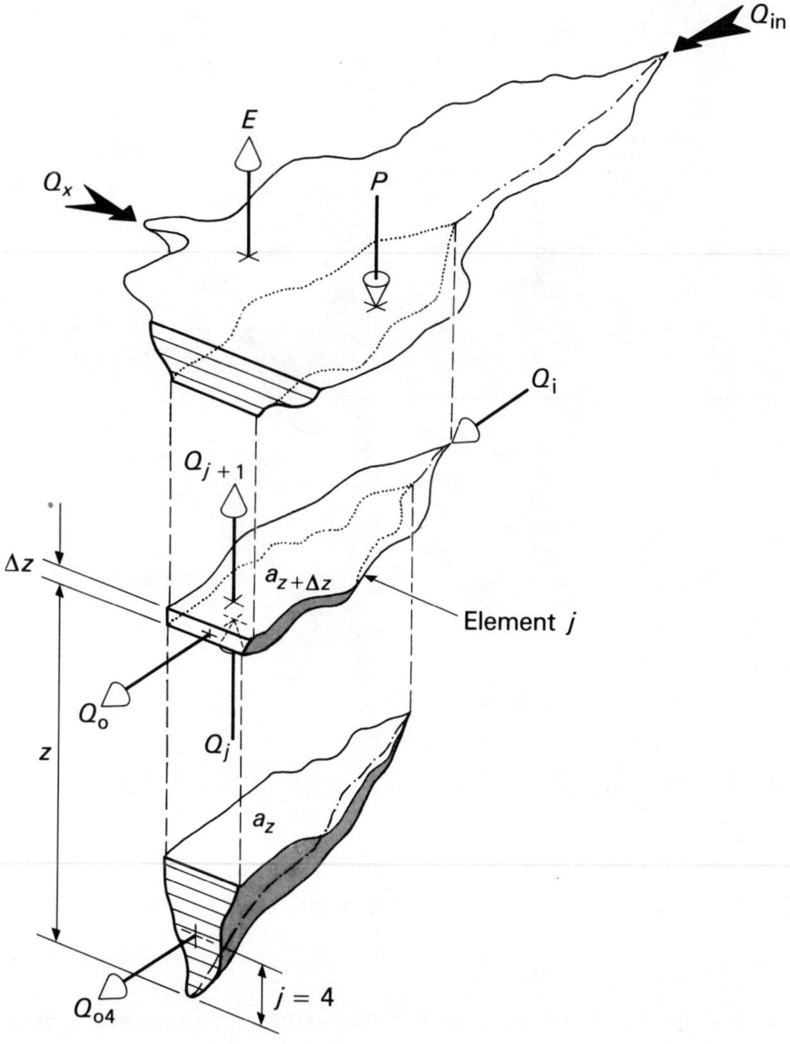

FIGURE 7.1 Conceptual representation of a stratified reservoir (after WRE, Inc., 1968). Q_{in}, inflow to reservoir; Q_x, local drainage; Q_i, advected flow to element j; Q_0, withdrawal from element j; Q_j, Q_{j+1}, vertically advected flows to element j; Q_{04}, withdrawal from reservoir at level $j = 4$; E, evaporation rate; P, precipitation rate.

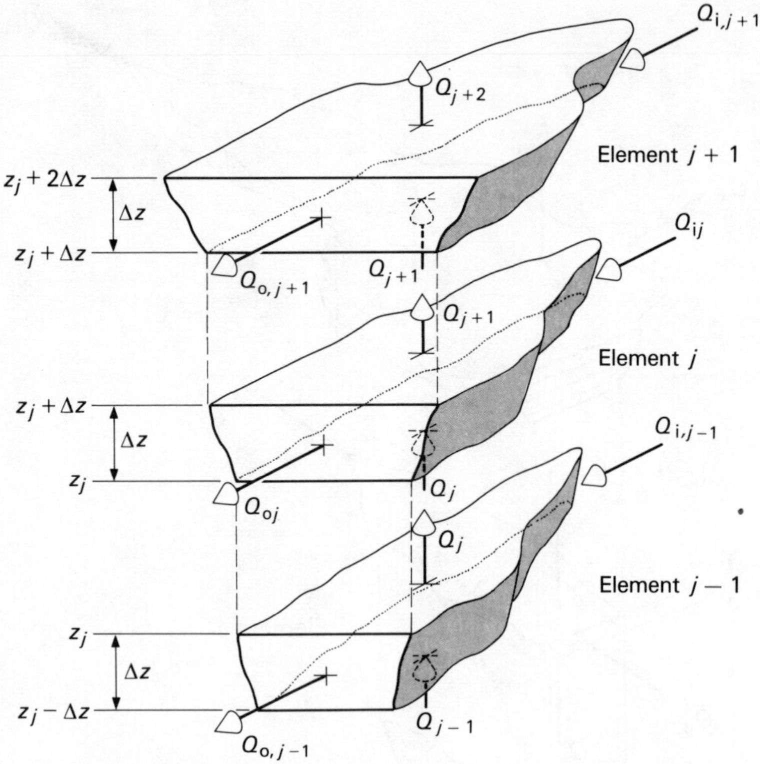

FIGURE 7.2 Mass continuity between adjacent reservoir elements (after WRE, Inc., 1968).

7.5. FORMULATION FOR TEMPERATURE PREDICTION

7.5.1. Conservation of Mass

Mass conservation for a volume element $\overline{V}_j$ within the impoundment is expressed by

$$\frac{\partial \overline{V}_j}{\partial t} = Q_{zj} - Q_{z,j+1} + Q_{ij} - Q_{oj}, \tag{7.3}$$

where

$\overline{V}_j$ $= \frac{1}{2}(a_z + a_{z+\Delta z})\Delta z$ is the volume of the jth element [L^3],
$a_z, a_{z+\Delta z}$ are the areas of horizontal planes bounding $\overline{V}_j$ [L^2],
Δz is the thickness of the element [L],
z is the depth, measured from the bottom [L],
j is the element index; $j = 1$ at the bottom,

Q_z is the vertical flow rate $[L^3 T^{-1}]$,
Q_i is the flow advected into the jth element in the horizontal plane $[L^3 T^{-1}]$,
Q_o is the flow advected out of the control volume in the horizontal plane $[L^3 T^{-1}]$.

$\partial \bar{V}_j / \partial t = 0$, except for the surface element.

7.5.2. Conservation of Heat Energy

Conservation of heat energy in volume element $\bar{V}_j$ is expressed by

$$\frac{\partial H_j}{\partial t} = (h_i - h_o + h_{sz})_j - (h_{wj} - h_{w,j+1}) - (h_{dj} - h_{d,j+1}), \qquad (7.4)$$

where

$H_j = c\rho V_j \Theta_j$ is the heat content of the jth element (J),
c is the specific heat $[J M^{-1} {}^\circ C^{-1}]$,
ρ is the density $[M L^{-3}]$,
Θ_j is the temperature (°C),
$h_i = c\rho Q_i \Theta_i$ is the heat advected by inflow $[J T^{-1}]$,
$h_o = c\rho Q_o \Theta_o$ is the heat withdrawn by outflow $[J T^{-1}]$,
$h_{sz} = \int_z^{z+\Delta z} q_{sz} a_z \, dz$ is the direct insolation $[J T^{-1}]$,
$q_{sz} = (1-\beta) q_{sn} e^{-\eta z}$ is the solar radiation flux at depth z $[J L^{-2} T^{-1}]$,
q_{sn} is the net solar radiation flux penetrating the surface $[J L^{-2} T^{-1}]$,
β is the ratio of radiation absorbed at the surface to the net incoming radiation,
η is the bulk extinction coefficient $[L^{-1}]$,
$h_{wj}, h_{w,j+1} = c\rho Q(z)\Theta_z$ represent the heat advected along the vertical axis $[J T^{-1}]$,
$h_{dj}, h_{d,j+1} = c\rho E_z a_z \partial\Theta/\partial z$ represent the heat diffused along the vertical axis $[J T^{-1}]$,
E_z is the coefficient of vertical diffusion $[L^2 T^{-1}]$.

7.5.3. Heat Budget for an Element

The heat energy budget for element $\bar{V}_j$ at temperature Θ_j is given by

$$\frac{\partial(\bar{V}_j \Theta_j)}{\partial t} = \underbrace{(Q_i \Theta_i - Q_o \Theta_o)_j}_{\text{local advection}} + \underbrace{\left(\frac{1}{c\rho}\int_z^{z+\Delta z} q_{sz} a_z \, dz\right)_j}_{\text{solar radiation}}$$

$$\underbrace{- (Q_z \Theta)_j + (Q_z \Theta)_{j+1}}_{\text{vertical advection}} + \underbrace{\left(Ea\frac{\partial\Theta}{\partial z}\right)_j - \left(Ea\frac{\partial\Theta}{\partial z}\right)_{j+1}}_{\text{vertical diffusion}}. \qquad (7.5)$$

Solution of (7.5) for $\Theta_j, j = 1, 2, 3, \ldots, n$, requires knowledge of $\bar{V}_j$, a, Q_i, Q_o, q_{sz}, and E. $\bar{V}_j(z)$ and $a(z)$ are physical characteristics that can be obtained from area–volume–elevation curves if Δz is fixed. The other terms may be evaluated as follows.

7.5.4. Evaluation of Heat Budget Terms

Local Advection

In the earliest versions of the one-dimensional temperature model the local inflow at temperature Θ_i was simply introduced into an element $\bar{V}_j$, where the condition $\Theta_j > \Theta_i > \Theta_{j-1}$ was satisfied. Outflow was considered to occur from an element spanning the vertical dimension of the outlet(s). In a later version for the US Environmental Protection Agency (EPA) (WRE, Inc., 1969) outflows were distributed uniformly over a layer estimated by Debler's (1959) criterion, which is related to the densimetric Froude number in the zone of the outlet. For withdrawal well below the thermocline sensitivity analysis indicated that a uniform distribution produced virtually equivalent results to those obtained when the outflow was given a pattern of Gaussian form. Since velocity patterns in the region of the outlet were actually poorly defined the added refinement of a nonuniform distribution was not considered justified.

Huber *et al.* (1972) have employed the theoretical relationship of Kao (1965) to estimate the thickness of the withdrawal layer:

$$\delta = 4.8\left(\frac{q^2}{g\epsilon}\right)^{1/4}, \tag{7.6}$$

where

δ is the thickness of a withdrawal layer of uniform velocity approaching a line sink [L],
q is the outflow per unit width of reservoir $[\mathrm{L}^2\,\mathrm{T}^{-1}]$,
ϵ $= (1/\rho)\,\mathrm{d}\rho/\mathrm{d}z\ [\mathrm{L}^{-1}]$.

They distribute the withdrawal in a Gaussian pattern over a layer assumed to be twice as thick as estimated by (7.6).

In the MIT model, inflows are also distributed in a Gaussian pattern defined by a standard deviation in the inflow velocity field.

Solar Radiation Flux

The net flux of solar radiation to the water column at the air–water interface, q_{sn}, is either measured directly or estimated from meteorological measurements. Procedures for estimating q_{sn} that have been incorporated into the WRE and MIT reservoir models, as well as the stream models, QUAL I and QUAL II

(Chapter 6), were developed by the Engineering Laboratory of the Tennessee Valley Authority (1972). Details of the estimation equations for the principal heat flux components are given in the appendix to Chapter 5.

The incremental supply of solar radiation to an element $\bar{V}_j$ in the water column is obtained by integrating $q_{sz} a_z$ between z and $z + \Delta z$. The heat energy supplied to the uppermost element ($j = n$) is the total heat flux across the air–water interface less that passing through the bottom of the element ($j = n - 1$), i.e.

$$\frac{\partial(\bar{V}_n \Theta_n)}{\partial t} = \frac{1}{c\rho}(\beta q_{\mathrm{sn}} + q_{\mathrm{at}} + q_{\mathrm{ws}} + q_{\mathrm{e}} + q_{\mathrm{h}})a_n$$
$$- \frac{1}{c\rho}[(1 - \beta)q_{\mathrm{sn}} \mathrm{e}^{-\eta z_n}]\, a_{n-1}, \tag{7.7}$$

where

$\bar{V}_n$ is the volume of the surface element [L^3],
Θ_n is the temperature of the surface element (°C),
a_n is the area of the reservoir surface [L^2],
a_{n-1} is the area at the bottom of the surface element [L^2],
z_n is the thickness of the surface element [L],
c is the specific heat [cal force^{-1}],
ρ is the density [force · T^2 L^{-4} or M L^{-3}],
q_{sn} is the net solar radiation heat flux [J L^{-2} T^{-1}],
q_{at} is the net atmospheric radiation heat flux [J L^{-2} T^{-1}],
q_{ws} is the water surface radiation heat flux [J L^{-2} T^{-1}],
q_{e} is the evaporation heat flux [J L^{-2} T^{-1}],
q_{h} is the sensible heat flux [J L^{-2} T^{-1}],
β is the ratio of absorbed to net incoming radiation,
η is the bulk light extinction coefficient.

For each element below the uppermost element in the water column the incremental solar radiation flux is given by

$$\frac{\partial(\bar{V}_j \Theta_j)}{\partial t} = \frac{1}{c\rho}(1 - \beta)\, q_{\mathrm{sn}} \int_{j-1}^{j} \mathrm{e}^{-\eta z_j}\, a_j \,\mathrm{d}z, \tag{7.8}$$

in which $1 < j < n - 1$. The extinction coefficient η may itself vary with depth, particularly within the epilimnion of eutrophic impoundments. For most practical purposes there is no net solar radiation flux to elements below the thermocline.

The appendix to Chapter 5 provides details of the estimation of the heat flux terms in (7.7) and (7.8).

7.5.5. Effective Diffusion

In addition to vertical advection due to flow imbalances at various levels in the impoundment, heat is transferred by diffusion, by the random motions of ambient turbulence, and by secondary currents not otherwise represented in the one-dimensional approximation. Collectively, these diffusion and dispersion mechanisms have been described as "effective diffusion" (HEC, 1974) by considering their combined net effects in heat transfer as analogous to the truly random process of molecular diffusion. The magnitude of the effective diffusion coefficient E_z varies widely in both time and space, depending on the physical circumstances of the water body being represented by the one-dimensional model. In laboratory experiments in the total absence of fluid turbulence, E_z may approach minimal levels in the range of $1 \cdot 10^{-6}$ to $1 \cdot 10^{-8}$ cm^2 s^{-1}, while in modeling of large lakes and reservoirs E_z may be in the range of $1 \cdot 10^{-4}$ to $1 \cdot 10^{-2}$ cm^2 s^{-1} (WRE, Inc., 1968). In these cases the coefficient is regarded as essentially empirical and must be derived from observations of the impoundment with due regard for the unknown mechanisms, including even characteristics of the model itself, e.g. "numerical mixing."

E_z is evaluated from field data by integrating the heat budget equation between the limits of the reservoir bottom ($z = 0$) and a specific elevation z, accounting for all explicitly defined heat transport or flux terms, i.e.

$$\bar{E}_z = \frac{1}{a_z \, \overline{\partial\Theta/\partial z}} \int_{t_1}^{t_2} \left(\int_0^z \Theta_z \, a_z \, dz - Q_a \Theta_a - Q_z \Theta_z - \frac{1}{c\rho} \int_0^z q_{sz} \, a_{z,1} \, dz \right) dt, \tag{7.9}$$

where

- $\bar{E}_z$ is the average effective diffusion coefficient at level z over the time interval $t_2 - t_1$ [$L^2 T^{-1}$],
- a_z is the horizontal area at level z [L^2],
- $\overline{\partial\Theta/\partial z}$ is the average temperature gradient at level z over the interval $t_2 - t_1$ [°C L^{-1}],
- Θ_z is the temperature at level z (°C),
- Q_a is the lateral advected flow, $Q_a(z, t)$[$L^3 T^{-1}$],
- Θ_a is the temperature of the lateral advected flow, $\Theta_a(z, t)$(°C),
- $Q_{z,1}$ is the vertical advected flow at level z_1, $Q_{z,1}(z, t)$[$L^3 T^{-1}$],
- q_{sz} is the shortwave insolation, $q_{sz}(z)$[J $L^{-2} T^{-1}$].

Analyses of temperature profiles in actual lakes and reservoirs have demonstrated that E_z varies widely with depth, with minima occurring in the region of the thermocline and near the bottom (Orlob and Selna, 1967, 1970; WRE, Inc., 1968). Highest values occur, of course, in the epilimnion and are clearly related to mixing induced by wind shear at the surface. At intermediate depths in the hypolimnion E_z may be several orders of magnitude greater than at the thermocline.

Various empirical representations of E_z have been used in one-dimensional temperature models, ranging from those derived by the method of (7.9) to merely applying a constant, independent of depth. Water Resources Engineers, Inc. (1969) derived a functional relationship between E_z and densimetric stability $(1/\rho)\,d\rho/dz$ that indicated a maximum of about $2\ \mathrm{cm^2\ s^{-1}}$ for densimetric stabilities above about $1 \cdot 10^{-6}\ \mathrm{m^{-1}}$. The Hydrologic Engineering Center, in its WQRRS model (HEC, 1974), recommends using a constant value of E_z for density gradients less than the critical, and lower values depending on the gradient within the zone of the thermocline. Experience has shown that in most instances minimum values two or three orders of magnitude greater than the molecular diffusion coefficient are appropriate, although several investigators have noted that simulation of heat transport deep in the reservoir is especially sensitive to the density gradient and to the numerical methods of calculation (Orlob and Selna, 1967; Harleman and Hurley, 1976). Explicit numerical techniques, in particular, often introduce numerical mixing effects that may be greater than those of the physical processes being approximated in the model (Bella and Dobbins, 1968).

Investigations of the development and erosion of the thermocline (Sundaram and Rehm, 1973; Spaulding and Svensson, 1976; Svensson, 1978) have led to improved descriptions of the mixing induced by wind shear at the water surface. From the classical mixing length concepts of Prandtl, Spaulding and Svensson obtained a description of E_z that takes the form:

$$E_z = C\rho \frac{K^2}{\epsilon}, \tag{7.10}$$

where

- C is an empirical constant,
- K is the turbulent kinetic energy,
- ϵ is the rate of turbulent kinetic energy dissipation per unit mass.

Values of K and ϵ are calculated from two transport equations derived from the Navier–Stokes equations. Comparison of results from laboratory flume experiments (under conditions of steady wind) with the model predictions indicates development and erosion of the vertical structure comparable with observations of lakes and reservoirs (Svensson, 1978). However, there may be some practical limitations in the use of this approach in long-term simulation at lake temperatures because of computational requirements (Bloss and Harleman, 1979).

The combined effect of mixing processes in small to medium-sized lakes on heat energy distribution has been successfully simulated using a total energy integration approach (Stefan and Ford, 1975; Imberger *et al.*, 1978). This modeling approach accounts for both potential and kinetic energy exchanges through a continuum of one-dimensional segments. Depending on their relative importance the energy budget can include terms for heat exchange through the air–water interface, convective mixing due to diurnal cooling, wind-induced turbulence, internal waves and seiches, and viscous damping. The Stefan–Ford model, known also as the Minnesota Lake Temperature Model (MLTM), has been applied with excellent results to a number of small dimictic lakes in the northern United States. DYRESM, the Imberger *et al.* model, has been applied to several medium-sized reservoirs, notably Wellington Reservoir in Australia. It is described more fully in section 7.13 and some representative results are illustrated in Figure 7.10.

An intermediate approach to representation of mixing processes in stratified reservoirs has been taken by Hurley-Octavio *et al.* (1977) and Bloss and Harleman (1979). These investigators have accounted for wind-induced entrainment across the thermocline by equating turbulent kinetic energy input from wind to potential energy increases due to mixing against the density gradient. Transient and dissipative effects of the entrainment process are included in an updated algorithm for the MIT Lake and Reservoir Model.

7.5.6. Dimictic Lakes: Simulation of the Freeze–Thaw Cycle

Lakes in the north or south temperate and polar regions experience two cycles of stratification–destratification annually. Inasmuch as the winter cycle, associated with the formation of an ice cover over the impoundment during extended periods of subzero temperature, is of special consequence from both water quality and ecological viewpoints it is desirable to include in temperature simulation models a capability to represent the freeze–thaw cycle.

The original WRE one-dimensional temperature model was modified to include the freeze–thaw cycle for applications to two impoundments on the border between the United States and Canada, Lake Koocanusa (Chen and Orlob, 1973) and Lake Ross (Norton and King, 1975). An additional term, accounting for the heat exchange accompanying the change of state of water from liquid to solid, was added to the heat budget equation. When the temperature in the water column is dropped to 4°C, a continuing net loss of heat at the surface reduces surface temperatures toward 0°C. When freezing temperatures are reached, further loss of heat, proportional to the latent heat of fusion, results in ice formation. Other heat transfer processes are modified in accordance with the restriction imposed by the ice sheet. In the spring, net warming results in melting of the ice until open water develops, whereupon the water column returns to isothermal conditions at the maximum density point, 4°C.

7.6. TECHNIQUES FOR SOLUTION OF ONE-DIMENSIONAL TEMPERATURE MODELS

7.6.1. Implicit Finite-Difference Method

The following solution technique, devised by Water Resources Engineers, Inc. (1968, 1969), is employed in the WRE and WQRRS versions of the one-dimensional temperature model. Equation 7.5 is formed into a finite-difference set for the discretized impoundment according to the notation of Figures 7.1 and 7.2. The rate of change of temperature Θ_j in the jth element is given by

$$\dot{\Theta}_j = \left[\frac{1}{V_j}\left(Q + \frac{Ea}{\Delta z}\right)_j\right]\Theta_{j-1} - \left[\frac{1}{V_j}\left(\frac{Ea}{\Delta z}\right)_j + \frac{1}{V_j}\left(\frac{Ea}{\Delta z}\right)_{j+1} + \frac{1}{V_j}(Q_{j+1} + Q_{oj} + \dot{V}_j)\right]\Theta_j + \left[\frac{1}{V_j}\left(\frac{Ea}{\Delta z}\right)_{j+1}\right]\Theta_{j+1} + \frac{1}{V_j}(Q_i\Theta_i)_j + \frac{1}{V_j}\left(\frac{q_{sz}a}{c\rho}\right)_j. \tag{7.11}$$

Written in matrix form, where the bracketed { } terms (vectors) are coefficients of Θ_z, (7.11) becomes

$$[\mathbf{I}]\{\dot{\boldsymbol{\Theta}}\} = [\mathbf{S}]\{\boldsymbol{\Theta}\} + \{\boldsymbol{P}\}, \tag{7.12}$$

where $[\mathbf{I}]$ is an identity matrix, $[\mathbf{S}]$ is a tridiagonal matrix of coefficients, and $\{\boldsymbol{P}\}$ is a vector of the known parameters and external heat soûrces and sinks.

The finite-difference equations are integrated by a linear acceleration method, wherein the incremental change in temperature is approximated by

$$\Theta_{t+\Delta t} = \Theta_t + \tfrac{1}{2}\Delta t\,(\dot{\Theta}_t + \dot{\Theta}_{t+\Delta t}) \tag{7.13}$$

or

$$\Theta_{t+\Delta t} = \tfrac{1}{2}\Delta t\dot{\Theta}_{t+\Delta t} + \tfrac{1}{2}\Delta t\dot{\Theta}_t + \Theta_t, \tag{7.14}$$

in which Θ and its derivative $\dot{\Theta}$ are known at time t. If the time subscripts are removed to simplify notation, the temperature of the jth element becomes

$$\Theta_j = \tfrac{1}{2}\Delta t\,\dot{\Theta}_j + b_j, \tag{7.15}$$

which in matrix form for the full equation set is

$$\{\boldsymbol{\Theta}\} = \tfrac{1}{2}\Delta t\,\{\dot{\boldsymbol{\Theta}}\} + \{\mathbf{b}\}, \tag{7.16}$$

where $\{\mathbf{b}\}$ is a matrix of Θ and its time derivative for time t.

Equation 7.16 is then substituted in eqn. 7.12, giving

$$[\mathbf{I}]\{\dot{\boldsymbol{\Theta}}\} = [\mathbf{S}]\{\mathbf{b}\} + \tfrac{1}{2}\Delta t\,[\mathbf{S}]\{\dot{\boldsymbol{\Theta}}\} + \{\boldsymbol{P}\}, \tag{7.17}$$

which may be simplified to

$$[\mathbf{S}^*]\,\{\dot{\boldsymbol{\Theta}}\} = \{\boldsymbol{P}^*\}, \tag{7.18}$$

where

$$[\mathbf{S}^*] = [\mathbf{I}] - \tfrac{1}{2}\Delta t\,[\mathbf{S}]$$

and

$$\{\boldsymbol{P}^*\} = [\mathbf{S}]\,\{\mathbf{b}\} + \{\boldsymbol{P}\}.$$

The computational procedure is as follows.

(1) Form $\{\mathbf{b}\}$ from initial conditions or the most recently computed values of $\{\boldsymbol{\Theta}\}$ and $\{\dot{\boldsymbol{\Theta}}\}$.
(2) From known values of coefficients and boundary conditions, determine for the end of the time step the values of $[\mathbf{S}]$, $\{\boldsymbol{P}\}$, $[\mathbf{S}^*]$, and $\{\boldsymbol{P}^*\}$.
(3) Solve for $\{\dot{\boldsymbol{\Theta}}\}$ from (7.18).
(4) Solve for $\{\boldsymbol{\Theta}\}$ from (7.16).
(5) Repeat for all succeeding time steps.

The steady state solution, i.e. $\dot{\Theta} = 0$, is readily obtained by solving

$$[\mathbf{S}]\,\{\boldsymbol{\Theta}\} = -\,\{\boldsymbol{P}\}, \tag{7.19}$$

a special case of (7.12).

The solution technique outlined above has been demonstrated as efficient in many practical applications. Its particular advantages are stability of solution and flexibility in adapting to temporal variations in boundary conditions.

7.6.2. Explicit Finite-Difference Method

An explicit solution technique has been used in the MIT model (Ryan and Harleman, 1971), wherein the equation set is solved in spatial order ($j = 1, 2, 3, \ldots, n$) for successive times ($t = 1, 2, 3, \ldots, k$). The method requires adherence to the more restrictive of two stability criteria,

$$\Delta t < \frac{1}{2}\frac{\Delta z^2}{E(z)} \tag{7.20}$$

or

$$\Delta t < \Delta z\,\frac{a(z)}{Q(z)}, \tag{7.21}$$

where $E(z)$ is the diffusion coefficient and $Q(z)/a(z)$ the vertical advection velocity. Selection of an appropriate Δt depends, therefore, on the desired spatial detail as well as on hydrological and operational conditions. Elements at reservoir levels near large-capacity bottom outlets often represent limiting conditions for choosing Δt and Δz.

The explicit method demands less computer storage capacity than the implicit technique, but this advantage may be offset partially by increased computational effort when the Δt required is smaller than that needed to describe the requisite temporal variations in temperature.

7.6.3. Finite-Element Method

Baca and Arnett (1976) have employed the finite-element method in a general water quality model that includes temperature simulation. Advantages of the method, which is also implicit in its treatment of the governing equations, are avoidance of numerical mixing and instability (inherent difficulties of explicit methods), flexibility in time-step selection, and adaptability to steep gradients that cause problems when linear approximations are made. The solution technique will be outlined in section 7.13.1.

7.7. APPLICATIONS OF ONE-DIMENSIONAL TEMPERATURE MODELS OF LAKES AND RESERVOIRS

There have been many practical applications of one-dimensional temperature models, far too numerous to describe in detail here. A few typical examples have been selected to indicate the general capabilities, as well as limitations, of these models.

7.7.1. Fontana Reservoir, North Carolina

One of the first attempts to simulate the annual cycle of a thermally stratified impoundment was an application of an early version of the WRE model to Fontana Reservoir in the TVA system. Data for the simulation were derived from a carefully planned and executed field survey by the TVA Engineering Laboratory.

Figure 7.3, derived from a report to the California Department of Fish and Game (WRE, Inc., 1968), compares simulated temperatures with observations from March to December 1966. Broken lines are isotherms predicted by the model, while full lines represent *in situ* observations. The model gives a credible representation of the distribution of temperature throughout most of the annual cycle, particularly during the period of strongest stratification. The pattern of thermocline formation is followed reasonably well by the model, although warming of strata near the surface in the spring was predicted to occur more

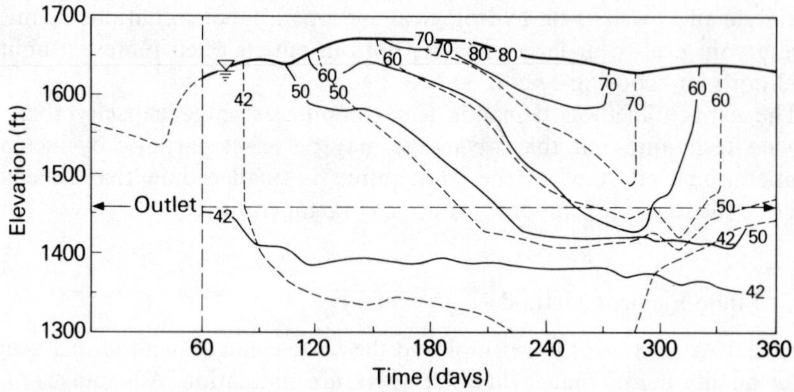

FIGURE 7.3 Computed (broken lines) and observed (full lines) thermal energy distribution in Fontana Reservoir using different diffusion coefficients (after WRE, Inc., 1968). The temperature (°F) is shown for each isotherm.

slowly than actually happened, possibly due to attenuation of shortwave insolation by turbid spring runoff not properly accounted for in the model. The uncertain ability of the model to represent convective mixing as a consequence of cooling during fall overturn is evident in the disparity between model and impoundment during late fall. It is exemplary, as well, of a general deficiency of this class of models in which the instability brought about by surface cooling is only crudely represented by numerical mixing driven by temperature anomalies in the upper strata. Overall, the simulation is a credible depiction of the annual thermal cycle for such a reservoir, rather typical of results expected with one-dimensional temperature models.

Comparable results for Fontana Reservoir employing the MIT model were reported by Harleman and Hurley (1976), as illustrated in Figure 7.4. In this example, the investigators examined the sensitivity of vertical heat transfer during thermocline development to the magnitude of the "diffusion coefficient." In Figure 7.4(a), corresponding to 27 April 1966, the assumption of molecular diffusion as the only mixing process results in a considerable disparity between model and impoundment. This difference is rectified somewhat by increasing the diffusion coefficient by two orders of magnitude, but it appears that in this period, at least, an even greater mixing effect might be necessary.

Later, in the early summer with a well established profile the model tends to produce more heat transfer in the region below the thermocline than was observed in the water body. Since these simulations were made largely for investigating the mechanics of internal heat transfer, no attempt was made to calibrate the model. Results, however, do show that the model is capable of reasonable representation of the gross processes of heat transfer in a thermally stratified deep reservoir.

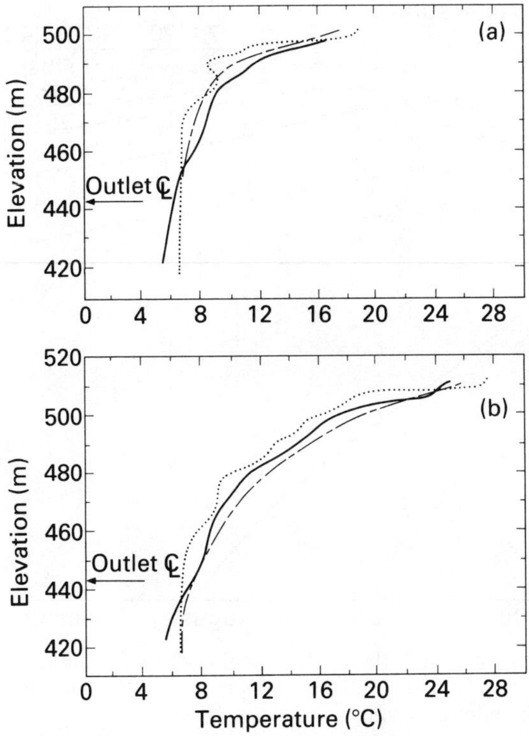

FIGURE 7.4 Comparison of predicted temperature profiles in Fontana Reservoir using different diffusion coefficients (after Harleman and Hurley, 1976). (a) 27 April 1966; (b) 22 June 1976. Full line, measured; dotted line, predicted (molecular diffusion); broken line, predicted (diffusion coefficient = 100 · molecular value).

7.7.2. Hungry Horse Reservoir, Montana

Results of a simulation of Hungry Horse Reservoir in Montana (Table 7.1) are presented in Figure 7.5, showing the capability of the one-dimensional temperature model to represent a strongly stratified reservoir (Fr = 0.0026) with a very shallow epilimnion. The thermocline developed late in the spring as the reservoir water level was rising, owing to flow of snowmelt into the bottom of the reservoir (4°C isotherm near the bottom). Apparently, because of higher turbidity in the waters of the impoundment, shortwave energy influx during the summer was confined to surface strata, the body of water below about 20 m remaining at less than 8°C throughout the year, even though surface temperatures exceeded 20°C in August. The agreement between model and impoundment is considered excellent for the most part, except that the model seems to predict fall cooling and associated convective mixing earlier than they actually occurred, illustrating once again the difficulty in simulating this phenomenon.

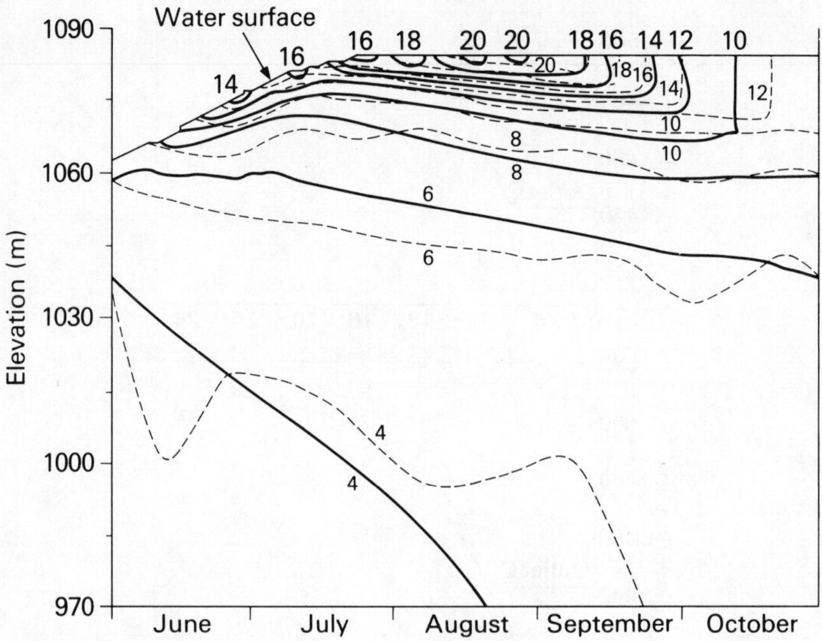

FIGURE 7.5 Simulated and observed thermal regimes for Hungry Horse Reservoir, 1965 (after WRE, Inc., 1969). Full lines, computed; broken lines, measured. The temperature (°C) is shown for each line.

7.7.3. Lake Päijänne, Finland

The model EPAECO (Gaume and Duke, 1975) was used to simulate the principal thermal cycle of Lake Päijänne in Central Finland (Kinnunen *et al.*, 1978) from late May, when temperatures were isothermal at 5.5°C, through the summer when surface temperatures reached about 20°C. Typical temperature profiles for June and August in three successive years are illustrated in Figure 7.6. Results are regarded as generally satisfactory, although there is evidence that surface heating occurs more rapidly in late spring in the model than in reality.

7.7.4. Ross Lake, Washington

The earliest versions of one-dimensional temperature models neglected dimictic lakes, which go through two distinctive periods of thermal stratification, one occurring during the period of ice cover. This capability was required in order to study Ross Lake on the Skagit River in the Cascade Mountains of Washington State. As a result of prolonged subzero air temperatures and the formation of ice cover, temperatures of surface strata during winter months drop to near

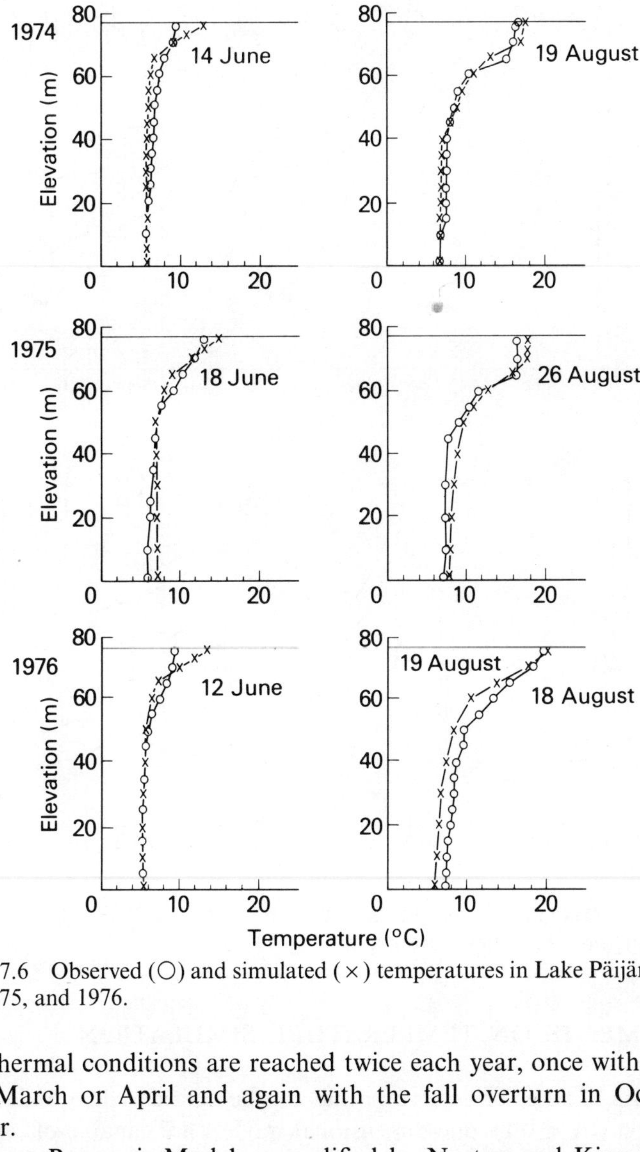

FIGURE 7.6 Observed (○) and simulated (×) temperatures in Lake Päijänne, Finland in 1974, 1975, and 1976.

zero. Isothermal conditions are reached twice each year, once with the spring thaw in March or April and again with the fall overturn in October and November.

The Deep Reservoir Model as modified by Norton and King (1975) was used to simulate the entire annual cycle of the lake and the predicted temperatures were compared with those recorded in monthly surveys during 1971, as illustrated in Figure 7.7. A random sampling of temperatures from model and lake indicates agreement to within $\pm$ 0.5°C throughout most of the year. The double cycle of stratification and subsequent mixing is faithfully simulated by the model.

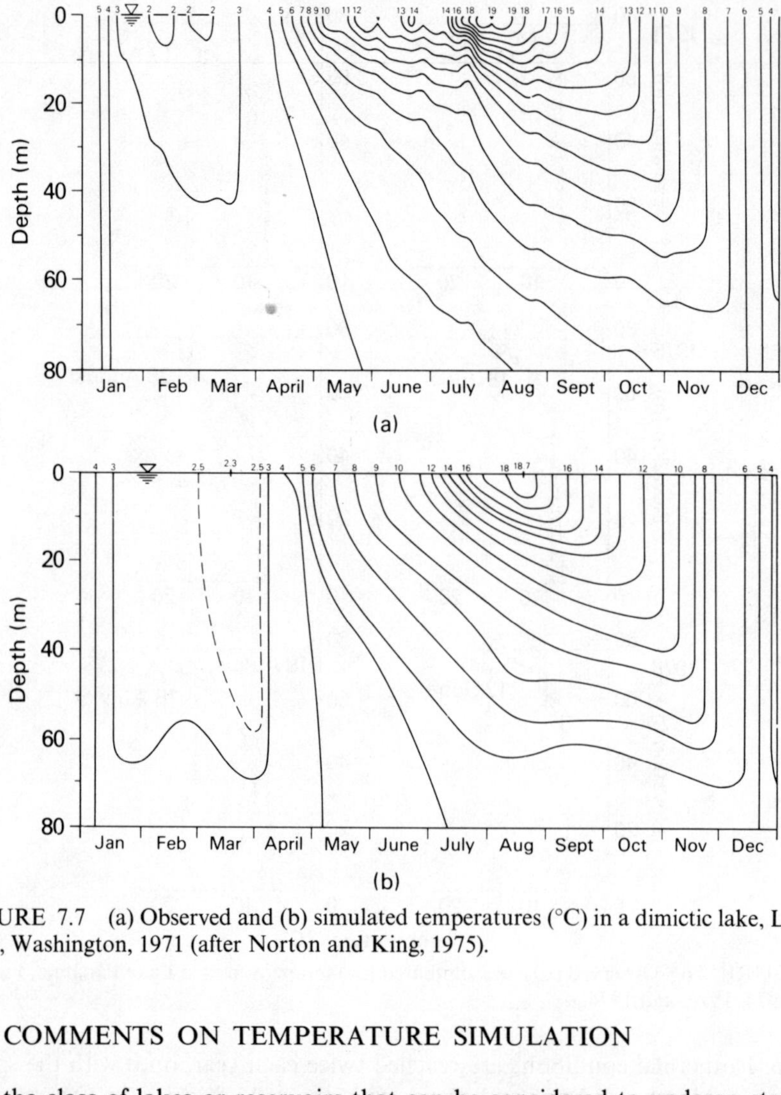

FIGURE 7.7 (a) Observed and (b) simulated temperatures (°C) in a dimictic lake, Lake Ross, Washington, 1971 (after Norton and King, 1975).

7.8. COMMENTS ON TEMPERATURE SIMULATION

For the class of lakes or reservoirs that can be considered to undergo strong stratification ($Fr \ll 0.1$), one-dimensional models are capable of representing the principal features of the annual thermal cycle, including even winter stratification under ice. Some difficulties are still experienced with such models in simulation of convective mixing because of cooling in the fall, although this deficiency does not appear to be serious, at least insofar as temperature simulation is concerned. When such models are employed to drive simulation of other state variables, water quality and biota for example, this weakness may become more of a limitation.

Another deficiency appears to lie in the characterization of mixing processes in the epilimnion after the onset of stratification, but this may be more of academic interest than of consequence from a practical water quality viewpoint. As noted in section 7.5.5, the effects of wind-induced mixing in the upper strata are currently under investigation, which holds promise of a more rigorous description of these processes. As regards water quality changes, it may be of greater importance to examine more critically what takes place in the region of the thermocline and below, inasmuch as these areas are likely to be the more sensitive ecologically.

7.9. APPROACH TO ONE-DIMENSIONAL MODELING OF WATER QUALITY AND ECOLOGY IN LAKES

Mathematical models for impoundments have evolved along two different lines that have gradually merged in recent years to provide a fairly comprehensive capability for simulation of water-quality–ecological relationships in deep, stratified water bodies. Those that may be treated by the one-dimensional conceptualization were presented earlier in this chapter. First, there was the rather logical extension of the one-dimensional stratified reservoir model of the WRE–MIT type to include vertical transport of quality constituents, both abiotic and biotic. These developments are well represented by the work of Chen and Orlob (1968), Markofsky and Harleman (1973), and Chen and Orlob (1975). Then there were the modeling activities that focused primarily on the ecosystem and nutrient balance of the lake, like the models of Vollenweider (1965, 1969) and the International Biological Program team of researchers at Rensselaer Polytechnic Institute (Park *et al.*, 1974a, b) who initiated the model series beginning with CLEAN, which was subsequently modified to deal with multisegment systems, e.g. MS CLEANER (Leung *et al.*, 1978). In both instances, models emerged that were essentially one-dimensional (although attempts have been made to extend them to systems that have higher dimensionality and deal with both water quality and ecology). Examples of such extensions of the basic concepts of modeling the water quality and ecology of lakes will be presented in Chapter 8. For the present, however, we will confine our discussion to the somewhat simpler one-dimensional case, using the general principles outlined in Chapters 3 and 4.

Because of the disciplinary predilections of the development teams, one engineering-oriented and the other concentrating more on biology, the two lines of one-dimensional water-quality–ecological model development tended to place different emphasis on two important and essential considerations in realistic presentation of the stratified water body. The WRE–MIT approach, founded as it was on considerations of thermal energy balance mechanisms and advective–diffusive transport along the vertical axis of a strongly stratified impoundment, concentrated more on vertical movement and distribution of the water quality state variables of traditional pollution control concern, e.g. DO, BOD, conservative substances, nutrients, coliforms, etc., and tended to treat the ecosystem in a more aggregated state. For example, algae, zooplankton, and fish were divided conceptually into simple groupings, rather than by species, age classes, life stages, etc. (Figure 4.1). The line of development represented by CLEAN, CLEANER, and MS CLEANER tended toward a more rigorous representation of biological phenomena at the sacrifice of some realism from the hydrological and hydrodynamic viewpoints (Figures 9.10 and 9.11). Both schools of model development have gradually combined, with each drawing from the other to provide more correct, yet practical, capability to model the one-dimensional class of impoundments.

7.9.1. Brief Review of Development

Zero-Dimensional and Two-Layer Models

Virtually all mathematical water-quality–ecological models have evolved from application of the law of mass conservation, supplemented by the principles of kinetics applicable to chemical and biological systems. Among the earliest such models of lake systems were those addressing the problem of nutrient balance in lakes. They were generally gross nutrient budget models that described the entire lake as a continuously stirred tank reactor (CSTR), fully mixed over the period of interest, often the full annual cycle (Vollenweider, 1965, 1969). At first, no particular attempt was made to discretize spatially and the models often treated only a single water quality constituent, e.g. nutrients considered to be important regulators of eutrophication, like carbon, nitrogen, and phosphorus. The Vollenweider approach was seized upon by other investigators (O'Melia, 1974; Snodgrass and O'Melia, 1975; Larsen and Mercier, 1975; Bella, 1976) in dealing more explicitly with nutrient and algal budgets in well mixed systems. Imboden (1974) and O'Melia (1974) adapted the approach to two-segment (epilimnion–hypolimnion) systems.

Transition to Multisegment Models

Among the early models for simulation of eutrophication of lakes, the one proposed by Chen and Orlob (1968) adopted the CSTR analogy to describe a set of interactive (coupled) mathematical relationships for production of algae from nutrient input to natural systems. This development took place, however, with the full expectation that the generalized set of mass balance equations that might evolve for the multiplicity of water quality and ecological compartments would be embedded directly in one-dimensional transport models, such as those already being operated to simulate the thermal structure of deep reservoirs. In fact, since the basic one-dimensional segmented model had already been constructed and tested for the temperature simulation problem (WRE, Inc., 1968, 1969), the transition to an operational water-quality–ecological model for this case was more easily accomplished. A similar transition was effected between the MIT temperature model (Huber *et al.*, 1972) and extensions to deal with DO–BOD balance in deep reservoirs (Markofsky and Harleman, 1973).

An important step in the evolution of water-quality–ecological models was the development of CLEAN, a model conceptualized by some 25 investigators of the Eastern Deciduous Forest Biome of the US International Biological Program (Park *et al.*, 1974a). At first, the model considered only a single spatial segment but as applications were made to real systems it was modified to accommodate at least two-segment systems, such as stratified lakes in which the epilimnion and hypolimnion could be treated as unique biological entities. In the version CLEANER, 20 state variables (mostly biological) were represented (Park *et al.*, 1974b; Scavia and Park, 1976). As Chapter 9 will show, it evolved further into a version (MS CLEANER) capable of representing the impoundment in multiple segments with up to 40 state variables (Leung *et al.*, 1978).

7.10. CONCEPTUAL REPRESENTATION OF ONE-DIMENSIONAL WATER-QUALITY–ECOLOGICAL MODELS

For this class of models our conceptual view of the water system is virtually identical to that of the vertically stratified, deep reservoir, depicted in Figure 7.1, wherein each control volume, a segment of the water column, is treated as a CSTR, i.e. as a zero-dimensional submodel. Thus, the physical, chemical, and biological interactions are all assumed to occur within the confines of a discrete volume element. Exchanges occur across the boundaries of such an element as a result of exogenous forces, e.g. advection, diffusion, insolation, sedimentation, etc., and the "states" of variables from time to time are governed by kinetic principles that must be defined as functions of space and time. Finally, all discrete elements are linked in space to provide a continuum in model form that represents the water body.

7.11. FORMULATION OF WATER-QUALITY–ECOLOGICAL EQUATIONS

The general form and notation for the water-quality–ecological equations resemble those set forth earlier in this chapter for one-dimensional temperature models. The general one-dimensional mass conservation equation for a volume element V_j is given by

$$\underbrace{\frac{\partial(V_jC_j)}{\partial t} = (Q_iC_i - Q_oC_o)_j}_{\text{local advection}} \underbrace{- (Q_zC)_j + (Q_zC)_{j+1}}_{\text{vertical advection}} + \underbrace{\left(Ea\frac{\partial C}{\partial z}\right)_j - \left(Ea\frac{\partial C}{\partial z}\right)_{j+1}}_{\text{vertical diffusion}}$$

$$+ \underbrace{C_j\frac{\partial V_j}{\partial t}}_{\text{volume change}} + \underbrace{V_j\frac{\mathrm{d}C}{\mathrm{d}t}}_{\text{process change}} \underbrace{\pm \quad S,}_{\text{sources and sinks}} \tag{7.22}$$

where C_j is the concentration of any constituent, abiotic or biotic, that moves with the fluid mass or may be transferred by a diffusive process in proportion to the gradient $\partial C/\partial z$. The total derivative $\mathrm{d}C/\mathrm{d}t$ represents all processes, other than advection, diffusion, and volume change, that act to modify C_j. Equations of this form can be written for each quality parameter of interest, either independent or coupled, conservative or nonconservative, through an appropriate formulation of $\mathrm{d}C/\mathrm{d}t$. Customarily, the process reactions exemplified by $\mathrm{d}C/\mathrm{d}t$ comprise first- or zero-order terms, the coefficients of which may be determined by either preset exogenous conditions or the values of other state variables.

7.11.1. Formulation for Oxygen Balance

In modeling the water quality of a stratified impoundment, perhaps the best illustration of the explicit formulation of an equation of type (7.22) is the mass balance equation for dissolved oxygen. For simplicity, the subscript j has been dropped:

$$\frac{\partial(V\mathrm{O})}{\partial t} = \underbrace{-A}_{\text{advection}} \quad \underbrace{+D}_{\text{diffusion}} \quad \underbrace{-\mathrm{O}\frac{\partial V}{\partial t}}_{\text{volume change}}$$

$$+ \underbrace{a_s K_2(\mathrm{O}^* - \mathrm{O})}_{\text{reaeration}} \underbrace{- K_1LV}_{\text{BOD}} \underbrace{- K_4 a_b (\mathrm{OS})^*}_{\text{benthic demand}}$$

$$\underbrace{- \beta_1\alpha_1(\mathrm{NH_3})V}_{\text{ammonia oxidation}} \quad \underbrace{- \beta_2\alpha_2(\mathrm{NO_2})V}_{\text{nitrite oxidation}} \quad \underbrace{- \beta_3\alpha_3(\mathrm{DET})V}_{\text{detritus oxidation}}$$

$$\underbrace{- K_b(r_P P)V}_{\text{biotic respiration, e.g. of algae}} \quad \underbrace{+ K_b\gamma(\mu_P P)V}_{\text{photosynthesis}} \quad \underbrace{\pm S_0,}_{\text{external sources and sinks}} \tag{7.23}$$

where

- O is the dissolved oxygen concentration $[\mathrm{M\,L^{-3}}]$,
- O* is the dissolved oxygen concentration at saturation $[\mathrm{M\,L^{-3}}]$,
- A is the advection rate $[\mathrm{M\,T^{-1}}]$,
- D is the diffusion rate $[\mathrm{M\,T^{-1}}]$,

a_s is the surface area of the uppermost element [L^2],
K_2 is the reaeration coefficient [$L\,T^{-1}$],
K_1 is the biochemical oxygen demand coefficient [T^{-1}],
L is the carbonaceous biochemical oxygen demand [$M\,L^{-3}$],
K_4 is the benthic sediment oxygen demand coefficient [T^{-1}],
a_b $= a_j - a_{j-1}$ is the bottom area associated with element j [L^2],
$(OS)^*$ is the organic sediment accumulation per unit of bottom area [$M\,L^{-2}$],
NH_3 is the ammonia nitrogen concentration [$M\,L^{-3}$],
NO_2 is the nitrite nitrogen concentration [$M\,L^{-3}$],
DET is the detritus concentration [$M\,L^{-3}$],
$\alpha_1, \alpha_2, \alpha_3$ are the stoichiometric equivalences with oxygen for ammonia, nitrite, and detritus [$M\,M^{-1}$],
$\beta_1, \beta_2, \beta_3$ are the decay coefficients for ammonia, nitrite, and detritus [T^{-1}],
P is the phytoplankton concentration [$M\,L^{-3}$],
r_P is the phytoplankton respiration rate [T^{-1}],
μ_P is the phytoplankton growth rate [T^{-1}],
K_b is the biota activity coefficient (dimensionless),
γ is the stoichiometric oxygenation factor for algal growth [$M\,M^{-1}$],
S_0 is the source or sink [$M\,T^{-1}$].

In eqn. 7.23 we perceive the complexity of interactions between water quality and the ecosystem. The first three terms represent the primary processes of the impoundment that affect oxygen transport, i.e. advection, diffusion, and change in water surface elevation. The next three terms represent the classical formulation of the Streeter–Phelps equation (Chapter 6) for oxygen sag. Nitrogenous demand for oxygen is represented in the oxidation of ammonia and nitrite. Oxidation of organic detritus and respiration of living forms, e.g. algae, account for additional decrements in the available resource, while photosynthesis adds oxygen to the pool in proportion to the production of new algal cell material. Sources and sinks of oxygen, exogenous to the element, are represented for completeness.

7.11.2. Formulation of the Equation Set

A complete formulation for an impoundment requires writing similar equations of the form of (7.22) for each state variable represented. Table 7.2, which has been adapted from the LAKECO subroutine of the WQRRS model (HEC, 1974; Chen *et al.*, 1975), illustrates the approach applied to a system involving 15 state variables. Figure 7.8 shows the relationships between the various constituents of the aquatic system.

TABLE 7.2 Mass Balance Equations for a One-Dimensional Water-Quality–Ecological Model of Stratified Impoundments.

Constituent Concentration $[\mathrm{M\,L^{-3}}]$	Mass Balance Equation
Conservative substances, C	$\dot{\overline{CV}} = A + D + C\dot{V}$
Biochemical oxygen demand, L	$\dot{\overline{LV}} = A + D - L\dot{V} - K_1 LV$
Detritus, DET	$\dot{\overline{(\mathrm{DET})V}} = A + D - (\mathrm{DET})\dot{V} - \underbrace{\beta_3(\mathrm{DET})V}_{\text{decay}} - \underbrace{a_b s_D(\mathrm{DET})V}_{\text{sedimentation}}$ $+ \underbrace{K_b \mu_Z ZV\left(\frac{1}{\eta_{ZA}} - 1\right) + F_Z\left(\frac{1}{\eta_{FZ}} - 1\right)}_{\text{zooplankton excreta}} + \underbrace{F_B\left(\frac{1}{\eta_{FB}} - 1\right)}_{\text{fish excreta}}$
Organic sediment, OS	$\dot{\overline{(\mathrm{OS})V}} = \underbrace{-K_s a_b(\mathrm{OS})^*V}_{\text{decay}} + \underbrace{a_b s_P PV}_{\text{phytoplankton}} + \underbrace{a_b s_D(\mathrm{DET})V}_{\text{detritus}}$
Carbon, C	$\dot{\overline{\mathrm{C}V}} = A + D - \mathrm{C}\dot{V} + \underbrace{a_s K_C(\mathrm{CO_2^*} - \mathrm{CO_2})V}_{\text{reaeration}} + \underbrace{\beta_3 c_D(\mathrm{DET})V}_{\text{detritus release}} + \underbrace{K_1 c_L LV}_{\text{BOD decay}} + \underbrace{K_b(c_P r_P P + c_Z r_Z Z + c_F r_F F + c_B r_B B)V}_{\text{respiration of biota}}$
Ammonia nitrogen, NH_3	$\dot{\overline{(\mathrm{NH_3})V}} = A + D - (\mathrm{NH_3})\dot{V} - \underbrace{\beta_1(\mathrm{NH_3})V}_{\text{decay}} + \underbrace{\Psi_D n_D(\mathrm{DET})V}_{\text{detritus release}} + \underbrace{\Psi_S n_S(\mathrm{OS})V}_{\text{organic sediment release}}$
Nitrite nitrogen, NO_2	$\dot{\overline{(\mathrm{NO_2})V}} = A + D - (\mathrm{NO_2})\dot{V} - \underbrace{\beta_2(\mathrm{NO_2})V}_{\text{decay}}$
Nitrate nitrogen, NO_3	$\dot{\overline{(\mathrm{NO_3})V}} = A + D - (\mathrm{NO_3})\dot{V} + \underbrace{\beta_2(\mathrm{NO_2})V}_{\text{nitrite oxidation}} - \underbrace{K_b n_P \mu_P PV}_{\text{phytoplankton uptake}}$
Phosphate phosphorus, PO_4	$\dot{\overline{(\mathrm{PO_4})V}} = A + D - (\mathrm{PO_4})\dot{V} + \underbrace{\Psi_D p_D(\mathrm{DET})V}_{\text{detritus release}} + \underbrace{\Psi_S p_S(\mathrm{OS})V}_{\text{organic sediment release}} - \underbrace{K_b p_P \mu_P PV}_{\text{phytoplankton uptake}}$
Coliform bacteria, CB	$\dot{\overline{(\mathrm{CB})V}} = A + D - (\mathrm{CB})\dot{V} - \underbrace{K_c(\mathrm{CB})V}_{\text{decay}}$
Algae, P	$\dot{\overline{PV}} = A + D - P\dot{V} + \underbrace{K_b(\mu_P - r_P)PV}_{\text{net growth}} - \underbrace{a_b s_P PV}_{\text{sedimentation}} - \underbrace{K_b \mu_Z ZV/\eta_{ZA}}_{\text{zooplankton grazing}}$

Zooplankton, Z	$\dot{\overline{ZV}} = A + D - Z\dot{V} + \underbrace{K_b(\mu_Z - r_Z - m_Z)ZV}_{\text{net growth}} - \underbrace{F_Z/\eta_F}_{\text{grazing by fish}}$
Benthic animals, B	$\dot{\overline{BV}} = \underbrace{K_b(\mu_B - r_B - m_B)B^* a_b}_{\text{net growth}} - \underbrace{F_B/\eta_B}_{\text{grazing by fish}}$
Nekton, F	$\dot{\overline{FV}} = \underbrace{K_b(\mu_F - r_F - m_F)FV}_{\text{net growth}} - \underbrace{hFV}_{\text{harvest}}$
Dissolved oxygen, O	$\dot{\overline{OV}} = A + D - O\dot{V} + \underbrace{a_s K_2(O^* - O)}_{\text{reaeration}} - \underbrace{K_1 LV}_{\text{BOD}} - \underbrace{K_s a_b(OS)V}_{\text{benthic demand}} - \underbrace{\beta_1\alpha_1(NH_3)V}_{\text{ammonia oxidation}} - \underbrace{\beta_2\alpha_2(NO_2)V}_{\text{nitrite oxidation}} - \underbrace{\beta_3\alpha_3(DET)V}_{\text{detritus demand}}$
	$- \underbrace{K_b(r_P P + r_Z Z + r_F F)V}_{\text{biotic respiration}} + \underbrace{K_b(\mu_P P)\gamma V}_{\text{photosynthesis}}$

Terms not previously defined:

B^*	is the saturation concentration of benthic animals per unit of bottom area [M L^{-2}],
$\mu_P, \mu_Z, \mu_B, \mu_F$	are the growth rates of phytoplankton, zooplankton, benthic animals, and nekton [T^{-1}],
r_P, r_Z, r_B, r_F	are the respiration rates of phytoplankton, zooplankton, benthic animals, and nekton [T^{-1}],
m_Z, m_B, m_F	are the mortality rates of zooplankton, benthic animals, and nekton [T^{-1}],
s_P, s_D	are the sedimentation rates of phytoplankton and detritus [L^{-2} T^{-1}],
Ψ	is the release rate of ammonia nitrogen [T^{-1}],
η_{ab}	is the efficiency of predator *a* in digesting prey *b*,
c	is the proportion of carbon,
n	is the proportion of nitrogen,
p	is the proportion of phosphate phosphorus,
h	is the harvest rate constant [T^{-1}],
K_s	is the rate of organic sediment decay per unit of bottom area [T^{-1} L^{-2}],
a_b	is the area of bottom sediment [L^2],
K_C	is the mass transfer rate of carbon as CO_2 per unit of surface area [T^{-1} L^{-2}],
a_s	is the area of the water surface [L^2].

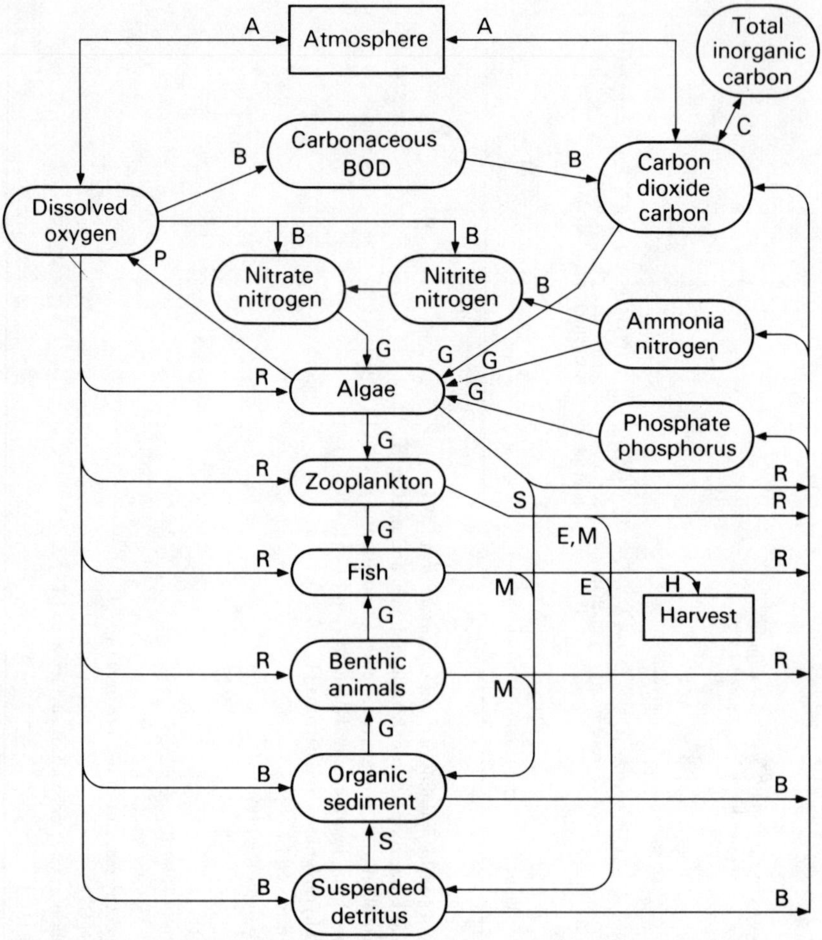

FIGURE 7.8 Water quality and ecological relationships in LAKECO (after HEC, 1978). A, aeration; B, bacterial decay; C, chemical equilibrium; E, excreta; G, growth; H, harvest; M, mortality; P, photosynthesis; R, respiration; S, settling.

7.12. TECHNIQUES FOR SOLUTION OF ONE-DIMENSIONAL WATER-QUALITY–ECOLOGICAL MODELS

7.12.1. Implicit Finite-Difference Method

The set of equations resulting from formulation of $j = 1, 2, \ldots, n$ equations for each of $i = 1, 2, \ldots, k$ state variables may be solved by an implicit technique like that outlined in section 7.6 for solution of the heat energy equations. This procedure has been used in LAKECO (Chen and Orlob, 1975) and WQRRS

(HEC, 1974) and in other versions of the original model, like EPAECO (Gaume and Duke, 1975).

7.12.2. Explicit Finite-Difference Method

Extensions of the MIT model to include water quality parameters like DO and BOD (Markofsky and Harleman, 1973) have followed the original solution procedure, a step-forward explicit method that requires observance of certain stability criteria (eqns. 7.20 and 7.21).

7.13. OTHER ONE-DIMENSIONAL WATER-QUALITY–ECOLOGICAL MODELS

Three additional one-dimensional water quality models are worthy of special mention in this chapter. They are the Finite-Element Model of Baca and Arnett (1976), a "dynamic" reservoir model developed by Imberger *et al.* (1978), and an adaptation of the Deep Reservoir Model by the US Army Waterways Experiment Station (Fontane and Bohan, 1974).

7.13.1. Finite-Element Model—Baca and Arnett

Baca and Arnett (1976) developed a water quality model for eutrophic lakes and reservoirs that, in addition to employing the finite-element method for forming and solving the basic differential equations, provides an improved representation of the nitrogen and phosphorus cycles in lakes.

Formulation of the basic mass balance equations in their model follows the general pattern outlined in Table 7.2, with a few important distinctions. These relate primarily to the nitrogen and phosphorus cycles and the recycling of these nutrients between the water column and the sediment.

Formulation of the Baca and Arnett model involves a set of $j = 1, 2, \ldots, n$ equations, corresponding to n reservoir slices, for each of $i = 1, 2, \ldots, k$ state variables, just as in LAKECO. These are solved by transforming the equations and applying the finite-element method of implicit solution. The method is summarized below.

The advection–diffusion equation can be written as a complex function of C:

$$L(C) = \frac{\partial C}{\partial t} + w\frac{\partial C}{\partial z} - \frac{\partial}{\partial z}\left(E_z\frac{\partial C}{\partial z}\right) + \lambda C - S, \tag{7.24}$$

where

C is the nutrient concentration $[\mathrm{M\,L^{-3}}]$,
w is the vertical advection velocity $[\mathrm{L\,T^{-1}}]$,
E_z is the diffusion coefficient $[\mathrm{L^2\,T^{-1}}]$,
λ is the decay (or growth) coefficient $[\mathrm{T^{-1}}]$,

and in vector notation as

$$\begin{Bmatrix} a \\ b \end{Bmatrix} = \frac{1}{\Delta z}\begin{bmatrix} z_{j+1} & -z_j \\ -1 & 1 \end{bmatrix}\begin{Bmatrix} C_j \\ C_{j+1} \end{Bmatrix}. \tag{7.25}$$

By employing the Galerkin method of weighted residuals, we can recast the governing model eqn. 7.24 in integral form:

$$\chi = \int_{\mathrm{R}} L(C) W_i \, \mathrm{d}z, \tag{7.26}$$

where W_i is an arbitrary set of weighting functions. If the approximating function W_j for $\tilde{C}$ is taken to be the same as W_j according to Galerkin's method, then

$$\tilde{C} = \sum_{j=1}^{k} C_j W_j \tag{7.27}$$

and

$$\tilde{C}(z, t) = [W_1, W_2]\begin{Bmatrix} C_j \\ C_{j+1} \end{Bmatrix}, \tag{7.28}$$

where the weighting functions are given by

$$W_1 = \frac{1}{\Delta z}(z_{j+1} - z) \tag{7.29}$$

and

$$W_2 = \frac{1}{\Delta z}(z - z_j). \tag{7.30}$$

Expanding (7.26) and integrating by parts results in

$$\chi = \int_{\mathrm{R}} \left(\frac{\partial C}{\partial t} W_i + w \frac{\partial C}{\partial z} W_i - E_z \frac{\partial C}{\partial z} \frac{\partial W_i}{\partial z} + \lambda C W_i - S W_i \right) \mathrm{d}z, \tag{7.31}$$

which can be partitioned so that

$$\chi = \sum_{j=1}^{n} \chi_j^{\mathrm{e}}, \tag{7.32}$$

where n is the number of subdomains. This may be expanded to give

$$\chi = \int_{z_1}^{z_2} [\,]\,\mathrm{d}z + \int_{z_2}^{z_3} [\,]\,\mathrm{d}z + \ldots + \int_{z_{m-1}}^{z_m} [\,]\,\mathrm{d}z. \tag{7.33}$$

Individual terms in the Galerkin functional eqn. 7.26 and the expanded form, eqn. 7.33, are approximated by

$$\frac{\partial C}{\partial t} = [W_1, W_2]\begin{Bmatrix} \dot{C}_j \\ \dot{C}_{j+1} \end{Bmatrix}, \tag{7.34}$$

$$\frac{\partial C}{\partial z} = \left[-\frac{1}{\Delta z}, \frac{1}{\Delta z} \right] \begin{Bmatrix} C_j \\ C_{j+1} \end{Bmatrix}, \tag{7.35}$$

and

$$\frac{\partial W_j}{\partial z} = \left[-\frac{1}{\Delta z}, \frac{1}{\Delta z} \right]. \tag{7.36}$$

Substituting these quantities into the Galerkin functional for a particular subdomain results in a set of integrals that can be evaluated by

$$\int_{z_j}^{z_{j+1}} W_1^{\gamma} W_2^{\beta} \, \mathrm{d}z = \frac{\gamma! \beta! \Delta z}{\gamma + \beta + 1}. \tag{7.37}$$

From these result specific expressions for the time derivative, advection, diffusion, decay, and the sources. Evaluation, summing, and gathering terms leads to a general matrix equation:

$$[\mathbf{I}]\{\dot{C}\} = [\mathbf{S}]\{C\} + \{\boldsymbol{R}\}, \tag{7.38}$$

in which $[\mathbf{I}]$ is an identity matrix, $[\mathbf{S}]$ is a tridiagonal matrix of the coefficients of C, and $\{\boldsymbol{R}\}$ is a vector of all known parameters and external sources of C. Solution of the equation for each quality constituent is accomplished sequentially according to the method outlined for (7.11), i.e. the procedure adopted for the WRE Deep Reservoir Model and for LAKECO (WRE, Inc., 1968, 1969; HEC, 1974).

The advantages of the finite-element technique cited by Baca and Arnett lie in avoidance of numerical mixing effects, stability of solution, flexibility in length of time step, and adaptability to steep gradients. Experience in use of the method in even more complex problems, involving hydrodynamic processes, seems to confirm these attributes (Gallagher *et al.*, 1973; King *et al.*, 1974).

7.13.2. Dynamic Reservoir Water Quality Model—Imberger *et al.*

Additional rigor in modeling the hydromechanics of small to medium-sized impoundments is provided in a new approach by Imberger *et al.* (1978; Spigel, 1978; Fischer *et al.*, 1979). A Lagrangian one-dimensional model is constructed by forming a set of turbulent kinetic energy budget equations. In Imberger's notation, the rate of change of mean potential energy in a control "slab" is given by

$$\frac{\mathrm{d}V}{\mathrm{d}t} = \Lambda_{\mathrm{K}} + \Lambda_{\mathrm{S}} - \Lambda_{\mathrm{T}} - \Lambda_{\mathrm{L}} - \Lambda_{\mathrm{D}}, \tag{7.39}$$

where

V is the potential energy,
Λ_{K} is the kinematic flux of turbulent kinetic energy imposed at the top surface,

Λ_S is the rate of production of turbulent kinetic energy by the shear across the bottom,

Λ_T is the rate of increase of turbulent kinetic energy in the slab (or column),

Λ_L is the rate of leakage of energy by radiation of internal waves into the quiescent fluid,

Λ_D is the total rate of energy dissipation by viscosity.

Drawing on research in oceanography, primarily that of Niiler (1975), Imberger transforms (7.39) into a simple entrainment law:

$$\frac{1}{q}\frac{\mathrm{d}h}{\mathrm{d}t} = \frac{C_K}{\mathrm{Ri}}, \tag{7.40}$$

where

q is the characteristic turbulent velocity scale,

h is the depth of the mixed layer,

C_K is a constant that determines the entrainment rate,

Ri $= \alpha\Delta\Theta gh/q^2$ is the Richardson number, in which α is the expansion coefficient for water and $\Delta\Theta$ is the temperature change.

The entrainment rate is estimated from observations of the impoundment in terms of two characteristic damping depths, h_0 and h_1, that determine the effectiveness of mixing by wind energy and convective mixing by mechanical buoyancy, respectively.

Effective turbulent mixing, which accounts for vertical transport of turbulent kinetic energy, is determined empirically by using the depth of the reservoir, the stability due to density differences from top to bottom, and a characteristic mixing time. A constant-flux model is proposed:

$$\epsilon(z) = \frac{\bar{\kappa}H^2}{T_m S}, \tag{7.41}$$

where

$\epsilon(z)$ is the effective turbulent mixing coefficient along the vertical (z) axis,

H is the depth of the reservoir,

T_m $= V/(P_1 + P_2)$ is the mixing time scale, in which P_1 and P_2 are rates of work of inflowing streams and wind, respectively,

S $= -(\mathrm{d}\rho/\mathrm{d}x)H/(\rho(0) - \rho(H))$ is the stability,

$\bar{\kappa}$ is the reservoir constant.

Application of the model requires estimation, from impoundment measurements, of four so-called "universal" constants: C_D, a drag coefficient of the entering streams; $\bar{\kappa}$, the constant of proportionality of the diffusion coefficient (eqn. 7.41); and h_0 and h_1. In demonstrating the model on Wellington Reservoir in Australia, Imberger *et al.* (1978) derived best values of the constants by simulation during representative periods when only one of the four constants was dominant. For the case examined C_D was found to be about $8.5 \cdot 10^{-3}$, $\bar{\kappa}$ was $4.8 \cdot 10^{-4}$, $h_1 = 5$ m, $h_0 = 20$ m, and ϵ varied from 0 to 10^{-4} m^2 s^{-1}.

The procedure for setting up the Imberger model is as follows. The reservoir is divided into slabs of arbitrary thickness for which the initial values of temperature, salinity, density, volume, etc. are known. Surface thermal energy, determined by a meteorological subroutine, is added according to an exponential penetration law at quarter-hour intervals up to one day, or when the surface slab has cooled by 3°C. A mixing routine then adjusts for instabilities to produce a new density profile, which is subsequently relaxed by the diffusion coefficient calculated according to (7.41). This is followed by adjustments for energy input by inflow and energy extraction by outflow. Volume continuity is maintained within each Lagrangian slab and the bounding planes are tracked as they move up and down. No convection is allowed between slabs.

In simulation of Wellington Reservoir with the model DYRESM (Dynamic Reservoir Model) a typical run of 100 days required 5.5 s per day of CPU time on a Cyber 73 computer. Preliminary results of the calibrated model give an excellent account of its capability to simulate changes in temperature and salinity in a medium-sized reservoir under rather complex operational conditions (Imberger, 1981).

7.13.3. Waterways Experiment Station Model

The Waterways Experiment Station (WES) of the US Army Corps of Engineers developed a one-dimensional model for the simulation of temperature and oxygen balance in deep reservoirs (Fontane and Bohan, 1974), known as WESTEX. It was patterned after an early version of the WRE Deep Reservoir Model, extended by Fogg and Fruh (1973) for use on Lake Lyndon B. Johnson in Texas, and designed to overcome an inherent difficulty of one-dimensional models: failure to represent correctly the actual detention time for flows entering when the reservoir is strongly stratified. Paradoxically, this problem is especially acute when the reservoir has a large volume relative to the rate of inflow, just the conditions that favor the one-dimensional temperature model. Under such conditions, organic loads entering the reservoir may distribute their demands for oxygen variably over space and time depending on BOD concentrations, temperature and relative density of inflow, and the time of flow through the reservoir. Clearly, the problem requires more than a simple one-dimensional model, wherein the loads are instantaneously distributed throughout the horizontal "slice" into which the load is introduced. The WES has devised a method whereby the times of flow through the reservoir can be estimated with the aid of a simple physical model of the stratified system. These times are then utilized in the simulation to give a more plausible description of the dissolved oxygen distribution along the reservoir and over the season. Comparison of simulated DO profiles in operating reservoirs with actual observations indicate that WESTEX can produce more realistic results than the unmodified one-dimensional model.

7.14. APPLICATIONS OF ONE-DIMENSIONAL WATER-QUALITY–ECOLOGICAL MODELS

While there have been many applications of one-dimensional water-quality–ecological models, since their first appearance in the mid-1970s, to deep lakes and reservoirs, relatively few cases have yielded sufficient data for comparative purposes. The best results seem to have been for simulation of dissolved oxygen balances, although some success has been demonstrated with nutrient budgets and primary productivity. Three examples have been selected to indicate the general capabilities of such models, as well as to point out certain inherent limitations.

7.14.1. Lake Washington, Washington

Figure 7.9 illustrates an early test of the model LAKECO on Lake Washington. The variation of dissolved oxygen concentration in time and space appears to be fairly well simulated, although certain details, notably the steeper gradients revealed in field observations, are missed. The model can characterize the moderate depletion of dissolved oxygen in the upper part of the hypolimnion during the late summer and early fall, as well as the progressive decline in DO levels at the bottom from early spring until the fall overturn. The initial and peak stages of phytoplankton growth, also simulated in the model, were represented satisfactorily, although during the fall observed algal biomass (as indicated by chlorophyll *a* concentrations) tended to be much higher than predicted (Chen and Orlob, 1975).

7.14.2. Wellington Reservoir, Australia

Figure 7.10 shows results of a simulation using DYRESM, the model of Imberger *et al.* (1978), in which both temperature and salinity (NaCl) distributions are compared with detailed observations in the field. The simulation begins with day 133, at which time the reservoir was slightly stratified thermally with a thermocline about 10 m above the bottom. Salinity at this time was almost uniform from top to bottom. Subsequently, the reservoir cooled (day 190) and there was a large influx of saline water. This inflow was apparently cooler and denser (owing to temperature and salinity) with the result that there was a large alteration in quality of the impounded water below about 13 m

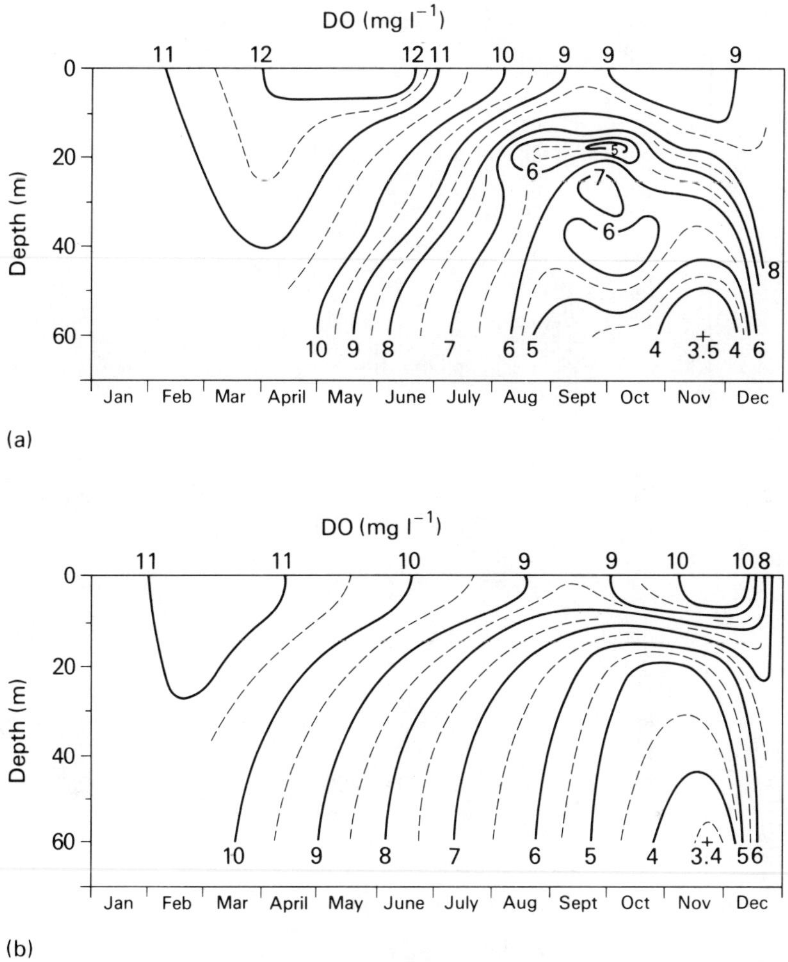

FIGURE 7.9 (a) Observed and (b) simulated dissolved oxygen concentrations in Lake Washington, 1963 (after Chen and Orlob, 1975).

from the bottom. These changes were fairly well simulated by the model, as indicated by the broken and dotted lines.

The subsequent warming period, illustrated by days 348 and 405, is closely followed by the model, especially in the region of the thermocline. Upward transport of salinity from lower strata is evident in reservoir data and seems to be followed well by the model. Simulation of the dynamics of the wind-mixed layer and internal mixing phenomena are still under investigation (Spigel and Imberger, 1980; Imberger, 1981).

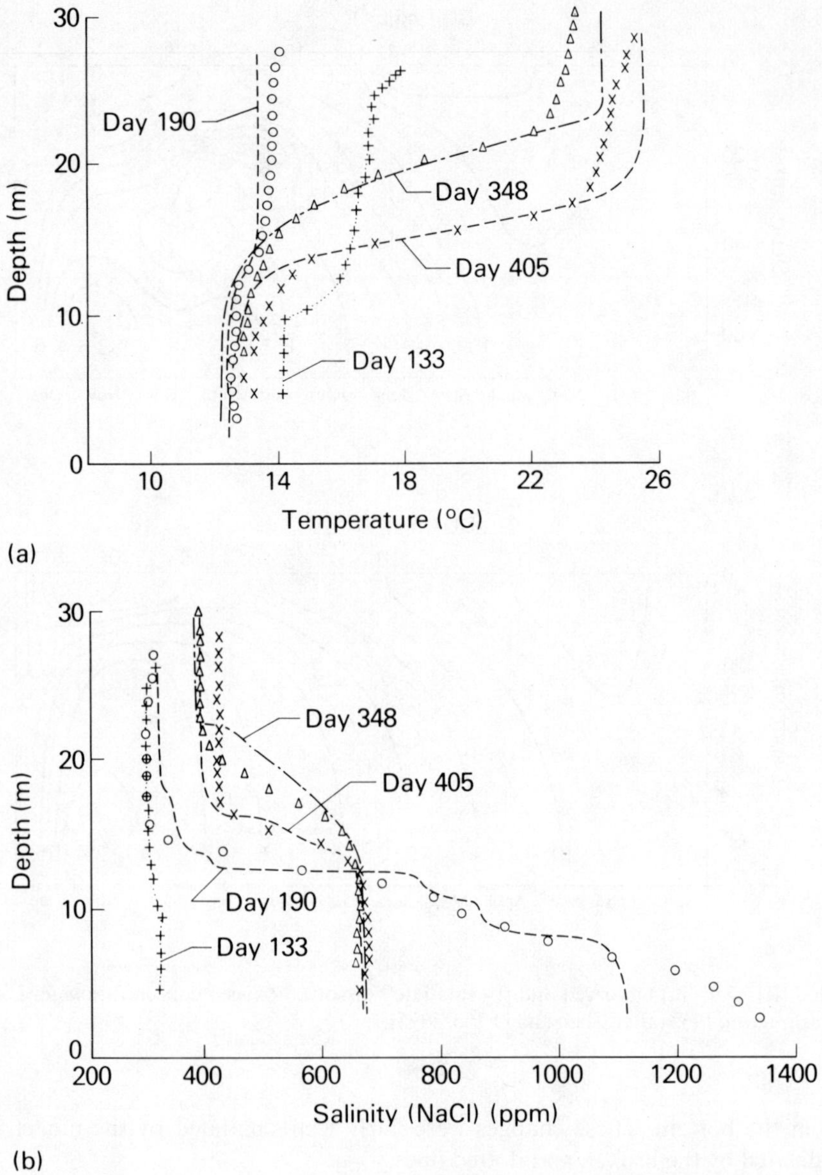

FIGURE 7.10 Simulation of (a) temperature and (b) salinity in Wellington Reservoir using DYRESM, and comparison with observations, 1975–76 (after Imberger *et al.*, 1978).

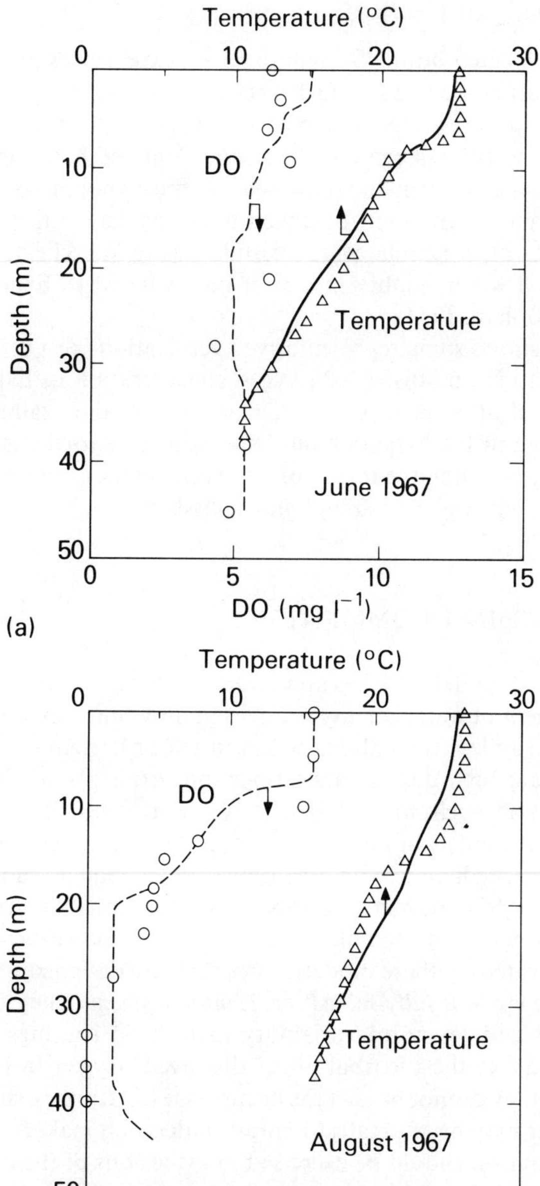

FIGURE 7.11 Observed (points) and simulated (lines) temperatures and dissolved oxygen concentrations in Lake Hartwell, Georgia, 1967 (after Fontane and Bohan, 1974).

7.14.3. Lake Hartwell, Georgia

This reservoir, located on the Savannah River, experiences an annual thermal cycle with the formation of a strong density gradient in the region of the thermocline that, in turn, restricts transport of oxygen into the hypolimnion. In addition, because the reservoir is large and summer flows are regulated by upstream projects, the demand imposed on the hypolimnion often depletes dissolved oxygen to the point of anaerobiosis by late summer or early fall. This system has been simulated successfully using WESTEX, the specialized one-dimensional water quality model of the Waterways Experiment Station (Fontane a Bohan, 1974).

Figure 7.11 shows some representative "verification" test results for simulation of temperature and dissolved oxygen concentration in Lake Hartwell for 1967. The model appears to give a fair account of the gradual depletion of dissolved oxygen in the hypolimnion. Especially noteworthy is the simulation of a steep DO gradient in the region of the thermocline, a characteristic of deep, stratified impoundments with a high nutrient supply.

7.15. CONCLUDING COMMENTS

One-dimensional models of lakes and reservoirs have proven to be useful tools for the assessment of water quality changes in impoundments that are small or medium-sized, say less than about 50 km in major horizontal dimension, and that experience a high degree of stratification. Attempts to simulate thermal energy changes in such water bodies have been generally successful and a variety of useful models are available for this purpose. Also, some results of oxygen balance prediction with one-dimensional models indicate that they can be used as tools to evaluate the impacts of operational or structural changes in deep reservoirs. Considerable care is required, however, in interpreting responses generated by these models, since they are designed to represent only *average conditions in a fully mixed, horizontal segment*. They cannot account for the spatial and temporal variability in organic loadings and biological activities that affect the distribution of dissolved oxygen in the real system. Consequently, they cannot predict the occurrence of extremes, such as anaerobic conditions, that may be of greatest interest to decision makers.

The same caution should be exercised in extensions of the one-dimensional conceptualization to simulate the aquatic ecosystem. While some one-dimensional models appear to provide reasonable accounts of gross nutrient budgets and primary production, they are increasingly unreliable in representing biological responses at the higher trophic levels. Such phenomena as patchiness, localized scavenging, and preferential grazing by motile predators cannot be described adequately in a one-dimensional conceptualization of the water body.

These limitations are highlighted here to emphasize the need for balance between the capabilities of the one-dimensional model and the complexities of the real world. For larger water bodies, for those that are not strongly stratified, and for cases where higher trophic levels are of major concern, we shall require more realistic models. These requirements suggest more complexity and rigor, multidimensional conceptualization, improved hydrodynamic representation, and more information from the field. They impose a challenge to modelers that the modelers seem to have been eager to accept. Evidence of this will be clearly demonstrated in subsequent chapters of this book.

REFERENCES

Baca, R. G., and R. C. Arnett (1976). A finite-element water quality model for eutrophic lakes. Finite Elements in Water Resources vol. 4, eds. W. G. Gray, G. Pinder and C. Brebbia (Plymouth, Devon: Pentech), pp. 125–147.

Baca, R. G., M. W. Lorenzen, R. D., Mudd, and L. V. Kimmel (1974). A generalized water quality model for eutrophic lakes and reservoirs. *Battelle Pacific Northwest Laboratories*, Report PB-256925.

Bella, D. A. (1976). Simulating the effect of sinking and vertical mixing in algal population dynamics. *Journal of Water Pollution Control Federation* 4(5), Part 2: R140–R152.

Bella, D. A., and W. E. Dobbins (1968). Difference modeling of stream pollution. *Proceedings of American Society of Civil Engineers, Journal of Sanitary Engineering Division* 94 (SA5), Paper 619, pp. 995–1016.

Bloss, S., and D. R. F. Harleman (1979). Effect of wind mixing on the thermocline formation in lakes and reservoirs. *MIT Department of Civil Engineering, R. M. Parsons Laboratory Technical Report* 249, 68 pp.

Burt, W. V. (1958). Heat budget terms for Middle Snake River Reservoir. *US Fish and Wildlife Service Report* 58-7, 23 pp.

Burt, W. V. (1960). A forecast of temperature conditions for the research behind the proposed Bruces Eddy Dam. *Water Research Associates, Inc., Corvallis, OR Report*, 16 pp.

Burt, W. V. (1963). Preliminary study on the predicted water changes in the Lower Snake River. *Water Research Associates, Inc., Corvallis, OR Report*, 16 pp.

Chen, C. W., M. Lorenzen, and D. J. Smith (1975). A comprehensive water-quality–ecologic model for Lake Ontario. *Report to Great Lakes Environment Research Laboratory, National Oceanic and Atmospheric Administration by Tetra Tech, Inc.*, 202 pp.

Chen, C. W., and G. T. Orlob (1968). A proposed ecologic model for a eutrophic environment. *Report to Federal Water Quality Agency, Southwest Region by Water Resources Engineers, Inc.*

Chen, C. W., and G. T. Orlob (1972). Predicting quantity effects of pumped storage. *Proceedings of American Society of Civil Engineers, Journal of Power Division* 98(PO1), Paper 8984.

Chen, C. W., and G. T. Orlob (1973). Ecologic study of Lake Koocanusa. *Report to US Army Corps of Engineers, District of Seattle by Water Resources Engineers, Inc.*, 63 pp.

Chen, C. W., and G. T. Orlob (1975). Ecologic simulation of aquatic environments. *Systems Analysis and Simulation in Ecology* vol. 3, ed. B. C. Patten (New York, NY: Academic Press), ch. 12, pp. 475–528.

Dake, J. M. K., and D. R. F. Harleman (1966). An analytical and experimental investigation of thermal stratification in lakes and ponds. *MIT Department of Civil Engineering, R. M. Parsons Laboratory Technical Report* 99, 271 pp.

Debler, W. E. (1959). Stratified flow into a line sink. *Proceedings of American Society of Civil Engineers, Journal of Engineering Mechanics Division* 85(EM3): 51–65.

Elder, R. A., and W. O. Wunderlich (1968). Evaluation of Fontana Reservoir field measurements. *Paper presented at ASCE Specialty Conference on Current Research into Effects of Reservoirs on Water Quality, Portland, OR.*

Ertel, H. E. (1954). Theorie der thermischen Sprungschicht in Seen. *Acta Hydrophysica* 1: 151–171.

Fischer, H. B., E. J. List, R. C. Koh, J. Imberger, and N. H. Brooks (1979). *Mixing in Inland and Coastal Water* (New York, NY: Academic Press), 483 pp.

Fogg, J. B., and E. G. Fruh (1973). A dissolved oxygen model for the hypolimnion of Lake Lyndon B. Johnson. *Center for Research in Water Resources, University of Texas, Austin, CRWR Report* 103, 90 pp.

Fontane, D. G., and J. P. Bohan (1974). Richard B. Russell Lake water quality investigation. *Waterways Experiment Station, Vicksburg, MS Technical Report* H-74-14, 26 pp.

Gallagher, R. H., J. A. Liggett, and S. T. K. Chan (1973). Finite-element shallow lake circulation analysis. *Proceedings of American Society of Civil Engineers, Journal of Hydraulics Division* 99(HY7), Paper 9855, pp. 1083–1096.

Gaume, A. M., and J. H. Duke, Jr. (1975). Computer program documentation for the reservoir ecologic model EPAECO. *Report to US Environmental Protection Agency, Washington, DC by Water Resources Engineers, Inc.*

Harleman, D. R. F., and K. A. Hurley (1976). Simulation of the vertical thermal structure of lakes under transient meteorological conditions. *Dynamics of Stratification and of Stratified Flows in Large Lakes: Proceedings of Workshop of the Committee on Lake Dynamics* (Windsor, Ontario: International Joint Commission, Regional Office), pp. 79–96.

Henderson-Sellers, B. H. (1978). Forced plumes in a stratified reservoir. *Proceedings of American Society of Civil Engineers, Journal of Hydraulics Division* 104(HY4): 487–501.

Huber, W. C., D. R. F. Harleman, and P. J. Ryan (1972). Temperature prediction in stratified reservoirs. *Proceedings of American Society of Civil Engineers, Journal of Hydraulics Division*, Paper 8839, pp. 645–666.

Hurley-Octavio, K. E., G. H. Jirka, and D. R. F. Harleman (1977). Vertical heat transport mechanisms in lakes and reservoirs. *MIT Department of Civil Engineering, R. M. Parsons Laboratory Technical Report* 227, 70 pp.

Hutchinson, G. E. (1957). *A Treatise on Limnology* vol. 1 (New York, NY: Wiley).

Hydrologic Engineering Center (1974) (revised 1978). WQRRS, generalized computer program for river–reservoir systems. *US Army Corps of Engineers, HEC, Davis, CA User's Manual* 401-100, 100A, 210 pp.

Imberger, J. (1981). Influence of stream salinity on reservoir water quality. *Agricultural Water Management* 4, 9 pp.

Imberger, J., J. Patterson, B. Hebbert, and I. Loh (1978). Dynamics of reservoirs of medium size. *Proceedings of American Society of Civil Engineers, Journal of Hydraulics Division* 104: 725–743.

Imboden, D. M. (1974). Phosphorus model for lake eutrophication. *Limnology and Oceanography* 19: 297–304.

Kao, T. W. (1965). The phenomenon of blocking in stratified flow. *Journal of Geophysical Research* 70(4): 815–822.

King, I. P., W. R. Norton, and K. R. Iceman (1974). A finite element model for two-dimensional flow. *Finite Elements in Flow Problems*, eds. J. T. Oden *et al.* (Huntsville, AL: University of Alabama, Huntsville Press), pp. 133–137.

Kinnunen, K., J. S. Neimi, and J. Eloranta (1978). Adaptation of the EPAECO model to a lake in Central Finland. *Modeling the Water Quality of the Hydrological Cycle. Proceedings of Baden Symposium, September. International Association of Hydrological Sciences Publication* 125, pp. 115–127.

Larsen, D. P., and H. T. Mercier (1975). Shagawa Lake recovery characteristics as depicted by predictive modeling. *Paper presented at American Institute of Biological Sciences Meeting, Corvallis, OR.*

Leung, K. K., R. A. Park, D. J. Desormeau, and J. Albanese (1978). MS CLEANER—An overview. *Proceedings of Conference on Modeling and Simulation, Pittsburgh, PA* (Troy, NY: Center for Ecological Modeling, Rensselaer Polytechnic Institute), 8 pp.

Markofsky, J., and D. R. F. Harleman (1973). Prediction of water quality in stratified reservoirs. *Proceedings of American Society of Civil Engineers, Journal of Hydraulics Division* 99(HY5), Paper 9730, pp. 729–745.

McEwen, G. F. (1929). A mathematical theory of the vertical distribution of temperature and salinity in water under the action of radiation, conduction, evaporation and mixing due to the resultant convection. *Scripps Institute of Oceanography Technical Bulletin* 2:197–306.

Niiler, P. O. (1975). Deepening of the wind-mixed layer. *Journal of Marine Research* 33:405–422.

Norton, W. R., and I. P. King (1975). Mathematical simulation of water temperature to determine the impact of raising an existing dam. *Progress in Astronautics and Aeronautics* 36:53–70.

O'Melia, C. R. (1974). Phosphorus cycling in lakes. *North Carolina Water Resources Research Institute, Raleigh Report* 97, 45 pp.

Orlob, G. T. (1965). A mathematical model of thermal stratification in deep reservoirs. *Conference of American Fisheries Society, Portland, OR*, 14 pp.

Orlob, G. T., and L. Selna (1967). Prediction of thermal energy distributions in deep reservoirs. *Proceedings of 6th Annual Sanitation and Water Resources Engineering Conference, Vanderbilt University, Nashville, TN*, pp. 64–77.

Orlob, G. T., and L. Selna (1970). Temperature variations in deep reservoirs. *Proceedings of American Society of Civil Engineers, Journal of Hydraulics Division* 96(HY2), Paper 7063.

Park, R. A., *et al.* (1974a). A generalized model for simulation of lake ecosystems. *US International Biological Program Technical Report.*

Park, R. W., D. Scavia, and N. L. Clesceri (1974b). CLEANER, the Lake George model. *US International Biological Program, Eastern Deciduous Forest Biome Contribution* 184, 13 pp.

Raphael, J. M. (1961). The effect of Wanapum and Priest Rapids Dams on the temperature of the Columbia River. *Public Utility District No. 2 of Grant County, Ephrata, WA Report*, 22 pp.

Raphael, J. M. (1962a). The effect of Wells and Rocky Reach Dams on the temperature of the Columbia River. *Public Utility District No. 2 of Grant County, Ephrata, WA Report*, 9 pp.

Raphael, J. M. (1962b). Prediction of temperature in rivers and reservoirs. *Proceedings of American Society of Civil Engineers, Journal of Power Division* 88(PO2):157–181.

Ryan, P. J., and D. R. F. Harleman (1971). Prediction of the annual cycle of temperature changes in a stratified lake or reservoir: Mathematical model and user's manual. *MIT Department of Civil Engineering, R. M. Parsons Laboratory Technical Report* 137.

Scavia, D., and R. A. Park (1976). A user's manual for the aquatic model CLEANER. *Rensselaer Fresh Water Institute at Lake George, Rensselaer Polytechnic Institute, Troy, NY Report* 76–16, 81 pp.

Snodgrass, W. J., and C. R. O'Melia (1975). A predictive phosphorus model for lakes—Sensitivity analysis and applications. *Environmental Science and Technology* 9:937–945.

Spaulding, D. B., and V. Svensson (1976). The development and erosion of the thermocline. *Heat Transfer and Turbulent Buoyant Convection* (Washington, DC: Hemisphere), pp. 113–122.

Spigel, R. H. (1978). Wind mixing in lakes. *Ph.D. Dissertation, University of California, Berkeley.*

Spigel, R. H., and J. Imberger (1980). The classification of mixed-layer dynamics in lakes of small to medium size. *Journal of Physical Oceanography* 10(7):1104–1121.

Stefan, H., and D. E. Ford (1975). Temperature dynamics in dimictic lakes. *Proceedings of American Society of Civil Engineers, Journal of Hydraulics Division*, Paper 11058, pp. 97–114.

Sundaram, T. R., C. C., Easterbrook, K. R. Piech, and G. Rudinger (1969). An investigation of the physical effects of thermal discharge into Cayuga Lake. *Cornell Aeronautical Laboratory, Buffalo, NY Report* VT-2616-0-2.

Sundaram, T. R., and R. G. Rehm (1973). The seasonal thermal structure of deep temperate lakes. *Telbes* 25:157–167.

Svensson, J. (1978). A mathematical model of the seasonal thermocline. *Department of Water Resources Engineering, Lund Institute of Technology, Lund, Sweden Thesis Report* 1002.

Tennessee Valley Authority (1972). Heat and mass transfer between a water surface and the atmosphere. *TVA Water Resources Research Engineering Laboratory, Norris, TN Report* 14, 123 pp.

Vollenweider, R. A. (1965). Calculation models of photosynthesis–depth curves and some implications regarding day rate estimates in primary production. *Memorie dell Istituto Italiano di Idrobiologia* 18 Suppl:425–457.

Vollenweider, R. A. (1969). Möglichkeiten und Grenzen elementarer Modelle der Stoffbilanz von Seen. *Archiv Hydrobiologie* 66:1–36.

Water Resources Engineers, Inc. (1968). Prediction of thermal energy distribution in streams and reservoirs. *Report to California Department of Fish and Game*, 90 pp.

Water Resources Engineers, Inc. (1969). Mathematical models for prediction of thermal energy changes in impoundments. *US Environmental Protection Agency Water Pollution Control Research Series* 16130 EXT, Contract 14-22-422. US Government Printing Office, Washington, DC, 157 pp.

CHAPTER 7: NOTATION

a area of horizontal plane through a reservoir
A advection rate
b average width of a reservoir
B benthic animal biomass concentration
d average depth of a reservoir
D diffusion rate
E effective diffusion rate
F nekton (fish) biomass concentration
h heat flow
H heat content of a volume element
K turbulent kinetic energy
K_1 BOD decay coefficient
K_2 reaeration coefficient
K_4 benthic demand coefficient
l length of a reservoir
L biochemical oxygen demand
m mortality rate
P phytoplankton biomass concentration
q discharge per unit width or depth; heat flux per unit area
Q_a lateral advected flow
r respiration rate coefficient
s sedimentation rate
S source or sink of constituent
U average flow-through velocity of a reservoir

w	vertical advection velocity
z	depth (above bottom of reservoir)
β	density gradient; ratio of absorbed to net incoming radiation
δ	thickness of withdrawal layer
ϵ	rate of energy dissipation per unit mass
η	bulk (light) extinction coefficient; digestion efficiency of predator
Θ	temperature
σ	stoichiometric oxygenation factor
Ψ	release rate.

8 Two- and Three-Dimensional Mathematical Models for Lakes and Reservoirs

M. Watanabe, D. R. F. Harleman, and O. F. Vasiliev

8.1. INTRODUCTION

A knowledge of the hydrophysical processes in lakes and reservoirs is important for the understanding of ecological processes, particularly (i) the transport of various materials such as nutrients needed by aquatic organisms, or pollutants; (ii) temperature distribution; (iii) population dynamics and transport of plankton; and (iv) distribution of water quality parameters such as DO, pH, BOD, and COD.

There is no universal model detailed enough to describe the entire physical and biochemical behavior of lakes. Simplifications must be introduced in order to formulate models and to make solutions possible. The degree of simplification, including averaging or the dropping of higher-order terms, necessarily reduces the generality of a mathematical model.

One-dimensional models can simulate the vertical (or longitudinal) behavior of water bodies. While the vertical structure is largely important for the analysis of lake behavior, from an ecological point of view vertical, one-dimensional models cannot predict the horizontal divergences and convergences of the wind-driven surface current, the return current in lake basins, the effects of the earth's rotation in large basins, or upwelling and downwelling. Patterns of distribution or circulation that extend over large space and time scales are often determined by the shape of the lake or by short-term strong disturbances (e.g. localized shear instabilities) that are characterized by much shorter space and time scales. This suggests the necessity for three-dimensional circulation models capable of simulating the important mechanisms over a wide range of scales (Mortimer, 1974).

Although the equations of fluid dynamics can describe motions that include all of these scales, practical difficulties prohibit (or make unnecessary) the use of the full equations of motion for real problems involving a wide variety of phenomena. Considerable effort and ingenuity have been expended to approximate these equations in order to obtain simpler equations and methods of solution.

The predictive power and, therefore, the usefulness of such models will be greatly dependent on the depth of understanding and recognition of those mechanisms that are physically important (excellent reviews have been written by Mortimer (1974), Boyce (1974), and Csanady (1975)) and also on well planned model verifications in actual lakes. In the last few years, hydrodynamic models for the calculation of lake circulation have become quite detailed and realistic and have contributed greatly to our understanding of lake behavior.

This chapter is concerned with the state of the art of two- and three-dimensional mathematical models for describing the hydrophysical behavior of lakes and reservoirs under natural influences (such as wind). In addition, mathematical models of cooling ponds or reservoirs under artificial heat loading from power plants will be discussed. The latter topic is relatively new, but significantly important for the understanding of eutrophication processes and for predicting other possible effects of excess heat loading on lake ecosystems. Artificial heat loads may change the entire flow and temperature fields and affect the rates of biochemical reactions.

In section 8.2, a brief discussion of some general characteristics of currents in lakes is followed by the three-dimensional, time-dependent hydrodynamic and heat exchange equations, boundary conditions, and parameters basic to all the models. These general equations are simplified, based on length and time scales significant for particular situations.

Mathematical models of wind-driven circulation can be classified broadly as horizontal two-dimensional models, two- and multi-layer models, Ekman-type models, vertical two-dimensional stratified flow models, and three-dimensional models.

Various models are compared in terms of their basic structure and assumptions, physical validity, and range of applicability. Brief reviews of numerical techniques, such as finite-difference and finite-element methods, and stability criteria are presented. The next section briefly reviews two- and three-dimensional water quality models in which hydrodynamic and water quality models are coupled. Because computer capacity is limited, integration of complex hydrodynamic circulation models and ecological models causes problems in terms of computation time, core memory, and cost. Simplifications based on physical reality are essential.

Following this, a critical survey is presented of two- and three-dimensional mathematical models for lakes, reservoirs, and cooling ponds, which require different simplifications and solution techniques.

8.2. GENERAL CHARACTERISTICS OF HYDROTHERMAL CIRCULATION MODELS AND GOVERNING EQUATIONS

8.2.1. Introduction

Circulation in lakes is primarily caused by wind shear acting on the water surface. Temperature or density differences produce some additional hydrodynamic effects that result in density currents. At the same time, density gradients affect currents and play an important role in stratified water bodies.

The geometry of a lake (or reservoir) has a significant influence on its circulation. Water circulation influenced by wind, density gradients, and many other factors is therefore complex and includes various motions associated with different length and time scales. Boyce (1974) summarized motions in lakes and their associated length and time scales, as shown in Table 8.1.

The relative importance of each force can be evaluated by selecting the basic scale of interest. Therefore, the comparison of various forces as they affect motions at these scales allows the simplification of the governing equations of motion by neglecting the less important terms. The differences between existing mathematical models are therefore related to the different scales that are significant for the specific conditions. In this section, the basic equations describing fluid motion and heat transfer in lakes or reservoirs are presented before each model is discussed in detail.

8.2.2. Governing Equations

The foundation for the development of models of fluid motion lies in the Navier–Stokes equations for the conservation of momentum. In addition, conservation laws for mass and energy are required. The complete set of equations is without formal, general solutions, yet methods have been devised to obtain solutions for particular cases (including numerical methods).

The three components of velocity, u, v, w, and the pressure p are decomposed according to the Reynolds conceptualization into mean and fluctuating components:

$$u = \bar{u} + u', \qquad v = \bar{v} + v', \qquad w = \bar{w} + w', \qquad p = \bar{p} + p'.$$

The following approximations are introduced in the Navier–Stokes equations.

(i) The Boussinesq approximation: the density variation in a water body is much smaller than the density itself; therefore, constant density can be used in the equation of motion, except in the term for the buoyancy force.

(ii) The hydrostatic approximation: in lakes and reservoirs, the vertical velocity component is usually small. Therefore, the vertical component of acceleration is much smaller than the gravitational acceleration.

If the expressions above are substituted into the Navier–Stokes equations and the time averages are taken, the Reynolds equations for a fluid body on a rotating earth are as follows:

$$\frac{\partial \bar{u}}{\partial t} + \frac{\partial}{\partial x}\,\bar{u}\bar{u} + \frac{\partial}{\partial y}\,\bar{u}\bar{v} + \frac{\partial}{\partial z}\,\bar{u}\bar{w} - f\bar{v} = -\frac{1}{\rho}\frac{\partial \bar{p}}{\partial x} + \nu\nabla^2\bar{u} - \frac{\partial}{\partial x}\,\overline{u'u'} - \frac{\partial}{\partial y}\,\overline{u'v'} - \frac{\partial}{\partial z}\,\overline{u'w'}, \qquad (8.1)$$

$$\frac{\partial \bar{v}}{\partial t} + \frac{\partial}{\partial x}\,\bar{v}\bar{u} + \frac{\partial}{\partial y}\,\bar{v}\bar{v} + \frac{\partial}{\partial z}\,\bar{v}\bar{w} + f\bar{u} = -\frac{1}{\rho}\frac{\partial \bar{p}}{\partial y} + \nu\nabla^2\bar{v} - \frac{\partial}{\partial x}\,\overline{v'u'} - \frac{\partial}{\partial y}\,\overline{v'v'} - \frac{\partial}{\partial z}\,\overline{v'w'}, \qquad (8.2)$$

$$0 = -g - \frac{1}{\rho}\frac{\partial \bar{p}}{\partial z}, \qquad (8.3)$$

and the continuity equation can be expressed as

$$\frac{\partial \bar{u}}{\partial x} + \frac{\partial \bar{v}}{\partial y} + \frac{\partial \bar{w}}{\partial z} = 0, \qquad (8.4)$$

where

- t is time,
- x, y, z are the coordinate axes ($+z$ is vertically upward),
- $\bar{u}, \bar{v}, \bar{w}$ are the mean velocities in the x, y, z directions,
- $\overline{u_i'u_j'}$ is the time-averaged turbulent eddy transport of momentum,
- g is the gravitational acceleration,
- ρ is the density,
- $\bar{p}$ is the pressure,
- f is the Coriolis parameter,
- ν is the (molecular) kinematic viscosity,
- $\nabla^2 = \partial^2/\partial x^2 + \partial^2/\partial y^2 + \partial^2/\partial z^2$.

Similarly, the equation of heat balance for a turbulent flow is

$$\frac{\partial \bar{T}}{\partial t} + \frac{\partial}{\partial x}\,\bar{T}\bar{u} + \frac{\partial}{\partial y}\,\bar{T}\bar{v} + \frac{\partial}{\partial z}\,\bar{T}\bar{w} = \chi\nabla^2 T - \frac{\partial}{\partial x}\,\overline{T'u'} - \frac{\partial}{\partial y}\,\overline{T'v'} - \frac{\partial}{\partial z}\,\overline{T'w'} + Q_{\mathrm{H}}, \qquad (8.5)$$

where χ is the (molecular) thermal diffusivity, $\overline{T'u_i'}$ is the time-averaged turbulent eddy transport of heat in direction x_i and Q_{H} is the heat source. Advective transport $\bar{T}\bar{u}_i$ and turbulent eddy transport $\overline{T'u_i'}$ depend on the state of flow.

TABLE 8.1 Classification of Motions in Lakes (according to Boyce, 1974). Letters in parentheses refer to the type of scale used. M, amplitude of motion; S, distance over which phenomenon varies significantly; P, period; T, time interval over which phenomenon varies significantly; C, wave speed; V, vertical particle velocity; H, horizontal particle velocity. The governing terms in the equations of motion and continuity are listed in column 6 according to the code: 1, time-dependent horizontal accelerations; 2, time-dependent vertical accelerations; 3, advective component of horizontal acceleration; 4, advective component of vertical acceleration; 5, Coriolis force; 6, pressure gradient force due to slope of free surface; 7, pressure gradient force due to slope of the thermocline; 8, pressure gradient force due to atmospheric pressure field; 9, variations in bottom topography; 10, wind energy/stress; 11, internal stresses arising from horizontal current shear; 12, internal stresses arising from vertical current shear; 13, friction against lake boundaries; 14, potential energy changes due to surface heating and cooling.

Phenomenon	Length Scale		Time Scale	Velocity Scale	Dynamics of Major Components	Notes
	Horizontal	Vertical				
Wind-driven surface gravitational waves	10 m (S)	1 m (M)	1 S (P)	10 m s^{-1} (C) 1 m s^{-1} (V, H)	1, 2, 6, 10	
Surface gravitational waves—seiches	100 km (S)	10 cm (M)	2–10 h (P)	2 cm s^{-1} (H)	1, 6, 9, 10	
Short, freely propagating internal waves	100 m (S)	2 m (M)	5 min (P)	2 cm s^{-1} (H)	1, 2, 7, 10	

Long, propagating internal waves steered by topography	10 km (S)	2 m (M)	1 day (T)	50 cm s^{-1} (C)	1, 5, 7, 9, 10	
Internal gravitational standing waves or seiches	10 km (S)	2 m (M)	16 h (P)	10 cm s^{-1} (H)	1, 5, 7, 10	
Surface wind drift	—	10 cm (S)	—	2 cm s^{-1} (H)	10, 12	
Coastal currents	10 km (S)	—	1 day (T)	10 cm s^{-1} (H)	All	Horizontal scale measured perpendicular to coast
Upwelling and downwelling	10 km (S)	10 m (M)	1 day (T)	<1 cm s^{-1} (V)	All	Horizontal scale measured perpendicular to coast
Wind-driven horizontal circulation	100 km (S)	100 m (S)	1 day (T)	10 cm s^{-1} (H)	All	
Basin-wide horizontal diffusion	100 km	—	1 month (T)	—	All	Integrated effect of all motions
Langmuir circulations, vertical mixing of epilimnion	—	10 m (S)	1 h (T)	1 cm s^{-1} (V)	1, 2, 3, 4, 10, 12, 14, and others	Dynamics not yet understood
Formation and decay of thermocline	—	10–100 m (S)	1 month (T)	—	10, 12, 14	

In addition, an equation of state defines the relationship between temperature and density:

$$\rho = \rho(T). \tag{8.6}$$

Initial conditions and boundary conditions have to be specified for the six unknowns $\bar{u}_i$, ρ, $\bar{p}$, and $\bar{T}$. Also, the eddy transport terms $\overline{u_i' u_j'}$ and $\overline{T' u_i'}$ need to be parameterized through the flow state variables.

8.2.2.1. Boundary Conditions

The specification of boundary conditions depends on the nature of the problem.

At the free surface

(1) A kinematic free-surface condition:

$$\left(\frac{\partial \eta}{\partial t} + \bar{u}\frac{\partial \eta}{\partial x} + \bar{v}\frac{\partial \eta}{\partial y} - \bar{w}\right)_{z=\eta} = 0, \tag{8.7}$$

where η is the elevation of the free surface;

(2) a dynamic free-surface condition for the pressure on the air–water interface:

$$\bar{p}(x, y, \eta, t) = p_a(x, y, t), \tag{8.8}$$

where p_a is the atmospheric pressure;

(3) wind stress on the air–water interface:

$$\boldsymbol{\tau}(x, y, \eta, t) = \boldsymbol{\tau}_s(x, y, t), \tag{8.9}$$

where $\boldsymbol{\tau}_s$ is the specified surface wind stress;

(4) surface heat flux:

$$q(x, y, \eta, t) = \phi_n(x, y, t), \tag{8.10}$$

where ϕ_n is the net heat flux through the surface.

At the bottom and side boundaries

(5) A kinematic condition:

$$\left(\bar{u}\frac{\partial h}{\partial x} + \bar{v}\frac{\partial h}{\partial y} - w\right)_{z=-h} = 0, \tag{8.11}$$

where h is the height from the datum to the bottom;

(6) no fluid motion or a specification of surface stresses in terms of near-surface velocities;

(7) a specification of either temperature or heat flux.

At the inflow or outflow boundaries

(8) Inflow and outflow velocities or volume flux are specified;
(9) inflow temperature is given by

$$T = f_T(x, y, z, t); \tag{8.12}$$

(10) outflow temperature satisfies the following relation (Vasiliev, 1978a,b):

$$\frac{\partial \bar{T}}{\partial t} + \bar{u}\frac{\partial \bar{T}}{\partial x} + \bar{v}\frac{\partial \bar{T}}{\partial y} + \bar{w}\frac{\partial \bar{T}}{\partial z} = Q_{\mathrm{H}}. \tag{8.13}$$

The rigid lid approximation can be considered as a special free-surface boundary condition that assumes no change in surface elevation. This assumption eliminates all surface waves; therefore, currents resulting directly from wind shear can be calculated.

The rigid lid approximation simplifies the kinematic condition at the free surface:

$$w = 0 \quad \text{at} \quad z = 0 \tag{8.14}$$

and the pressure can be set to an arbitrary value at any point. However, the rigid lid does not act like a solid boundary with respect to friction and shear. Therefore, in the rigid lid approximation pressure differences are transmitted instantaneously everywhere in the water body instead of propagating with the celerity of gravity waves.

Since the propagation of gravity waves is very rapid compared with the response of the wind-driven current, this approximation has little effect on the response of currents. The advantage of the rigid lid approximation is that it results in a simpler computational process and the ability to use longer time steps. Besides the boundary conditions, the initial conditions should be given: the distributions of velocity and temperature in the water body and the free-surface position at an initial time.

8.2.3. Turbulent Shear and Diffusion

8.2.3.1. Turbulence Models

The covariances $\rho\overline{u_i' u_j'}$ in (8.1)–(8.3), which result from time averaging over the scale of the turbulent fluctuations, represent an effective shear acting in direction x_i on a plane perpendicular to x_j. In order to solve the governing equations, these covariances need to be specified. This specification has been the subject of different theories of turbulence, starting with Reynolds and Boussinesq in the later part of the last century. A useful summary of these theories and their practical limitations has been given by Vreugdenhil (1973).

A complete and accurate description of turbulence is complicated by the fact that different mechanisms govern the turbulent fluctuations (measured by its kinetic energy e):

$$e = \frac{1}{2}(\overline{u'^2} + \overline{v'^2} + \overline{w'^2}).$$

These are (Schlichting, 1968): (a) the generation of turbulence as a result of gradients in the mean motion, (b) the transfer of turbulence from more turbulent to less turbulent zones, and (c) the dissipation of turbulence by internal friction between fluid elements. Thus, successful descriptions are usually limited to cases in which a local equilibrium persists between these three mechanisms, such as at a cross section of uniform pipe or open-channel flow.

Attempts to quantify the turbulent shear stresses range from the eddy viscosity concept to direct transport equations for the shear stresses. All hypotheses involve empirical coefficients and need to be verified with experiments. According to Boussinesq, the turbulent shear stress τ_{ij}^{t} can be formulated as

$$\tau_{ij}^{t} = -\rho\overline{u_i' u_j'} = \rho\epsilon_i\left(\frac{\partial \bar{u}_i}{\partial x_j} + \frac{\partial \bar{u}_j}{\partial x_i}\right), \tag{8.15}$$

completely in analogy to the laminar (molecular) shear stress. The coefficient ϵ_i in (8.15) is termed the kinematic eddy viscosity and is represented as a vector signifying the directional dependence. (This is a simplification because the eddy viscosity should be a higher-order tensor, as shown by Hinze (1959; also Monin and Yaglom, 1965).)

8.2.3.2. Diffusion Models

The covariances $\overline{u_i' T'}$ represent an effective transport of heat by the random, turbulent displacement of lumps of fluid and the subsequent exchange with adjacent fluid masses. The specification of $\overline{u_i' T'}$ presents some difficulty because temperature, by virtue of the equation of state (8.6), also affects the density of the fluid. On the other hand, most research on turbulent diffusive transport $\overline{u_i' c'}$ has been focused on passive (inert) scalar quantities of small concentrations c, which therefore do not change the density of the ambient fluid.

Similarly to the turbulent shear models, two approaches have been followed for diffusion models, namely the eddy diffusivity concept and the direct specification of $\overline{u_i' T'}$ by means of the equations for higher-order statistical moments. Discussion will concentrate on the eddy diffusivity method since there are computational restrictions with the other method. The turbulent diffusive transport of heat is expressed by

$$q_i^{t} = -\overline{u_i' T'} = K_{Ti}\frac{\partial \bar{T}}{\partial x_i}, \tag{8.16}$$

where K_{Ti} is the eddy diffusivity, in analogy to the Fourier law of heat conduction by molecular diffusion.

8.2.4. Eddy Viscosity and Eddy Diffusivity

8.2.4.1. Eddy Viscosity

Unlike the molecular viscosity, the eddy viscosity ϵ_i is not simply a fluid property, but rather a function of the flow conditions. An approach to describe this dependence is the mixing length concept, first proposed by Prandtl, and used by Taylor, von Karman, and others. The mixing length concept is, in principle, an empirical relation:

$$\epsilon \approx l^2 \frac{\mathrm{d}\bar{u}}{\mathrm{d}z}, \tag{8.17}$$

where l is the characteristic (mixing) length, $\mathrm{d}\bar{u}/\mathrm{d}z$ is the mean velocity gradient perpendicular to the flow, and z is the distance from the bottom. This concept has been useful for describing certain equilibrium shear flows, in particular for uniform open-channel flow in which the vertical eddy viscosity can be written as

$$\epsilon_z = u_* l, \qquad l = \kappa z(1 - z/H), \tag{8.18}$$

where

u_* $=(\tau_0/\rho)^{1/2}$ is the shear velocity,
τ_0 is the bottom shear,
κ $= 0.4$ is the von Karman constant,
H is the water depth.

Another approach can be made by using turbulence energy e and a length scale of turbulence:

$$\epsilon \approx l e^{1/2}. \tag{8.19}$$

This requires an additional transport equation for e (derived from the Navier–Stokes equations), involving higher-order covariances (statistical moments) and requiring more empirical coefficients. The equation of the turbulent energy balance is discussed in detail by Monin and Yaglom (1965).

A more complex turbulence closure model is based on two transport equations: one for e and the other for the turbulent energy dissipation ϵ (Jones and Launder, 1973; Launder, 1976). These more elaborate turbulence models require substantial computational effort. Individual models have usually been verified with a limited class of experiments (mostly boundary layer types). Examples of application of turbulent models of the latter types will be given in section 8.3.5.

Effects of density stratification

Turbulence in the presence of a density stratification is discussed by Phillips (1969). A stable stratification hampers vertical displacement of fluid. Thus, for turbulence to be maintained there must be a sufficiently great destabilizing supply of turbulent kinetic energy to overcome the stabilizing effect of the potential energy of stable density stratification. The parameter that characterizes these relative effects is the Richardson number:

$$\mathrm{Ri} = \frac{(g/\rho)\partial\rho/\partial z}{(\partial u/\partial z)^2}, \tag{8.20}$$

where z is the coordinate in the direction of g and u is the horizontal velocity. The Richardson number is a local parameter and varies with depth as the density gradient $\partial\rho/\partial z$ and velocity gradient $\partial u/\partial z$ change. Large values of Ri reflect a high degree of stability and reduced vertical transport of momentum (shear). A theoretical condition for turbulence to be maintained is

$$\mathrm{Ri} < \tfrac{1}{4}. \tag{8.21}$$

An elementary derivation of this criterion, including critical remarks about its significance, is given by Long (1970; also Monin and Yaglom, 1965).

Little is known in exact, quantitative terms about the dependence of the vertical eddy viscosity coefficient ϵ_z on the Richardson number. Karelse (1974) compared available data with a variety of empirical formulas proposed by different investigators and discovered considerable scatter. A characteristic empirical formula is one by Munk and Anderson (1948):

$$\epsilon_z = \epsilon_{z_0}(1 + 10\,\mathrm{Ri})^{-1/2}, \tag{8.22}$$

where ϵ_{z_0} is the eddy viscosity under nonstratified conditions.

In many instances, when the stratification is significant and a distinct interface exists, it is advantageous to make use of the discretely layered structure and apply one- or two-dimensional stratified flow theory with more or less well defined interfacial friction factors.

8.2.4.2. Eddy Diffusivity

Eddy diffusivities K_i are again strongly dependent on flow conditions. The ratio between the eddy viscosity and the eddy diffusivity is termed the turbulent Prandtl number:

$$\mathrm{Pr} = \frac{\epsilon_i}{K_i}. \tag{8.23}$$

The Prandtl number is about unity for wall turbulence such as the shear flow considered in (8.17) and (8.18), so that these equations are directly applicable for the vertical diffusivity K_z.

Diffusion in open water, such as in lakes or coastal zones, is in contrast to open-channel shear flow in which the diffusive transport can be effectively related to the mean flow characteristics. Oceanic or lake turbulence represents a spectrum of different eddies resulting from the breakdown of large-scale circulations in shore zones and perturbation by wind and waves. Attempts to analyze this situation have shown that the horizontal diffusive transport K_h depends on the length scale l of the phenomenon. The most widely used formula is the four-thirds power law:

$$K_h = A_D l^{4/3}, \tag{8.24}$$

where A_D is the dissipation parameter. Useful summaries of lake and ocean diffusion data have been provided by Yudelson (1967), Okubo (1968), and Ozmidov (1968).

Effects of density stratification

Density stratifications have a pronounced effect on the vertical turbulent diffusivity K_z. Furthermore, K_z is more influenced by stratification than is the eddy viscosity ϵ_z. Available laboratory and field data on the relation of K_z/ϵ_z (the inverse of the Prandtl number) to the local Richardson number have been summarized by several investigators. For example, the results of Kullenburg (1974) are shown in Figure 8.1. Since eddy diffusion coefficients are more readily determined from tracer diffusion experiments, qualitative relations such as those in Figure 8.1 are of great practical significance, as they allow one to estimate the eddy viscosity coefficients, which are difficult to determine directly.

There have also been many attempts to relate K_z to the local Richardson number. A typical formula is that of Munk and Anderson (1948):

$$K_z = K_{z_0}(1 + \tfrac{10}{3}\,\mathrm{Ri})^{-3/2}, \tag{8.25}$$

where K_{z_0} is the eddy diffusivity without stratification, indicating a reduced value of K_z for large Ri. Further compilations of data and comparisons of different formulas have been made by Okubo (1962), Koh and Fan (1970), and Karelse (1974).

8.2.5. Surface Shear Expressions

The turbulent shear stress τ_s exerted at the water surface is usually expressed as

$$\tau_s = C_w \rho_a |\mathbf{W}| \mathbf{W}, \tag{8.26}$$

where $\mathbf{W}$ is the wind velocity measured at a certain height, usually 10 m, ρ_a is the air density, and C_w is a "wind stress coefficient." The coefficient depends on the surface roughness, which in turn is dependent on wind velocity. C_w is also dependent on the fetch, the stability of the air mass, the relative temperatures of air and water, and the topography and roughness upwind of the water body. In general, the roughness of the air–water interface is created by a great variety of waves differing in height, shape, and phase velocity. Moreover, the waves are subjected to continuous and irregular changes.

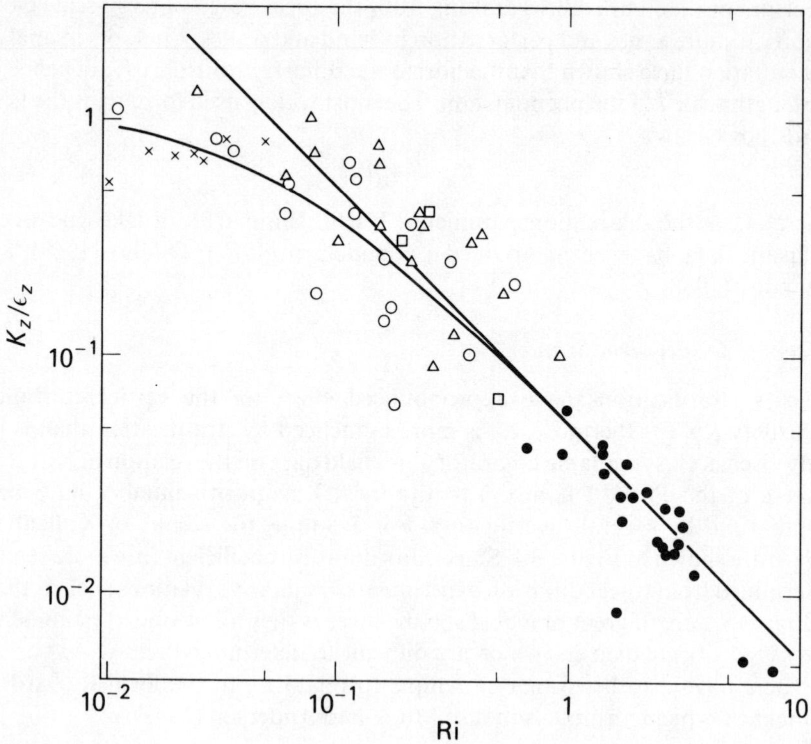

FIGURE 8.1 Ratio of the vertical diffusion coefficient and the vertical momentum transfer coefficient plotted against the Richardson number. The straight line represents (K_z/ϵ_z) Ri = 0.05. The curve represents Ellison theory with the critical flux Richardson number equal to 0.005. Experimental data are from: □ Bowden, ○ and △ Ellison and Turner, ● Lofquist, and × Swinbank (after Kullenburg, 1974).

Depending on wind speed in general, two regions can be distinguished from available data (Wu, 1969): (i) the region of roughness establishment, in which surface roughness increases with wind velocity, and (ii) the region of established roughness, in which surface roughness reaches an equilibrium value. Table 8.2, taken from Wu (1969), summarizes different expressions for the surface roughness $C_w = C_{10}$ (wind velocities in meters per second measured at 10 m) as a function of wind speed, sea state, or Beaufort number.

8.2.6. Surface Heat Flux

Several thermophysical processes, such as radiation, conduction, and evaporation, determine the heat transfer through the water surface. Detailed descriptions of these processes were presented in Chapter 5.

TABLE 8.2 Wind Stress Coefficients C_{10} as Functions of Wind Conditions (from Wu, 1969).

<table>
<tr><td>Type of Wind</td><td colspan="3">Breeze</td><td colspan="5">Light Wind</td><td colspan="5">Strong Wind</td></tr>
<tr><td>Wind speed W_{10} (m s^{-1})</td><td></td><td>1</td><td>3</td><td>5</td><td>7   9</td><td>11</td><td>13</td><td>15</td><td>18</td><td>21</td><td>24</td><td>27</td><td>30</td></tr>
<tr><td>Sea state</td><td>0</td><td>1</td><td>2</td><td>3</td><td colspan="3">4</td><td colspan="2">5</td><td>6</td><td>7</td><td>8</td><td>9</td></tr>
<tr><td>Beaufort number</td><td>0</td><td>1</td><td>2</td><td>3</td><td>4</td><td>5</td><td>6</td><td>7</td><td>8</td><td>9</td><td>10</td><td>11</td><td>12</td></tr>
<tr><td>Boundary layer flow conditions</td><td colspan="3">Aerodynamically smooth flow</td><td colspan="2">Transition</td><td colspan="8">Aerodynamically rough flow</td></tr>
<tr><td>Region of surface roughness</td><td colspan="3"></td><td colspan="5">Region of roughness establisment</td><td colspan="5">Region of established roughness</td></tr>
<tr><td>Surface roughness k (cm)</td><td colspan="3"></td><td colspan="2">$0.03 < k < 0.3$</td><td colspan="3">$0.3 < k < 1.5$</td><td colspan="5">8</td></tr>
<tr><td>Wind stress formula for C_{10}</td><td colspan="2">$\left(\frac{\kappa}{\ln(9.190/\nu)}\right)^2$</td><td colspan="6">$\left(\frac{\kappa}{\ln(\gamma s/0.0156 C_{10} u_{10}^2)}\right)^2$</td><td colspan="5">$2.6 \cdot 10^{-3}$</td></tr>
<tr><td>Approximate wind stress formula</td><td>$\frac{0.00125}{u_{10}^{1/5}}$</td><td colspan="7">$0.5 W_{10}^{1/2} \cdot 10^{-3}$</td><td colspan="5">$2.6 \cdot 10^{-3}$</td></tr>
</table>

In the analysis of the heat balance for water bodies the net surface heat flux ϕ_n is usually introduced as a boundary condition directly at the free surface. This is only approximately correct inasmuch as radiation absorption does not occur simply at the air–water interface but rather in a water layer close to the surface, the thickness of the layer being dependent on the absorptive characteristics, such as turbidity. This approximation is sufficiently accurate for many applications. In specific applications, such as the thermal stratification of lakes, the actual radiation absorption should be accounted for in the model formulation.

The general form of ϕ_n is nonlinearly dependent on the water surface temperature. The equation may be linearized over a certain range of surface temperatures upon introduction of an equilibrium temperature T_E. This is the water surface temperature at which, under given meteorological conditions, the net surface heat flux is zero. A linear approximation can be defined as follows:

$$\phi_n = -K_H(T_S - T_E), \tag{8.27}$$

where K_H denotes the heat transfer coefficient valid for a certain range of surface temperatures T_S. This linearization of surface heat transfer has an advantage only with analytical solutions or simple steady state models. Since both K_H and T_E are functions of time, direct computation of ϕ_n is recommended for transient numerical models (Ryan and Harleman, 1973).

Plate and Wengefeld (1979) made a survey of the transport processes for momentum, heat, and mass at a lake surface. The survey includes, in particular, the description of the formation of boundary layers near the free surface and the analysis of wave generation.

8.3. MODELS OF WIND-DRIVEN CIRCULATION

Although circulation and heat exchange in lakes or reservoirs can be expressed in general by the equations of motion, the continuity equation, and the thermodynamic energy equation, modelers wish to isolate the particular phenomena that they want to simulate and to retain only the corresponding terms in the governing equations.

Existing circulation models can be categorized into the following groups, based on the relevant assumptions and simplifications:

(1) horizontal two-dimensional circulation (single-layer) models,
(2) two- and multi-layer models,
(3) Ekman-type models,
(4) vertical two-dimensional stratified flow models, and
(5) three-dimensional models.

Each type of model has its own advantages and applicability. The derivation of the governing equations and limitations of the models will now be discussed.

8.3.1. Vertically Averaged Two-Dimensional Circulation Models (Single-Layer Models)

Single-layer models are formed by averaging the three-dimensional equations over the depth, with bottom and surface boundary conditions. The result of these manipulations is a set of reduced two-dimensional equations, which are comparatively easy to analyze and require relatively little computer time. These models do not give details of the vertical variation of the flow and calculate only the total mass transport.

The vertically averaged formulation was originally developed in the theory of long waves and in open-channel hydraulics. The first numerical model of storm surges in the North Sea based on such a formulation was developed by Hansen (1956). Platzman (1963) applied a similar model to several of the Great Lakes, Reid and Bodine (1968) to the Gulf of Mexico, and Heaps (1969) to the North Sea. Later, this approach was used to simulate circulation in shallow, vertically well mixed systems by Leenderste (1967), Cheng and Tung (1970), Simons (1971), Cheng (1972), Abbott *et al.* (1973), Gallagher *et al.* (1973), and Wang and Connor (1975).

The principal limitation of vertically averaged models is that they do not consider the effects of velocity and density variations in the vertical direction, though these details are often necessary for the complete understanding of the flow characteristics and for an accurate description of dependent problems in ecological applications (for example, when there are bottom return currents in wind-induced lake circulation).

However, the two-dimensional model can be adequate for the consideration of pronounced unsteady flows in shallow water bodies and might be useful for preliminary investigations of flows. This type of model is also the basis for multilayer models. The basic equations for this model are obtained by integrating the equations of motion and continuity from the bottom, $z = -h(x, y)$, to the surface, $z = \eta(x, y)$. (For convenience, hereinafter the bar indicating mean velocity will be omitted.)

The integrated continuity equation is

$$\int_{-h}^{\eta} \frac{\partial u}{\partial x}\, dz + \int_{-h}^{\eta} \frac{\partial v}{\partial y}\, dz + w|_{\eta} - w|_{-h} = 0. \tag{8.28}$$

According to the Leibnitz rule,

$$\begin{aligned} &\frac{\partial}{\partial x}\int_{-h}^{\eta} u\, dz - u|_{\eta}\frac{\partial \eta}{\partial x} + u|_{-h}\frac{\partial(-h)}{\partial x} + \frac{\partial}{\partial y}\int_{-h}^{\eta} v\, dz \\ &\quad - v|_{\eta}\frac{\partial \eta}{\partial y} + v|_{-h}\frac{\partial(-h)}{\partial y} + w|_{\eta} - w|_{-h} = 0. \end{aligned} \tag{8.29}$$

If the kinematic conditions (8.7) and (8.11) at the surface and bottom, respectively, are used, the following integrated continuity equations can be obtained:

$$\frac{\partial \eta}{\partial t} + \frac{\partial}{\partial x} UH + \frac{\partial}{\partial y} VH = 0, \tag{8.30}$$

where the averaged velocities are defined as:

$$U = \frac{1}{H}\int_{-h}^{\eta} u \, dz, \qquad V = \frac{1}{H}\int_{-h}^{\eta} v \, dz,$$

and $H = h + \eta$. The following assumptions are introduced for momentum equations.

(i) The hydrostatic approximation is imposed in the vertical direction because the horizontal scale $L \gg H$:

$$\frac{\partial p}{\partial z} = -\rho g.$$

(ii) The Boussinesq approximation is introduced, whereby the density in all terms is replaced by the constant mean density ρ_0.

(iii) Horizontal exchange of momentum is neglected, on the assumption that it is much less significant than other processes such as bottom and surface friction, etc.†

(iv) The distribution of horizontal velocities is almost uniform along the vertical.

With these assumptions, the momentum equations are integrated from $z = -h$ to $z = \eta$, resulting in

$$\frac{\partial U}{\partial t} + U\frac{\partial U}{\partial x} + V\frac{\partial U}{\partial y} - fV = -\frac{\partial}{\partial x}\left(g\eta + \frac{p_a}{\rho_0}\right) + \frac{1}{\rho_0 H}(\tau_{sx} - \tau_{bx}), \tag{8.31}$$

$$\frac{\partial V}{\partial t} + U\frac{\partial V}{\partial x} + V\frac{\partial V}{\partial y} + fU = -\frac{\partial}{\partial y}\left(g\eta + \frac{p_a}{\rho_0}\right) + \frac{1}{\rho_0 H}(\tau_{sy} - \tau_{by}). \tag{8.32}$$

The bottom shear stresses depend on the character of the flow and on the vertical velocity distribution, particularly in the zone near the bottom, which is in turn related to the horizontal pressure gradients. As an example, for steady wind-driven flows in lakes with well developed bottom return currents, vertically averaged velocities U and V can be negligible in comparison with the actual velocities. Therefore, the evaluation of bottom shear stresses is not easy in the vertically integrated model. Usually, these stresses are parameterized

† Consideration of the problem taking into account horizontal momentum exchange can be found, e.g., in Simons (1980).

through U and V, which can be more or less substantiated only for the so-called translation wave motion (in which the particle velocity is the same, or nearly so, at all points along the vertical):

$$\tau_{bx} = c_f \rho_0 (U^2 + V^2)^{1/2} U, \qquad \tau_{by} = c_f \rho_0 (U^2 + V^2)^{1/2} V, \tag{8.33}$$

where c_f is a resistance coefficient. Several empirical expressions, the Darcy–Weisbach, Chezy, and Manning formulas, were originally derived from measurements of steady flow in channels or pipes:

$$\begin{aligned} c_f &= f_0/8 && \text{Darcy–Weisbach} \\ c_f &= g/C^2 && \text{Chezy} \\ c_f &= n^2 g/H^{1/3} && \text{Manning } (C = H^{1/6}/n). \end{aligned} \tag{8.34}$$

C is a Chezy coefficient and f_0 (or λ) is a friction coefficient in the Darcy–Weisbach formula for head losses in pipes. Values of Manning's n are known for fully developed, rough turbulent flow, which normally prevails in lakes.

The nonlinear bottom friction can produce strong effects at large velocities and small depths, since it has a quadratic dependence on velocity and an inverse quadratic dependence on depth.

There are many computational methods for solving the vertically integrated momentum and continuity equations numerically with appropriate initial and boundary conditions. These methods differ according to: (a) whether they solve governing equations directly for the "primitive variables" U, V, and p† or instead solve for the vorticity and stream function; (b) the type of time integration scheme, such as implicit or explicit; and (c) the kind of spatial differencing, such as finite-difference or finite-element methods. An effective numerical method to solve the problems being considered is a splitting method. Different aspects of the usage of this method for problems of atmospheric and ocean dynamics have been considered by Marchuk (1974, 1975) and Marchuk and Zalesny (1974). A complete review of computational techniques applicable to the equations describing the circulation in water basins is far beyond the scope of this chapter. For the interested reader there are reviews by Lick (1976), Cheng *et al.* (1976), Sündermann (1979), and Simons (1980).

Types of Variable

Most circulation models use primitive variables U, V, and p. Primitive-variable models tend to be more accurate for similar computational labor probably because the vorticity–stream-function approach requires finite-difference approximations to more critical derivatives than does the primitive-variable approach.

---

† The pressure p is represented in the momentum equations 8.31 and 8.32 by the surface elevation η.

Time-Differencing Procedures

In the numerical calculation it is desirable to use space and time steps that are as large as possible and yet consistent with physical meaning. However, other restrictions on the allowable steps are related to the numerical methods used and to their accuracy and stability. The basic equations 8.31 and 8.32 describe the propagation of long waves in a shallow water body, and the effects of various time integration schemes on the accuracy of solution are important for numerical stability.

Detailed discussions of computational (or numerical) stability in connection with time-variable circulation problems have been presented by Fischer (1959, 1965), Platzman (1963), Kasahara (1965), Gates (1968), and Simons (1973). For computing unsteady flows in shallow water bodies, one of the most simple and familiar methods is based on the scheme of centered time differences. For centered differences in the space and time domains, the stability criterion of Courant-Friedrichs and Levy is imposed:

$$\Delta t < \frac{\Delta x}{(2gH)^{1/2}}. \tag{8.35}$$

This condition indicates that the time step is related to the time for a surface gravity wave (of speed $c = (gH)^{1/2}$) to travel a distance Δx.

The nonlinear acceleration terms and Coriolis term may cause some numerical instability; Lilly (1961), Simons (1973), and others have discussed improved methods for the evaluation of these terms.

Space-Differencing Procedures

It is impossible to discuss space-resolving techniques completely independently from the time-resolving procedures described above. Usually, each of them is designed in close relation to a particular technique for another dimension. Thus, the scheme of centered time differences is commonly used together with the staggered grid in space, in which U, V, and p are computed at different grid points.

One advantage of the staggered grid is that it allows the use of central differences in space and time, which is desirable for accuracy and numerical stability while keeping the number of variables low and the variables themselves partly uncoupled. However, there are problems in representing the physical boundaries properly and special treatment is necessary to avoid errors and instability. These problems have remained in later modeling efforts.

For marine applications (circulation in shallow seas) Hansen (1956) solved the vertically integrated equations by the finite-difference method, using the variables η, U, and V on a staggered grid in space and time.

Leendertse (1967) discussed the numerical stability and accuracy of the alternative-direction implicit (ADI) method in its application to the single-layer circulation model. The treatment of the nonlinear terms causes severe problems and time-centered differences cannot be used for the convective terms in the momentum equation.

Heaps (1969) used a staggered grid in space with velocities at the same points for the linearized and vertically integrated dynamic equations. Care was taken to center the differences in space and an explicit time integration scheme was used.

For the Great Lakes, Simons (1971) developed a finite-difference model based on the vertically integrated equations, including horizontal eddy viscosity, using space- and time-staggered grids to avoid problems with the convective terms. Several variations on the treatment of bottom friction and convective terms were tried.

Finite-difference methods normally make use of orthogonal grids. These may cause some difficulties in the representation of the geometry of natural water bodies. To suit complex boundaries better, other types of mesh are applied that can be used even with finite-difference schemes (Bauer, 1979). An unsteady, two-dimensional, depth-averaged model was developed for long, shallow reservoirs by Tatom and Waldrop (1978). The model uses orthogonal curvilinear coordinates and considers reservoir inflows and outflows, cooling water circulation, and surface wind stress. Irregular meshes, e.g. of triangular shape, are more often used with another effective numerical method, the finite-element method.

Gallagher *et al.* (1973) analyzed steady wind-driven circulation for shallow lakes using the rigid lid approximation (eqn. 8.14). Full advantage of the freedom of varying the grid was not taken in the examples given, but the possibility of using existing general-purpose finite-element programs was emphasized.

Taylor and Davis (1972), Cheng (1972), Grotkop (1973), Norton *et al.* (1973), and King *et al.* (1973) developed finite-element models for application in a number of different cases. Wang and Connor (1975) devised a finite-element model for near-coastal circulation and discussed the fundamental transformation of the original vertically integrated equations to the so-called weighted residual form. This method was applied to simulate wind- and tidally driven circulation in Massachusetts Bay.

Cheng (1972) developed a finite-element method for wind circulation in Lake Erie. Linearized momentum equations, which were obtained by neglecting inertia terms and introducing linear bottom shear, were written in the form of a stream function. The finite-element grid consisted of 516 three-node triangular elements and 308 nodal points. This gave a reasonably accurate spatial resolution in the domain of Lake Erie. Figure 8.2 shows the finite-element grid and the mean circulation streamlines driven by a linear wind distribution (along the lake surface).

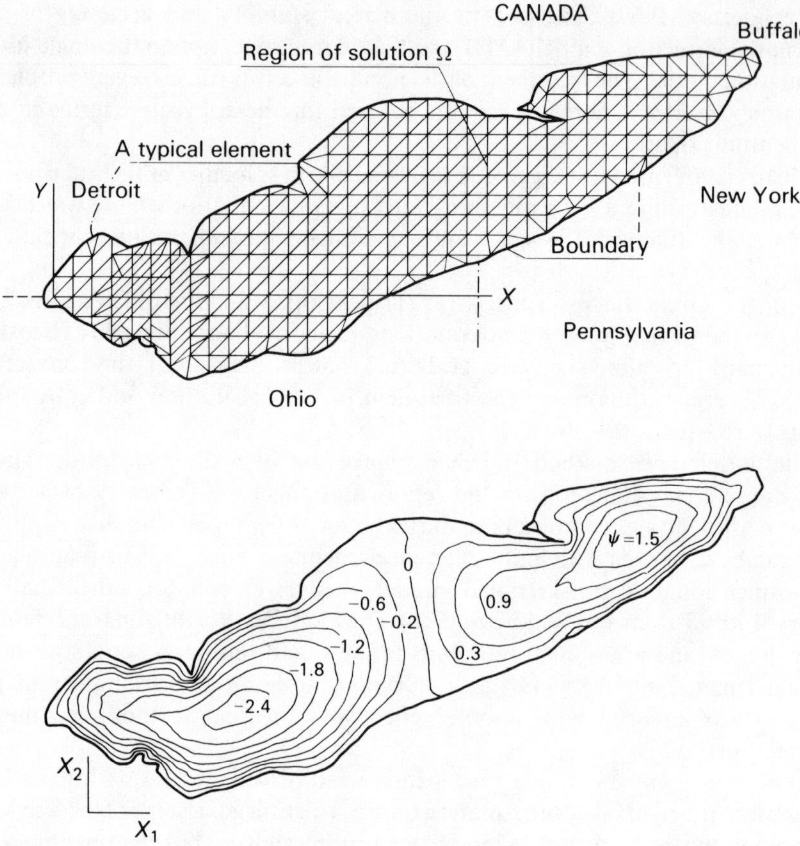

FIGURE 8.2 Finite-element grid and two-celled mean circulation driven by a quadratic wind stress distribution: application to Lake Erie. ψ is the stream function.

8.3.2. Two- and Multi-Layer Models

In this type of model, the water body is represented by layers in each of which the density is assumed to be constant. The thickness of each layer is variable and the layers move in response to free-surface and internal waves. This type of model, discussed by Simons (1973, 1980), Wang and Connor (1975), and Cheng *et al.* (1976), is useful for strong stratification with little interfacial mixing. There are relatively few applications of this type of model to lakes or reservoirs. A four-layer model was used by McNider and O'Brien (1973); however, the most common version of moving-interface models is a two-layer scheme representing the epilimnion and hypolimnion.

The vertically integrated equations for continuity and momentum are derived by integrating three-dimensional equations over each constant-density layer.

Momentum is transmitted between layers by interfacial stresses and by mass transfer (entrainment) through the interfaces. Wang and Connor (1975) treated the mass transfer across the interface in the continuity equation as follows:

$$\frac{\partial H_k}{\partial t} + \frac{\partial}{\partial x} U_k H_k + \frac{\partial}{\partial y} V_k H_k = q_k - w_k + \frac{\rho_{k-1}}{\rho_k} w_{k-1}, \tag{8.36}$$

where

$$H_k = \eta_k - \eta_{k-1}$$

is the thickness of layer k,

$$U_k = \frac{1}{H_k} \int_{\eta_{k-1}}^{\eta_k} u_k \, dz, \qquad V_k = \frac{1}{H_k} \int_{\eta_{k-1}}^{\eta_k} v_k \, dz, \qquad q_k = \int_{\eta_{k-1}}^{\eta_k} S \, dz,$$

and S is the source (or sink) strength; w_k, w_{k-1} are the relative normal velocities at the interface between layers k and $k - 1$, that is, w_k is a net entrainment or mixing velocity between these layers. Vertically integrated momentum equations for each layer are obtained similarly to those in the single-layer formulation, with hydrostatic and Boussinesq approximations.

Bottom shear stress and surface wind stress are determined with the same relationships that are used in single-layer formulations. The idealization of constant-density layers cannot represent explicitly the mass and momentum exchanges in the transition region between layers. Wang and Connor (1975) included a shear stress and a velocity of entrainment at the interface to simulate these processes; therefore, their functional dependence on the mean flow variables must be specified.

Interfacial shear stress is related to the square of the velocity differences of the layers:

$$\begin{aligned} \tau_{xk} &= \rho_k C[(U_k - U_{k+1})^2 + (V_k - V_{k+1})^2]^{1/2}(U_{k+1} - U_k) \\ \tau_{yk} &= \rho_k C[(U_k - U_{k+1})^2 + (V_k - V_{k+1})^2]^{1/2}(V_{k+1} - V_k), \end{aligned} \tag{8.37}$$

where C is an interfacial shear stress coefficient that depends on the Reynolds and densimetric Froude numbers. Derivations of governing equations for multilayer models are also discussed by Simons (1973). Wang and Connor (1975) have applied this model to a two-layer flow that was investigated experimentally by Hyden (1974). The multilayer models present difficulties in practical application when upwelling or downwelling effects are significant; use of this type of model is limited to strongly stratified conditions with little interfacial mixing.

8.3.3. Ekman-Type Models

The approach initiated by Welander (1957) is based on the earlier work of Ekman, who investigated the rotational effects of the earth on oceanic circulation. The approach is specifically designed for wind-driven currents. If the Rossby number, Ro = (inertial terms)/(rotational terms) = u/fL, is small and the hydrostatic approximation is valid, the nonlinear convective terms can be neglected. It is also assumed that the role of horizontal turbulent exchange is small in comparison with the vertical exchange; that is, horizontal shearing motions (such as in zones near the shore) are not considered. The governing equations are, therefore:

$$\frac{\partial u}{\partial t} - fv = -\frac{1}{\rho}\frac{\partial p}{\partial x} + \epsilon_z \frac{\partial^2 u}{\partial z^2} \tag{8.38}$$

$$\frac{\partial v}{\partial t} + fu = -\frac{1}{\rho}\frac{\partial p}{\partial y} + \epsilon_z \frac{\partial^2 v}{\partial z^2}, \tag{8.39}$$

$$\frac{\partial u}{\partial x} + \frac{\partial v}{\partial y} + \frac{\partial w}{\partial z} = 0. \tag{8.40}$$

This model can be categorized as an Ekman-type model. The class of models allows one to solve the governing equations through a combination of analytical and numerical solutions. At least for the steady state case, the vertical velocity distribution is determined analytically in terms of the prescribed wind stresses and the unknown pressure gradients. The pressure gradients must be obtained numerically through the depth-integrated equations.

Liggett and Hadjitheodorou (1969) applied the Ekman-type model to simulate a steady state wind-driven circulation in a rectangular lake with variable depth. Applications of the Ekman-type models have been demonstrated by Gedney and Lick (1970, 1972) and by Bonham-Carter and Thomas (1973) in their studies of wind-driven circulations in Lake Erie and Lake Ontario.

Liggett and Hadjitheodorou (1969) considered steady state on the basis of the following assumptions. (i) The vertical eddy viscosity is constant throughout. (ii) The no-slip condition is imposed at the bottom boundary:

$$u = v = w = 0 \quad \text{at} \quad z = -h(x, y). \tag{8.41}$$

(iii) The rigid lid approximation is applied at the surface and the wind stress is prescribed as follows:

$$\epsilon_z \frac{\partial u}{\partial z} = \tau_{sx}, \qquad \epsilon_z \frac{\partial v}{\partial z} = \tau_{sy} \quad \text{on} \quad z = 0. \tag{8.42}$$

From the momentum equations with boundary conditions, the general solution for velocity is obtained analytically as a function of the imposed wind stress, the bottom topography, and the pressure distribution (which is unknown):

$$W = u + \mathrm{i}v = \frac{\sinh \lambda(h + z)Q}{\lambda \cosh \lambda h} - \mathrm{i}\left(\frac{\cosh \lambda z}{\cosh \lambda h} - 1\right)P, \tag{8.43}$$

where

$$Q = \tau_{sx} + \mathrm{i}\tau_{sy}, \qquad P = \frac{\partial p}{\partial x} + \mathrm{i}\frac{\partial p}{\partial y}, \qquad \lambda = \left(\frac{\mathrm{i}}{E_v}\right)^{1/2},$$

and $E_v = \epsilon_z/fh^2$ is the vertical Ekman number. If the general solution is integrated vertically and a stream function is introduced that satisfies the vertically integrated continuity equation, a Poisson-type equation for a stream function is derived:

$$\nabla^2\Psi = a\frac{\partial \Psi}{\partial x} + b\frac{\partial \Psi}{\partial y} + c, \tag{8.44}$$

where *a*, *b*, and *c* are functions of the wind stress, Ekman number, and local depth. This equation, with proper boundary conditions, is solved using finite-difference methods. The numerical results for the stream function give the pressure distribution to complete the general solution for the velocity, eqn. 8.43.

The steady state model was applied to an artificial basin of rectangular planform, shown in Figure 8.3. A vector diagram of velocities in the test lake at the surface and at two depths is also shown for when a south wind is blowing. The highest velocities on the surface occur near the boundaries and are in the downwind direction, and the return flow occurs in the deeper, central portion of the lake in the upwind direction. Rao and Murty (1970) presented a similar Ekman-type model and applied it to Lake Ontario.

Gallagher *et al.* (1973) applied a finite-element method to the steady state Ekman-type model described by eqns. 8.43 and 8.44. The same method was applied for an unsteady state in shallow, homogeneous lakes by Liggett (1969). Lee and Liggett (1970; Liggett and Lee, 1971) also applied the Ekman-type model for two-layer stratified flow conditions. The solutions obtained from the Ekman-type model are applied to each layer separately. Additional boundary conditions between the layers, consisting of equating velocities at the interface, and continuity of pressure and shears across the interface, are imposed as follows:

$$u_1 = u_2, \qquad v_1 = v_2, \qquad w_1 = w_2, \tag{8.45}$$

$$p_1 = p_2, \qquad \epsilon_{z_1}\frac{\partial u_1}{\partial z} = \epsilon_{z_2}\frac{\partial u_2}{\partial z}, \qquad \epsilon_{z_1}\frac{\partial v_1}{\partial z} = \epsilon_{z_2}\frac{\partial v_2}{\partial z}, \tag{8.46}$$

where the subscripts 1 and 2 refer to the upper and lower layers, respectively. These boundary conditions, together with a kinematic condition, lead to the interface specifications.

This type of model has certain limitations in its application to real water bodies: (i) simplifications of governing equations are necessary for

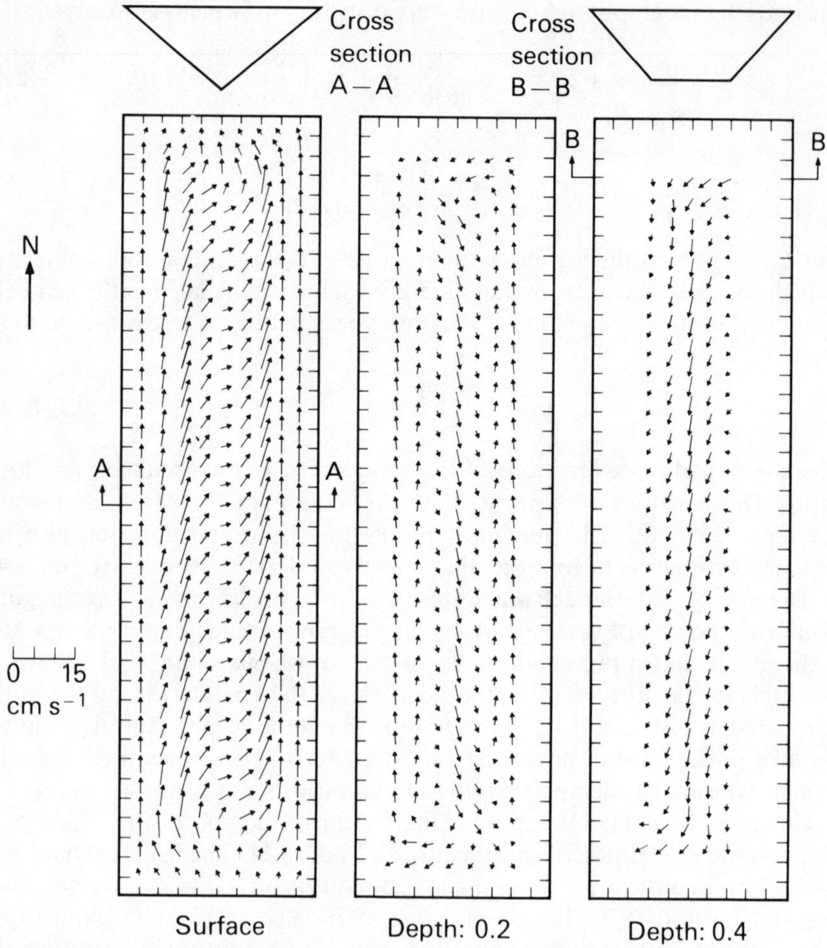

FIGURE 8.3 Horizontal velocities in the test lake under a south wind: at the surface and at 0.2 and 0.4 of the total depth (source: Liggett and Hadjitheodorou, 1969).

analytical solutions, such as neglecting the convective terms and horizontal diffusion terms; (ii) the assumption of constant vertical eddy viscosity may not be realistic for stratified flows; and (iii) the analytical solutions are quite complex for a multilayer system.

8.3.4. Laterally Averaged Two-Dimensional Flow Models

If the lateral dimension of the water body is sufficiently small compared with the longitudinal dimension, the system can be approximated as two-dimensional (in the longitudinal–vertical plane). This type of model may be applicable in reservoirs with strong through flows, but is less important for natural lakes (another field of application is in estuarine hydrodynamics).

Withdrawal patterns, i.e. withdrawal "layers," are primarily affected by the degree of stratification in the immediate vicinity of the outlet. This selective withdrawal influences the distributions of flow and temperature within the entire impoundment. The resulting patterns of flow and temperature are essentially interdependent and they, in turn, govern the distributions in space and time of other water quality characteristics of the impoundment.

A notable example is Lake Roosevelt behind Grand Coulee Dam in Washington. In this reservoir the period of maximum runoff corresponds roughly with the period of greatest insolation. The result is a pronounced longitudinal temperature gradient as well as vertical stratification. So-called "tilted isotherms," planes of equal temperature inclined downstream, are evident.

The two-dimensional formulation of inhomogeneous fluid motion, including vertical and longitudinal distributions of density and velocity, can be obtained by averaging the momentum and heat balance equations in the lateral direction (Vasiliev *et al.*, 1973, 1974; Vasiliev, 1978a,b). After averaging, the governing equations are as follows (the hydrostatic approximation is used for the vertical momentum equation):

$$\frac{\partial u}{\partial t} + u\frac{\partial u}{\partial x} + w\frac{\partial u}{\partial z} = -g\frac{\partial}{\partial x}\left(\eta + \int_z^\eta \frac{\rho - \rho_0}{\rho_0}\,\mathrm{d}z\right) + \frac{1}{b}\frac{\partial}{\partial z}\left(b(v + \epsilon_z)\frac{\partial u}{\partial z}\right) - \frac{\sigma}{b}\tau, \tag{8.47}$$

$$\frac{\partial(bu)}{\partial x} + \frac{\partial(bw)}{\partial z} = 0, \tag{8.48}$$

$$\frac{\partial T}{\partial t} + u\frac{\partial T}{\partial x} + w\frac{\partial T}{\partial z} = \frac{1}{b}\frac{\partial}{\partial z}\left(b(\chi + K_z)\frac{\partial T}{\partial z}\right) + \frac{1}{\rho_0 c_0}\left(\phi - \frac{\sigma}{b}q_n\right), \tag{8.49}$$

$$\sigma = \sum_{i=1}^{2}\left[1 + \left(\frac{\partial b_i}{\partial x}\right)^2 + \left(\frac{\partial b_i}{\partial z}\right)^2\right]^{1/2}, \tag{8.50}$$

$$\tau = \left(\frac{\lambda}{8}\right)|u|u,$$

where

$\rho_0 c_0$ is the product of average density and specific heat,
b $= b_1 + b_2$ is the channel width,
$b_1(x, z), b_2(x, z)$ characterize the shape of the lateral surface of the channel,
η is the free-surface elevation,
v is the coefficient of molecular viscosity
ϵ_z is the coefficient of turbulent viscosity,

χ is the coefficient of molecular heat diffusivity,
K_z is the coefficient of turbulent heat diffusivity,
τ is the frictional resistance of the lateral surface,
λ is the frictional resistance coefficient,
q_n is the heat flux through the lateral surface (per unit surface area),
ϕ is the heat source or the volumetric heat influx (section 8.2.2).

King *et al.* (1973) applied a finite-element method to two-dimensional stratified flow. The vertically two-dimensional model was applied to a laboratory experiment at the Waterways Experiment Station. Numerical results were compared with measurements from the vicinity of a broad-crested weir. This model was also applied to Lower Granite Reservoir by Norton *et al.* (1973).

A similar model was developed by Edinger and Buchak (1978) and applied to Sutton Reservoir, West Virginia. Results of the test simulations indicated four possible circulation regimes. These depend on inflow and outflow rates, inflow density, and surface heat exchange.

8.3.5. Three-Dimensional Models

The most advanced models can be found in this group, in which a three-dimensional discretization of the water body is constructed by using a spatial Eulerian grid and the unknown variables are computed at each node for successive time steps. The development of high-speed computers has facilitated remarkable progress in this type of modeling (Simons, 1972, 1973, 1974, 1975, 1980; Bennett, 1974; Vasiliev and Kvon, 1977; Wittmiss, 1979; Tjomsland, 1979; Bauer, 1979; Raney *et al.*, 1979; Tsvetova, 1979).

The governing equations for three-dimensional models of a thermally stratified turbulent flow were given in section 8.2.2. The equations can be solved by the straightforward application of conventional numerical techniques, whereby the continuous derivatives are replaced by finite differences.

A number of models of lake circulation have been developed by Simons, as previously noted. Adapting the concepts of modeling utilized in the three-dimensional representation of atmospheric circulation, Simons has modeled several of the Great Lakes, as well as Lake Vänern in Sweden. Calibration and verification tests using field data have shown that these models simulate reasonably well the patterns of circulation induced by wind and affected by temperature changes and Coriolis forces. The approach employed is illustrated by Simons' model VÄNERN (Simons *et al.*, 1977), which was an adaptation of his earlier multilevel model for the Great Lakes.

A three-dimensional finite-difference model can be visualized as a sequence of fixed but permeable levels. Simons (1980) discussed the advantages of this interpretation of a finite-difference model as follows: (i) the general circulation problem can be reduced to a quasi-two-dimensional problem; (ii) the model equations are obtained by vertical integration over each layer instead of applying the equations at given levels, and this procedure ensures that the conservative character of the original differential equations and the relevant volume integrals is preserved.

The equations of motion, heat conservation, and continuity developed by Simons are:

Motion

$$\frac{\partial u}{\partial t} + \nabla \cdot (Vu - \epsilon_h \nabla u) + \frac{\partial}{\partial z}\left(wu - \epsilon_z \frac{\partial u}{\partial z}\right) = fv - \frac{\partial}{\partial x}(P + Q) \quad (8.51)$$

$$\frac{\partial v}{\partial t} + \nabla \cdot (Vv - \epsilon_h \nabla v) + \frac{\partial}{\partial z}\left(wv - \epsilon_z \frac{\partial v}{\partial z}\right) = -fu - \frac{\partial}{\partial y}(P + Q) \quad (8.52)$$

Heat conservation

$$\frac{\partial T}{\partial t} + \nabla \cdot (VT - K_h \nabla T) + \frac{\partial}{\partial z}\left(wT - K_z \frac{\partial T}{\partial z}\right) = 0 \quad (8.53)$$

Continuity

$$\frac{\partial w}{\partial z} = -\frac{\partial u}{\partial x} - \frac{\partial v}{\partial y}, \quad (8.54)$$

where

u, v, w are velocity components along the x, y, z axes,
V is the horizontal component of the velocity vector,
T is the temperature,
ϵ_h, ϵ_z are the horizontal and vertical eddy viscosities,
K_h, K_z are the horizontal and vertical heat diffusivities,
∇ is the horizontal gradient operator,
f is the Coriolis parameter.

Motion is induced by Coriolis forces fv and fu and by barotropic and baroclinic pressure components P and Q:

$$P = g\eta + \frac{p_a}{\rho_0} \quad (8.55)$$

$$Q = g \int_z^\eta \frac{\Delta\rho}{\rho_0} \, dz, \quad (8.56)$$

where

η is the free-surface elevation,
g is the gravitational acceleration,

p_a is the atmospheric pressure at the air–water interface,
ρ_0 is the mean water density,
$\Delta\rho$ is the density difference.

Vertical velocities are computed from the horizontal flow divergence through application of the continuity equation, with the lower boundary condition that there can be no flow normal to the bottom. Integration of the same equation along the vertical axis results in a determination of the free-surface elevation, which, in turn, determines the barotropic pressure function (8.55). Surface wind stresses, bottom friction, and heat fluxes at the surface and bottom comprise boundary conditions for fluxes of momentum and heat. At lateral boundaries, where normal components of velocity must vanish, tangential velocities are stipulated as required by "slip" or "no slip" assumptions that fit a particular model.

Numerical solutions of the equations presented above are obtained by finite differencing on a staggered grid. The vertical structure of the model, which fixes the computational scheme, comprises a series of arbitrarily fixed levels at which vertical velocities, stresses, and vertical fluxes are calculated. Temperatures and currents are defined as averages for the intermediate layers. The effects of free-surface oscillations on the computation of internal flows are minimized by reducing the layered system of equations to a single equation for vertically integrated flow and a set of equations for the shears between adjacent moving layers. Integration over time is essentially explicit, except that the Coriolis term is treated implicitly. Simons (1973) provided other details of the solution technique and numerical approximations. He applied the above-mentioned multilevel model to three cases.

(i) The model was applied by Simons (1974) to Lake Ontario in spring and early summer. The predicted water levels, currents, and temperatures were compared with observations made shortly after the passage of tropical storm Agnes during the latter part of June 1972. Figure 8.4 indicates that the currents follow the wind on the shallow, northern side of the lake, whereas the resulting pressure gradient returns the water masses in the deeper parts of the lake. Figure 8.5 compares observed and predicted water levels during the storm. The model simulation of the observed circulation pattern is substantially correct. For the period of weak stratification the model is found to be most sensitive to parameters related to the vertical flux of momentum. However, satisfactory simulations of observed water levels and currents require a wind stress coefficient considerably larger than those obtained from direct measurements.

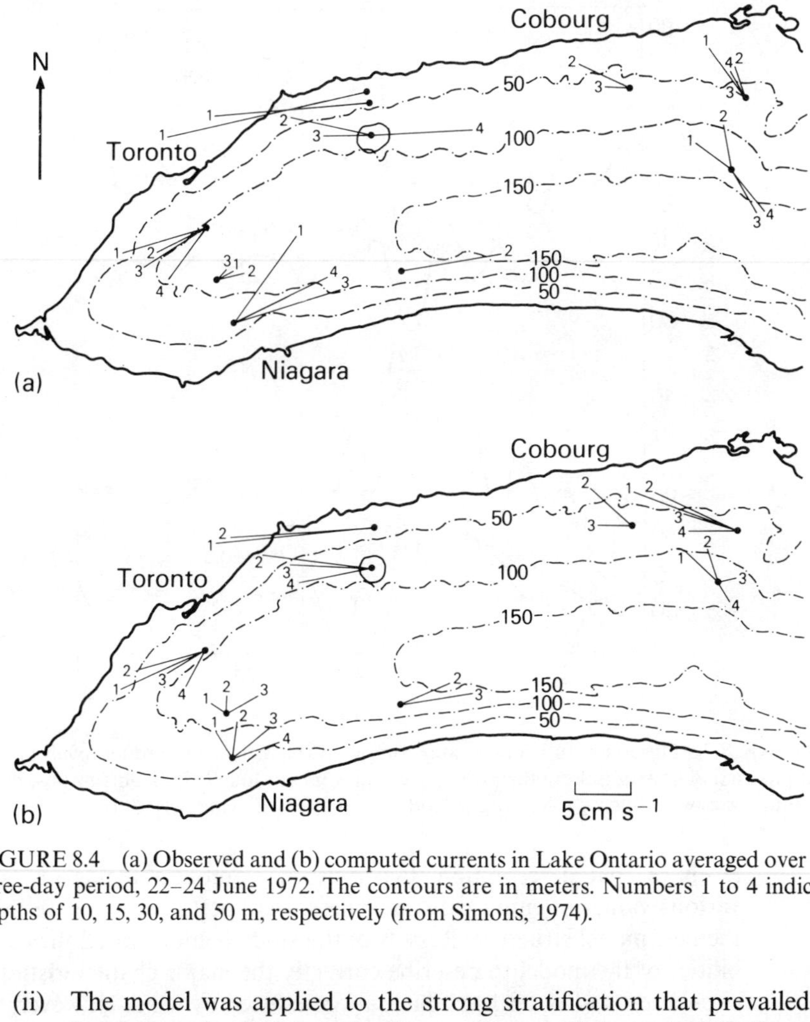

FIGURE 8.4 (a) Observed and (b) computed currents in Lake Ontario averaged over the three-day period, 22–24 June 1972. The contours are in meters. Numbers 1 to 4 indicate depths of 10, 15, 30, and 50 m, respectively (from Simons, 1974).

(ii) The model was applied to the strong stratification that prevailed in Lake Ontario during and after a storm on 9 August 1972 (Simons, 1975). Good agreement between predicted and observed data was obtained for water levels, as shown in Figure 8.6. The model results shown in Figure 8.7 neglect heat diffusion and include only heat transport by advection. There is good agreement between model and observations with regard to the wave-like pattern of temperature in the deeper parts of the lake.

(iii) The model was applied by Simons *et al.* (1977) to Lake Vänern in Sweden. Verification checks were made for two distinct conditions, one when the lake was essentially homogeneous and another when the lake was thermally stratified. Sensitivity tests, involving alternative

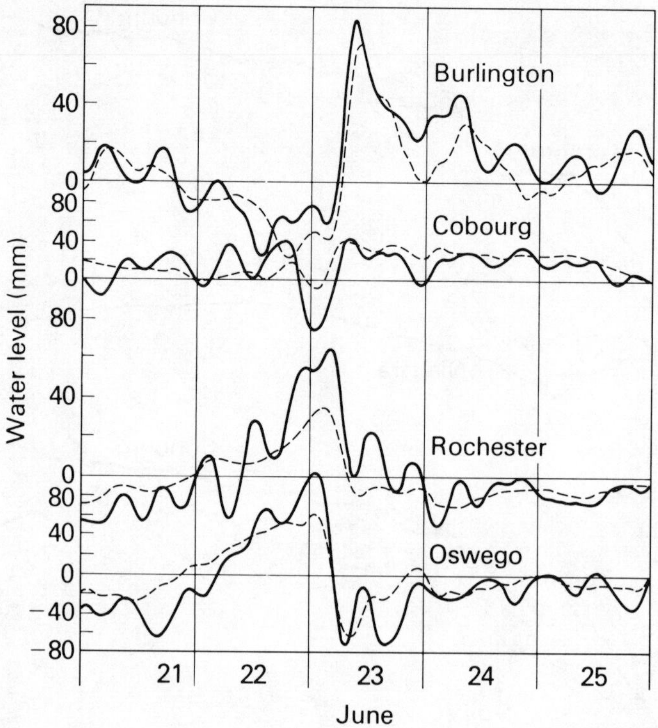

FIGURE 8.5 Observed (full curves) and computed (broken curves) water levels at four stations on the shore of Lake Ontario during storm Agnes in June 1972 (from Simons, 1974). All time series were filtered to remove periodic components having periods of less than 5 h.

levels of spatial resolution, two different computational grids, and various values of empirical coefficients, were performed to determine the best model structure. Results of the study demonstrated the capabilities of the model to describe correctly the major characteristics of circulation and water level fluctuation induced by wind. However, the tests indicated that attempts to improve the results of verification by increasing the resolution (using a smaller grid size, more layers, shorter time steps, a refined computational scheme) tended toward a level of diminishing returns against the added time and cost of computation. The most sensitive parameters were the empirical mixing coefficients and the eddy viscosities. These, in turn, through their influence on current structure, controlled the ability of the model to simulate temperature changes. Heat diffusion, *per se*, was of little consequence. A general conclusion derived by Simons and his colleagues from these experiments was that "very expensive, non-linear, high-resolution models" are not justified for dealing with slowly varying lake-wide circulations.

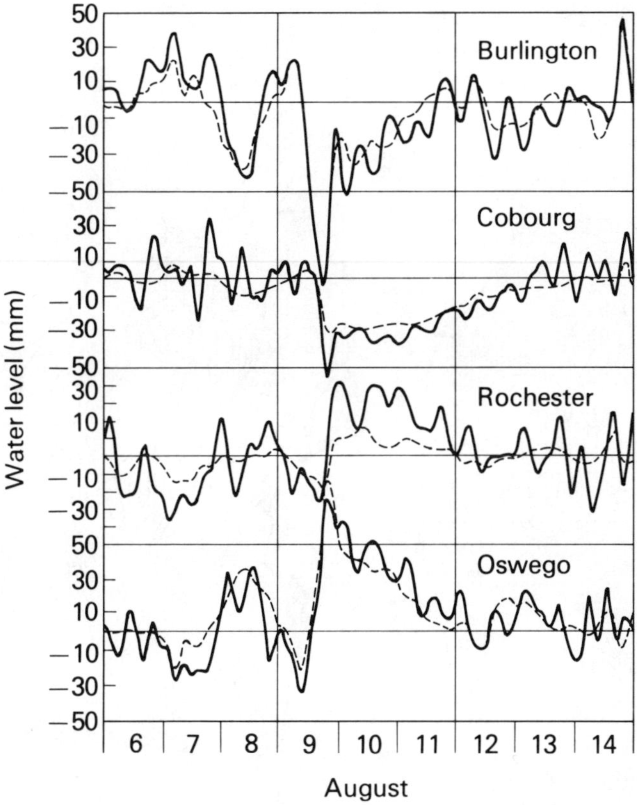

FIGURE 8.6 Observed (full curves) and computed (broken curves) water levels at four stations on the shore of Lake Ontario during and after a storm on 9 August 1972 (from Simons, 1975).

A similar model was also developed by Bennett (1974). This model employed a rigid lid approximation to eliminate the short-term dynamic response of the water surface and used a stream function to ensure continuity. As a result, this model allows a computational time step of the order of an hour, whereas the Simons model required a time step of a few minutes. However, this advantage is offset by the requirement of iterative procedures to solve the pressure equation. Thus, instead of more time steps, one needs more iterations and so requires about the same computer time.

Most circulation models have used constant eddy diffusivities for computation of the momentum equations. Vasiliev and Kvon (1977), Vasiliev (1978a,b), and Kvon (1979a,b) developed more comprehensive circulation models, including a one- or two-equation turbulence closure model with two parameters. Using the Boussinesq approximation and the hydrostatic condition and

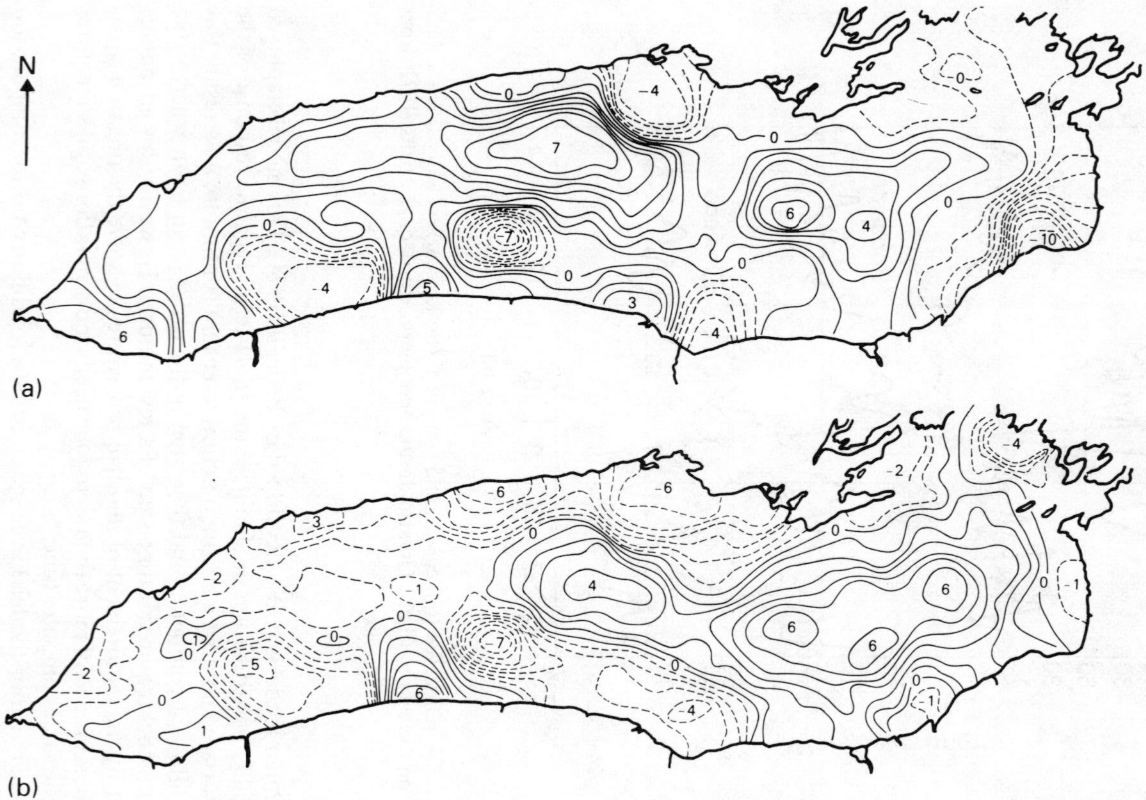

FIGURE 8.7 (a) Observed and (b) computed temperature changes (°C) in Lake Ontario during 2–5 August 1972, corresponding to the second layer of the model (between 10 and 20 m) (from Simons, 1975).

eliminating horizontal turbulent exchange, they obtained the following governing equations for a temperature-stratified turbulent flow:

$$\frac{\partial u}{\partial t} + u\frac{\partial u}{\partial x} + v\frac{\partial u}{\partial y} + w\frac{\partial u}{\partial z} - fv = -g\frac{\partial}{\partial x}\left[\eta + \int_z^{z_0}\left(\frac{\rho - \rho_0}{\rho_0}\right)\mathrm{d}z\right] + \frac{\partial}{\partial z}\left(\epsilon\frac{\partial u}{\partial z}\right), \tag{8.57}$$

$$\frac{\partial v}{\partial t} + u\frac{\partial v}{\partial x} + v\frac{\partial v}{\partial y} + w\frac{\partial v}{\partial z} + fu = -g\frac{\partial}{\partial y}\left[\eta + \int_z^{z_0}\left(\frac{\rho - \rho_0}{\rho_0}\right)\mathrm{d}z\right] + \frac{\partial}{\partial z}\left(\epsilon\frac{\partial v}{\partial z}\right), \tag{8.58}$$

$$\frac{\partial u}{\partial x} + \frac{\partial v}{\partial y} + \frac{\partial w}{\partial z} = 0, \tag{8.59}$$

$$\frac{\partial T}{\partial t} + u\frac{\partial T}{\partial x} + v\frac{\partial T}{\partial y} + w\frac{\partial T}{\partial z} = \frac{\partial}{\partial z}\left(K_{\mathrm{T}}\frac{\partial T}{\partial z}\right), \tag{8.60}$$

$$\frac{\rho - \rho_0}{\rho_0} = -\beta(T - T_0).$$

On the lateral boundaries, either the velocity component normal to the boundary is zero, or it must be specified. At the free surface, $z = z_0$.

$$\frac{\partial z}{\partial t} = w, \qquad \epsilon\frac{\partial \boldsymbol{u}}{\partial z} = \frac{\boldsymbol{\tau}}{\rho_0}, \qquad c_p\rho_0 K_{\mathrm{T}}\frac{\partial T}{\partial z} = -K(T - T_{\mathrm{E}}); \tag{8.61}$$

and at the bottom, $z = h(x, y)$:

$$w = u\frac{\partial z_0}{\partial x} + v\frac{\partial z_0}{\partial y}, \qquad \epsilon\frac{\partial \boldsymbol{u}}{\partial z} = K_l|\boldsymbol{u}|\boldsymbol{u}, \qquad \frac{\partial T}{\partial z} = 0, \tag{8.62}$$

where $\boldsymbol{u} = (u, v)$, $\boldsymbol{\tau} = (\tau_x, \tau_y)$, K_l is the coefficient of the bottom tangential stress, T_{E} is the equilibrium temperature, and z_0 is the average free-surface elevation. Vasiliev *et al.* (1973) determined the turbulent exchange coefficients, ϵ and K_{T}, from the turbulent kinetic energy e and a length scale of turbulence, L, as follows:

$$\epsilon = e^{1/2}Lf(e^{1/2}L/v), \qquad K_{\mathrm{T}} = \alpha_{\mathrm{T}}\epsilon,$$

$$\frac{\partial e}{\partial t} = \frac{\partial}{\partial z}\left(k_{\mathrm{e}}\frac{\partial e}{\partial z}\right) + (\epsilon - v)\left[\left(\frac{\partial u}{\partial z}\right)^2 + \left(\frac{\partial v}{\partial z}\right)^2\right](1 - \alpha_{\mathrm{T}}\mathrm{Ri}) - ck_{\mathrm{e}}(e/L^2), \tag{8.63}$$

where k_{e} is the coefficient of total exchange for turbulent energy transfer and c is a constant. The Richardson number Ri and the ratio α_{T} of the turbulent exchange coefficients for heat and momentum transfer are given by:

$$\mathrm{Ri} = -\frac{(g/\rho_0)\partial\rho/\partial z}{(\partial u/\partial z)^2 + (\partial v/\partial z)^2} \tag{8.64}$$

$$\alpha_{\mathrm{T}} = \alpha_0 \frac{(1 + 10\,\mathrm{Ri})^{1/2}}{[1 + (10/3)\,\mathrm{Ri}]^{3/2}}, \tag{8.65}$$

where α_0 is a constant. The model was solved numerically by means of an implicit finite-difference scheme.

Later, Kvon applied the two-equation turbulence model founded on two transport equations for the turbulent energy and the turbulent energy dissipation (Jones and Launder, 1973; Launder, 1976). On this basis he developed the flow model with a slip condition at the bottom. The model was used by Kvon (1979) in the consideration of a three-dimensional temperature-stratified flow in a water body.

8.3.6. Concluding Comments

A well developed capability exists for simulating vertically averaged two-dimensional circulation and water level fluctuations in shallow, well mixed lakes and impoundments. Single-layer circulation models solved by finite-difference methods have been successfully applied to shallow lakes, such as Lake Erie, producing sufficient information in the form of depth-averaged velocities to drive compatible water quality models. Added flexibility in providing local detail and in fitting irregular boundaries and topography is available in the form of finite-element and finite-difference models, both of which have been demonstrated in a number of practical cases. Documentation of vertically averaged two-dimensional circulation models is generally adequate for transferring this technology to new users. The state of the art of modeling circulation in the horizontal two-dimensional formulation is fairly well advanced.

Another type of simulation technique, which is also of two-dimensional character, has been developed relatively recently for deep, oblong reservoirs and lakes when density stratification takes place. Though even vertical one-dimensional models can be helpful in many of these cases, the vertical two-dimensional models allow much better resolution of the problem and can include the simulation of withdrawal patterns, "tilted isotherms," flow patterns in the upstream reach of a reservoir, etc. However, experience with the application of these models is less than for the horizontal two-dimensional models.

The state of the art of mathematical modeling of circulation in large lakes is exemplified by the three-dimensional models of Simons. This type of model can be visualized by a sequence of fixed but permeable levels, except for the free surface and the bottom. Therefore, a generalized system of model equations is reduced to a quasi-two-dimensional form. Such models have been applied to many lakes and, therefore, are well documented for other users.

Conceptually, all of the most important physical mechanisms that govern water movement are included in the models, although it is generally recognized that model performance is most sensitive to empirically determined coefficients of mixing. A more comprehensive approach to determine the coefficient of turbulent exchange, as discussed by Vasiliev (1978) and by Kvon (1979a,b), is desirable. Modeling efforts are likely to be extended in the direction of better characterization of mechanisms of turbulent transport along both the vertical and horizontal axes.

8.4. TWO- AND THREE-DIMENSIONAL WATER QUALITY AND ECOLOGICAL MODELS

In the one-dimensional models for stratified impoundments, hydrodynamic considerations are much simplified. A continuity equation for the vertical direction is used to determine the advection caused by inflows and outflows at different elevations. The vertical transport of heat is governed by advection and diffusion and by internal heating due to shortwave solar radiation. For deep impoundments that are dominated by inflows and outflows, as opposed to diffusion and internal heating, these models are generally reliable (Orlob and Selna, 1967; Huber and Harleman, 1968; Ryan and Harleman, 1971; Hurley-Octavio *et al.*, 1977).

Simulation of the annual temperature cycle in such water bodies has been accomplished with reasonably good agreement between mathematical model and impoundment. However, extension of the advection–diffusion approach to include other nonconservative parameters (for example, dissolved oxygen, BOD, nutrients) that are not distributed in the impoundment by the same mechanisms as heat energy raises questions as to the appropriateness of the one-dimensional approximation (Markofsky and Harleman, 1971, 1973; Chen and Orlob, 1972). This concern is further accentuated for long, narrow or broad, shallow impoundments, in which temperature or concentration gradients may develop either longitudinally or laterally. Various two- and three-dimensional water-quality–ecological models have been developed to take into account horizontal or vertical resolution. There are three broad categories in two- and three-dimensional water quality modeling: (a) compartment models, (b) a network of one-dimensional channel models, and (c) two- and three-dimensional models. In the following discussion the emphasis will be on the coupling of the hydrothermal and water quality components; however, the details of the water quality and ecological components will be omitted since they are discussed in Chapters 3, 4, and 9.

8.4.1. Compartment Models

By ignoring hydrodynamics completely and concentrating on the kinetics of water quality and biota, modelers have been able to construct models that describe gross changes in the mass balance of well mixed water bodies. The impoundment is treated as an interconnected system of continuously stirred tank reactors (CSTR) to give added spatial dimensions and to account in part for temporal changes. This approach has been adopted in extending laboratory idealizations of chemical and biological kinetics to two-dimensional systems in which net transport between reaction cells is derived either by independent simulation of hydrodynamics or by field measurements.

Snodgrass and O'Melia (1975) developed a model that mass-balances both particulate and ortho-phosphorus for a two-compartment lake under two sets of seasonal conditions. The lake is treated as a single, well mixed compartment in the winter and as two compartments (epilimnion and hypolimnion) in the summer. The model is formulated in mass balance form as a CSTR for epilimnetic and hypolimnetic particulate phosphorus. The CSTR equations contain vertical transport coefficients for the exchange of mass across the thermocline.

The Snodgrass–O'Melia model has been applied by the developers to predict average phosphorus concentrations in lakes with a wide range of detention times. Comparisons of predicted and observed values show "excellent" agreement (O'Melia, 1974). The model is recommended as a tool for prediction of permissible phosphorus loadings in lakes. A similar model for a two-layer lake was developed by Imboden (1974).

Larsen and Mercier (1975) applied the Snodgrass–O'Melia model to Lake Shagawa, Minnesota, observing that the model underestimated the amount of epilimnetic phosphorus. They developed and applied a three-compartment epilimnetic model, a simplified version of models developed by Baca *et al.* (1974) and Thomann *et al.* (1975). The distinctive features of this model were the inclusion of algae as a sink for soluble reactive phosphorus and the conversion of particulate phosphorus to the soluble form so that it would be available for algal growth.

Detailed models of the eutrophication process have been proposed and several have been implemented, particularly for Lakes Erie and Ontario. A noteworthy example is the development of phytoplankton models for Lake Ontario (Thomann *et al.*, 1975). Three basic models, each with a different level of detail, were developed and tested. The model Lake 1 simulated the impoundment as a three-layer system—epilimnion, hypolimnion, and benthos—and concentrated on phytoplankton and zooplankton dynamics. Lake 2 provided additional vertical resolution with seven layers and also considered temperature, chemistry, and sediment interactions. Lake 3, which was developed only to the preliminary stage, provided additional spatial resolution with up to seven layers and 67 segments, and accommodated 10 to 15 variables. A future model is being planned with up to 5000 compartments (segments · variables).

Each of the models is founded on mass conservation and kinetic principles. A mass balance equation is written for each constituent and each segment.

Solution over a suitable time horizon with appropriate time steps results in the required space–time description of all variables. The models are driven by circulations developed from field observations or from models, such as those of Simons (1973), that have been applied to Lake Ontario.

Verification runs with Lake 1 indicated generally good agreement with measurements of such constituents as chlorophyll *a*, zooplankton, carbon, four forms of nitrogen, and phosphorus. Lake 2 was studied analytically but was not tested against the lake itself. Lake 3 was tested preliminarily against selected observations but was not verified.

The plankton model was combined with a three-dimensional circulation model of Lake Ontario by Simons (1976). The simulations were compared with three-dimensional data on Lake Ontario collected during 1972. Typical spatial differences between deep-water and near-shore zones appeared reasonably well reproduced in the model.

A water quality model to predict phytoplankton production in western Lake Erie was developed by Di Toro *et al.* (1975). The basis of the model is a set of mass conservation equations that relate the variables to each other. The model includes effects of biological phenomena (predator–prey relationships), chemical reactions (nitrification), and other interactions that provide nutrients necessary for phytoplankton growth. Figure 8.8 indicates the seven spatial segments used in the model.

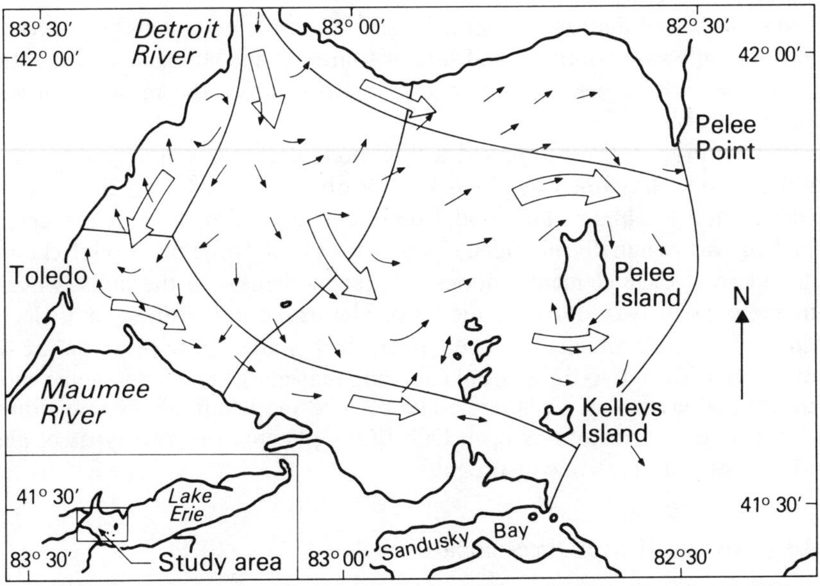

FIGURE 8.8 Circulation pattern in western Lake Erie, showing prevailing current directions and spatial segments of the water quality model (Di Toro *et al.*, 1975).

Rates of flow from one compartment to the other were assigned on the basis of observed and computed flow patterns. The magnitude of mixing of flows between adjacent segments was calibrated by use of a conservative tracer, in this case chloride concentration. The model was calibrated by adjusting parameters that specified the internal kinetics of the seven dependent variables for the period 1967–1970. Initial values were derived from the literature or laboratory and "fine-tuned" until agreement between model and lake was considered acceptable.

8.4.2. Network of One-Dimensional Channel Models

Network models were originally developed for estuarine systems by Chen and Orlob (1972), Dailey and Harleman (1972), and Najarian and Harleman (1975). If lakes and reservoirs are connected to each other and they are narrow and long, a network approximation is advantageous.

In the Chen and Orlob model the water body is subdivided into discrete, fully mixed volume units, called "nodes." Nodes are characterized by surface area, depth, volume, and side slopes. All water quality parameters that characterize the system are associated with nodes. Nodes are interconnected by channels or "links," which are defined by length, width, cross-sectional area, hydraulic radius, and a friction factor. Water is constrained to flow from one node to another through the defined channels, advecting and diffusing water quality constituents between nodes.

Mass balance equations are applied to these components, including advection, diffusion, inflow and outflow, and sink or source terms. Mass balance equations are coupled with hydrodynamic equations to produce the response of water quality variables.

Patterson *et al.* (1975) applied a link–node dynamic water quality model, similar to one developed by Chen and Orlob (1972), to Green Bay in northeastern Lake Michigan. The model represents a shallow lake or estuary (for which it was originally intended) as a network of one-dimensional channels (links) and storage elements (nodes). Water movement in the links is usually simulated by a hydrodynamic model of identical configuration or is derived from field measurements. In the Green Bay model, current structure was simulated with a two-dimensional, orthogonal-mesh, finite-difference model patterned after that of Leendertse (1970). The model simulates temperature, coliforms, four forms of nitrogen, DO, BOD, phosphorus, two types of algae, and several conservative constituents.

8.4.3. Two- and Three-Dimensional Models

In lakes or reservoirs, spatial variations of nutrients and plankton or other living organisms are strongly associated with the development of thermoclines in the vertical direction and horizontal variations between shore zones and deep water. In such lakes or reservoirs, two- or three-dimensional models are necessary to simulate the behavior of the whole system. Usually two- or three-dimensional ecological models are applied and they consist of horizontal or vertical arrangements of volume elements within each of which the state variables change with time according to a set of biochemical reaction equations. Each volume element is coupled by water movements that are simulated by two- or three-dimensional hydrothermal models.

Lam and Simons (1976) devised an advection–diffusion model of Lake Erie for the transport and dispersion of conservative substances, e.g. mineral salts. The governing equation for the model is a statement of mass conservation for a layer bounded by two horizontal planes 1 and 2 (upper and lower layer, respectively):

$$\frac{\partial}{\partial t}(h_1 c) = -\nabla \cdot (V h_1 c) + \nabla \cdot (K_h h_1 \nabla c) + (wc)_2 - (wc)_1 + \left(K_z \frac{\partial c}{\partial z}\right)_1 - \left(K_z \frac{\partial c}{\partial z}\right)_2 + S_e, \tag{8.66}$$

where

- c is the concentration of a conservative substance,
- h_1 is the local depth of a model layer,
- V is the horizontal component of velocity,
- K_h, K_z are the horizontal and vertical eddy diffusion coefficients,
- w is the vertical water displacement perpendicular to thc bounding surface,
- S_e represents all external sources.

The model is driven by Simons' multilayer lake circulation model, which produces the time-varying quantities V and w. In a two-layer model vertical transport occurs only at the intermediate bounding plane, i.e. at the top of the bottom layer and at the bottom of the upper layer. For a vertically mixed (homogeneous) system, $w = 0$ and $\partial c/\partial z = 0$, so the right-hand side of (8.66) reduces to the first two terms and the source term.

Lam and Simons (1976) applied their model to the simulation of chloride distributions in Lake Erie for conditions corresponding to summer and fall, 1970 and gained generally satisfactory results. Studies were made to determine the sensitivity of the models to various parameters and parameter levels.

The model developed by Lam and Simons can be applied to simulate non-conservative water quality parameters, including nutrients. Lam and Halfon (1978) developed a two-compartment phosphorus model coupled with a hydrodynamic model and applied it to Lake Superior. The hydrodynamic model was developed by Simons (1973) and described in section 8.3.5. A horizontal grid of size $\Delta x = \Delta y = 10$ km divides the lake so that there are 734 surface points.

The vertical structure consists of four layers, separated at 10, 30, and 40 m below the surface. The ecological model consists of two variables; the transformation from one variable (soluble reactive phosphorus) to the other (particulate phosphorus) is controlled by a primary production submodel. The two-variable model was first calibrated with time-averaged data on the assumption that the lake was spatially homogeneous. The model was then coupled to the transport processes computed by the hydrodynamic model.

The primary production submodel is a function of some physical factors, e.g. water temperature, sunlight, albedo, day length, and water turbidity. Figures 8.9(a) and (b) show the observed soluble reactive phosphorus distributions in June and September, respectively. The June observations were used as initial conditions for the model. Figures 8.9(c) and (d) show the predicted results of soluble reactive phosphorus for September without and with the transport processes, respectively. The strong interaction of the physical transport processes with the biochemical model is demonstrated by Figure 8.9(d).

The agreement between observed values (Figure 8.9(b)) and predicted values (Figure 8.9(d)) is reasonable. Similar models were applied to Lake Ontario by Simons (1976) and by Simons and Lam (1978). The investigation utilized the data base accumulated during the 1972 International Field Year on Lake Ontario. They demonstrated that the effects of water transport processes on biochemical processes are comparable in magnitude with other physical processes.

Chen and Smith (1979) developed a three-dimensional ecological–hydrodynamic model for Lake Ontario. The model includes mass balance equations for 15 different classes of biotic and abiotic substances. The biological compartments (algae, zooplankton, and fish) were substantially expanded to give a

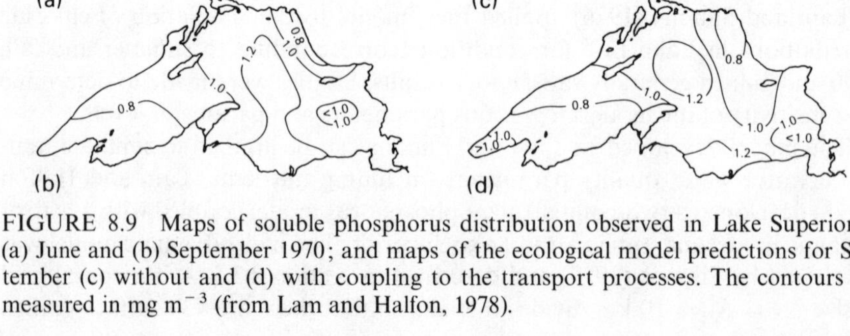

FIGURE 8.9 Maps of soluble phosphorus distribution observed in Lake Superior in (a) June and (b) September 1970; and maps of the ecological model predictions for September (c) without and (d) with coupling to the transport processes. The contours are measured in mg m^{-3} (from Lam and Halfon, 1978).

more complete picture of the ecological behavior of the lake. For example, the algae compartment was enlarged to include four groups of phytoplankton and attached algae; four groups of zooplankton were included and four groups of fish in three life stages (adult, young, and eggs or larvae) were represented.

The Chen–Smith model of Lake Ontario was designed to give a three-dimensional quality characterization of the impoundment by dividing it into layers and segments, i.e. hydraulic elements, for which separate mass balance equations can be written. Implementation of the model requires an independent determination of the flow field in three dimensions and estimates of empirical diffusion coefficients.

The model was tested against a representation of Lake Ontario that included 41 surface elements and seven vertical layers of varying thickness, a total of 209 elements. Hydraulic inputs were derived from a hydrodynamic model attributed to Bennett (1974). The hydrodynamic model is of orthogonal form, with 715 surface elements and seven layers. It is wind-driven and considers densimetric (temperature) effects on circulation. Integration of the hydrodynamic output to provide flows for the quality model was accomplished by a special interface program.

The model was simulated for the year 1972, with an hourly time step for hydrodynamics and a daily time step for water quality and ecology. No rigorous verification was attempted; however, the correspondence of the model results with some limited available water quality data has generally been good.

The 1979 version of the model required about 60 k decimal words of core storage and 9 s of Univac 1108 CPU time per day of simulation ($\Delta t = 1$ day). By way of contrast, the hydrodynamic program requires 130 k of storage and 120 s of CPU time per day of simulation.

8.4.4. Concluding Comments

The phytoplankton productivity models of Di Toro *et al.* (1975) and Thomann *et al.* (1975) and the water-quality–ecological models of Chen (1970; Chen and Orlob, 1972) are fairly representative of the state of the art of ecological modeling of impoundments, at least from an engineering viewpoint. However, the very nature of the modeling exercise that tends to aggregate, average, and smooth over biological subtleties leaves the more rigorous aquatic biologist somewhat disconcerted. He would prefer to concentrate efforts on a more correct representation of biological interactions, the kinetics of varying life stages, shifts in grazing preferences, ecological instabilities, and the like. Some trade-offs are necessary, simply because the model is an approximation of the real system. These seem to have occurred in the engineer-developed models by gross simplification of the aquatic ecosystem and in the biologist-developed models by simplifying the circulatory and exchange processes of the impoundment. Somewhere between these extremes probably lies the best practical ecological model of a eutrophic lake or reservoir. It seems from our review that the capability to model these systems has probably outstripped our understanding of them. At best, calibrations are rather rough and verifications more so. Part of the reason is a lack of good data from the field, but there has also been some over-zealousness on the part of modelers. Sensitivity testing, which has only been occasionally employed, will no doubt reveal, as it has for models of lake circulation, that some simplification in ecosystem description is justified. If this can be accomplished while still satisfying the biologist that his science has not been unduly compromised, some useful models can probably be produced.

8.5. MATHEMATICAL MODELS OF COOLING IMPOUNDMENTS

Lakes, reservoirs, or shallow artificial impoundments are frequently utilized to dissipate excess heat from power plants. These heated condenser water discharges may alter the hydrothermal, chemical, and ecological environment of the water body and it is necessary to have techniques for the prediction of such impacts.

Cooling impoundments dissipate waste heat by direct transfer between the water surface and the atmosphere. A portion of the heat is dissipated by radiation while the remainder is transferred by evaporation and conduction. The water that is lost in the evaporative process must be replaced by "makeup" water. This is supplied to the impoundment either by natural inflow or by pumping from an adjacent water body. The average makeup flow must exceed the average rate of evaporative water loss so that the content of dissolved solids in the cooling water can be controlled. The difference between the makeup and the loss is the "blow-down" or discharge from the impoundment. Cooling impoundments may be classified into two types: cooling lakes and cooling ponds. A *cooling lake* is an existing or man-made water body that impounds a stream or river (i.e. an "on-stream" reservoir) and provides cooling as part of a circulating water system. In the terminology of certain regulatory agencies, this is called a "recirculating cooling water body" and thermal and water quality standards apply both within the impoundment and for downstream discharges. A *cooling pond* is an artificial impoundment that does not intercept a stream or river (i.e. an "off-stream" or "perched" pond). This is a closed, circulating water system and, except for the blow-down to an adjacent natural body, is not normally subject to temperature or other water quality standards.

Both types of cooling impoundment have a number of advantages over forced- or natural-draft evaporative cooling towers. Because of the lower pumping head and heat rejection temperature, power production is more efficient; the thermal inertia of cooling impoundments reduces the diurnal meteorological temperature fluctuations associated with towers; and their ability to store water on a seasonal basis reduces or eliminates the demand for makeup water during periods of low flow and increases siting flexibility in areas that are short of water. Disadvantages include land cost and availability for off-stream ponds and environmental constraints in siting on-stream impoundments. In addition, the use of cooling impoundments has been constrained by a lack of confidence in the ability to predict hydrothermal performance and environmental impacts. The disadvantages associated with land cost and availability for off-stream cooling ponds may sometimes be offset by the requirement to purchase land in order to obtain "rights" to water usage.

The analysis of thermal discharges in large bodies of water, typical of once-through systems, has been considerably aided by the development of near- and far-field zone models. This conceptualization is possible because the far-field advective and diffusive processes are weakly coupled to the buoyant jet-induced mixing in the near-field zone. In contrast, transport processes in a cooling impoundment are dominated by currents induced by the cooling water discharge and intake. The development and verification of mathematical models for the hydrothermal structure and performance of cooling ponds and lakes are summarized below.

8.5.1. Classification of Cooling Impoundments

Existing and proposed cooling impoundments display a wide diversity in physical features and thermal loading. Thermal loading is defined as the waste heat rejected by the cooling water system in thermal megawatts per hectare of water surface. Figure 8.10(a) shows a proposed *cooling lake* having an area of 620 ha and a thermal loading of 2.4 $MW_t\ ha^{-1}$ (Lake Merom, Indiana). Cooling lakes, formed by impoundment of a natural water course, generally have an irregular shape and relatively large depth (6–30 m). Long, dead-end side arms are frequently found. Cooling lakes can offer advantages as multipurpose sites by accommodating facilities for hydroelectric power, water supply, and recreation. Environmental regulations generally provide quantitative limits on maximum temperatures or temperature rises within the lake or otherwise set ecological guidelines on the impact of the induced heat load.

Figure 8.10(b) shows an off-stream *cooling pond* that does not impede the flow of a natural stream (Powerton, Illinois). Cooling ponds are usually more shallow than cooling lakes and, in addition to the external dikes forming the pond, they may contain internal baffles and dikes to channel flow from discharge to intake. The pond shown in Figure 8.10(b) has an area of 575 ha, a thermal loading of 4.2 $MW_t\ ha^{-1}$, and a mean depth of 3.4 m. Water quality regulations usually do not apply to cooling ponds. Therefore, the ponds are usually more heavily loaded (by a factor of 2) than cooling lakes and the major design consideration is optimization of thermal performance to achieve minimal intake temperatures and high thermal inertia.

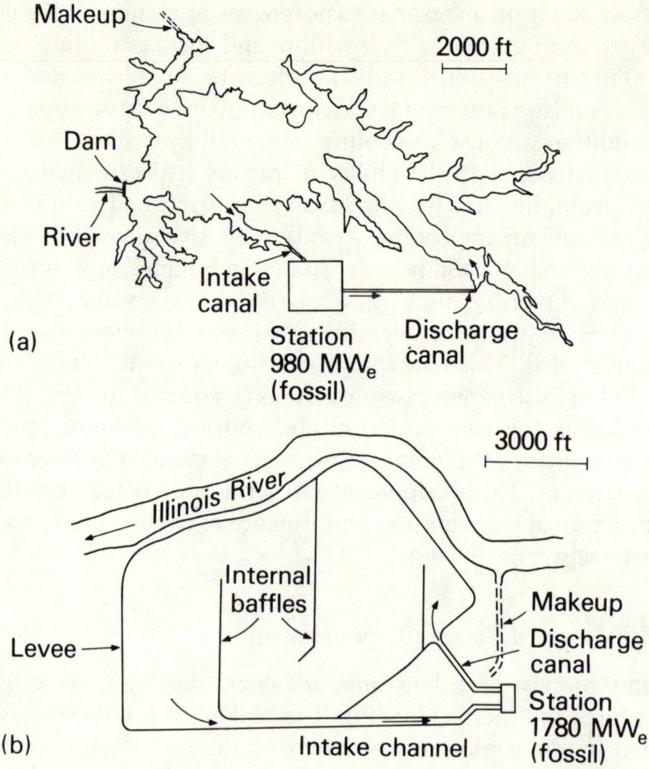

FIGURE 8.10 Types of cooling impoundment: (a) a cooling lake, Lake Merom, Indiana; (b) a cooling pond at Powerton, Illinois (from Jirka and Harleman, 1979).

Table 8.3 (Jirka and Harleman, 1979) compares the physical characteristics and thermal loadings of some cooling ponds and lakes in the United States.

8.5.2. Distinction Between Stratified and Vertically Mixed Impoundments

The classification into cooling lakes and cooling ponds is useful for legal and environmental assessment purposes. However, it does not suffice for defining the internal hydrothermal structure of the impoundment. In order to develop mathematical models for use in design and prediction of performance, it is necessary to have an additional classification that distinguishes between stratified and vertically mixed impoundments. This depends on the thermal loading, the impoundment depth and shape, and the design and location of intake and discharge structures.

All natural water bodies have a tendency to stratify thermally under the action of solar and atmospheric radiation. In a cooling impoundment of average

TABLE 8.3 Comparison of Physical Characteristics for Typical Cooling Lakes and Cooling Ponds in the United States.

Impoundment	Area (ha)	Electricity Generating Capacity (MW_e)	Waste Heat (MW)	Thermal Loading ($MW\ ha^{-1}$)
Cooling lakes				
Lake Anna, VA	5200	3784 (nuclear)	7600	1.5
Clinton Lake, IL	1960	1982 (nuclear)	3750	1.9
Gibbons Creek Res., TX	920	896 (fossil)	1170	1.3
Lake Merom, IN	620	980 (fossil)	1465	2.4
Lake Robinson, NC	900	135 (fossil) +730 (nuclear)	1750	1.9
Lake Sanchris, IL	865	1232 (fossil)	1930	2.2
Sutherland Reservoir, NE	855	1300 (fossil)	2040	2.4
Cooling ponds				
Braidwood, IL	1015	2200 (nuclear)	4520	4.5
Collins, IL	805	2520 (fossil)	3074	3.8
Dresden, IL	510	1600 (nuclear)	2678	5.3
La Salle, IL	825	2156 (nuclear)	4362	5.3
Powerton, IL	575	1780 (fossil)	2437	4.2

size, the kinetic energy input due to the flow-through induced by the condenser water input and withdrawal is usually large enough to destroy the natural stratification. Instead, an artificial stratification may be created because of the forced temperature gradient between the discharge and the condenser intake. Based on earlier work of Watanabe and Harleman (1977), Jirka and Harleman (1979) proposed a dimensionless "pond number" P as a measure of whether a cooling impoundment will be stratified or vertically mixed. The pond number is defined as

$$P = \left(\frac{f_i}{4} \frac{Q_0^2}{\beta \Delta T_0 g H^3 W^2} D_v^3 \frac{L}{H}\right)^{1/4}, \tag{8.67}$$

where

f_i is the interfacial quadratic-law friction factor,
Q_0 is the condenser flow rate,
ΔT_0 is the condenser temperature rise,
β is the coefficient of thermal expansion,
g is the gravitational acceleration,
H is the average pond depth,
W is the average pond width (or flow path width in a pond with internal baffles),
D_v is the dilution ratio for vertical entrance mixing at the plant discharge,
L is the pond length following the mean flow-through path.

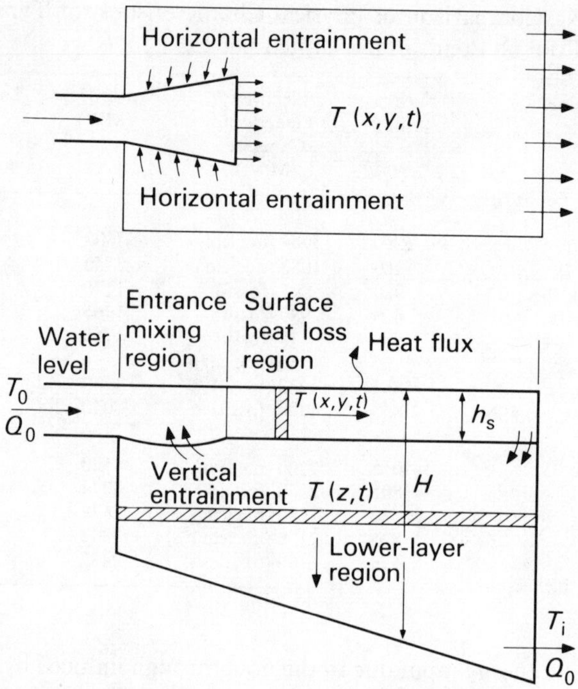

FIGURE 8.11 Diagram of a stratified cooling impoundment.

A stratified cooling impoundment is shown schematically in Figure 8.11. A derivation of the pond number is given by Jirka *et al.* (1978). Physically, P represents the ratio of the thickness of the upper layer, h_1, to the total depth H. If the density of the upper layer, $\rho_1(x)$, is assumed to vary linearly with x because of surface heat dissipation, then $h_1(x) = h_s =$ constant and

$$P = \frac{h_s}{H}. \tag{8.68}$$

In the calculation of pond numbers, the following quantities are useful: for typical Reynolds numbers, $f_i = 0.01$ (field) and $f_i = 0.10$ (laboratory) (Jirka *et al.*, 1975); $\beta\Delta T_0 = \Delta\rho_0/\rho_2$, where $\beta = 0.0002\ ^\circ\mathrm{C}^{-1}$ at 20 °C; D_v is a function of the densimetric Froude number of the discharge channel:

$$D_v = 1.2\mathrm{Fr}'_0 - 0.2, \tag{8.69}$$

where

$$\mathrm{Fr}'_0 = \frac{U_0}{(\beta\Delta T_0 g a_0^{1/2})^{1/2}} \tag{8.70}$$

and a_0 is the channel exit area if the discharge is along a wall and half the exit area for a symmetrical discharge. Jirka *et al.* (1978) recommend a minimum value of $D_v = 1.5$ for rectangular discharges and $D_v = 1.2$ for special, low-mixing discharge structures with radial guide vanes. The higher the entrance mixing parameter, the less likely it is that the pond will be stratified.

Inspection of available field and laboratory data indicates that cooling impoundments in which $P \leq 0.3$ are well stratified "deep" ponds. On the other hand, if $P \geq 0.7$ the impoundments are vertically fully mixed "shallow" ponds. Mathematical models for the prediction of hydrothermal performance are available for these two cases and will be discussed below. Impoundments in the range $0.3 < P < 0.7$ are partially stratified, having no distinct surface layer and exhibiting variable degrees of vertical stratification throughout the pond. The analysis of this class of ponds is difficult, but this range of pond numbers can be avoided by proper choice of design parameters.

Currents in cooling ponds result from three mechanisms: through flow (generated by pumping), density differences, and wind stresses. The relative magnitudes of density and through-flow currents are particularly important and are closely linked to the thermal structure of cooling ponds; density currents usually prevail in deep ponds, while shallow ponds may be governed by the through flow, eddies in the discharge zone, and flow separations at constrictions or around baffles. Under certain conditions, wind-induced currents may be stronger than either through-flow or density currents. However, strong winds are intermittent and surface heat loss during these periods is enhanced.

Deep, stratified impoundments are less subject to short-circuiting between the discharge and condenser intake than shallow, vertically mixed ponds. In a stratified pond, buoyancy acts to spread the flow over the surface of the impoundment, including dead-end side arms (Brocard *et al.*, 1977). The thermal performance of shallow, vertically mixed ponds is highly influenced by the pond geometry. Short-circuiting and the generation of large ineffective eddies should be minimized by the use of interior baffles.

8.5.3. Vertically Mixed Cooling Ponds

Cooling impoundments that are characterized by pond numbers of 0.7 or greater are classed as vertically mixed ponds. Thus, the vertically averaged wind-driven circulation models discussed in section 8.3.1 should be applicable in combination with an appropriate heat transport equation. However, the existence of longitudinal temperature gradients and flows generated by the intake and discharge (which usually dominate the wind-induced currents) warrants a special category of models.

Yeh *et al.* (1973) developed a two-dimensional model of a vertically mixed cooling pond. They made three major assumptions in their transient model. (a) The flow field is in steady state and the momentum equations contain only

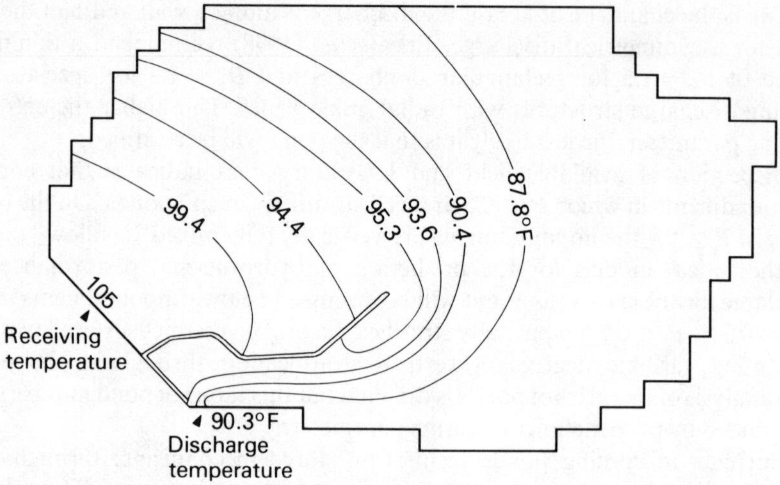

FIGURE 8.12 Surface temperature distribution in a vertically mixed cooling pond, as predicted by Yeh *et al.* (1973).

pressure and linearized friction terms, leading to a potential flow solution in analogy to that for viscous flows. (b) There is no entrance mixing, thus reducing the flow field to a pure source–sink motion. (c) The transient heat transport equation neglects lateral diffusion and is solved in a Lagrangian sense, i.e. following the stream tubes as given from the potential flow solution. This fully mixed model was applied to a cooling pond with a retention time of about eight days. No analysis or field data have been provided to ascertain that the pond under study was indeed fully mixed. Results for the predicted temperature field are given in Figure 8.12. The temperature pattern shows large lateral temperature gradients at the power plant intake. This is contradictory to usual observations and could be attributed to the neglect of convective momentum and lateral heat diffusion terms.

Vasiliev (1978a) presented a model of a vertically averaged cooling pond with quadratic bottom shear and surface wind shears. Both longitudinal and lateral heat diffusion were included. At the inflow boundary, volume and temperature flux conditions were imposed. No numerical results are available.

The importance of well designed internal baffles or dikes in vertically mixed ponds is shown in Figure 8.13. Figure 8.13(a) shows surface isotherms in the Powerton, Illinois pond and indicates recirculation eddies in each of the pond compartments. In the Dresden, Illinois pond (Fig. 8.13(b)) the flow is basically longitudinal. The lateral temperature variations (higher temperatures at the centerline of each channel and lower temperatures at the boundaries) indicate a longitudinal dispersion effect in the heat transport.

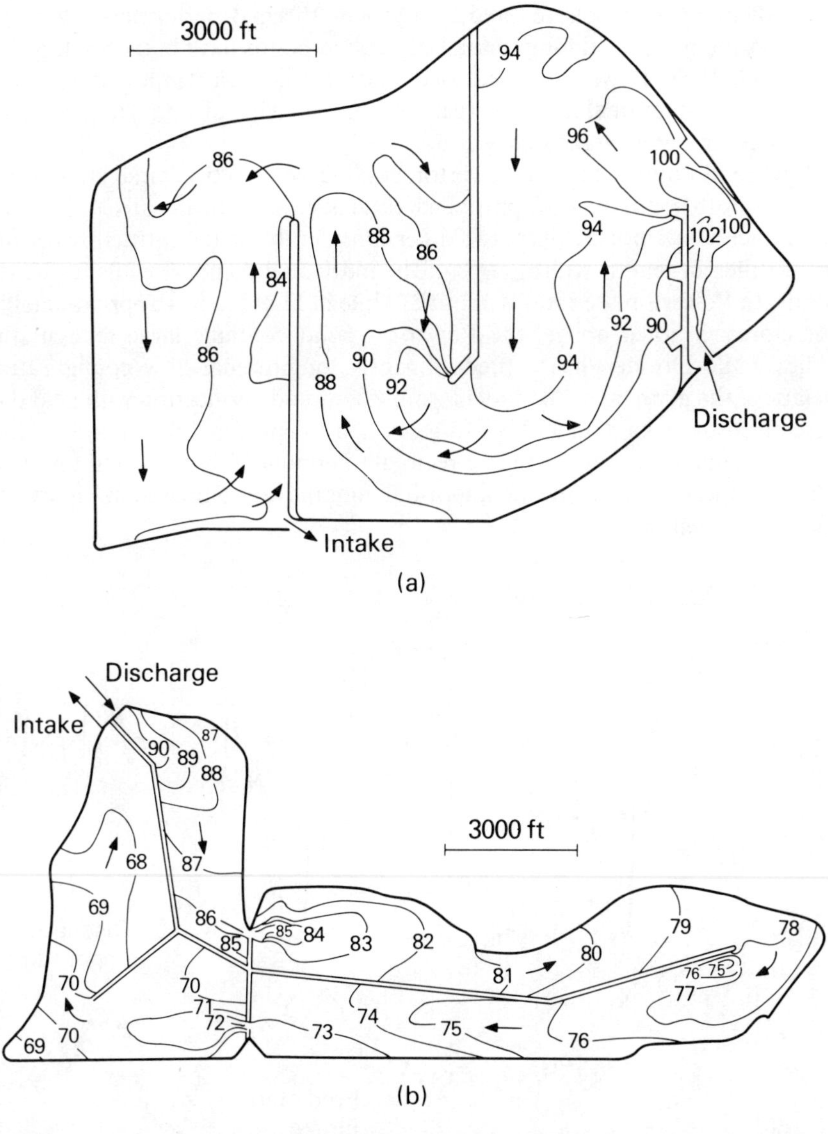

FIGURE 8.13 Surface isotherms (°F) for two shallow cooling ponds: (a) Powerton, IL, with predominant lateral recirculation behavior, and (b) Dresden, IL, with predominant longitudinal dispersion behavior (from Jirka and Harleman, 1979).

Mathematical models to predict transient thermal performance for both recirculating ponds and longitudinal dispersion ponds have been developed by Jirka *et al.* (1978). Power plant heat loads and meteorological inputs are averaged over the computational time step (usually between 3 h and 1 day) for prediction of condenser water intake temperatures.

The pond number for the Powerton cooling pond is 0.6, while for Dresden $P = 0.8$. Both ponds may be approximately classed as vertically mixed; however, the mathematical models for the two differ considerably in their structure because of the different length: width ratios of the individual compartments in the two ponds. In Powerton the ratio is about 2, while in Dresden it is approximately 6. Therefore, as noted above, the Powerton pond contains large recirculating eddies while Dresden has a predominantly longitudinal flow configuration. Details of the mathematical model formulations and verification with field data were given by Jirka *et al.* (1978). In general, a longitudinal flow pond exhibits better thermal performance than a recirculating pond. Observed and predicted intake temperatures during one-and-a-half months are compared in Figure 8.14 for the Dresden pond.

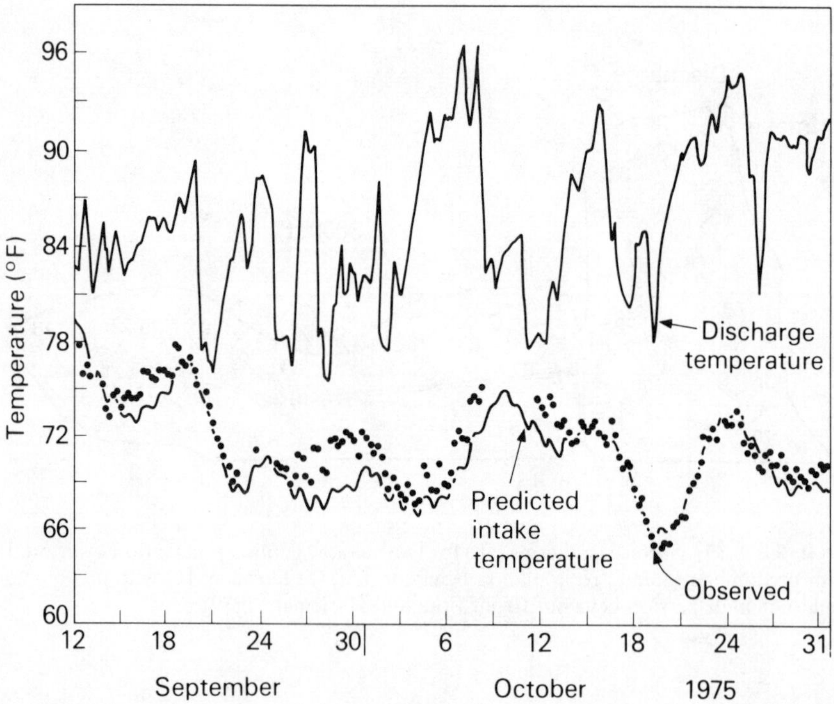

FIGURE 8.14 Predicted and measured intake temperatures for the Dresden, IL, cooling pond during September and October 1975 (from Jirka *et al.*, 1978). Mean error, 0.8°F; standard deviation, 1.0°F.

8.5.4. Stratified Cooling Ponds

An extensive survey of techniques for the hydrothermal performance of stratified cooling impoundments has been made by Ryan and Harleman (1973). The important hydrodynamic and heat transport features of stratified cooling lakes are as follows. Heated water enters the cooling lake through a discharge channel; entrance mixing occurs as a result of vertical and lateral entrainment into the discharge "jet"; beyond the entrance mixing region there exists a *stratified surface layer* that is warmer than the underlying water, and *surface heat loss* plays a dominant role in gradually decreasing the temperature of this layer. Ultimately, at distances far from the discharge, the water in the surface layer down-wells and enters the *subsurface region*. In this region the water is advected downward owing to pond through flow. A skimmer wall may be used at the intake to provide *selective withdrawal* from the cooler lower layers. The water from the lower layers finally enters the *power plant intake*. This process occurs in a continuous, closed-cycle fashion as the condenser water is discharged with a temperature rise equal to ΔT_0 above that at the intake. Cooling impoundments exhibit highly transient behavior because of variations in the power plant loading and diurnal and seasonal meteorological fluctuations.

The stratified cooling impoundment model developed by Ryan and Harleman (1973) explicitly accounts for the three-dimensional hydrodynamic and heat transport features described above and illustrated in Figure 8.11. This model is an extension of an earlier one-dimensional vertical reservoir model (Huber and Harleman, 1968; Ryan and Harleman, 1971). The general applicability of this model has been established through comparison with field data, notably in the TVA system (Wunderlich, 1973; Parker *et al.*, 1975).

Ryan and Harleman (1973) used the buoyant jet model developed by Stolzenbach *et al.* (1972) as the basis for determining the entrance mixing and the thickness of the stratified surface layer, h_s. Thus, the upper-layer thickness is a function only of the discharge channel geometry and densimetric Froude number. The surface heat loss and, therefore, the temperature in the upper layer are computed in terms of the fractional surface area but not explicitly in terms of x–y position. Because of the large depth of the subsurface zone (in comparison with the surface layer thickness), horizontal flows are neglected and this zone is assumed to be horizontally stratified.

The intake flow under the skimmer wall takes water selectively from certain layers, depending on the magnitude of the vertical density gradient. The thickness of the withdrawal layer is computed using relationships developed by Kao (1965). River inflows and reservoir outflows can be included at any pond depth, and entrance mixing or withdrawal layer considerations are included in these cases.

The transient pond behavior is computed by assuming essentially quasi-steady state conditions within each time step. Quasi-steady state conditions that apply to components such as entrance mixing, layer thickness, etc. are taken to be constant within the time step. For this assumption to be correct, it is necessary that the time step be longer than the characteristic time for dynamic changes, which is essentially the impoundment length divided by the internal wave speed. A time step of about one day is appropriate for this reason and is short enough to allow the study of transient behavior due to plant loading and meteorology.

Ryan and Harleman (1973) applied the model to laboratory experiments and field data. Satisfactory agreement was found in most cases, including surface temperature distribution, vertical profiles, and intake temperatures. As an example, Figure 8.15 shows a comparison of predicted and measured intake temperatures over one year for the Hazelwood, Australia cooling pond, for which $P = 0.17$; thus, it is well stratified (Jirka *et al.*, 1978). The overall accuracy is of the order of 1 °C. Another comparison, in Figure 8.16, shows the vertical temperature distribution in Lake Norman, North Carolina, with heated discharges from the Marshall power station. This cooling lake also has river through flow, which is included in the model. In summary, the model by Ryan and Harleman (1973) seems a reliable tool for transient prediction of stratified impoundments.

The Ryan and Harleman model was extended by Watanabe *et al.* (1975) to take into account horizontal two-dimensional flow and temperature distributions in the surface layer. Since it is difficult to solve both near-field jet phenomena and far-field stratified flow by using one model, the model is constructed by integrating separate models for each region.

The major components of the mathematical model (Figure 8.11) are an entrance mixing region and a surface layer heat loss region with horizontal temperature gradients overlying a region with a vertical temperature gradient. The entrance mixing region creates a strong lateral recirculation that significantly affects the temperature distribution in the far field. In contrast to the Ryan and Harleman (1973) model, which assumed that the surface layer thickness was a function only of the discharge channel geometry and densimetric Froude number, the Watanabe *et al.* (1975) model includes far-field effects, interfacial friction, density differences between the surface and the lower layers, and pond geometry. The surface layer thickness h_s is given by eqns. 8.67 and 8.68.

The vertically integrated two-dimensional momentum and heat transport equations for the surface layer are solved by finite-element methods. Since the response of the stratified cooling pond is relatively steady over a short time interval, the transient hydraulic behavior and heat distribution for the surface layer of the cooling pond are approximated by solving the steady state momentum and heat equations sequentially with new boundary conditions and new meteorological conditions at each time step. The lower-layer formulation is the same as the Ryan–Harleman model. The model has been tested in the laboratory (Watanabe and Harleman, 1977) and applied to Lake Anna, Virginia (Jirka *et al.*, 1977), the site of a large nuclear power station.

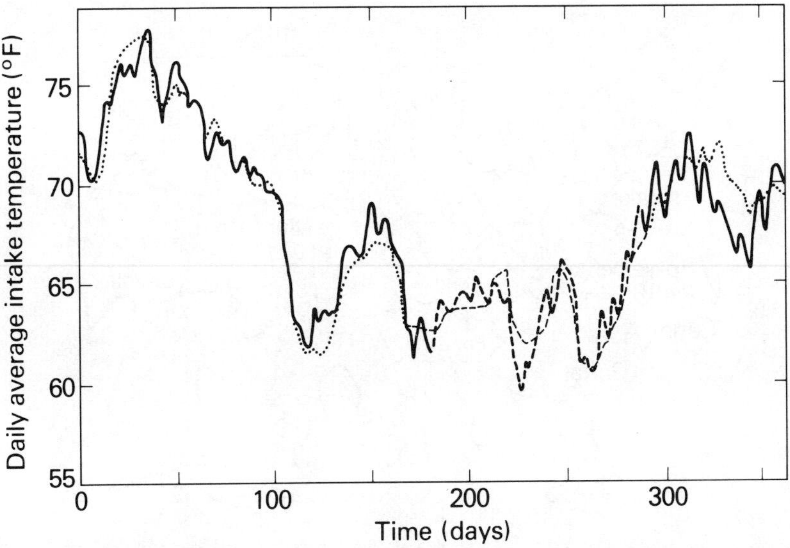

FIGURE 8.15 Predicted and measured intake temperatures of the Hazelwood, Australia cooling pond in 1969 (from Ryan and Harleman, 1973). Full line, mean measured intake temperature; bold broken line, mean measured temperature at skimmer wall opening; dotted line, predicted intake temperature; light broken line, predicted temperature at skimmer wall opening.

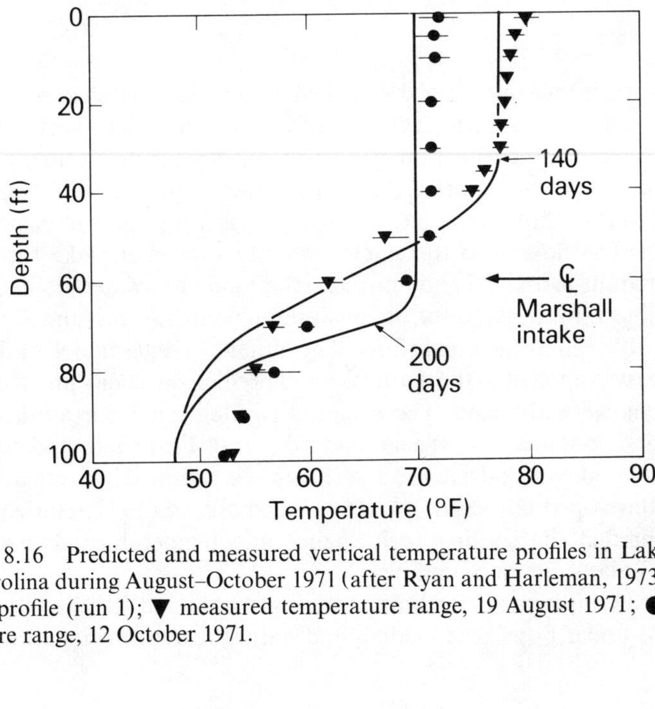

FIGURE 8.16 Predicted and measured vertical temperature profiles in Lake Norman, North Carolina during August–October 1971 (after Ryan and Harleman, 1973). Full line, predicted profile (run 1); ▼ measured temperature range, 19 August 1971; ● measured temperature range, 12 October 1971.

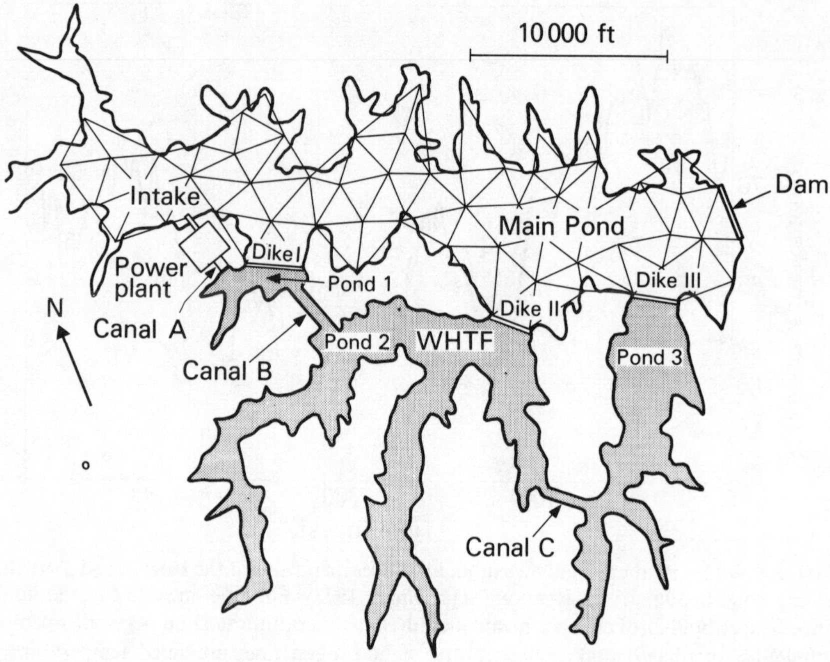

FIGURE 8.17 The combined cooling pond (WHTF) and cooling lake of Lake Anna, Virginia.

Lake Anna is an example of a stratified cooling lake, as shown in Table 8.3. Actually it is a combined cooling pond and cooling lake. The heated condenser water is discharged into a portion of the lake that is separated from the main lake by dikes, as shown in Figure 8.17. A large fraction of the total waste heat is dissipated in this "hot pond" enclosure, which is not subject to environmental constraints. The flow from this enclosure enters the main lake through submerged conduits in dike III and returns to the condenser water intake. The main lake is subject to constraints on maximum temperatures and temperature increases above natural conditions. The finite-element model of Watanabe *et al.* (1975) was applied to the main lake and predicted velocity and temperature distributions were obtained. The enclosed cooling pond portion is thermally stratified and contains several dead-end side arms. Flows into and out of these arms occur as buoyancy-driven counterflows. A segmented hydrothermal model has been developed (Brocard *et al.*, 1977; Jirka *et al.*, 1977; Harleman *et al.*, 1978) to predict the cooling performance and temperature changes for the combined Lake Anna impoundment.

Predicted temperatures for Lake Anna were obtained for a ten-year simulation period under both heat loading and natural (preoperational) conditions.

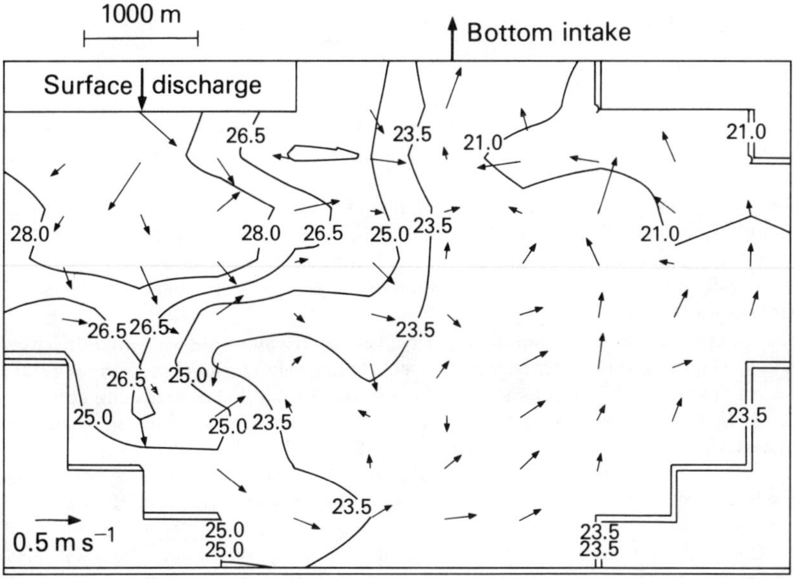

FIGURE 8.18 Surface velocity and temperature (°C) distributions in the First Ekibastuz Thermal Power Plant.

The results from such long-term simulations can be processed to demonstrate environmental effects such as induced temperatures and induced excess (above natural) temperatures.

The model presented in section 8.3.5, developed by Vasiliev *et al.* (1974), was modified through the elimination of the Coriolis term and the nonlinear inertial term, and applied to the cooling reservoir of the First Ekibastuz Thermal Power Plant (surface area 18.5 km^2, volume $0.86 \cdot 10^8$ m^3, average depth 4.6 m) in the USSR. Figure 8.18 indicates the predicted behavior of the currents and the temperature distribution on the water surface under steady state conditions (Vasiliev, 1978b).

8.5.5. Concluding Comments

The state of the art of predicting the transient performance of cooling impoundments by means of one-, two-, and three-dimensional mathematical models is fairly well developed. Observations of small-scale cooling impoundments under controlled laboratory conditions are useful in the development and verification of mathematical models. However, because of the different requirements for hydrodynamics, stratified flows, and surface heat dissipation and because laboratory meteorology is essentially steady state, the direct transfer of laboratory measurements to prediction of the behavior of an impoundment is not recommended.

REFERENCES

Abbott, M. B., A. Danesgaard, and G. S. Rodenhuis (1973). System 21, Jupiter. *Journal of Hydraulic Research* 11(1): 1–28.

Baca, R. G., M. W. Lorenzen, R. D. Mudd, and L. V. Kimmel (1974). A generalized water quality model for eutrophic lakes and reservoirs. *Battelle Pacific Northwest Laboratories Report* PB-256925.

Bauer, S. W. (1979). Three-dimensional simulation of time-dependent elevations and currents in a homogeneous lake of arbitrary shape using an irregular-grid finite-difference model. *Hydrodynamics of Lakes*, eds. W. H. Graf and C. H. Mortimer (Amsterdam: Elsevier), pp. 267–276.

Bennett, J. R. (1974). On the dynamics of wind-driven lake currents. *Journal of Physical Oceanography* 4(3): 400–414.

Bonham-Carter, G., and J. H. Thomas (1973). Numerical calculation of steady wind-driven currents in Lake Ontario and the Rochester embayment. *Proceedings of 16th Conference on Great Lakes Research* (Ann Arbor, MI: International Association of Great Lakes Research), pp. 640–662.

Boyce, F. M. (1974). Some aspects of Great Lakes physics of importance to biological and chemical processes. *Journal of Fisheries Research Board of Canada* 31(5): 689–730.

Brocard, D. N., G. H. Jirka, and D. R. F. Harleman (1977). A model for the convective circulation in side arms of cooling lakes. *MIT Department of Civil Engineering, R. M. Parsons Laboratory Technical Report* 223.

Chen, C. W. (1970). Concepts and utilities of ecologic model. *Proceedings of American Society of Civil Engineers, Journal of Sanitary Engineering Division* 96(SA5): 1085–1097.

Chen, C. W., and G. T. Orlob (1972). Ecologic simulation for aquatic environments. *Report for Office of Water Resources Research, US Department of Interior by Water Resources Engineers, Inc.*

Chen, C. W., and D. J. Smith (1979). Preliminary insights into a three-dimensional ecological–hydrodynamic model. *Perspectives on Lake Ecosystem Modeling*, eds. D. Scavia and A. Robertson (Ann Arbor, MI: Ann Arbor Science), pp. 249–279.

Cheng, R. T. (1972). Numerical integration of lake circulation around island by the finite-element method. *International Journal of Numerical Methods in Engineering* 5: 103.

Cheng, R. T., T. M. Powell, and T. M. Dillon (1976). Numerical models of wind-driven circulation in lakes. *Applied Mathematical Modelling* 1(3): 141–159.

Cheng, R. T., and C. Tung (1970). Wind-driven lake circulation by the finite-element method. *Proceedings of 13th Conference on Great Lakes Research* (Ann Arbor, MI: International Association of Great Lakes Research), pp. 891–903.

Csanady, G. T. (1975). Hydrodynamics of large lakes. *Annual Review of Fluid Mechanics* 7: 357–386.

Dailey, J. E., and D. R. F. Harleman (1972). Numerical model for the prediction of transient water quality in estuary networks. *MIT Department of Civil Engineering, R. M. Parsons Laboratory Technical Report* 158.

Di Toro, D. M., D. J. O'Connor, R. V. Thomann, and J. L. Mancini (1975). Phytoplankton–zooplankton–nutrient interaction model for western Lake Erie. *Systems Analysis and Simulation in Ecology* vol. 3, ed. B. C. Patten (New York, NY: Academic Press), ch. 11.

Edinger, J. E., and E. D. Buchak (1978). Reservoir longitudinal and vertical implicit hydrodynamics. *Proceedings on Environmental Effects of Hydraulic Engineering Works, Knoxville, TN* (International Association of Hydraulics Research), pp. 160–169.

Fischer, G. (1959). Ein numerisches Verfahren zur Errechnung von Windstau und Gezeiten in Randmeeren. *Tellus* 11(1): 60–76.

Fischer, G. (1965). A survey of finite-difference approximations to the primitive equations. *Monthly Weather Review* 93(1): 1–10.

Gallagher, R. H., J. A., Liggett, and S. K. T. Chan (1973). Finite-element shallow lake circulation analysis. *Proceedings of American Society of Civil Engineers, Journal of Hydraulics Division* No. HY7: 1083–1095.

Gates, W. L. (1968). A numerical study of transient Rossby waves in a wind-driven homogeneous ocean. *Journal of Atmospheric Science* 25(1):3–22.

Gedney, R. T., and W. Lick (1970). Numerical calculation of the steady state wind-driven currents in Lake Erie. *Proceedings of 13th Conference on Great Lakes Research* (Ann Arbor, MI: International Association of Great Lakes Research), pp. 829–838.

Gedney, R. T., and W. Lick (1972). Wind-driven currents in Lake Erie. *Journal of Geophysical Research* 77:2714–2723.

Grotkop, G. (1973). Finite-element analysis of long-period water waves. *Computer Methods in Applied Mechanics and Engineering* 2(2):147–157.

Hansen, W. (1956). Theorie zur Errechnung des Wasserstandes und der Strömungen in Randmeeren nebst Auwendungen. *Tellus* 8(3):287–300.

Harleman, D. R. F., G. H. Jirka, D. N. Brocard, K. A. Hurley-Octavio, and M. Watanabe (1978). Transient simulation of cooling lake performance under heat loading from the North Anna Power Station. *Proceedings of 2nd Conference on Waste Heat Management and Utilization, Miami, FL.*

Heaps, N. S. (1969). A two-dimensional numerical sea model. *Philosophical Transactions, Series A* 265(1160):93–137.

Hinze, J. O. (1959). *Turbulence* (New York, NY: McGraw-Hill).

Huber, W. C., and D. R. F. Harleman (1968). Laboratory and analytical studies of thermal stratification of reservoirs. *MIT Department of Civil Engineering, R. M. Parsons Laboratory Technical Report* 112.

Hurley-Octavio, K. A., G. H. Jirka, and D. R. F. Harleman (1977). Vertical heat transport mechanisms in lakes and reservoirs. *MIT Department of Civil Engineering, R. M. Parsons Laboratory Technical Report* 227.

Hyden, H. (1974). Water exchange in two-layer stratification waters. *Proceedings of American Society of Civil Engineers, Journal of Hydraulics Division* 100(HY3):345–361.

Imboden, D. M. (1974). Phosphorus model for lake eutrophication. *Limnology and Oceanography* 19(2):297–304.

Jirka, G. H., G. Abraham, and D. R. F. Harleman (1975). An assessment of techniques for hydrothermal prediction. *MIT Department of Civil Engineering, R. M. Parsons Laboratory Technical Report* 203.

Jirka, G. H., D. N. Brocard, K. A. Hurley-Octavio, M. Watanabe, and D. R. F. Harleman (1977). Analysis of cooling effectiveness and transient long-term simulations of a cooling lake (with application to the North Anna Power Station). *MIT Department of Civil Engineering, R. M. Parsons Laboratory Technical Report* 232.

Jirka, G. H., and D. R. F. Harleman (1979). Cooling impoundments: Classification and analysis. *Proceedings of American Society of Civil Engineers, Journal of Environmental Engineering Division* 105(EE2):291–309.

Jirka, G. H., M. Watanabe, K. A. Hurley-Octavio, C. F. Cerco, and D. R. F. Harleman (1978). Mathematical predictive models for cooling ponds and lakes. Part A: Model development and design considerations. *MIT Department of Civil Engineering, R. M. Parsons Laboratory Technical Report* 238.

Jones, W. P., and B. E. Launder (1973). The calculation of low Reynolds number phenomena with a two-equation model of turbulence. *International Journal of Heat and Mass Transfer* 16(6):1119–1130.

Kao, T. W. (1965). The phenomenon of block in stratified flow. *Journal of Geophysical Research* 70(4):815–822.

Karelse, M. (1974). Vertical exchange coefficients in stratified flows. *Momentum and Mass Transfer in Stratified Flows*, ed. H. N. C. Breusers. *Delft Hydraulic Laboratory Report* R.880, ch. 2.

Kasahara, A. (1965). On certain finite-difference methods for fluid dynamics. *Monthly Weather Review* 93(1):27–31.

King, I. P., W. R. Norton, and G. T. Orlob (1973). Finite-element model for two-dimensional density-stratified flow. *Water Resources Engineers, Inc. Report to Office of Water Resources Research.*

Koh, R. C. T., and L. N. Fan (1970). Mathematical models for the prediction of temperature distributions resulting from the discharge of heated water in large bodies of water. *US Environmental Protection Agency Research Series* No. 16130 DWO.

Kullenberg, G. (1974). An experimental and theoretical investigation of the turbulent diffusion in the upper layer of the sea. *Copenhagen University, Institute for Physical Oceanography Report* 25.

Kvon, V. I. (1979a). Hydrothermal computation. *Izvestiya Akademia Nauk SSSR. Energetika i Transport* No. 5 (in Russian).

Kvon, V. I. (1979b). A temperature-stratified flow in a running flow water body. *Meteorologiia i Gidrologiia* No. 6: 74–79 (in Russian).

Lam, D. C. L., and E. Halfon (1978). Model of primary production, including circulation influences in Lake Superior. *Journal of Applied Mathematical Modeling* 2(1): 30–40.

Lam, D. C. L., and T. J. Simons (1976). Numerical computations of advective and diffusive transports of chloride in Lake Erie during 1970. *Journal of Fisheries Research Board of Canada* 33(3): 537–549.

Larsen, D. P., and H. T. Mercier (1975). Shagawa Lake recovery characteristics as depicted by predictive modeling. In: Water Quality Criteria Research of the US Environmental Protection Agency. *US Environmental Protection Agency, Corvallis, OR Report* EPA 660/3-76-079.

Launder, B. E. (1976). Comments on improved form of the low Reynolds number k–ϵ turbulence models, by G. H. Hoffman. *Physics of Fluids* 19(5): 765–766.

Lee, K. K., and J. A. Liggett (1970). Computation for circulation in stratified lakes. *Proceedings of American Society of Civil Engineers, Journal of Hydraulics Division* No. HY10: 2089–2115.

Leendertse, J. J. (1967). Aspects of a computational model for long-period water-wave propagation. *Rand Corporation, Santa Monica, CA Memorandum* RM-5294-PR.

Leendertse, J. J. (1970). A water quality simulation model for well mixed estuaries and coastal seas: vol. I, Principles of computation. *Rand Corporation, Santa Monica, CA Memorandum* RM-6230-RC.

Lick, W. (1976). Numerical models of lake currents. *US Environmental Protection Agency Environmental Research Laboratories, Duluth, MN Report* EPA 600/3-76-020.

Liggett, J. A. (1969). Unsteady circulation in shallow homogeneous lakes. *Proceedings of American Society of Civil Engineers, Journal of Hydraulics Division* No. HY4: 1273–1288.

Liggett, J. A., and C. Hadjitheodorou (1969). Circulation in shallow homogeneous lakes. *Proceedings of American Society of Civil Engineers, Journal of Hydraulics Division* 95(HY2): 609–620.

Liggett, J. A., and K. K. Lee (1971). Properties of circulation in stratified lakes. *Proceedings of American Society of Civil Engineers, Journal of Hydraulics Division* No. HY1.

Lilly, D. K. (1961). A proposed staggered-grid system for numerical integration of dynamic equations. *Monthly Weather Review* 89(3): 59–65.

Long, R. R. (1970). A theory of turbulence in stratified fluids. *Journal of Fluid Mechanics* 42(2): 349–365.

Marchuk, G. I. (1974). *Numerical Solution of the Problems of Atmospheric and Ocean Dynamics* (Leningrad: Gidrometizdat) (in Russian).

Marchuk, G. I. (1975). *Methods of Numerical Mathematics* (New York, NY: Springer).

Marchuk, G. I., and V. B. Zalesny (1974). *Numerical Model of the Global Ocean Circulation* (Siberian Branch, USSR Academy of Sciences) (in Russian).

Markofsky, M., and D. R. F. Harleman (1971). A predictive model for thermal stratification and water quality in reservoirs. *MIT Department of Civil Engineering, R. M. Parsons Laboratory Technical Report* 134.

Markofsky, M., and D. R. F. Harleman (1973). Prediction of water quality in stratified reservoirs. *Proceedings of American Society of Civil Engineers, Journal of Hydraulics Division* 99(HY5): 729–745.

McNider, R. T., and J. J. O'Brien (1973). A multilayer transient model of coastal upwelling. *Journal of Physical Oceanography* 3(3): 258–273.

Monin, A. S., and A. M. Yaglom (1965). *Statistical Fluid Mechanics: Mechanics for Turbulence* vol. 1 (Moscow: Nauka; English translation, 1971, by Massachusetts Institute of Technology).

Mortimer, C. H. (1974). Lake hydrodynamics. *Mitteilungen der Internationalen Vereinigung Limnologie* 20:124–197.

Munk, W. H., and E. R. Anderson (1948). Notes on the theory of the thermocline. *Journal of Marine Research* 1.

Najarian, T. O., and D. R. F. Harleman (1975). A real-time model of nitrogen-cycle dynamics in an estuarine system. *MIT Department of Civil Engineering, R. M. Parsons Laboratory Technical Report* 204; also *Progress in Water Technology* (1977) 8:323–345.

Norton, W. R., I. P. King, and G. T. Orlob (1973). A finite-element model for Lower Granite Reservoir. *Water Resources Engineers Inc., Walnut Creek, CA Technical Report.*

Okubo, A. (1962). A review of theoretical models for turbulent diffusion in the sea. *Journal of Ocean Society of Japan*, 20th anniversary volume, 1:286–320.

Okubo, A. (1968). A new set of oceanic diffusion diagrams. *Chesapeake Bay Institute, Johns Hopkins University Technical Report* 38.

O'Melia, C. R. (1974). Phosphorus cycling in lakes. *North Carolina Water Resources Research Institute, Raleigh Report* 97.

Orlob, G. T., and L. G. Selna (1967). Progress report on development of a mathematical model for prediction of temperatures in deep reservoirs—Phase 3: Castle Lake investigation. *Water Resources Engineers, Inc., Walnut Creek, CA Technical Report.*

Ozmidov, R. V. (1968). *Horizontal Turbulence and Turbulent Exchange in the Ocean* (Moscow: Nauka).

Parker, F. L., B. A. Benedict, and C. Tsai (1975). Evaluation of mathematical models for temperature prediction in deep reservoirs. *US Environmental Protection Agency, National Environmental Research Center, Corvallis, OR Report* EPA 660/3-75-038.

Patterson, D. J., E. Epstein, and J. McEvoy (1975). Water pollution investigations: Lower Green Bay and Lower Fox River. *Report to US Environmental Protection Agency*, No. 68-01-1572.

Phillips, J. O. (1969). *The Dynamics of the Upper Ocean* (London: Cambridge University Press).

Plate, E. J., and P. Wengefeld (1979). Exchange processes at the water surface. *Hydrodynamics of Lakes*, eds. W. H. Graf and C. H. Mortimer (Amsterdam: Elsevier), pp. 277–301.

Platzman, G. W. (1963). The dynamic prediction of wind tides on Lake Erie. *Meteorological Monographs* 4(26):1–43.

Raney, D. C., D. L. Durham, and H. L. Butler (1979). Application of a three-dimensional storm surge model. *Hydrodynamics of Lakes*, eds. W. H. Graf and C. H. Mortimer (Amsterdam: Elsevier), pp. 77–86.

Rao, D. B., and T. S. Murty (1970). Calculation of the steady-state wind-driven circulations in Lake Ontario. *Archiv für Meteorologie, Geophysik, und Bioklima, Serie* A 19(2):195–210.

Reid, R. O., and B. R. Bodine (1968). Numerical model for storm surges in Galveston Bay. *Proceedings of American Society of Civil Engineers, Journal of Waterways and Harbors Division* 94(WW1):33–37.

Ryan, P. J., and D. R. F. Harleman (1971). Prediction of the annual cycle of temperature changes in a stratified lake or reservoir: Mathematical model and user's manual. *MIT Department of Civil Engineering, R. M. Parsons Laboratory Technical Report* 137.

Ryan, P. J., and D. R. F. Harleman (1973). An analytical and experimental study of transient cooling pond behavior. *MIT Department of Civil Engineering, R. M. Parsons Laboratory Technical Report* 161.

Schlichting, H. (1968). *Boundary Layer Theory* (New York, NY: McGraw-Hill).

Simons, T. J. (1971). Development of numerical models of Lake Ontario. *Proceedings of 14th Conference on Great Lakes Research* (Ann Arbor, MI: International Association of Great Lakes Research), pp. 654–669.

Simons, T. J. (1972). Development of numerical models of Lake Ontario, Part 2. *Proceedings of 15th Conference on Great Lakes Research* (Ann Arbor, MI: International Association of Great Lakes Research), pp. 655–672.

Simons, T. J. (1973). Development of three-dimensional numerical models of the Great Lakes. *Canada Center for Inland Waters, Burlington, Ontario Science Series* No. 12.

Simons, T. J. (1974). Verification of numerical models of Lake Ontario: Part I. Circulation in spring and early summer. *Journal of Physical Oceanography* 4(4):507–523.

Simons, T. J. (1975). Verification of numerical models of Lake Ontario: Part II. Stratified circulations and temperature changes. *Journal of Physical Oceanography* 5(1):98–110.

Simons, T. J. (1976). Analysis and simulation of spatial variations of physical and biochemical processes in Lake Ontario. *Journal of Great Lakes Research* 2(2):215–233.

Simons, T. J. (1980). Circulation models of lakes and inland seas. *Canadian Bulletin of Fish and Aquatic Sciences* 203, 146 pp.

Simons, T. J., L. Funkquist, and J. Svensson (1977). Application of a numerical model to Lake Vänern. *Swedish Meteorology and Oceanography Institute Report* RH09.

Simons, T. J., and D. C. L. Lam (1978). Nutrient exchange between volume elements in large lakes. *Modeling the Water Quality of the Hydrological Cycle. Proceedings of Baden Symposium, September. International Association of Hydrological Sciences Publication* 125, pp. 185–191.

Snodgrass, W. J., and C. R. O'Melia (1975). Predictive model for phosphorus in lakes. *Environmental Science and Technology* 9(10):937–944.

Stolzenbach, K. D., E. E. Adams, and D. R. F. Harleman (1972). A user's manual for three-dimensional heated surface discharge computations. *MIT Department of Civil Engineering, R. M. Parsons Laboratory Technical Report* 156.

Sündermann, J. (1979). Numerical modelling of circulation in lakes. *Hydrodynamics of Lakes*, eds. W. H. Graf and C. H. Mortimer (Amsterdam: Elsevier), pp. 1–29.

Tatom, F. B., and W. R. Waldrop (1978). Curvilinear unsteady two-dimensional depth-averaged hydrodynamic model. *Proceedings of Conference on Environmental Effects of Hydraulic Engineering Works, Knoxville, TN* (International Association of Hydraulics Research), pp. 329–339.

Taylor, C., and J. Davis (1972). Tidal and long-wave propagation—A finite-element approach. *Department of Civil Engineering, University College of Swansea Report* C/R/189.

Thomann, R. V., D. M. Di Toro, R. P. Winfield, and D. J. O'Connor (1975). Mathematical modeling of phytoplankton in Lake Ontario. *US Environmental Protection Agency, National Environmental Research Center, Corvallis, OR Report* EPA 660/3-75-005.

Tjomsland, T. (1979). Application of a mathematical hydrodynamic model to Lake Mjosa. *Hydrodynamics of Lakes*, eds. W. H. Graf and C. H. Mortimer (Amsterdam: Elsevier), pp. 41–49.

Tsvetova, H. A. (1979). Numerical modelling of the dynamics of Lake Baikal. *Professional Paper* PP-79-4 (Laxenburg, Austria: International Institute for Applied Systems Analysis).

Vasiliev, O. F. (1978a). Three-dimensional numerical model for hydrothermal analysis of cooling ponds. *Thermal Effluent Disposal from Power Generation*, ed. Z. P. Zarić (New York, NY: McGraw-Hill), pp. 115–132.

Vasiliev, O. F. (1978b). Two-dimensional numerical models for hydrothermal analysis of water bodies. *Thermal Effluent Disposal from Power Generation*, ed. Z. P. Zarić (New York, NY: McGraw-Hill), pp. 133–149.

Vasiliev, O. F., and V. I. Kvon (1977). Numerical modeling of flows and heat exchange in cooling ponds. *Proceedings of 16th Congress of International Association for Hydraulic Research, Baden-Baden* vol. 2, pp. 1–8.

Vasiliev, O. F., V. I. Kvon, and R. T. Chernishova (1973). Mathematical modelling of the thermal pollution of a water body. *Proceedings of the 15th Congress of the International Association for Hydraulic Research, Istanbul* vol. 2, pp. 129–136.

Vasiliev, O. F., V. I. Kvon, and R. T. Chernishova (1974). Thermally stratified flow in a water body of oblong form. *Gidroteknicheskoi Stroitel'stvo* No. 4 (in Russian).

Vreugdenhil, C. B. (1973). Computational methods for the vertical distribution of flow in shallow water. *Delft Hydraulics Laboratory Report* W152.

Wang, J. D., and J. J. Connor (1975). Mathematical modeling of near-coastal circulation. *MIT Department of Civil Engineering, R. M. Parsons Laboratory Technical Report* 200.

Watanabe, M., and D. R. F. Harleman (1977). A mathematical model of a stably stratified cooling pond. *Proceedings of 16th Congress of International Association for Hydraulic Research, Baden-Baden* vol. 2, pp. 67–74.

Watanabe, M., D. R. F. Harleman, and J. J. Connor (1975). Finite-element model for transient two-layer cooling pond behavior. *MIT Department of Civil Engineering, R. M. Parsons Laboratory Technical Report* 202.

Welander, O. (1957). Wind action on a shallow sea: Some generalizations of Ekman's theory. *Tellus* 9(1):45–52.

Wittmiss, J. (1979). Application of a transient mathematical model to Lake Kösen. *Hydrodynamics of Lakes*, eds. W. H. Graf and C. H. Mortimer (Amsterdam: Elsevier), pp. 31–40.

Wu, J. (1969). Wind stress and surface roughness at air–sea interface. *Journal of Geophysical Research* 74(2):444–445.

Wunderlich, W. (1973). Water temperature prediction model for deep reservoirs. *Tennessee Valley Authority, Water Resources Management Methods Staff, Technical Report* A-2.

Yeh, G. T., A. P. Verma, and F. H. Lai (1973). Unsteady temperature prediction for cooling ponds. *Water Resources Research* 9(6):1555–1563.

Yudelson, J. M. (1967). A survey of ocean diffusion studies and data. *W. M. Keck Laboratory for Hydraulics and Water Resources, California Institute of Technology Technical Memorandum* 67-2.

CHAPTER 8: NOTATION

A_D	dissipation parameter
b	channel width
c	concentration of a conservative substance
c_f	resistance coefficient
C	interfacial shear stress coefficient; Chezy coefficient
C_w	wind stress coefficient
D_v	dilution ratio
e	kinetic energy (per unit mass)
E_v	vertical Ekman number
f_0 or λ	friction coefficient (Darcy–Weisbach)
h	height from datum to the bottom
H	depth of water
K	heat diffusivity
K_H	heat transfer coefficient
l	mixing length
L	length scale of turbulence; pond length
n	Manning coefficient
P	pond number
Q_H	heat source
S_e	external sources
u_*	shear velocity
U, V	average velocities
w_k	net entrainment or mixing velocity between layers k and $k + 1$

W	wind velocity; average pond width
β	coefficient of thermal expansion
ϵ	eddy viscosity
η	elevation of free surface
κ	von Karman constant
ϕ_n	net heat flux through surface.

9 Ecological Modeling of Lakes

S. E. Jørgensen

9.1. BACKGROUND

9.1.1. Water Quality and Ecological Models of Lakes

Chapter 4 surveyed the modeling of ecological processes in streams and lakes. As elucidated in Figure 4.1, the various submodels must be coupled to make a total ecosystem model. This has already been demonstrated for models of streams in Chapter 6.

Nutrient enrichment of lakes has been of major concern in pollution studies, so it is not surprising that most of the working ecological models of lakes focus on the problem of eutrophication. The eutrophication models concentrate on the kinetics of water quality and biota and are therefore, in structure, ecological models.

The eutrophication of rivers is generally not a problem, because of the short retention time of water. The major problem in rivers is the concentration of dissolved oxygen, which depends on the algae producing oxygen, on the nitrogen cycle consuming oxygen, and on other chemical–biological constituents. The direct relationship between eutrophication and algal concentration implies, however, that ecological models have found a much wider use in lake management than in river management.

The purpose of this chapter is not to give all specifications of mathematical models and programs, but to describe some models that are characteristic of a reasonably wide range of those actually in use. Many are mentioned in the review (sections 9.2.2 and 9.2.3) and the total systems of equations for five models are shown in the appendix.

9.1.2. Brief Review of Development

Among the earliest models of lake systems were those that addressed the problem of nutrient balance. These nutrient budget models simulated the entire lake as a mixed tank reactor (Vollenweider, 1965, 1969). Later, this approach was seized upon by other investigators dealing more explicitly with nutrient and algal budgets in lakes.

Mathematical models of lakes have evolved along two different lines. First, there was the extension of the zero-dimensional model to one-, two-, and three-dimensional models (Chapters 7 and 8); then there were the modeling activities that focused primarily on a better and more detailed description of the chemical–biological processes (Park *et al.*, 1974; Chen and Orlob, 1975; Jørgensen, 1976; and others).

The line represented by CLEAN, CLEANER, and MS CLEANER demonstrates that the development has been toward inclusion of more biological constituents, linked by increasingly detailed description and interaction of biological phenomena. The latest development includes adaptation and biological feedback mechanisms (Straškraba, 1976, 1979; Jørgensen *et al.*, 1978; Park *et al.*, 1979). The tendency is to build more realism into the models by using more constraints.

9.2. ECOLOGICAL LAKE MODELS

9.2.1. Coupling of Submodels

Over the past decade there has been a clear evolution toward more and more complex models, not only by adding more constituents but also by making more complex links (adaptation, feedback, self-organization) between the submodels. Ten models have been chosen to demonstrate different levels of complexity and of coupling. The presentation of the models is taken from Park *et al.* (1979) with the authors' permission. Figure 9.1 shows the symbols used in the model descriptions.

Figure 9.2 shows the model of Chen and Orlob (1975). Light and nutrients are represented by saturation kinetics and temperature by the exponential van't Hoff equation (eqn. 1, Table 4.12). Nutrient limitation is set by the minimum of the nitrogen, phosphorus, and silicon limitations, while the contributions of light, nutrient, and temperature limitations are multiplicative. Light extinction is a function of depth and of detritus and phytoplankton concentrations. A Michaelis–Menten expression describes grazing. Di Toro *et al.* (1977) use Steele's (1962) equation for light and temperature limitations; grazing and phytoplankton mortality vary linearly with temperature (Figure 9.3).

In the model of Kremer and Nixon (1978) (Figure 9.4) the limitations are multiplied, but the functions are the same as in the model of Di Toro *et al.*

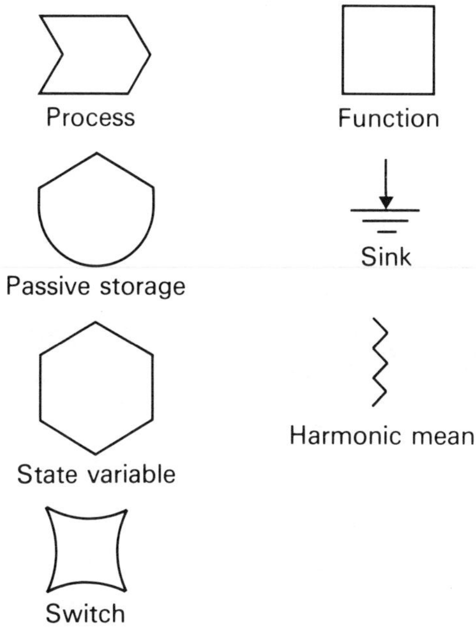

FIGURE 9.1 Symbols used in Figures 9.2–9.11.

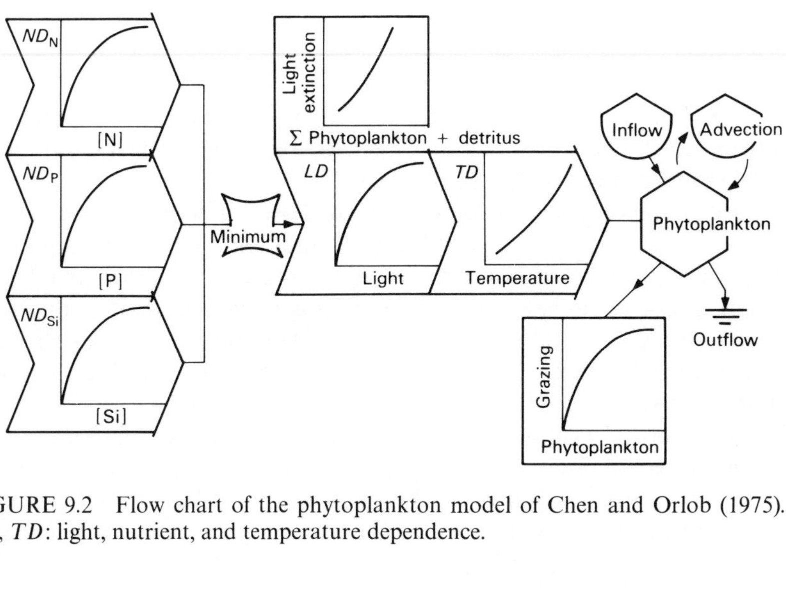

FIGURE 9.2 Flow chart of the phytoplankton model of Chen and Orlob (1975). *LD*, *ND*, *TD*: light, nutrient, and temperature dependence.

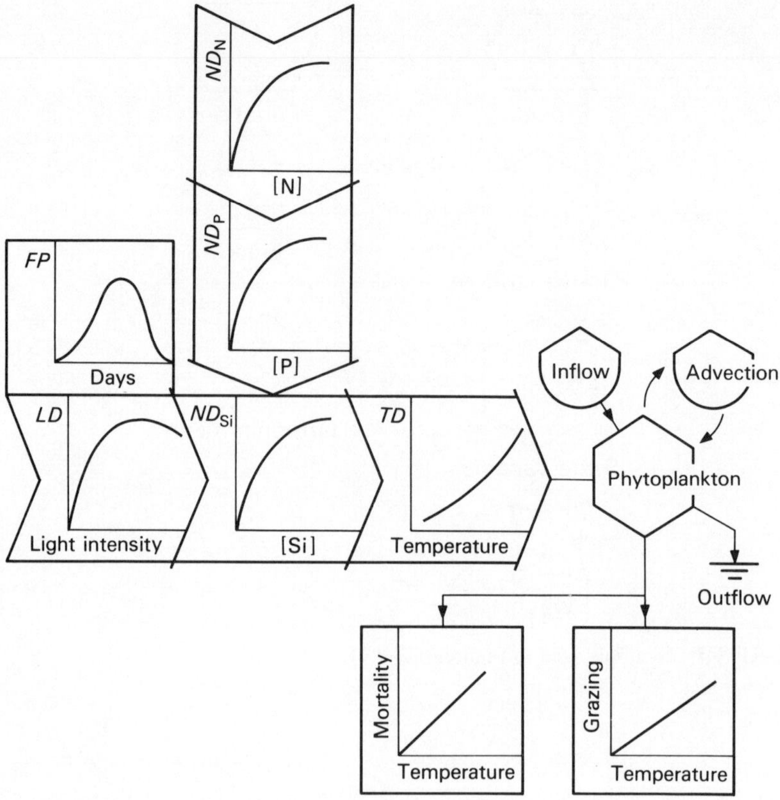

FIGURE 9.3 Flow chart of the phytoplankton model of Di Toro *et al.* (1977). *FP*: factor dependent on time.

Diurnal variations of light and a three-day running average for adaptation to light are included. The model was developed for a marine system, but is directly applicable to lake systems as well.

The model proposed by Straškraba (1976) is presented in Figure 9.5. Light limitation is based on the work of Steel (1972); the limiting factors are multiplicative; the Stokes law is applied to sinking, and grazing accounts for prey saturation. Light extinction considers particulate organic matter, dissolved organic matter, suspended sediment, and phytoplankton biomass. A noticeable feature is the temperature response of phytoplankton growth with an adaptive optimum temperature (see also Table 4.12).

The model of Lehman *et al.* (1975) (Figure 9.6) provides for intracellular storage of nutrients (discussed in section 4.2.2). Michaelis–Menten expressions for limitations are used multiplicatively. Mortality is dependent on the number of days when the growth is suboptimal (section 4.2.4) and sinking is constant.

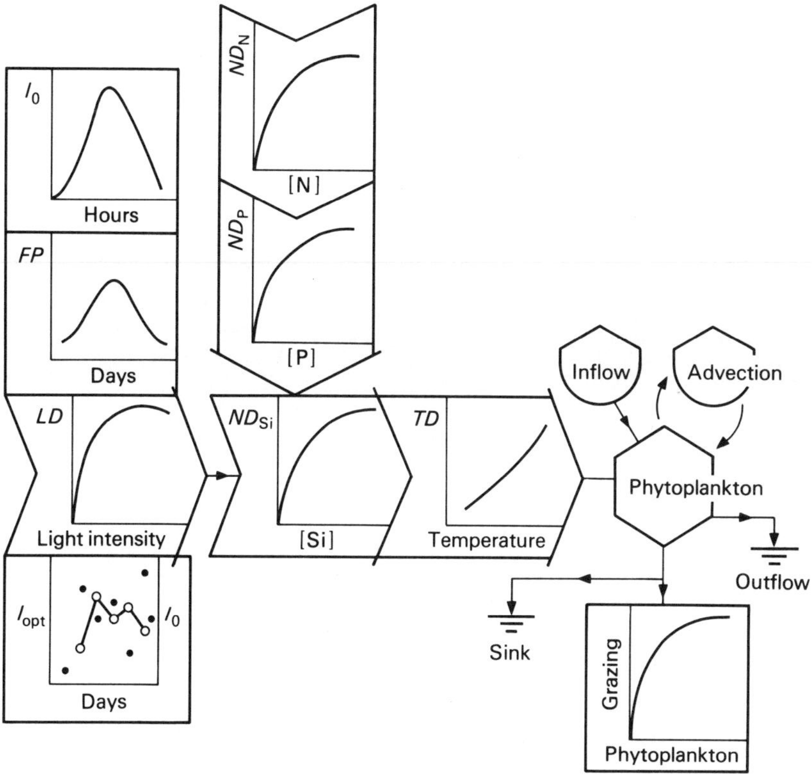

FIGURE 9.4 Flow chart of the phytoplankton model of Kremer and Nixon (1978).

Nyholm's (1978) model (Figure 9.7) takes intracellular storage of nutrient into account (see Table 4.1 and discussion in section 4.2.2). The nutrient limitation terms are combined as resistors in series, the so-called harmonic mean, suggested by Bloomfield *et al.* (1974). Light inhibition and adaptation are incorporated. The van't Hoff equation is used for temperature response and Michaelis–Menten kinetics for grazing.

The model of Scavia *et al.* (1976) uses Michaelis–Menten expressions for light and nutrient limitations and the minimum is used as the limiting factor (Figure 9.8), which is then multiplied with the nonlinear temperature factor of Bloomfield *et al.* (1974). Equations 2 in Table 4.3 are used for grazing and are modified by temperature and prey preference. Equation 5 of Table 4.6 is used to express sinking. Respiration is a linear function of temperature and mortality a function of temperature and nutrient limitation.

The model of Jørgensen (1976) and Jørgensen *et al.* (1978) in Figure 9.9 considers independent nutrient cycles and uses eqns. 14 of Table 4.1 to describe

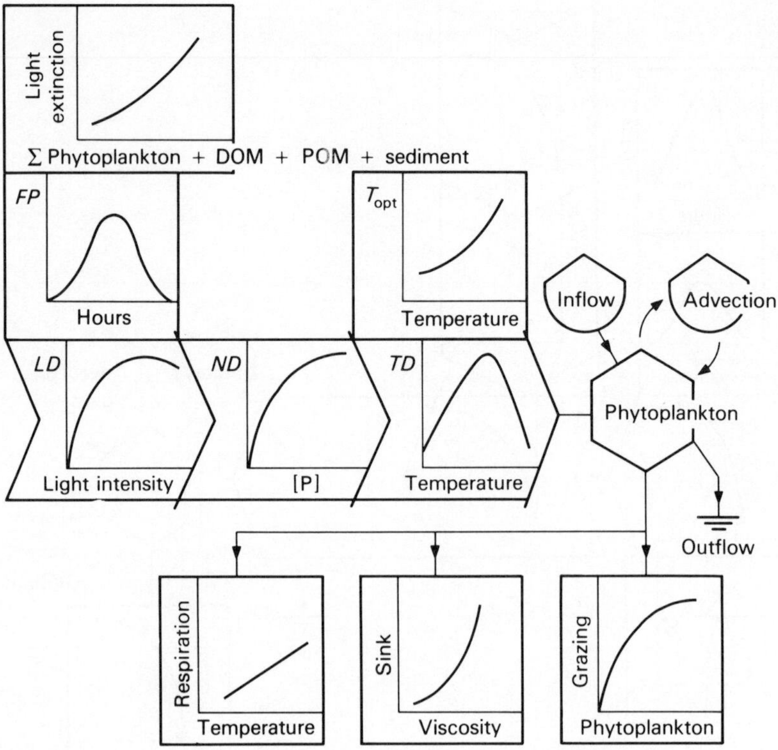

FIGURE 9.5 Flow chart of the phytoplankton model of Straškraba (1976).

algal growth. Light inhibition and adaptation are included, and photosynthesis determines the intracellular concentration of carbon (photosynthesis being the uptake of CO_2). In addition, respiration (release of CO_2) influences this concentration, but is controlled by the phytoplankton concentration, temperature, and the concentration of intracellular carbon. Light extinction accounts for phytoplankton, detritus, and zooplankton concentrations. Algal growth is controlled by the intracellular concentrations of nutrients (N, P, and C) and the temperature. A multiplicative expression is used. Outflow, inflow, and grazing are included; the latter is represented by a Michaelis–Menten expression and a temperature function. Sinking is controlled by viscosity and a rather complex sediment model for description of nutrient release is applied.

Figure 9.10 presents the model CLEANER (Youngberg, 1977). It uses Steele's (1962) equation for light limitation; the harmonic mean of Bloomfield *et al.* (1974) to combine nutrient limitations; the nonlinear response of growth to temperature; and grazing with saturation kinetic formulation, nonlinear temperature dependence, and prey preference. Respiration, mortality, and

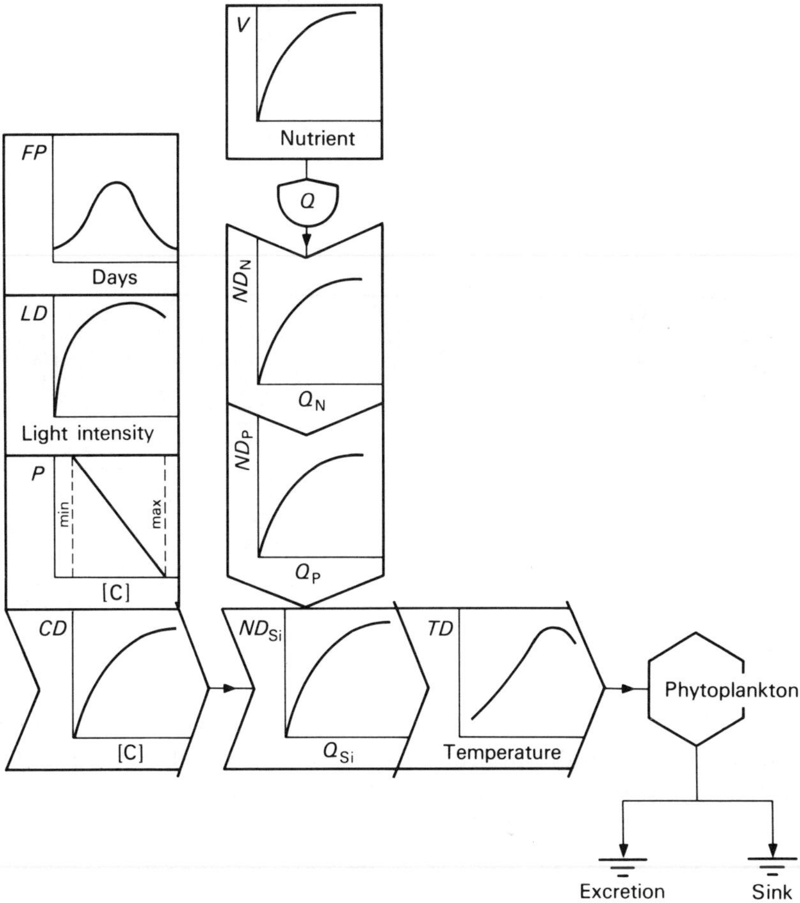

FIGURE 9.6 Flow chart of the phytoplankton model of Lehman *et al.* (1975).

sinking are temperature-dependent and nutrient limitation influences the mortality (Scavia *et al.*, 1976).

MS CLEANER (Figure 9.11) is the most complex of the models presented in this section. It is claimed that because the parameters have physical and biological meaning (Desormeau, 1978; Park *et al.*, 1979a) extensive calibration is not needed in order to apply the model to a new site. Intracellular storage of nutrients is used and the uptake processes for phosphorus and nitrogen are represented by a Michaelis–Menten kinetic (eqns. 14 of Table 4.1). Uptake is, furthermore, a function of light and temperature. Single-nutrient limitation is determined by comparing the N:P ratio with thresholds (Droop, 1974; Rhee, 1978).

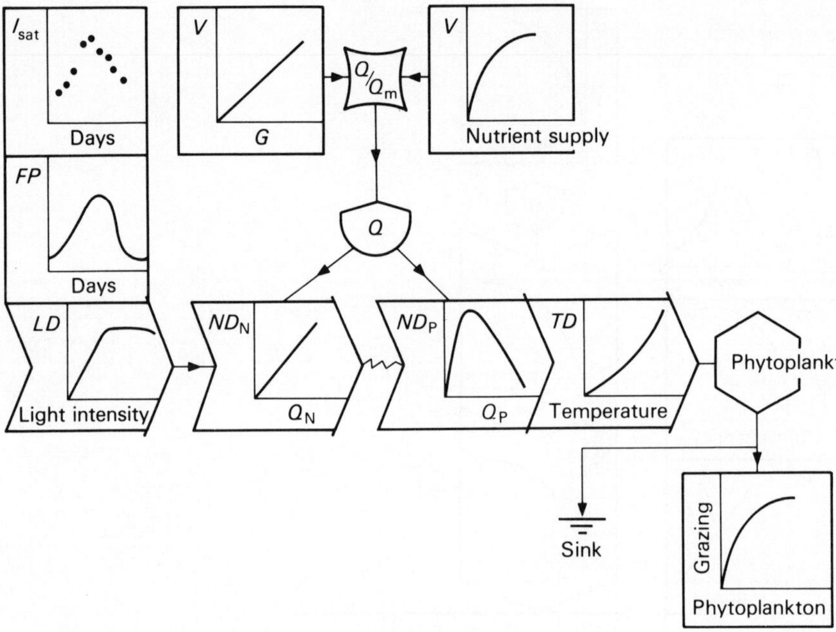

FIGURE 9.7 Flow chart of the phytoplankton model of Nyholm (1978).

Light limitation provides for adaptation to varying light intensities (Groden, 1977) and photosynthesis is a linear function of chlorophyll concentration, which, in phytoplankton, is an exponential function of light intensity. Light inhibition is represented by Steele's (1974) equation, and Smith's (1936) equation is used below the inhibitory level. Light extinction is a function of dissolved organic matter, particulate organic matter, and phytoplankton concentrations.

Temperature limitation does not control photosynthesis directly, but is used to determine the light saturation and nutrient uptake rate. The equation developed by Groden (1977) is applied, and temperature adaptation is described by using the empirical function suggested by Straškraba (1976). Grazing differentiates between saturation kinetic feeding, exhibited by copepods and fish, and feeding at a uniform rate, exhibited by cladocerans that filter at a constant rate. Grazing is also a function of temperature and prey preference (Park *et al.*, 1974).

9.2.2. Nutrient Budget Models

Vollenweider (1969, 1975) has suggested a simple nutrient budget model that considers input, output, and net loss to the sediments (plus a correction factor

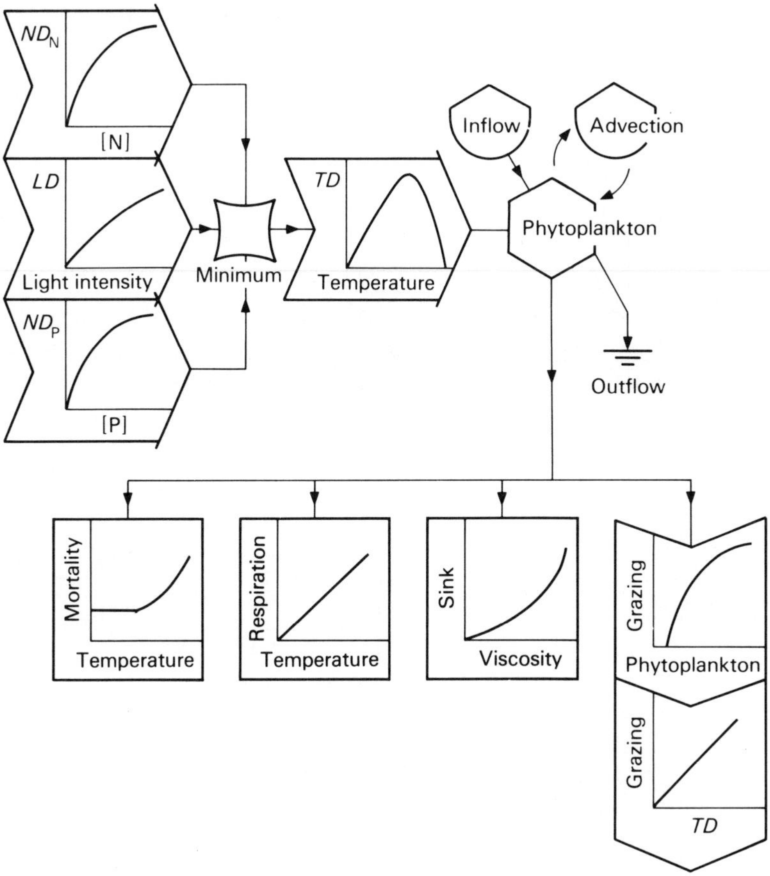

FIGURE 9.8 Flow chart of the phytoplankton model of Scavia *et al.* (1976).

for stratified lakes):

$$\frac{d[P]}{dt} = l_P - (\rho_w - \sigma_P)[P], \tag{9.1}$$

where

[P] is the phosphorus concentration in the lake $[M\,L^{-3}]$,
l_P is the phosphorus supply rate $[M\,L^{-3}\,T^{-1}]$,
ρ_w is the hydraulic washout coefficient $[T^{-1}]$,
σ_P is the sedimentation rate constant $[T^{-1}]$,
t is time.

This equation is valid if the lake is well mixed, its volume is constant, the outflow is at a concentration equivalent to that of the lake and is equal to the inflow,

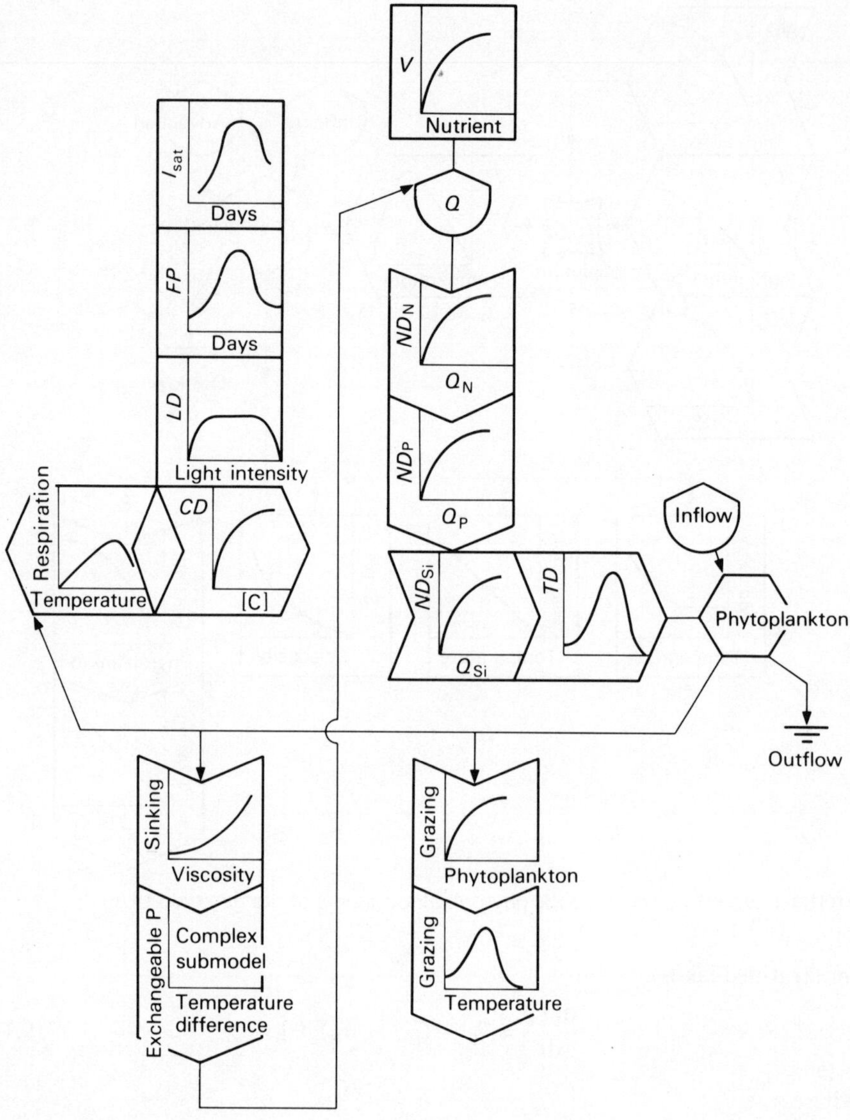

FIGURE 9.9 Flow chart of the phytoplankton model of Jørgensen (1976) and Jørgensen *et al.* (1978).

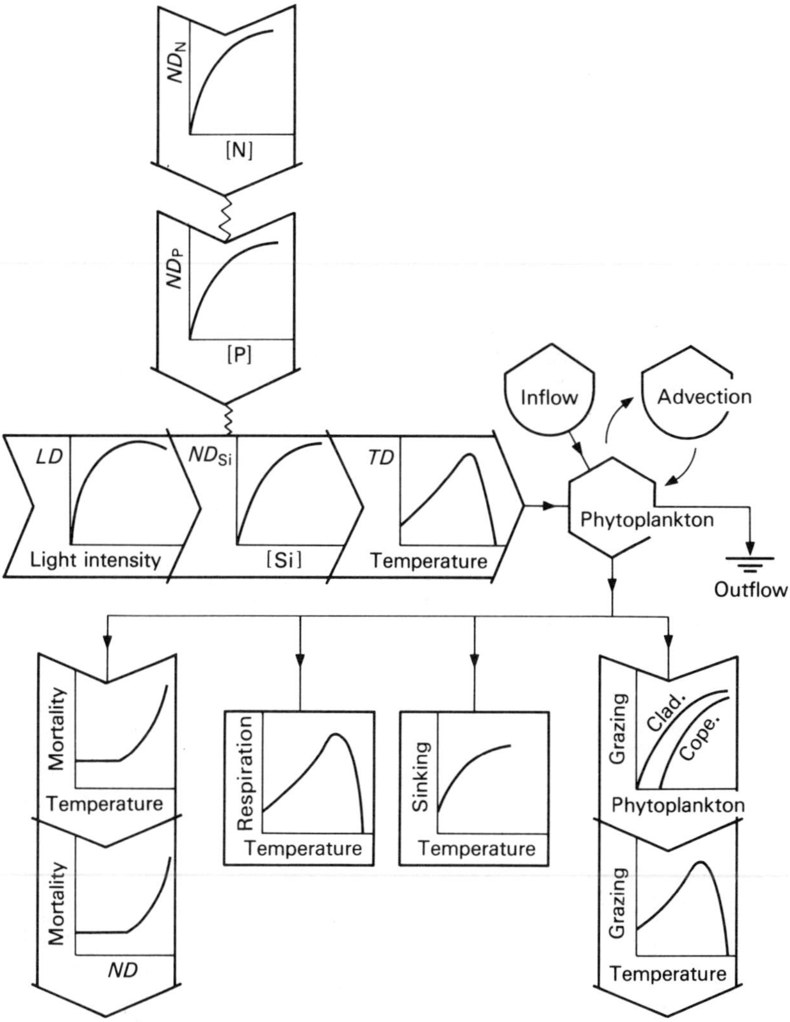

FIGURE 9.10 Flow chart of the CLEANER model of Youngberg (1977). Clad.: cladocerans; Cope.: copepods.

and there is no net supply of phosphorus from sediment. Provided that l_P, ρ_w, and σ_P are time-independent this equation can be solved:

$$[P(t)] = [P_0]\exp[-(\rho_w - \sigma_P)t] + \frac{l_P}{\rho_w + \sigma_P}\{1 - \exp[-(\rho_w - \sigma_P)t]. \quad (9.2)$$

Figure 9.12 shows the Vollenweider plot, which is based upon the above-mentioned considerations.

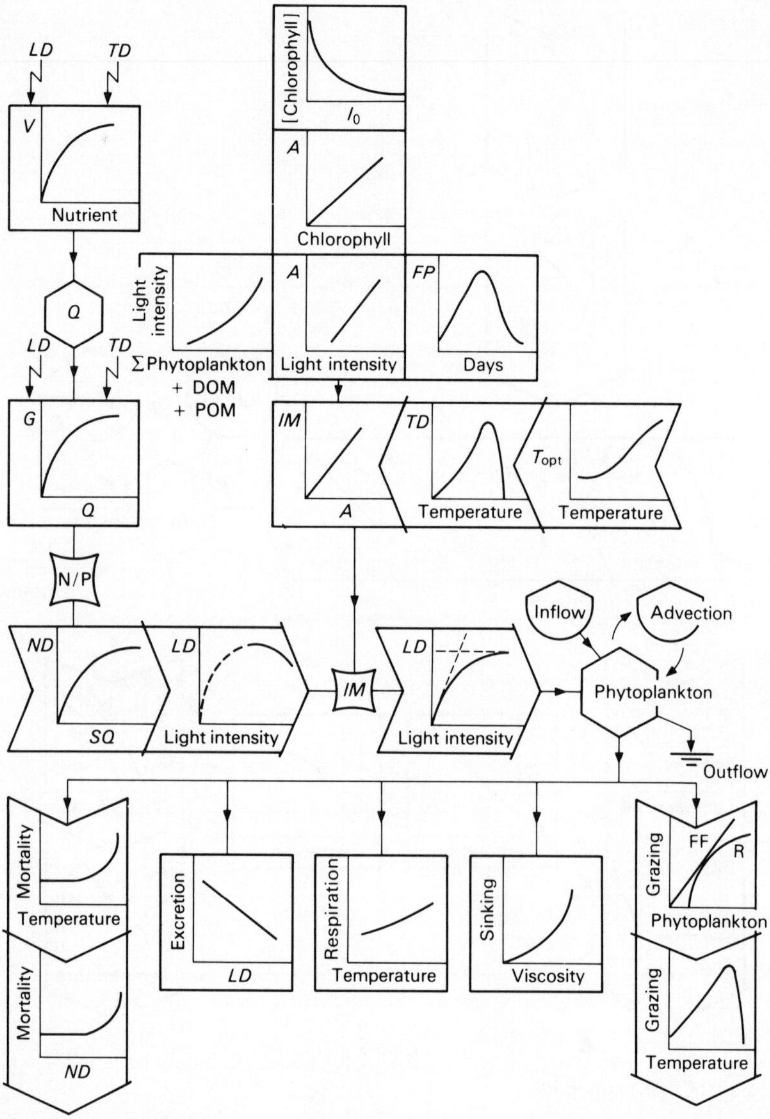

FIGURE 9.11 Flow chart of the phytoplankton submodel of MS CLEANER. FF, filter feeders; R, raptors; *LD*, *ND*, *TD*: light, nutrient, and temperature dependence; *IM*, intermediate quantity; *G*, intermediate factor; *Q*, flow; *SQ*, rate of nutrient uptake; *FP*, factor dependent on time (days).

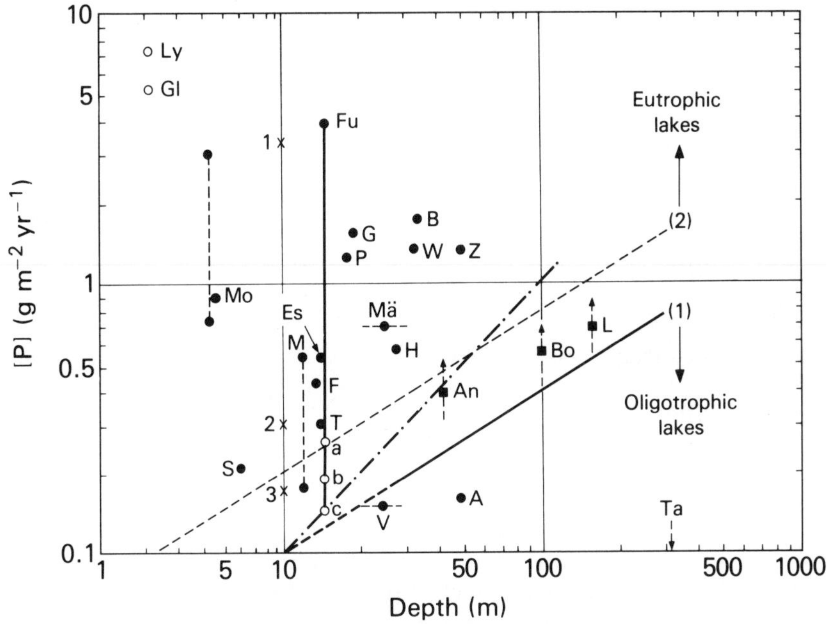

FIGURE 9.12 Vollenweider plot of phosphorus concentrations: a, b, and c correspond to removal of 90, 95, and 99%, respectively, of the phosphorus input to Furesø. Results for 1972 are shown for: Gl, Glumsø Sø (Jørgensen *et al.*, 1973); Ly, Lyngby Sø (Smith, 1973); Es, Esrum Sø; G, Greifensee; P, Pfäffiker See; B, Baldegger See; W, Lake Washington; Z, Zürich See; Mä, Mälaren; H, Hallwiler See; Bo, Bodensee; A, Ägeerisee; V, Vänern; Fu, Furesø; F, Furesø (1954); M, Lake Mendota; T, Türlersee; S, Lake Sebasticook; Mo, Lake Moses; An, Lac d'Annecy; L, Lac Léman; Ta, Lake Tahoe. (1) and (2) are the limits of the two lake conditions.

Imboden (1974) suggested a two-compartment model for phosphorus content. The model considers a stratified lake and includes input, output, and exchange between hypolimnion and epilimnion, as well as sediment exchange. Four coupled differential equations for dissolved and particulate phosphorus are applied. The model has been improved (Imboden and Gächter 1978; Imboden, 1979) by describing nutrient and biomass concentrations as continuous functions of time and depth and by replacing the first-order kinetic by Michaelis–Menten kinetics. O'Melia (1974) and Snodgrass and O'Melia (1975) developed a similar model, but did not include release of phosphorus from the sediment; however, depth-dependent rates of turbulent diffusion were considered.

Larsen *et al.* (1974) found that the Vollenweider and Snodgrass–O'Melia models underestimated the actual amount of epilimnetic phosphorus, when applied to Lake Shagawa in Minnesota. They then applied a slightly more complex model consisting of a three-compartment epilimnetic model, which includes algae as a sink for soluble reactive phosphorus and conversion of particulate phosphorus to the soluble form. The basic equations for this model are:

$$\frac{\mathrm{dPA}}{\mathrm{d}t} = \mathrm{MYMAX}(T) \cdot \mathrm{LIGHT} \cdot \frac{\mathrm{PS}}{\mathrm{KP} + \mathrm{PS}} \cdot \mathrm{PA} - (\mathrm{CONR1} + \mathrm{SETTL1} + \rho_{\mathrm{w}}) \cdot \mathrm{PA} \tag{9.3}$$

$$\frac{\mathrm{dPS}}{\mathrm{d}t} = \frac{\mathrm{PSIN}}{\mathrm{VE}} - \mathrm{MYMAX}(T) \cdot \mathrm{LIGHT} \cdot \frac{\mathrm{PS}}{\mathrm{KP} + \mathrm{PS}} \cdot \mathrm{PA} + \mathrm{CONR2} \cdot \mathrm{PP} + \rho_{\mathrm{w}} \mathrm{PS}, \tag{9.4}$$

$$\frac{\mathrm{dPP}}{\mathrm{d}t} = \frac{\mathrm{PPIN}}{\mathrm{VE}} + \mathrm{CONR1} \cdot \mathrm{PA} - (\mathrm{CONR2} - \mathrm{SETTL2} + \rho_{\mathrm{w}}) \cdot \mathrm{PP}, \tag{9.5}$$

where

PA is the concentration of algal phosphorus $[\mathrm{M\,L^{-3}}]$,
LIGHT is the fractional reduction of MYMAX(T) in the epilimnion due to the availability of light,
MYMAX(T) is the maximum specific growth rate of phytoplankton as a function of temperature $[\mathrm{T^{-1}}]$,
KP is the half-saturation constant for phosphorus $[\mathrm{M\,L^{-3}}]$,
CONR1 is the rate constant for conversion of algal phosphorus to particulate phosphorus $[\mathrm{T^{-1}}]$,
CONR2 is the rate constant for conversion of particulate phosphorus to soluble phosphorus $[\mathrm{T^{-1}}]$,
PP is the concentration of particulate (non-algal) phosphorus $[\mathrm{M\,L^{-3}}]$,
PPIN is the rate of supply of particulate phosphorus to the epilimnion $[\mathrm{M\,T^{-1}}]$,
PS is the concentration of soluble phosphorus $[\mathrm{M\,L^{-3}}]$,
PSIN is the rate of supply of soluble phosphorus to the epilimnion $[\mathrm{M\,T^{-1}}]$,
SETTL1 is the rate constant for settling of algal phosphorus (corresponding to a settling velocity of 0.02 m $\mathrm{day^{-1}}$),
SETTL2 is the rate constant for settling of non-algal particulate phosphorus (corresponding to a settling velocity of 0.04 m $\mathrm{day^{-1}}$),
T is the temperature,
VE is the volume of the epilimnion $[\mathrm{L^3}]$.

Lorenzen *et al.* (1976) developed a model consisting of two differential equations only, one for soluble phosphorus and one for exchangeable phosphorus in the sediment:

$$\frac{\mathrm{dPS}}{\mathrm{d}t} = \frac{\mathrm{PSIN}}{\mathrm{VL}} + \frac{K_2 \cdot \mathrm{AREA} \cdot \mathrm{PSED}}{\mathrm{VL}} - \frac{K_1 \cdot \mathrm{AREA} \cdot \mathrm{PS}}{\mathrm{VL}} - \frac{Q}{\mathrm{VL}} \cdot \mathrm{PS}$$

$$\frac{\mathrm{dPSED}}{\mathrm{d}t} = \frac{K_1 \cdot \mathrm{AREA} \cdot \mathrm{PS}}{\mathrm{VS}} - \frac{K_2 \cdot \mathrm{AREA} \cdot \mathrm{PSED}}{\mathrm{VS}} - \frac{K_1 K_3 \cdot \mathrm{AREA} \cdot \mathrm{PS}}{\mathrm{VS}}, \tag{9.6}$$

where

AREA is the lake surface area $[\mathrm{L}^2]$,
K_1 is the rate of transfer of phosphorus to the sediment $[\mathrm{L\,T}^{-1}]$,
K_2 is the rate of transfer of phosphorus from the sediment $[\mathrm{L\,T}^{-1}]$,
K_3 is the fraction of total phosphorus input to the sediment that is not available for exchange,
PSED is the total concentration of exchangeable phosphorus in the sediment $[\mathrm{M\,L}^{-3}]$,
VS is the sediment volume $[\mathrm{L}^3]$,
Q is the outflow $[\mathrm{L}^3\,\mathrm{T}^{-1}]$,
VL is the lake volume $[\mathrm{L}^3]$.

The purpose of the model is to predict long-term changes in lakes that have undergone significant changes in loading rates. PSIN is therefore understood as the annual loading of PS, Q the annual outflow, and K_1 and K_2 are measured in m yr^{-1}.

The equations can be solved analytically and the steady state solution of PS is

$$\mathrm{PS}_\infty = \frac{\mathrm{PSIN}}{Q + K_1 K_3 \cdot \mathrm{AREA}}. \tag{9.7}$$

A characteristic feature of this model is that, in spite of its simplicity, it considers the sediment-accumulated phosphorus and that only a fraction of the total phosphorus input to the sediment is available for the exchange process. More complex models do not include this important property of the phosphorus in the sediment, although it is of great importance for the long-term changes in lakes because a substantial part of the phosphorus in a lake system is accumulated in the sediment.

The parameters of this model are estimated by the following procedure. When reasonably good data on loading rates and average aqueous and sediment concentrations are known:

(1) K_3 is estimated.
(2) Since

$$K_1 K_3 = \frac{\mathrm{PSIN} - \mathrm{PS}_\infty \cdot Q}{\mathrm{PS} \cdot \mathrm{AREA}}, \tag{9.8}$$

K_1 can be calculated.

(3) K_2 is calculated from

$$K_2 = \frac{PS_\infty}{PSED_\infty} \cdot K_1(1 - K_3), \tag{9.9}$$

as the ratio of steady state aqueous to sediment phosphorus concentrations is given by (analytical solution)

$$\frac{PS_\infty}{PSED_\infty} = \frac{K_2}{K_1} \frac{1}{1 - K_3}. \tag{9.10}$$

The model was used on Lake Washington by applying data from 1941–1950 to calculate a consistent set of model constants, based upon $K_3 = 0.6$. K_3 can be found on the basis of sediment analysis (a more detailed examination of mud–water exchange of phosphorus was reported by Kamp-Nielsen, 1975). The observations during 1955–1970, which showed that the phosphorus loading increased up to 1964 and decreased thereafter, were well predicted by the model. However, $K_3 = 0.5$ gave a better result (Figure 9.13).

Lappalainen (1975) improved Vollenweider's approach by considering the state of a lake as a function of lake volume, discharge, and phosphorus input. In this model a regression equation that relates the net sedimentation of phosphorus and the oxygen concentration of the hypolimnion is determined. The model includes a relationship between the sedimentation of phosphorus and

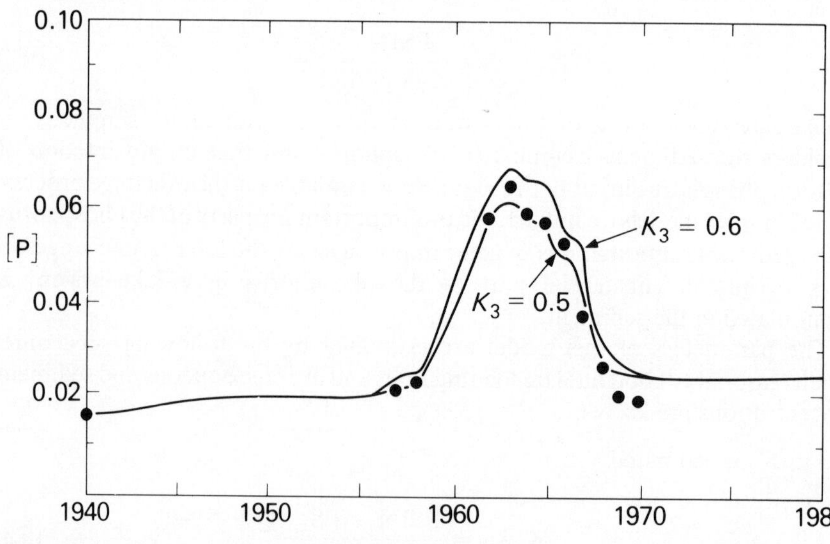

FIGURE 9.13 Calculated and observed (●) annual average total phosphorus concentrations (mg l^{-1}) in Lake Washington.

volume, discharge, and phosphorus input. This sedimentation submodel and the regression expression were used to construct a model for the prognosis of the oxygen concentration of the hypolimnion, which is used to determine the boundary phosphorus input, comparable with loads given earlier in the literature.

Jolankai and Szöllösi-Nágy (1978) constructed a simple model of algal dissolved reactive phosphorus to describe eutrophication in the Bay of Keszthely, Lake Balaton. The model is based on the specific condition that in Lake Balaton the main direct source of phosphorus is the sediment. The model considers three phosphorus state variables: phosphorus in water, in sediment, and in algae. Temperature and light limitations are not considered.

9.2.3. Multiconstituent Models

The models mentioned above are often described as nutrient budget models. They all deal with eutrophication in gross terms, i.e. with only one or two layers, long time scales, and only one governing nutrient. These models give no attention to short-term hydrodynamics, ecosystem dynamics, or spatial resolution of quality changes within the impoundment. The result is that such models have only limited utility in predicting impoundment response to pollution control strategies, particularly those that may be either of local impact or short-term.

More comprehensive models of eutrophication have been proposed and several have been implemented. The models are based upon the principle of mass conservation. The general mass conservation equation, exemplified below for a water element j in a one-dimensional model, is applied:

$$\frac{\partial(V_j C_j)}{\partial t} = \underbrace{-Q_j C_j + Q_{j+1} C_{j+1}}_{\text{advection}} + \underbrace{\left(E\frac{\partial C}{\partial Z}\right)_j - \left(E\frac{\partial C}{\partial Z}\right)_{j+1}}_{\text{diffusion}} + \underbrace{C_j \frac{\partial V_j}{\partial t}}_{\text{volume change}}$$

$$\underbrace{\pm V_j \frac{dC}{dt}}_{\text{sources and sinks}}. \qquad (9.11)$$

Thomann *et al.* (1975) developed three models with different levels of detail. Model 1 conceptualizes the impoundment as a three-layer system (epilimnion, hypolimnion, and benthos) and considers phytoplankton and zooplankton dynamics. Model 2 provides additional vertical resolution with seven layers and considers temperature, chemistry, and sediment interactions. Model 3 is far more comprehensive as it provides additional spatial resolution with up to seven layers, 67 segments, and 15 state variables. A future model with as many as 5000 compartments (segments · state variables) is envisioned.

Validation of model 1 indicates generally good agreement with observations for such state variables as chlorophyll *a*, zooplankton, carbon, four forms of nitrogen, and phosphorus. Unfortunately model 2 was only studied analytically and was not validated against observations, and model 3 was only tested preliminarily against selected observations from the field but was not validated.

The Chen–Orlob model (1975) was extended by Chen *et al.* (1975) for simulation of Lake Ontario. It included mass balance equations for 15 different classes of biotic and abiotic substances. Four groups of phytoplankton, four groups of zooplankton, and four groups of fish in three life stages were represented. Furthermore, the model divided the lake into segments and layers, for which separate mass balance equations could be written. Details of the model are given in the appendix. Implementation of the model required an independent determination of the flow field in three dimensions and estimates of empirical diffusion coefficients. The hydraulic inputs were derived from a hydrodynamic model attributed to Bennett (1974). Details of the hydrodynamic model are not available, except that it is of orthogonal form with 715 surface elements and seven layers. It is wind-driven and considers effects of temperature on circulation. Successful test simulations for the months of 1972 were carried out with 41 surface elements and seven vertical layers of varying thickness, a total of 209 elements. Results indicated reasonable performance, but no validation was carried out.

The 1979 version of this lake model required 9 s of Univac 1108 CPU time per day of simulation ($\Delta t = 1$ day), which is little compared with the hydrodynamic program, which required 120 s of CPU time per day of simulation. The Chen–Orlob model, in a version known as EPAECO, has also been applied to a Finnish lake (Kinnunen *et al.*, 1978).

Patten *et al.* (1975) developed the Lake Texoma Cove model, which attempts to consider every conceivable aspect of the cove ecology. Its most distinguishing feature is that it is an ecologist's model—an ecosystem description drawn by a relatively large number of scientists under conditions of prolonged, intensive interaction. The model is further distinguished by its linear construction (although a few minor nonlinearities occur). The forcing functions are time, temperature, solar radiation, rain, wind, current, and water level.

The primary producer submodel contains nine state variables: small, medium, and large phytoplankton, blue-green algae, floating algal mats, attached algae, aufwuchs, and submergent and emergent vascular plants. The zooplankton submodel includes those animals retained by a 64 μm plankton net, excluding fish larvae and dipteran larvae. Pupal zooplankton was divided into two groups: small (less than 0.75 mm) and large (0.75 mm or greater). The benthic invertebrate submodel consists of suspension feeders, deposit feeders, predators, and scavengers. The vertebrate submodel includes larval fish, fingerlings, filter-feeding fish, bottom-feeding fish, minnow-like fish, carnivorous fish, turtles, and herbivorous and carnivorous harvesters. In the decomposer submodel nine compartments were selected: dissolved organic matter, particulate organic matter, plant carcasses, animal carcasses, organic sediment, nitrate and nitrite nitrogen, phosphorus, carbon dioxide, and dissolved oxygen. Together, the submodels treat 33 state variables. A connectivity matrix for the state variables is shown in Figure 9.14. As indicated by the authors, the model cannot be considered as having been validated. No time-series data were at hand for comparison with model results. The model represents one extreme by giving a comprehensive mathematical description of the ecology.

Di Toro *et al.* (1975) developed a phytoplankton–zooplankton–nutrient interaction model for western Lake Erie. They considered seven segments and seven state variables: inorganic phosphorus, phytoplankton, nitrate nitrogen, ammonia nitrogen, organic nitrogen, zooplankton, and organic phosphorus. Thus the model has 49 compartments. The advective flows are established primarily by the inflow of the Detroit River and its passage through the Western Basin into the Central Basin (Figure 8.8, p. 311). In addition to advective flow, it is necessary to assess the magnitude of mixing of the flow between adjacent segments. This is accomplished by use of conservative tracer, the chloride ion.

The Monod kinetic is used to express the rates of growth of phytoplankton and zooplankton, while the transformations from organic nitrogen to ammonia and from organic phosphorus to orthophosphate are described by first-order temperature-dependent kinetics. The constants and their temperature-dependences are found initially by the use of available laboratory experimental data to set the probable ranges of constants (Di Toro *et al.*, 1971). Then detailed comparisons are made between observed and computed data in order to fine-tune the value (calibration).

The model was finally validated using a composite set of data from the year 1930. This year was chosen for two reasons: (i) there is a significant data base and (ii) the limnological conditions that existed in 1930 were far removed from those in 1970. Reasonably good agreement with observed data resulted for all systems. However, there was some discrepancy between the observed and the predicted zooplankton concentrations.

Bierman (1976) constructed a 14-compartment model of inner Saginaw Bay (Figure 9.15). The idea was to apply a spatially simplified version of the model, which would be incorporated at a later stage in a spatially segmented version. A unique feature of the model is that all growth is considered to be a two-step process, involving separate nutrient uptake and all synthesis mechanisms. This is in accordance with an increasingly large body of experimental evidence indicating that the mechanisms of nutrient uptake and all growth are actually quite distinct (Dugdale, 1967; Fuhs, 1969; Eppley and Thomas, 1969; Fuhs *et al.*, 1971; Caperon and Meyer, 1972a, b; Droop, 1973; Halmann and Stiller, 1974; Nyholm, 1975).

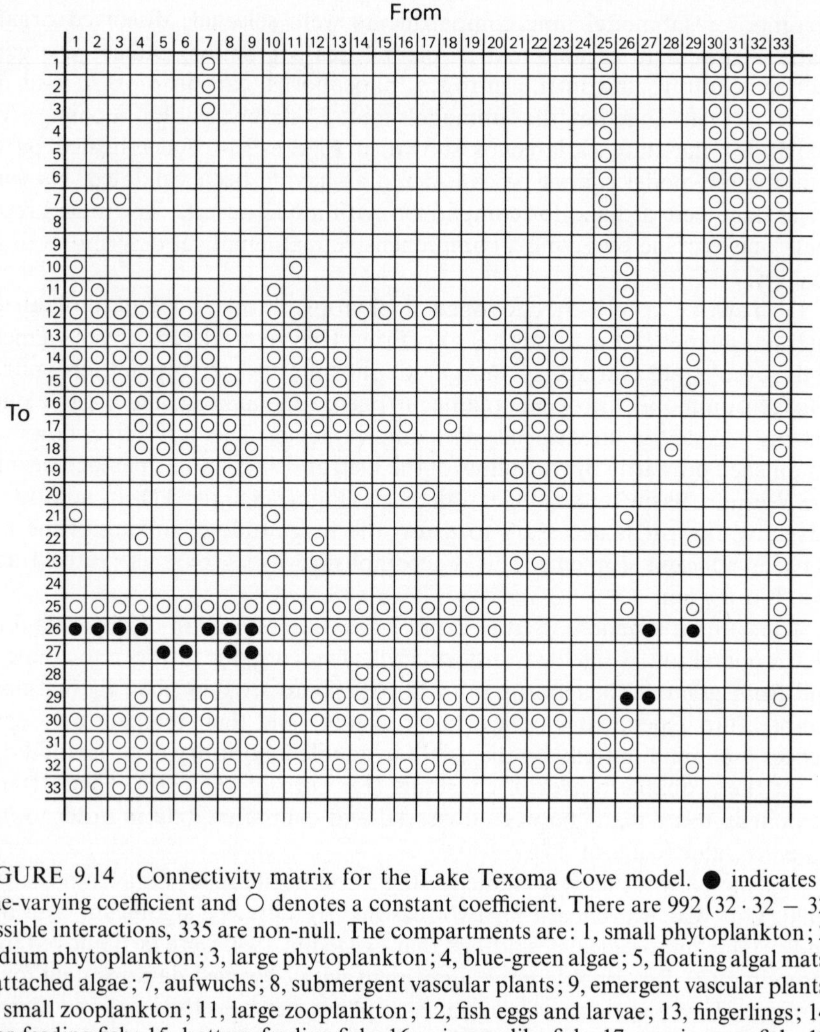

FIGURE 9.14 Connectivity matrix for the Lake Texoma Cove model. ● indicates a time-varying coefficient and ○ denotes a constant coefficient. There are 992 $(32 \cdot 32 - 32)$ possible interactions, 335 are non-null. The compartments are: 1, small phytoplankton; 2, medium phytoplankton; 3, large phytoplankton; 4, blue-green algae; 5, floating algal mats; 6, attached algae; 7, aufwuchs; 8, submergent vascular plants; 9, emergent vascular plants; 10, small zooplankton; 11, large zooplankton; 12, fish eggs and larvae; 13, fingerlings; 14, filter-feeding fish; 15, bottom-feeding fish; 16, minnow-like fish; 17, carnivorous fish; 18, turtles; 19, herbivorous vertebrate harvesters; 20, carnivorous vertebrate harvesters; 21, suspension-feeding invertebrates; 22, deposit-feeding invertebrates; 23, invertebrate predators; 24, invertebrate scavengers; 25, dissolved organic matter; 26, particulate organic matter; 27, plant carcasses; 28, animal carcasses; 29, organic sediment; 30, nitrate and nitrite nitrogen; 31, phosphorus; 32, carbon dioxide; 33, dissolved oxygen.

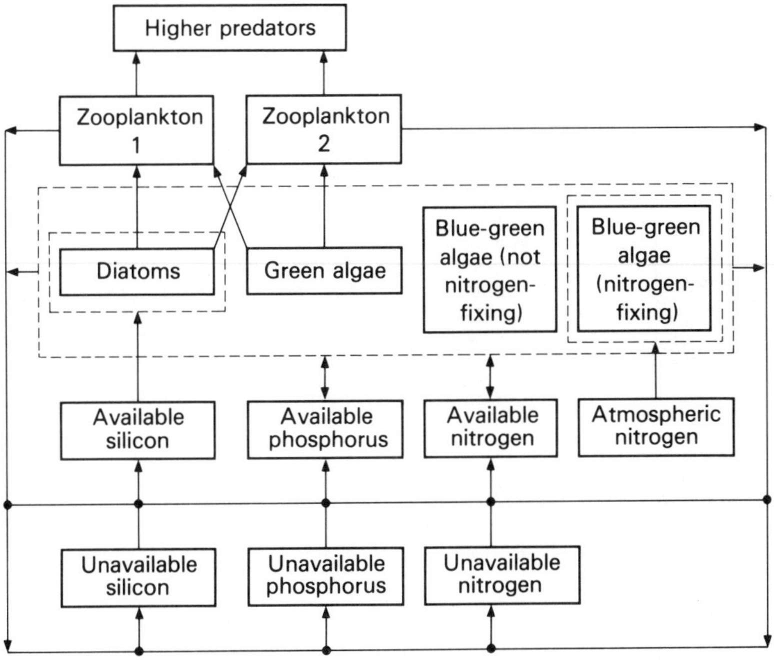

FIGURE 9.15 Principal constituents of the Saginaw Bay eutrophication model.

The specific phosphorus uptake in the Bierman model is described as a function of the balance between extracellular and intracellular dissolved phosphorus:

$$\text{uptake rate} = \text{maximum uptake rate} \cdot f(T)f(I) \cdot \left(\frac{1}{1 + \text{PKI} \cdot \text{PCA}} - \frac{1}{1 + \text{PKI} \cdot \text{PCM}}\right) \tag{9.12}$$

$$\text{PCA} = \text{PCAMIN} \cdot \exp(\text{PC}/\text{PO} - 1), \tag{9.13}$$

where

PKI is the uptake constant [L^3 mol^{-1}],
PC is the total amount of phosphorus per cell (PCA plus internal storage) (mol cell^{-1}),
PO is the minimum stoichiometric level of total phosphorus per cell (mol cell^{-1}),
PCAMIN is the minimum value of PCA [mol L^{-3}],
PCA is the internal dissolved phosphorus concentration [mol L^{-3}],
PCM is the external dissolved phosphorus concentration [mol L^{-3}].

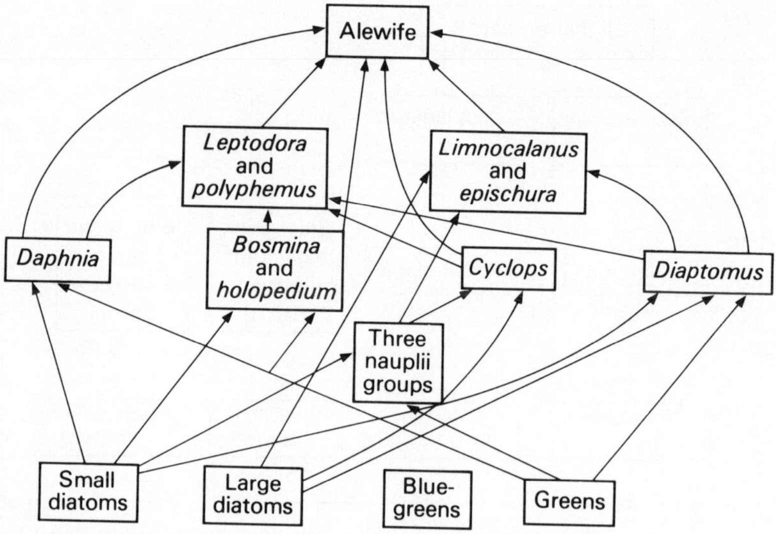

FIGURE 9.16 Principal constituents of the Lake Michigan food web model.

An identical approach is used for the nitrogen uptake. The state variable equations of this model are given in the appendix.

One of the interesting features of the work of Bierman is that the model has been used to examine which factors are significant for blue-green algae blooms. It is often important to control the blooms, since these algae cause formation of a blue-green foam on the water surface and produce toxic substances. Freedom from grazing and lower sinking rates seem to be the most important of the competitive advantages, although blue-green algae have a slower maximum growth rate than all other classes of algae. Relatively small differences in sinking rates between blue-green algae and other classes showed a significant effect. Furthermore, it seems essential that blue-green algae have a higher phosphorus uptake rate (approximately twice as high) and reach half-saturation at significantly lower extracellular phosphorus concentrations than other phytoplankton classes.

Canale *et al.* (1976) used a complex food web model for Lake Michigan; the food web is shown in Figure 9.16. The model distinguishes between raptors, selective filterers, and nonselective filterers and uses different equations for the feeding rates of the three classes of zooplankton. For raptors:

$$\text{feeding rate} = \text{MYZMAX} \cdot f(T) \frac{\sum_{\text{prey}} C_1}{\sum_{\text{prey}} C_1 + \text{KZ}}, \tag{9.14}$$

where

$\sum_{\text{prey}} C_1$ is the sum of concentrations of all states that can serve as food for raptor state Z,

MYZMAX is the maximum growth rate of zooplankton [T^{-1}],
KZ is the half-saturation concentration for grazing.

Selective filterers obtain food by grazing, but have developed an ability to vary their filtering rate. A proposed equation for this mechanism is:

$$\text{feeding rate} = \text{MYZMAX} \cdot f(T) \cdot \frac{\text{KMFM} \cdot \text{PHYT} + \text{KFLM}}{\text{PHYT} + \text{KFLM}} \cdot \text{PHYT}, \quad (9.15)$$

where

KFLM is the food level [$M\,L^{-3}$],
KMFM is the minimum filtering rate multiplier,
PHYT is the concentration of phytoplankton [$M\,L^{-3}$].

A formulation of the feeding mechanism of nonselective filterers is:

$$\text{feeding rate} = \text{MYZMAX} \cdot f(T) \sum_{\text{prey}} C_1. \quad (9.16)$$

The nonselective filterers cannot lower their filtering rate when the plankton content of filtered water increases. Therefore, they operate below maximum possible efficiency. A simple but unconfirmed formulation suggested by Canale *et al.* is:

$$\text{assimilation efficiency} = \text{A11N} \frac{\text{A24}}{\sum_{\text{prey}} C_1 + \text{A24}} \quad (9.17)$$

$$= \text{CON1} \frac{\text{CON2}}{\text{PHYT} + \text{CON2}}, \quad (9.18)$$

where A11, A24, CON1, and CON2 are constants.

For the other classes, however, a constant assimilation efficiency that is related to the growth rate is used:

$$\text{growth rate} = \text{assimilation efficiency} \cdot \text{feeding rate}. \quad (9.19)$$

The rate at which state z is consuming prey in state k is, in general, given by

$$\text{predation of } k \text{ by } z = \text{preference of } z \text{ for } k \cdot \text{feeding rate} \cdot C_z. \quad (9.20)$$

The formulation of the preference factor involves a simple assumption: the preference of z for k is proportional to the product of the electivity of z for k and the concentration of k. The electivity α_k^z is defined as the fraction of the diet of z that would be composed of food species k, if all food species were present in equal concentrations:

$$\text{preference of } z \text{ for } k = \frac{\alpha_k^z C_k}{\sum_i \alpha_i^z C_1}. \quad (9.21)$$

The web of electivities incorporated into the lower food web model is shown in Figure 9.17; Table 9.1 presents the state of the system.

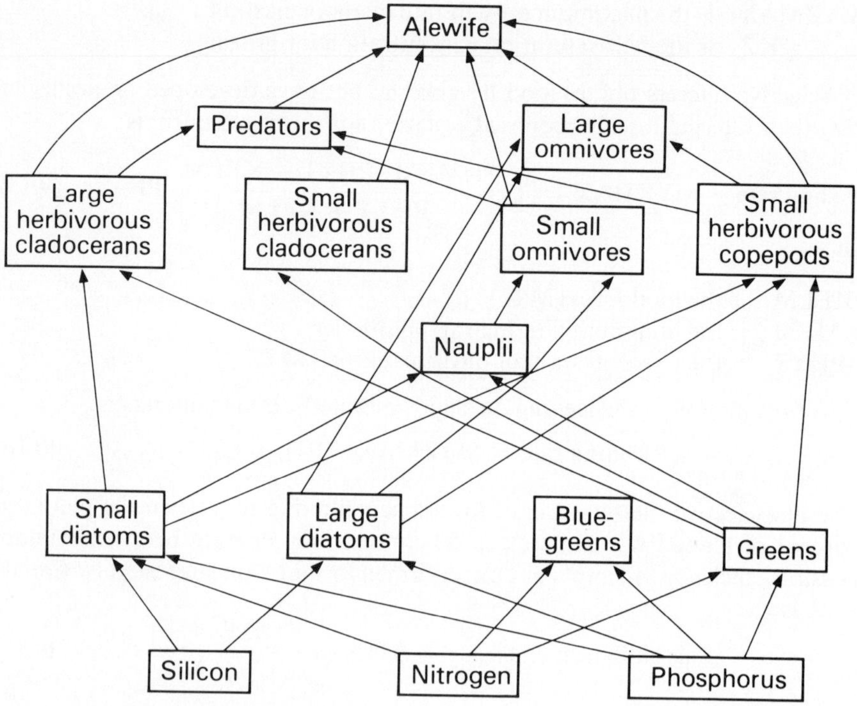

FIGURE 9.17 Web of electivities.

The model applied by Canale *et al.* has 25 water quality variables and two vertical layers. The model is calibrated, but cannot be considered validated. Simulations were conducted with the calibrated model to examine its behavior when phosphorus concentration and alewife predation became higher than normal.

A multispecies model has also been developed by Ikeda *et al.* (1979). The approach of parameter estimation in this model is interesting, as the main objective of the model is confined to determining the parameter values that generate a similar behavior of plankton dynamics rather than finding a best fit to the observed data. This probably indicates the state of the art of multispecies models: the model can be calibrated to obtain a similar behavior, but most data sets are not sufficient in number and accuracy to allow a very good fit between model and observed values.

Jørgensen (1976) applied a two-stage algal growth process in his lake model and compared the results of this model with a model based upon general Monod kinetics. It was concluded that the Jørgensen model gave a better description of the system response to seasonal variation in nutrient loading than the generally applied Monod kinetic. Another characteristic feature of

TABLE 9.1 The State of the System.

Description	Classification	
Leptodora and *Polyphemus*	Raptors	Zooplankton
Cyclops		
Cyclops nauplii	Selective filterers	
Diaptomus nauplii		
Limnocalanus and *Epischura* nauplii		
Diaptomus		
Limnocalanus and *Epischura*		
Daphnia	Nonselective filterers	
Bosmina and *Holopedium*		
Small diatoms		Phytoplankton
Large diatoms		
Blue-greens		
Greens		
Detrital nitrogen	Nitrogen	Epilimnion nutrients
Dissolved organic nitrogen		
Ammonia		
Nitrate		
Detrital phosphorus	Phosphorus	
Dissolved organic phosphorus		
Dissolved inorganic phosphorus		
Detrital silicon	Silicon	
Dissolved silicon		
Total inaccessible nitrogen		Nutrient mass held in hypolimnion and sediments
Total inaccessible phosphorus		
Total inaccessible silicon		

the model is that it contains a more complex submodel for the exchange of nutrient between sediment and water than was applied in the other models mentioned. The model distinguishes between exchangeable and nonexchangeable phosphorus and describes the release of phosphorus as a two-step process:

$$\begin{array}{c}\text{exchangeable}\\ \text{phosphorus}\end{array} \rightarrow \begin{array}{c}\text{interstitial}\\ \text{phosphorus}\end{array} \rightarrow \begin{array}{c}\text{orthophosphorus}\\ \text{in water phase}\end{array},$$

with different rate constants for aerobic and anaerobic conditions. Simulation of reduced phosphorus loading demonstrated that the inclusion of this more complex sediment submodel gives a more pronounced response than the use of a simple first-order kinetic for phosphorus release, which is used in most models, if the sediment is considered at all. This is not surprising as a substantial part of the sediment phosphorus cannot be exchanged. The submodel used for nutrient release is in accordance with comprehensive studies of these processes carried out on sediment cores from several lakes (Jørgensen *et al*., 1975; Jacobsen and Jørgensen, 1975; Kamp-Nielsen, 1975).

The Jørgensen model also considers denitrification. Its application to a Swedish lake study (Södra Bergundasjön) required the modeling of nitrogen fixation (Jørgensen *et al.*, 1981, 1982). The model, which is a one-layer model with 17 state variables in the last version (Jørgensen, 1976), has been validated in studies of two Danish lakes, Glumsø Sø and Lyngby Sø, with good results. An examination of several versions of the model is published in the same paper by Jørgensen *et al.* (1978). It concludes that distinguishing between two classes of zooplankton and the introduction of a biologically active layer on top of the sediment are not important for the validation of the model, and that the use of daily measured irradiance data rather than average values is essential. The two models compared are shown in the appendix.

Richey (1977) developed a phosphorus model of Castle Lake, California. The model includes the following state variables: dissolved inorganic P, dissolved organic P, phytoplankton P, bacterial P, zooplankton P, polyphosphate P, detrital P, ferric P, and sediment P. The changes of processes in pools over depth and time are considered. The distribution, formation of complexes, and precipitation of phosphate species as functions of pH and iron concentration are modeled by the following system of equations:

$$[PO_4^{3-}] = P_T(1 + [H^+]10^{-12.3} + [H^+]10^{-19.5} + [H^+]^3 10^{-21.7})$$
$$[HPO_4^{2-}] = P_T(1 + [H^+]10^{-12.3} + [H^+]10^{-7.2} + [H^+]^2 10^{-9.4})$$
$$[H_2PO_4^-] = P_T(1 + [H^+]10^{-2.2} + [H^+]10^{-7.2} + [H^+]^2 10^{-19.5})$$
$$[H_3PO_4] = P_T(1 + [H^+]10^{-2.2} + [H^+]^2 10^{-9.4} + [H^+]^3 10^{-21.7})$$
$$[HPO_4S] = 10^{-11}[H^+][Fe^{3+}] \qquad (HPO_4 \text{ removed})$$
$$[HPO_4P] = [HPO_4S][HPO_4^{2-}] \qquad (HPO_4 \text{ percentage removed})$$
$$[DIP] = 3.1 \cdot 10^7(P_T - [HPO_4^{2-}])(1 - [HPO_4S][HPO_4^{2-}])$$
$$(\text{phosphate left in solution}).$$

The model CLEANER (Scavia and Park, 1975; Scavia *et al.*, 1976) was originally applied to Lake George, but has been developed for application to a variety of impoundments, including Sarasota Lake (Florida), Loch Leven (Tayside, Scotland), several Scandinavian lakes, Lake Balaton (Hungary), and lakes in Italy and Czechoslovakia. A new version of the model, MS CLEANER, includes 40 state variables and provisions for horizontal and vertical transport. It can be applied to littoral, pelagic, and profundal zones. A typical solution for a stratified reservoir with 34 state variables requires about 63 s of Univac 1110 CPU time for simulation of an annual cycle. The model seems to be applicable to many case studies that require calibration of only a very limited number of parameters.

Nyholm (1978) developed an ecological model for shallow lakes, using seven state variables: phytoplankton, phosphorus in phytoplankton, nitrogen in phytoplankton, available phosphorus, available nitrogen, detrital phosphorus, and detrital nitrogen. The model does not include sediment nitrogen

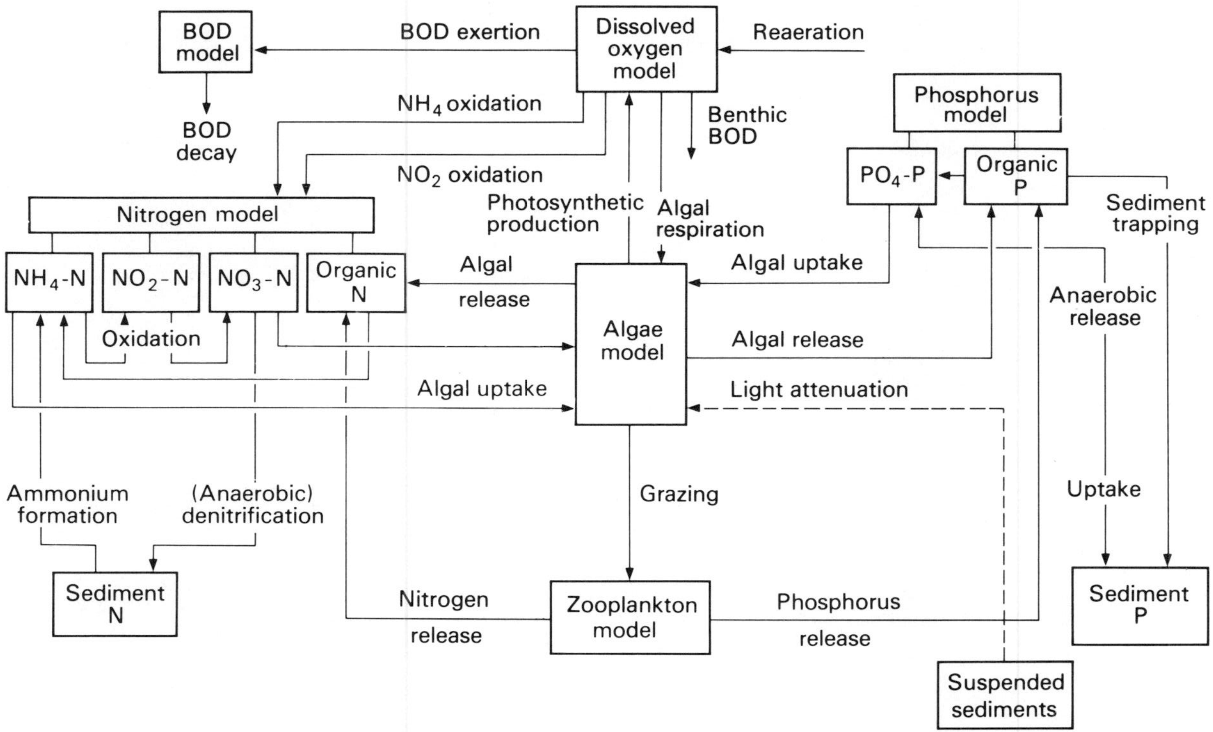

FIGURE 9.18 Processes considered in the limnological model of Baca and Arnett (1976).

TABLE 9.2 Lake Model Characteristics.

Model	Number of State Variables per Layer or Segment	Nutrients Considered	Number of Segments	Number of Dimensions (D) or Layers (L)	Constant Stoichiometrics (CS) or Independent Nutrient Cycle (NC)	Calibrated (C) and/or Validated (V)	Number of Case Studies in Literature
Vollenweider	1	P(N)	1	1L	CS	C + V	Many
Imboden	2	P	1	2L, 1D	CS	C + V	3
O'Melia	2	P	1	1D	CS	C	1
Larsen	3	P	1	1L	CS	C	1
Lorenzen	2	P	1	1L	CS	C + V	1
Thomann 1	8	P, N, C	1	3L	CS	C + V	1
Thomann 2	10	P, N, C	1	7L	CS	C	1
Thomann 3	15	P, N, C	67	7L	CS	—	1
Chen and Orlob	15	P, N, C	Many	7L	CS	C	At least 2
Patten	33	P, N, C	1	1L	CS	C	1
Di Toro	7	P, N	7	1L	CS	C + V	1
Bierman	14	P, N, Si	1	1L	NC	C	1
Canale	25	P, N, Si	1	2L	CS	C	1
Jørgensen	17	P, N, C	1	1–2L	NC	C + V	3
CLEANER	40	P, N, C, Si	Many	Many L	CS	C	Many
Nyholm	7	P, N	1–3	1–2L	NC	C + V	13

and sediment phosphorus, but describes the release of sediment nutrient as a function of temperature and sedimentation rates. The model accounts for variable internal storage of nutrients in the cell and describes the specific growth rate as a function of intracellular nutrient level (eqns. 13, Table 4.1).

The same model has been used to simulate 13 lakes in Denmark, some of which have been described by two or more compartments. Most parameters in the model were common to all simulations; a few key parameters were allowed to vary within a narrow range and if deviations from this range were necessary a qualitative explanation could usually be given. The simulations generally gave reasonably good results, especially for lakes with short retention times. The reason for the poorer agreement between model and observations for lakes with longer retention times is probably that the model describes the internal reactions of the lakes very crudely. It is of importance to have a good submodel of the sediment–water exchange of nutrient when the internal loading of nutrient is dominating, i.e. for lakes with long retention times.

The model of Baca and Arnett (1976) includes a comprehensive phosphorus and nitrogen cycle with a few distinctions, related primarily to the recycling of these nutrients between sediment and water. The model is presented in Figure 9.18 and the governing equations are shown in part (5) of the appendix.

Parker (1978) introduced a model that takes into account the spatial heterogeneity of phytoplankton and dissolved nutrient by nutrient–phytoplankton interaction. Dubois (1975) and Parker (1976) coupled the dynamics of predator–prey populations with the consequences of physical processes in a turbulent medium. Halfon and Lam (1978) simulated the spatial movements of phosphorus by using the computed currents from a three-dimensional hydrodynamic model. A biological submodel describes the phosphorus dynamics and primary production in each grid cell of 20 km · 20 km in Lake Superior.

As this survey shows, the lake models in operation cover a wide range of complex circumstances, although the list is not complete. Table 9.2 summarizes some of the model characteristics that can be found in the literature.

9.3. SOME ILLUSTRATIVE CASE STUDIES

9.3.1. Application of Case Studies

During the preparation of this survey of models, it was found that although a great number of aquatic models are described in the literature, only a limited number of lake models have been calibrated and validated. Jørgensen (1978) has paid attention to this problem and encouraged studies in which ecological models are tested with field data to determine their applicability to environmental management. Two case studies have been selected to illustrate the applicability of ecological lake models.

9.3.2. Övre Heimdalsvatn, Norway

The lake has a surface area of 0.775 km^2 and is 3 km long; the maximum depth is 13 m and the mean depth 4.7 m. The lake is subalpine and ultra-oligotrophic. It is a natural lake and is a good example, because the retention time is 2 months but can be as short as 2–3 days during the spring runoff. A version of CLEANER without intracellular storage of nutrients was used but it predicted a phytoplankton peak that was 3 weeks before the observed peak. However, MS CLEANER yielded an excellent fit to the data with an intracellular nutrient submodel. These observations are in accordance with the experiences published by Jørgensen (1976).

This case study shows that MS CLEANER is capable of modeling the combined effect of light, temperature, and nutrient in- and outflow. Figure 9.19 illustrates this capability for phytoplankton and zooplankton; these results

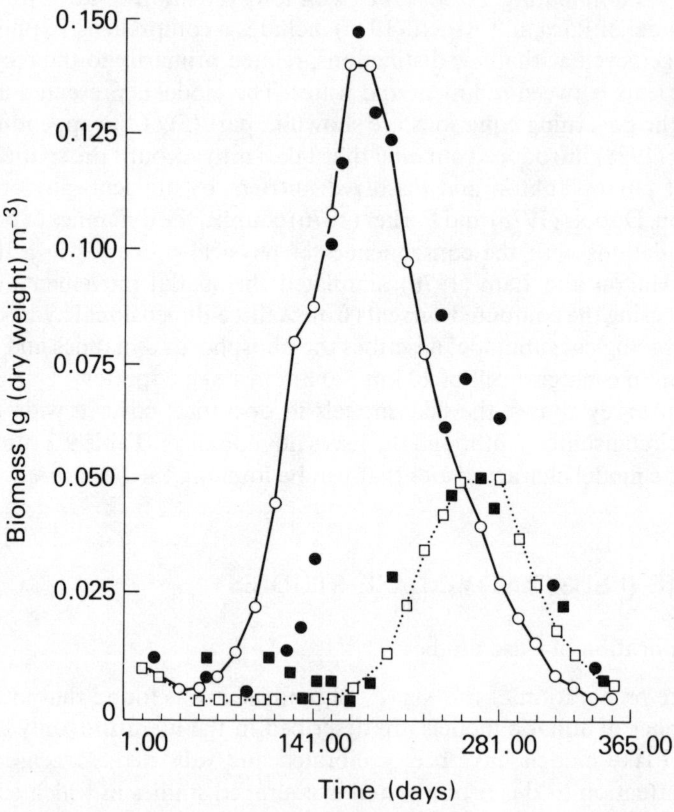

FIGURE 9.19 Simulation of biomass in Övre Heimdalsvatn, Norway, 1972. The results are for predicted (○) and observed (●) phytoplankton and for predicted (□) and observed (■) zooplankton (from Desormeau, 1978). Unpublished data are by courtesy of P. Larsson.

were obtained after the calibration of two parameters only. For all other parameters general values were applied.

9.3.3. Glumsø Sø, Denmark

Glumsø Sø is a small (250 000 m^2), shallow lake (maximum depth 2 m) with advanced eutrophication. Four versions of the Jørgensen model (details are given in part (4) of the appendix) were calibrated for the 1973–74 data and validated for the 1974–75 data. The validation was based on a comparison of soluble phosphorus, soluble nitrogen, phytoplankton concentration, zooplankton concentration, and productivity. The validation shows that version II is preferable to version I.

The details in the appendix cover both versions. Version II uses measured irradiance data while version I uses average data based on measurements over the last two decades. Version III, which accounts for light adaptation, and version IV, which considers diurnal variation of irradiance, gave poorer validation than version II. The result of the validation is summarized in Table 9.3, which shows Y and the correlation coefficient for comparison of predicted and measured values:

$$Y = \left(\frac{\sum_{i=1}^{n} (x_t^i - x_m^i)^2 / \bar{x}_m}{n} \right)^{1/2}, \tag{9.22}$$

x_t being the predicted values of the model, x_m the measured data, $\bar{x}_m$ the mean of the x_m, and n the number of compared pairs of figures. Figures 9.20–9.24 illustrate the validation by comparison of predicted and measured values.

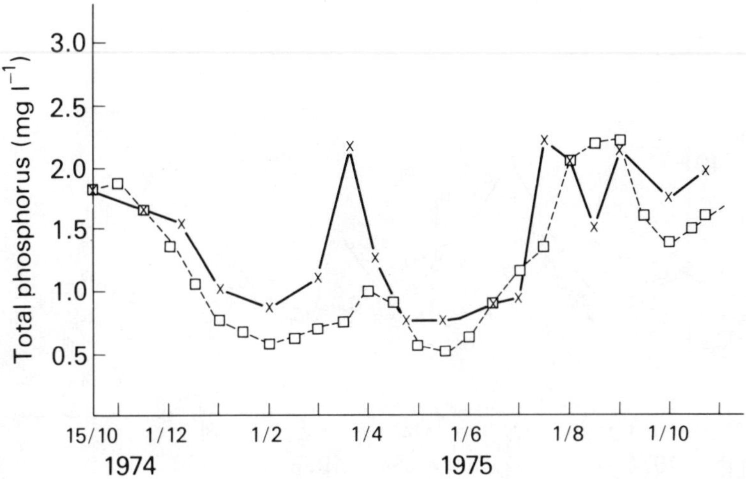

FIGURE 9.20 Concentration of total phosphorus in Glumsø Sø: observed data, ×; calculated data, □.

TABLE 9.3 Results of the Validation of Four Versions of the Jørgensen Model Applied to Glumsø Sø, Denmark.

Model Version	Y	Correlation Coefficient
I	0.574	0.78
II	0.420	0.79
III	0.644	0.74
IV	0.768	0.63

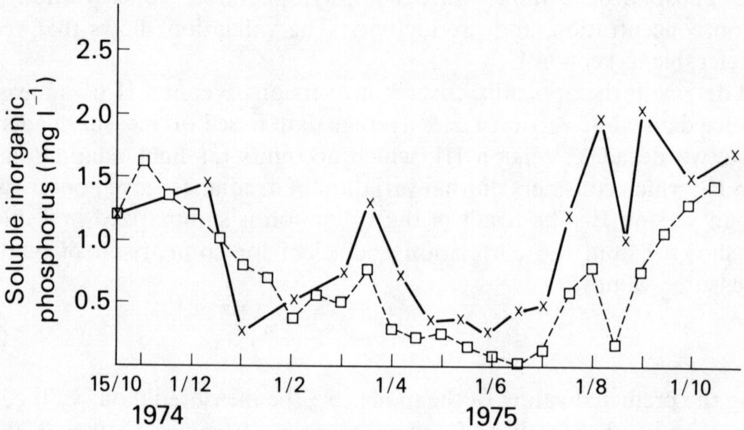

FIGURE 9.21 Concentration of soluble inorganic phosphorus in Glumsø Sø: observed data, ×; calculated data, □.

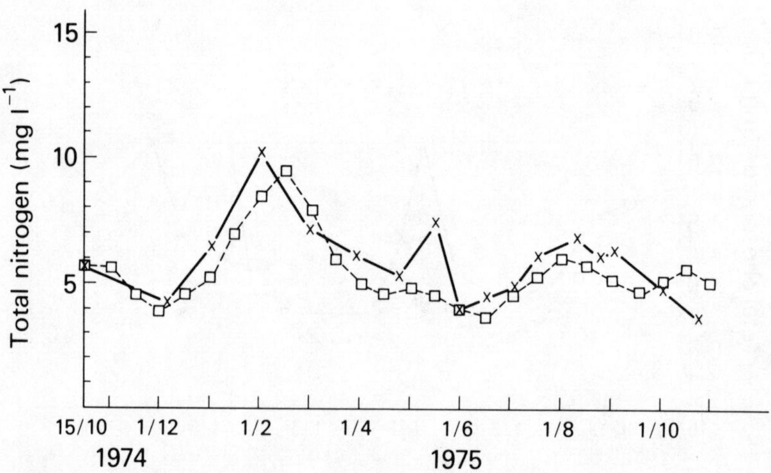

FIGURE 9.22 Concentration of total nitrogen in Glumsø Sø: observed data, ×; calculated data, □.

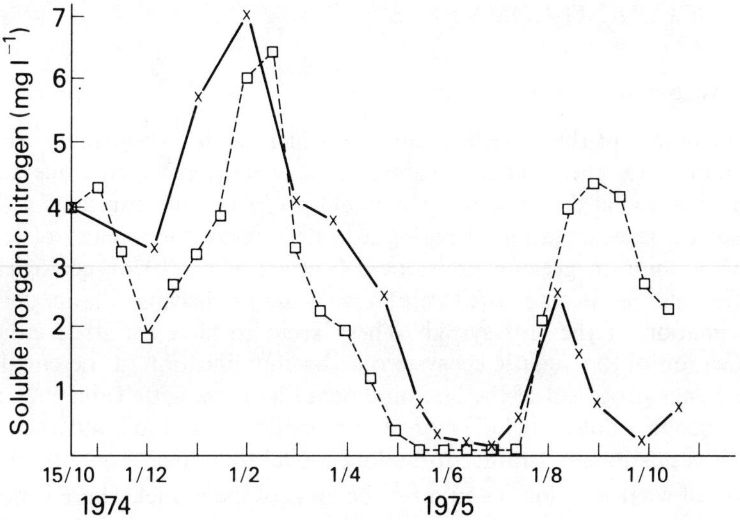

FIGURE 9.23 Concentration of soluble inorganic (nitrate and ammonia) nitrogen in Glumsø Sø: observed data, ×; calculated data, □.

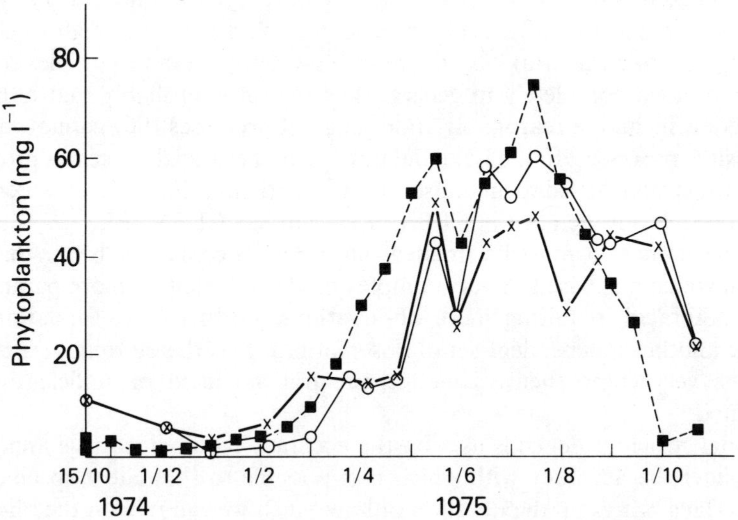

FIGURE 9.24 Concentration of phytoplankton in Glumsø Sø: observed data, × (on the basis of suspended matter, 1–80 μm) and ○ (on the basis of chlorophyll); calculated data, □.

9.4. CONCLUDING COMMENTS

9.4.1. Discussion and Conclusions

The very nature of the modeling exercise, which tends to aggregate, average, and smooth over the biological subtleties, leaves the more rigorous aquatic biologist somewhat disconcerted. He would prefer to concentrate efforts on a more correct representation of biological interactions, the kinetics of varying life stages, shifts in grazing preferences (Canale *et al.*, 1976), ecological instabilities, etc. Some trade-offs are necessary simply because the model is an approximation of the real system. These seem to have occurred either by simplification of the aquatic ecosystem or by simplification of the circulatory and exchange processes of the impoundment (compare with Table 9.2). However, the scope is not to include more and more details and to build as complex a model as possible, but rather to build a model that gives a quantitative description of what is in focus—to meet the aims of the model. There is no such thing as a general ecological lake model, but in every case study the goals and the resources available must be balanced so that the right model can be selected (see also Chapter 2).

The experience with CLEANER and MS CLEANER, however, has shown that some generality exists. These models have been applied in many case studies and have obviously a certain general applicability, although some case studies (e.g. Lake Balaton) have demonstrated that even these complex models cannot be used completely in general. The reason is probably that although the models include a reasonably wide range of processes they cannot include all possible processes. In the Lake Balaton study the crucial process is probably the resuspension of sediment caused by wind stirring the body of water, and this process is not included in CLEANER or in MS CLEANER.

In this context it must be stressed that a model cannot be better than the data on which it is based. A very complex model will contain more parameters to be calibrated, requiring more observations. Furthermore, validation will require another independent set of observations. It is therefore not surprising that the very comprehensive models are not validated or sufficiently well calibrated.

Model structure depends also on the accuracy required, so it is important to consider the accuracy with which it is possible to simulate a specific ecosystem. This brings up the question of how much we can rely on the observed data. It is assumed that ecological observations normally will have a standard deviation of 10–25%, which must be taken into consideration when the accuracy of the model is estimated.

Many lake models have been calibrated to fit limnological data collected with the usual frequency of two sets of measurements per month. Such data can hardly be used to describe the dynamics of the system, as demonstrated in

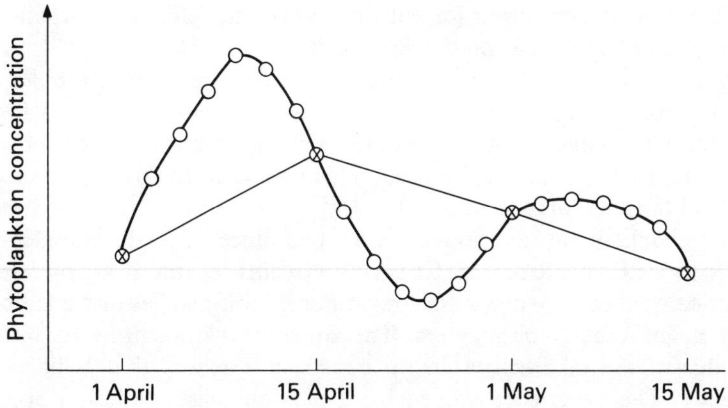

FIGURE 9.25 Comparison of measuring frequencies: × every second week; ○ every second day. The two frequencies give completely different pictures of phytoplankton dynamics, and will lead to different calibration results.

Figure 9.25. The data that are used for calibration should always be carefully collected with a frequency corresponding to the dynamics of the calibrated system or subsystem (Jørgensen, 1979; Jørgensen *et al.*, 1981). Only little work has been carried out to compare the results of different models, but it seems more important to include a detailed description of the nutrient uptake by phytoplankton and the sediment–water exchange of nutrient than to implement details of the higher trophic levels (Jørgensen, 1976; Jørgensen *et al.*, 1978; Nyholm, 1978). Nevertheless, acceptable validation results were obtained by several studies. This does not imply that the models can be used generally; rather, they were selected properly to solve specific problems.

A determination of sufficient model complexity enters the modeling process at two stages (Beck, 1978):

(1) during the initial stage, when the analyst must choose a certain level of complexity before attempting to verify the model against field data; and
(2) during the final phases, when the analyst must decide whether the model has been verified and has sufficient complexity for its intended application.

Jørgensen and Mejer (1977) suggested the use of a quantitative index for the selection of model complexity. The idea is to use a concept of sensitivity for identifying the model structure. Basically it is an inverse "submodel sensitivity," called the ecological buffer capacity, which measures the influence (sensitivity) that additional suggested submodels have on a particular state variable (e.g.

phytoplankton concentration for eutrophication models) to see whether anything is changed by increasing the complexity.

Tapp (1978) examined and compared the use of simple and complex eutrophication models. He concluded that simple models can be used for first-approximation analysis, but where data exist to establish a basis for a more complex model these should be used. This conclusion is in accordance with the state of the art (Jørgensen, 1979).

Lake modeling has developed along two lines. The development from CLEAN to CLEANER to MS CLEANER seems to aim at a general model that can be used on any new case study after minor changes and a calibration of only a few crucial parameters. The other development is to start with knowledge of the lake ecosystem to construct a model that is balanced in complexity. The experience gained from previous case studies is applied in selecting the model for each new study. A general uncritical application of a complex model seems dangerous, but if the models are used with some critical sense and the necessary modifications and calibrations are carried out from case to case, it seems a proper course to follow. Improvements will evidently result from this development, as the experiences gained steadily are included in the models.

However, our knowledge of ecosystem processes is limited and submodels included in a complex model may be valid in one case study but give wrong results in many others. This is avoided by using the other strategy, whereby the characteristic features of the modeled ecosystem form the basis for a more or less specific model, which will contain far fewer parameters and be more certain in the sense that the model components are selected in accordance with knowledge of the ecosystem. The amount of data required for calibration of the general model seems to be higher than for the more specific model provided that the more general model is used critically, which implies that it must be calibrated and to a certain extent validated. Blake and Gentil (1979) published an interesting sensitivity analysis of a discrete lake model, and concluded that a considerable amount of data is needed to calibrate a model for prediction purposes.

The conclusion of this discussion is that both approaches should be encouraged. Both developments will probably lead to a better understanding of the lake ecosystem and to better predictive models with a wider applicability.

There are impoundment models available today that are able to give an acceptable quantitative description of the eutrophication process. Except for some special cases, where phosphorus (or nitrogen) is limiting and a more simple model might be sufficient, it seems that a management model must include the main biogeochemical mass flows, including the sediment–water exchange processes (Jørgensen and Mejer, 1977), as well as a reasonably accurate hydrodynamic description. Furthermore, algae succession can be described fairly accurately (Ahlgren 1973a,b; Bierman, 1976; Park *et al.*,

1979b), although more knowledge of parameter values for different algal species or classes is desirable.

9.4.2. Future Research Needs

Several problems are still unsolved, and further research is needed. In particular, it is necessary (1) to examine the sensitivity of the models to the use of different submodels; (2) to gain more experience in the application of models that include ecology as well as hydrodynamics; and (3) to validate the existing models, preferably over a period that includes a change of loading.

Present-day models are based upon an ecological structure of lake ecosystems that might limit the possibility of using the models as predictive tools, since the impact of pollution might require the development of a different ecological structure. Future models will probably attempt to include such changes in ecological structure. Some promising but primary approaches have already been published, using catastrophe theory (Duckstein *et al.*, 1977; Dubois, 1979), thermodynamics (Jørgensen and Mejer, 1979; Mejer and Jørgensen, 1979; Mauersberger, 1979), or self-organization and adaptation (Straškraba, 1979). The next generation of models will probably contain some basic ecological principles evolved from these efforts, which implies that changes in structure can be predicted.

APPENDIX. FIVE LAKE MODELS

(1) Model by Bierman

State variables

PSA(L), NSA(L)	moles of phosphorus and nitrogen per milligram cell dry weight for phytoplankton, L,
PCM, NCM, SCM	moles of phosphorus, nitrogen, and silicon per liter of solution,
$A(L)$	milligrams dry weight per liter of phytoplankton, L,
$Z(K)$	milligrams dry weight per liter of zooplankton, K,
TOP, TON, TOS	moles of total unavailable phosphorus, nitrogen, and silicon per liter of solution.

For each state variable the model equations are written in the form of a mass balance differential equation:

rate of change of state variable = rate of change due to water circulation, Q
+ rate of change due to interactions in system volume, V

State variable equations

For each phytoplankton type, L, the rate of change of intracellular phosphorus is given by:

$$V\frac{\mathrm{d}A(L)\cdot \mathrm{PSA}(L)}{\mathrm{d}t} = Q(\mathrm{ABD}(L)\cdot \mathrm{PSABD}(L)\cdot A(L)\cdot \mathrm{PSA}(L)) + VA(L)\cdot \mathrm{R1PM}(L)\cdot f(T)f(I)\cdot 0.322\cdot 10^{-4}(\mathrm{mol\ mg}^{-1}) \cdot\left(\frac{1}{1+\mathrm{PK1}(L)\cdot \mathrm{PCA}(L)} - \frac{1}{1+\mathrm{PK1}(L)\cdot \mathrm{PCM}}\right) - VA(L)\cdot \mathrm{PSA}(L)\cdot\left(\mathrm{RAGRZD}(L)+\mathrm{RLYS}(L)\cdot T\cdot \mathrm{TCROP} + \frac{\mathrm{ASINK}(L)}{\mathrm{DEPTH}}\right). \tag{A1.1}$$

Expanding by the chain rule for derivatives gives:

$$V\frac{\mathrm{d}A(L)\cdot \mathrm{PSA}(L)}{\mathrm{d}t} = V\left(A(L)\frac{\mathrm{dPSA}(L)}{\mathrm{d}t} + \mathrm{PSA}(L)\frac{\mathrm{d}A(L)}{\mathrm{d}t}\right). \tag{A1.2}$$

For each phytoplankton type the rate of change of biomass is given by:

$$V\frac{\mathrm{d}A(L)}{\mathrm{d}t} = Q(\mathrm{ABD}(L) - A(L)) + VA(L)\left(\mathrm{SPGR}(L) - \mathrm{RAGRZD}(L) - \mathrm{RLYS}(L)\cdot T\cdot \mathrm{TCROP} - \frac{\mathrm{ASINK}(L)}{\mathrm{DEPTH}}\right). \tag{A1.3}$$

In (A1.3) it is assumed that the contribution to the derivative due to changes in intracellular phosphorus is negligible.

Setting the right-hand sides of (A1.1) and (A1.2) equal and substituting (A1.3) gives the following equation for the state variable PSA(L):

$$VA(L)\frac{\mathrm{dPSA}(L)}{\mathrm{d}t} = Q\cdot \mathrm{ABD}(L)\cdot(\mathrm{PSABD}(L) - \mathrm{PSA}(L)) + VA(L)\cdot \mathrm{R1PM}(L)\cdot f(T)f(I)\cdot 0.322\cdot 10^{-4} \cdot\left(\frac{1}{1+\mathrm{PK1}(L)\cdot \mathrm{PCA}(L)} - \frac{1}{1+\mathrm{PK1}(L)\cdot \mathrm{PCM}}\right) - VA(L)\cdot \mathrm{PSA}(L)\cdot \mathrm{SPGR}(L). \tag{A1.4}$$

An identical approach is used for the state variable NSA(L).

The equation for phosphorus concentration PCM is

$$V\frac{\mathrm{dPCM}}{\mathrm{d}t} = Q(\mathrm{PCMBD} - \mathrm{PCM}) - Vf(T)f(I)\cdot 0.322\cdot 10^{-4}$$

$$\cdot \sum_{L=1}^{\mathrm{Naspec}}\left[\mathrm{R1PM}(L)\cdot A(L)\left(\frac{1}{1+\mathrm{PK1}(L)\cdot\mathrm{PCA}(L)}\right.\right.$$

$$\left.\left.-\frac{1}{1+\mathrm{PK1}(L)\cdot\mathrm{PCM}}\right)\right] + V\sum_{L=1}^{\mathrm{Naspec}}[(\mathrm{RLYS}(L)\cdot TA(L)\cdot\mathrm{TCROP}$$

$$+\,\mathrm{RAEXC}(L)\cdot A(L))(\mathrm{PSA}(L) - \mathrm{PSAMIN}(L))]$$

$$+\,V\cdot\mathrm{RDCMP}\cdot T\cdot\mathrm{TOP} + \mathrm{WPCM}. \qquad \text{(A1.5)}$$

The equation for nitrogen concentration NCM is functionally identical to the equation for PCM.

The equation for silicon concentration SCM is

$$V\frac{\mathrm{dSCM}}{\mathrm{d}t} = Q(\mathrm{SCMBD} - \mathrm{SCM})$$

$$-\,V\sum_{\mathrm{Diatoms}}(A(L)\cdot\mathrm{SPGR}(L)\cdot\mathrm{SSA}(L))$$

$$+\,V\cdot\mathrm{RDCMP}\cdot T\cdot\mathrm{TOS} + \mathrm{WSCM}. \qquad \text{(A1.6)}$$

The equation for the concentration Z of zooplankton K is

$$V\frac{\mathrm{d}Z(K)}{\mathrm{d}t} = Q(\mathrm{ZBD}(K) - Z(K)) + VZ(K)(\mathrm{RZ}(K) - \mathrm{ZDETH}(K)). \qquad \text{(A1.7)}$$

The equation for the total unavailable phosphorus concentration TOP is

$$V\frac{\mathrm{dTOP}}{\mathrm{d}t} = Q(\mathrm{TOPBD} - \mathrm{TOP})$$

$$+\,VT\cdot\mathrm{TCROP}\sum_{L=1}^{\mathrm{Naspec}}(\mathrm{RLYS}(L)\cdot A(L)\cdot\mathrm{PSAMIN}(L))$$

$$+\,V\sum_{K=1}^{\mathrm{Nzspec}}(\mathrm{RZPEX}(K)\cdot Z(K))$$

$$-\,V\cdot\mathrm{TOP}\left(\mathrm{RDCMP}\cdot T + \frac{\mathrm{TOSINK}}{\mathrm{DEPTH}}\right) + \mathrm{WTOP}. \qquad \text{(A1.8)}$$

The equations for TON and TOS are functionally identical to (Al.8).

Process rate equations

The specific growth rate of phytoplankton L, SPGR(L), is equal to the minimum of the following three functions:

$$\text{maximum growth rate} \cdot f(T)f(I)\{1 - \exp[-0.693(\text{P/PO} - 1)]\},$$
$$\text{maximum growth rate} \cdot f(T)f(I) \cdot (\text{N} - \text{NO})/(\text{KNCELL} + \text{N} - \text{NO}),$$
$$\text{maximum growth rate} \cdot f(T)f(I) \cdot \text{SCM}/(\text{KSCM} + \text{SCM}).$$

The rate of growth of zooplankton K is expressed by:

$$\text{RZ}(K) = \text{RZMAX}(K) \cdot \text{ZASSIM}(K) \cdot \sum_{L=1}^{\text{Naspec}} \frac{\text{ZEFF}(K, L) \cdot A(L)}{\text{KZSAT}(K, L) + \sum_{L=1}^{\text{Naspec}} \text{ZEFF}(K, L) \cdot A(L)}.$$

The rate at which phytoplankton L is ingested by zooplankton is:

$$\text{RAGRZD}(L) = \sum_{K=1}^{\text{Nzspec}} \frac{\text{RZMAX}(K) \cdot Z(K) \cdot \text{ZEFF}(K, L)}{\text{KZSAT}(K, L) + \sum_{L=1}^{\text{Naspec}} \text{ZEFF}(K, L) \cdot A(L)}.$$

The rate at which phytoplankton L is excreted by zooplankton (used to calculate the amount of phosphorus excreted to the pool of available phosphorus) is:

$$\text{RAEXC}(L) = \sum_{K=1}^{\text{Nzspec}} \frac{(1 - \text{ZASSIM}(K)) \cdot \text{RZMAX}(K) \cdot \text{ZEFF}(K, L) \cdot Z(K)}{\text{KZSAT}(K, L) + \sum_{L=1}^{\text{Naspec}} \text{ZEFF}(K, L) \cdot A(L)}.$$

The rate at which phosphorus is excreted to the unavailable pool by zooplankton K is:

$$\text{RZPEX}(K) = (1 - \text{ZASSIM}(K)) \cdot \text{RZMAX}(K) \cdot \sum_{L=1}^{\text{Naspec}} \frac{\text{ZEFF}(K, L) \cdot A(L) \cdot \text{PSAMIN}(L)}{\text{KZSAT}(K, L) + \sum_{L=1}^{\text{Naspec}} \text{ZEFF}(K, L) \cdot A(L)}.$$

Miscellaneous functions

$$f(T) = \Theta^{T-20},$$

where $\Theta = 1.07$ for diatoms, 1.08 for greens, and 1.10 for blue-greens.

$$T = \text{TMAX}\left[0.50 - 0.50\ \text{SINE}\left(\frac{6.23 \cdot \text{TIME} + \phi}{360}\right)\right],$$

where ϕ is chosen such that SINE = 0 on 1 November.

$$f(I) = \frac{1}{ke \cdot \text{DEPTH}}[\exp(-\alpha 1) - \exp(-\alpha 0)],$$

where

$$\alpha 1 = \frac{\mathrm{Ia}}{\mathrm{Is}} \exp(-ke \cdot \mathrm{DEPTH}),$$

$$\alpha 0 = \frac{\mathrm{Ia}}{\mathrm{Is}},$$

$$ke = 1.9/\text{Secchi depth} + 0.17 \cdot \mathrm{TCROP}$$
$$= 0.633 + 0.17 \cdot \mathrm{TCROP},$$

Ia = 2000 foot-candles, Is is the irradiance at the surface, and the photoperiod is 0.50.

Notation

A	is the phytoplankton concentration (mg dry wt l^{-1}),
ASINK	is the phytoplankton sinking rate (m day^{-1}),
BD	is a suffix that refers to the boundary value of a particular variable,
$f(I)$, $f(T)$	are the light intensity and temperature reduction factors,
KNCELL	is the intracellular half-saturation constant for nitrogen-dependent growth (mol $cell^{-1}$),
KSCM	is the Michaelis constant for silicon-dependent growth (mol l^{-1}),
KZSAT(K, L)	is the half-saturation concentration of phytoplankton L for grazing by zooplankton K,
Naspec	is the number of phytoplankton species,
Nzspec	is the number of zooplankton species,
P, N	are the moles of phosphorus and nitrogen per phytoplankton cell,
PCA, NCA	are the intracellular phosphorus and nitrogen concentrations (mol l^{-1}(cell volume)$^{-1}$),
PCM, NCM, SCM	are the nutrient concentrations of phosphorus, nitrogen, and silicon in solution (mol l^{-1}),
PSA, NSA	are phosphorus and nitrogen storage in phytoplankton cells (mol (mg dry wt)$^{-1}$),
PK1, NK1	are affinity constants for phosphorus and nitrogen uptake mechanisms (l mol^{-1}),
PO, NO	are the minimum stoichiometric levels of phosphorus and nitrogen per phytoplankton cell (mol $cell^{-1}$),
Q	is the water circulation rate [L^3 day^{-1}],
R1PM, R1NM	are the maximum phosphorus and nitrogen uptake rates (day^{-1}),

RAEXC(L)	is the rate at which phytoplankton L is excreted by zooplankton (day^{-1}),
RAGRZD(L)	is the rate at which phytoplankton L is grazed (ingested) by zooplankton (day^{-1}),
RDCMP	is the decomposition rate from unavailable to available nutrient pools (day^{-1} °C^{-1}),
RLYS	is the algal death rate [(day °C mg l^{-1})$^{-1}$],
RZ	is the zooplankton specific growth rate (day^{-1}),
RZMAX	is the zooplankton maximum ingestion rate (day^{-1}),
RZPEX(K), RZNEX(K), RZSEX(K)	are phosphorus, nitrogen, and silicon excreted by zooplankton K to the unavailable nutrient poll [mol(mg zooplankton)$^{-1}$ day^{-1}],
SPGR	is the phytoplankton specific growth rate (day^{-1}),
SSA	is the silicon stoichiometry for diatoms [mol(mg dry wt)$^{-1}$],
T	is the temperature (°C),
TCROP	is the total phytoplankton biomass (mg dry wt l^{-1}),
TOP, TON, TOS	are the concentrations of unavailable phosphorus, nitrogen, and silicon (mol l^{-1}),
TOSINK	is the sinking rate of nonliving organic material (m day^{-1}),
V	is the system volume,
WPCM, WNCM, WSCM	are the external point loading rates of available phosphorus, nitrogen, and silicon (mol day^{-1}),
WTOP, WTON, WTOS	are the external point loading rates of unavailable phosphorus, nitrogen, and silicon (mol day^{-1}),
Z	is the zooplankton concentration (mg dry wt l^{-1}),
ZASSIM	is the zooplankton assimilation efficiency,
ZEFF(K, L)	is the efficiency of zooplankton K in ingesting phytoplankton, L
ZDETH	is the zooplankton death rate (day^{-1}).

(2) Model by Canale *et al.*

Summary of the model equations

$$\dot{c}_z = [\text{growth}]_z - \begin{bmatrix}\text{predation by other}\\ \text{zooplankton}\end{bmatrix}_z - \begin{bmatrix}\text{predation}\\ \text{by alewife}\end{bmatrix}_z - \begin{bmatrix}\text{respiration}\\ \text{loss}\end{bmatrix}_z - \underset{\substack{\text{natural}\\ \text{death rate}\\ \text{for copepods}}}{\text{A}14_z(t)} \cdot c_z \qquad (z = 1, 2, \ldots, 9)$$

$$\dot{c}_p = [\text{growth}]_p - \begin{bmatrix}\text{predation}\\ \text{loss}\end{bmatrix}_p - \begin{bmatrix}\text{respiration}\\ \text{loss}\end{bmatrix}_p - \begin{bmatrix}\text{sinking}\\ \text{loss}\end{bmatrix}_p$$

$$(p = 10, 11, 12, 13)$$

$$\dot{c}_{14} = \text{NCR} \cdot \sum_z \left(1 - \begin{bmatrix}\text{assimilation}\\ \text{efficiency}\end{bmatrix}_z\right)\begin{bmatrix}\text{eating}\\ \text{rate}\end{bmatrix}_z c_z + \text{NCR} \cdot \sum_z \begin{bmatrix}\text{natural}\\ \text{death rate}\end{bmatrix}_z$$
$$- \text{A18} \cdot Tc_{14} - \text{A23} \cdot \text{SINK}(t) \cdot c_{14}$$

$$\dot{c}_{15} = \text{NCR} \cdot \sum_p \begin{bmatrix}\text{respiration}\\ \text{loss}\end{bmatrix}_p + \text{NCR} \cdot \text{A21} \cdot \sum_z \begin{bmatrix}\text{respiration}\\ \text{loss}\end{bmatrix}_z + \text{A18} \cdot Tc_{14}$$
$$- \text{A20} \cdot Tc_{15} + \text{LOAD}_{15}$$

$$\dot{c}_{16} = \text{NCR} \cdot (1 - \text{A21}) \cdot \sum_z \begin{bmatrix}\text{respiration}\\ \text{loss}\end{bmatrix}_z + \text{A20} \cdot Tc_{15} - \text{A22} \cdot Tc_{16}$$
$$- \text{NCR} \cdot \left(\frac{\text{ANH3} \cdot c_{16}}{\text{ANH3} \cdot c_{16} + (1 - \text{ANH3})c_{17}}\right) \sum_p [\text{growth}]_p + \text{LOAD}_{16}$$

$$\dot{c}_{17} = \text{A22} \cdot Tc_{16} - \text{NCR} \cdot \left(\frac{(1 - \text{ANH3})c_{17}}{\text{ANH3} \cdot c_{16} + (1 - \text{ANH3})c_{17}}\right) \sum_p [\text{growth}]_p$$
$$+ \text{LOAD}_{17}$$

$$\dot{c}_{23} = \text{A23} \cdot \text{SINK}(t) \cdot c_{14} + \text{NCR} \cdot \sum_p \begin{bmatrix}\text{sinking}\\ \text{loss}\end{bmatrix}_p \cdot \text{VOLEP}$$
$$+ (\text{LOAD}_{15} + \text{LOAD}_{16} + \text{LOAD}_{17}) \cdot \text{VOLHY}$$

$$\dot{c}_{18} = \text{PCR} \cdot \sum_z \left(1 - \begin{bmatrix}\text{assimilation}\\ \text{efficiency}\end{bmatrix}_z\right)\begin{bmatrix}\text{eating}\\ \text{rate}\end{bmatrix}_z c_z + \text{PCR} \cdot \sum_z \begin{bmatrix}\text{natural}\\ \text{death rate}\end{bmatrix}_z$$
$$- \text{A17} \cdot Tc_{18} - \text{A23} \cdot \text{SINK}(t) \cdot c_{18}$$

$$\dot{c}_{19} = \text{PCR} \cdot \sum_p \begin{bmatrix}\text{respiration}\\ \text{loss}\end{bmatrix}_p + \text{A17} \cdot Tc_{18} - \text{A19} \cdot Tc_{19} + \text{LOAD}_{19}$$

$$\dot{c}_{20} = \text{PCR} \cdot \sum_z \begin{bmatrix}\text{respiration}\\ \text{loss}\end{bmatrix}_z + \text{A19} \cdot Tc_{19} - \text{PCR} \cdot \sum_p [\text{growth}]_p + \text{LOAD}_{20}$$

$$\dot{c}_{24} = \text{A23} \cdot \text{SINK}(t) \cdot c_{18} + \text{PCR} \cdot \sum_p \begin{bmatrix}\text{sinking}\\ \text{loss}\end{bmatrix}_p \cdot \text{VOLEP}$$
$$+ (\text{LOAD}_{19} + \text{LOAD}_{20}) \cdot \text{VOLHY}$$

$$\dot{c}_{21} = \text{SCR} \cdot \sum_{\text{diatoms}} \begin{bmatrix}\text{respiration}\\ \text{loss}\end{bmatrix}_p + \begin{bmatrix}\text{predation}\\ \text{loss}\end{bmatrix}_p - \text{A16} \cdot Tc_{21}$$
$$- \text{A23} \cdot \text{SINK}(t) \cdot c_{21}$$

$$\dot{c}_{22} = \text{A16} \cdot Tc_{21} - \text{SCR} \cdot \sum_{\text{diatoms}} [\text{growth}]_p + \text{LOAD}_{22}$$

$$\dot{c}_{25} = \text{A23} \cdot \text{SINK}(t) \cdot c_{21} + \text{SCR} \cdot \sum_{\text{diatoms}} \begin{bmatrix}\text{sinking}\\ \text{loss}\end{bmatrix}_p \cdot \text{VOLEP}$$
$$+ \text{LOAD}_{22} \cdot \text{VOLHY}.$$

Model coefficients

Table I Zooplankton-Related Coefficients.

Symbol	Definition	Unit	Value Used
$A7_{lept.-poly.}$	Maximum snatching rates at 20°C	$\frac{\text{mg food-C}}{\text{mg z-C day}}$	0.70
$A7_{cyc.}$			0.43
$A7_{cyc.\ n}$	Maximum filtering rates at 20°C	$\frac{1}{\text{mg z-C day}}$	2.6
$A7_{diap.\ n}$			6.5
$A7_{lim.-ep.\ n}$			5.2
$A7_{diap.}$			1.0
$A7_{lim.-ep.}$			1.25
$A7_{daph.}$	Filtering rates at 20°C		4.0
$A7_{bosm.-holo.}$			3.5
$KFOOD_{lept.-poly.}$	Half-saturation food level for raptors	mg food-C l^{-1}	0.2
$KFOOD_{cyc.}$			0.2
A9	Minimum filtering rate multiplier	None	0.1
A10	Food level where multiplier is (1 + A9)/2	mg food-C l^{-1}	0.2
A11R	Assimilation efficiency of raptors	$\frac{\text{mg z-C}}{\text{mg food-C}}$	0.4
A11S	Assimilation efficiency of selectives		0.7
A11N	Maximum assimilation efficiency of nonselectives		0.8
A24	Half effective food level for nonselectives	mg food-C l^{-1}	0.2
$A12_{adults}$	Respiration rates at 20°C	day^{-1}	0.06
$A12_{nauplii}$			0.04
$A14_{cyc.}$	Natural death rates for copepods	day^{-1}	0.005
$A14_{diap.}$			0.005
$A14_{lim.-ep.}$			0.003

Table II Phytoplankton-Related Coefficients.

Symbol	Definition	Unit	Value Used
$A1_{sm.diatoms}$	Maximum growth rates at 20°C	day^{-1}	2.1
$A1_{lg.diatoms}$			2.0
$A1_{blue-greens}$			1.6
$A1_{greens}$			1.9
$IS_{sm.diatoms}$	Optimum light intensities	ly day^{-1}	225
$IS_{lg.\ diatoms}$			225
$IS_{blue-greens}$			600
IS_{greens}			160
$A6_{sm.\ diatoms}$	Maximum sinking rates	day^{-1}	0.05
$A6_{lg.\ diatoms}$			0.03
$A6_{blue-greens}$			0
$A6_{greens}$			0.02
KN	Michaelis constants for phytoplankton growth	mg nutrient l^{-1}	0.015
KP			0.0025
KS			0.030
DEPTH	Depth of euphotic zone	m	20
A3	Respiration rate at 20°C for phytoplankton	day^{-1}	0.03

Table III Nutrient-Related Coefficients.

Symbol	Definition	Unit	Value Used
NCR, PCR, SCR	Nutrient-to-carbon ratios	$\frac{\text{mg nutrient}}{\text{mg C}}$	0.2
A18	Detrital nitrogen → organic nitrogen	$\text{day}^{-1}\,°\text{C}^{-1}$	0.001
A20	Organic nitrogen → ammonia	$\text{day}^{-1}\,°\text{C}^{-1}$	0.0012
A22	Ammonia → nitrate	$\text{day}^{-1}\,°\text{C}^{-1}$	0.008
A17	Detrital phosphorus → organic phosphorus	$\text{day}^{-1}\,°\text{C}^{-1}$	0.01
A19	Organic phosphorus → inorganic phosphorus	$\text{day}^{-1}\,°\text{C}^{-1}$	0.01
A16	Detrital silicon → dissolved silicon	$\text{day}^{-1}\,°\text{C}^{-1}$	0.0015

Symbol	Definition	Unit	Value Used
A23	Maximum detrital sinking rate	day^{-1}	0.05
A21	Fraction of zooplankton-respired nitrogen that is organic	None	0.7
Q	Net flow through Lake Michigan	l day^{-1}	$1.37 \cdot 10^{11}$
VOLEP	Volume of Lake Michigan epilimnion	l	$1.218 \cdot 10^{15}$
VOLHY	Volume of the hypolimnion	l	$3.654 \cdot 10^{15}$
ANH_3	Concentration of ammonium ion		
LOAD	Input		

(3) Model by Chen and Orlob

Ecological model formulations

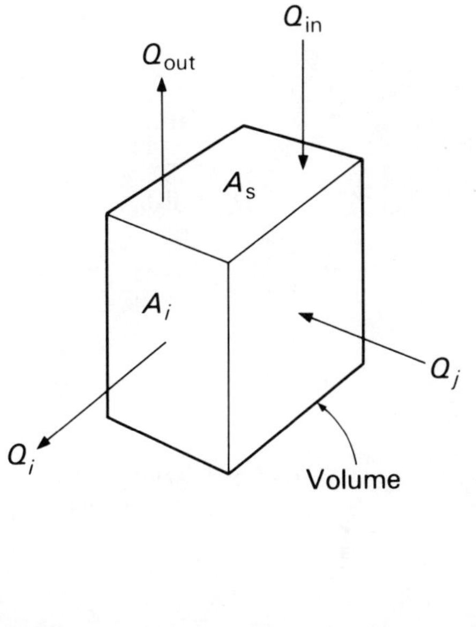

$$\begin{pmatrix} C_1 \\ C_2 \\ C_3 \\ \vdots \end{pmatrix} = \begin{pmatrix} \text{TDS} \\ \text{BOD} \\ \text{DO} \\ \text{TEMP} \\ \vdots \\ \text{ALGAE} \\ \text{ZOO} \\ \text{FISH} \end{pmatrix}$$

General mass balance equation for abiotic substances:

$$\frac{dVC_1}{dt} = \underset{\text{advection}}{\sum Q_1 C_1} + \underset{\text{diffusion}}{\sum EA \frac{dC_1}{dz}} + \underset{\text{input}}{\sum Q_{in} C_{in}} - \underset{\text{output}}{\sum Q_{out} C_1} \pm \underset{\text{settling}}{S_1 V C_1}$$

$$\pm \underset{\text{reaeration}}{K_r A_3 (C_1 - C_1^*)} - \underset{\text{decay}}{K_{d,1} V C_1} \pm \underset{\substack{\text{transformation} \\ NH_3 \to NO_2 \to NO_3}}{K_{d,2} V C_2} - \underset{\substack{\text{uptake} \\ \text{byproduct}}}{\sum \mu_3 V C_3 F_{3,1}}$$

$$+ \underset{\text{respiration release}}{\sum P_3 V C_3 F_{3,1}}.$$

General mass balance equation for biota:

$$\frac{dVC_1}{dt} = \sum Q_1 C_1 + \sum EA \frac{dC_1}{dz} + \sum Q_{in} C_{in} - \sum Q_{out} C_1 + (\mu_1 - R_1 - S_1 - M) V C_1 - \underset{\text{grazing}}{\mu_2 V C_2 F_{2,1}}.$$

Phytoplankton (algae):

$$\mu_1 = \hat{\mu}\theta^{T-20} \frac{L}{K_L + L} \frac{C}{K + C} \frac{N}{K_N + N} \frac{P}{K_P + P},$$

$$S_1 = \frac{S_t}{S_o} \qquad R_1 = r\theta^{T-20} \qquad \mu_2, C_2 = \text{zooplankton}.$$

Zooplankton:

$$\mu_1 = \hat{\mu}\theta^{T-20} \frac{\text{algae}}{K_A + \text{algae}},$$

$$M_1 = \alpha + \beta \cdot \text{toxicity} \qquad R_1 = r\theta^{T-20} \qquad \mu_2, O_2 = \text{fish}.$$

Fish:

$$\mu_1 = \hat{\mu}\theta^{T-20} \frac{\text{zooplankton}}{K_Z + \text{zooplankton}},$$

$$M_1 = \alpha + \beta \cdot \text{toxicity} \qquad R_1 = r\theta^{T-20} \qquad \mu_2, C_2 = \text{harvest}.$$

(4) Model by Jørgensen

Table IV Equations, Model II.

The model has 17 state variables:

CC	Carbon in algal cells ($g\ m^{-3}$),
FNF	Proportion of nitrogen in fish,
FNZ	Proportion of nitrogen in zooplankton,
FPF	Proportion of phosphorus in fish,
FPZ	Proportion of phosphorus in zooplankton,
NC	Nitrogen in algal cells ($g\ m^{-3}$),
ND	Nitrogen in detritus ($g\ m^{-3}$),
NSED	Nitrogen in sediment ($g\ m^{-3}$),
NS	Soluble nitrogen ($g\ m^{-3}$),
PB	Phosphorus released biologically from sediment ($g\ m^{-3}$),
PC	Phosphorus in algal cells ($g\ m^{-3}$),
PD	Phosphorus in detritus ($g\ m^{-3}$),
PE	Exchangeable phosphorus in sediment ($g\ m^{-3}$),
PHYT	Phytoplankton biomass ($g\ m^{-3}$),
PI	Phosphorus in interstitial water ($g\ m^{-3}$),
PS	Soluble phosphorus ($g\ m^{-3}$),
ZOO	Zooplankton biomass ($g\ m^{-3}$).

Level I, the differential equations:

$$\mathrm{dCC}/\mathrm{d}t = (\mathrm{UC} - \mathrm{RC})\cdot \mathrm{PHYT} - (\mathrm{SA} + \mathrm{GZ}/Y + Q/V)\cdot \mathrm{CC}$$

$$\mathrm{dFNF}/\mathrm{d}t = (\mathrm{PRED}/Y)\cdot(\mathrm{FNZ} - \mathrm{FNF})$$

$$\mathrm{dFNZ}/\mathrm{d}t = \mathrm{MYZ}\cdot(\mathrm{FNA} - \mathrm{FNZ})$$

$$\mathrm{dFPF}/\mathrm{d}t = (\mathrm{PRED}/Y)(\mathrm{FPZ} - \mathrm{FPF})$$

$$\mathrm{dFPZ}/\mathrm{d}t = \mathrm{MYZ}\cdot(\mathrm{FPA} - \mathrm{FPZ})$$

$$\mathrm{dNC}/\mathrm{d}t = \mathrm{UN}\cdot \mathrm{PHYT} - (\mathrm{SA} + \mathrm{GZ}/Y + Q/V)\cdot \mathrm{NC}$$

$$\mathrm{dND}/\mathrm{d}t = L\cdot \mathrm{GZ}\cdot \mathrm{NC} + \mathrm{MZ}\cdot \mathrm{NZOO} + L\cdot \mathrm{PRED}\cdot \mathrm{NFISH} - (\mathrm{KDN} + \mathrm{SD} + Q/V)\cdot \mathrm{ND} + \mathrm{QNDIN}$$

$$\mathrm{dNSED}/\mathrm{d}t = (\mathrm{SA}\cdot \mathrm{NC} + \mathrm{SD}\cdot \mathrm{ND} - \mathrm{NREL})/\mathrm{AE}$$

$$\mathrm{dNS}/\mathrm{d}t = \mathrm{KDN}\cdot \mathrm{ND} + \mathrm{RZ} + \mathrm{NZOO} + \mathrm{PRED}\cdot \mathrm{NFISH} + \mathrm{NREL} - \mathrm{UN}\cdot \mathrm{PHYT} + \mathrm{QNSIN} - (Q/V + \mathrm{DENIT})\cdot \mathrm{NS}$$

$$\mathrm{dPB}/\mathrm{dt} = \mathrm{QSED}/\mathrm{AB} - \mathrm{QBIO} - \mathrm{QDSORP}$$

$$\mathrm{dPC}/\mathrm{d}t = \mathrm{UP}\cdot \mathrm{PHYT} - (\mathrm{SA} + \mathrm{GZ}/Y + Q/V)\cdot \mathrm{PC}$$

$$\mathrm{dPD}/\mathrm{d}t = L\cdot \mathrm{GZ}\cdot \mathrm{PC} + \mathrm{MZ}\cdot \mathrm{PZOO} + L\cdot \mathrm{PRED}\cdot \mathrm{PFISH} - (\mathrm{KDP} + \mathrm{SD} + Q/V)\cdot \mathrm{PD} + \mathrm{QPDIN}$$

$$\mathrm{dPE}/\mathrm{d}t = [(12/29)\cdot \mathrm{SA}\cdot \mathrm{PC} - \mathrm{QSED} + \mathrm{SD}\cdot \mathrm{PD}]/\mathrm{AE} - \mathrm{KE}\cdot \mathrm{PE}$$

$$\mathrm{dPHYT}/\mathrm{d}t = (\mathrm{CDR} - \mathrm{SA} - \mathrm{GZ}/Y - Q/V)\cdot \mathrm{PHYT}$$

$$\mathrm{dPI}/\mathrm{d}t = (\mathrm{AE}/\mathrm{AI})\cdot \mathrm{KE}\cdot \mathrm{PE} - \mathrm{QDIFF}/\mathrm{AI}$$

$$\mathrm{dPS}/\mathrm{d}t = \mathrm{KDP}\cdot \mathrm{PD} + \mathrm{RZ}\cdot \mathrm{PZOO} + \mathrm{PRED}\cdot \mathrm{PFISH} - \mathrm{UP}\cdot \mathrm{PHYT} + \mathrm{QDIFF} + {} + \mathrm{QPSIN} - (Q/V)\cdot \mathrm{PS} + \mathrm{AB}\cdot(\mathrm{QBIO} + \mathrm{QDSORP})$$

$$\mathrm{dZOO}/\mathrm{d}t = (\mathrm{MYZ} - \mathrm{RZ} - \mathrm{MZ} - Q/V)\cdot \mathrm{ZOO} - \mathrm{PRED}\cdot \mathrm{FISH}/Y$$

QNDIN (QPDIN) and QNSIN (QPSIN) represent the inflows of detrital and soluble nitrogen (phosphorus)

Level II, rates:

$$\mathrm{CDR} = \mathrm{CDR}_{max}\cdot \mathrm{FT1}\cdot \mathrm{FN3}\cdot \mathrm{FC3}\cdot \mathrm{FP3}$$

$$\mathrm{GZ} = \mathrm{MYZ}\cdot \mathrm{FZP}$$

$$\mathrm{KDN} = \mathrm{KDN}_{10}\cdot \mathrm{FT3}$$

$$\mathrm{KDP} = \mathrm{KDP}_{10}\cdot \mathrm{FT3}$$

$$\mathrm{KE} = \mathrm{KE}_{20}\cdot \mathrm{FT2}$$

$$\mathrm{MYZ} = \mathrm{MYZ}_{max}\cdot \mathrm{FPH}\cdot \mathrm{FT1}$$

(*continued over*)

Table IV (*continued*)

$$\mathrm{NREL} = \mathrm{FTS}\cdot(\mathrm{KREL}\cdot\mathrm{NSED} + 0.08)/(1000D)$$
$$\mathrm{PRED} = \mathrm{PRED_{max}}\cdot\mathrm{FT1}\cdot\mathrm{FZ}$$
$$\mathrm{QBIO} = 0.563\cdot\mathrm{FT6}\cdot(\mathrm{PB}/1800)/(1000\cdot\mathrm{DB})$$
$$\mathrm{QDIFF} = \mathrm{FT4}\cdot[1.21\cdot(\mathrm{PI} - \mathrm{PS}) - 1.70]/(1000D)$$
$$\mathrm{QDSORP} = (0.60\ \lg \mathrm{PS} - 2.27)/(1000\cdot\mathrm{DB})$$
$$\mathrm{QSED} = \min(\mathrm{SA}\cdot\mathrm{PC}\cdot 5.06\cdot 10^{-3})$$
$$\mathrm{RC} = \mathrm{RC_{max}}\cdot\mathrm{FC4}\cdot\mathrm{FT1}$$
$$\mathrm{RZ} = \mathrm{RZ_{max}}\cdot\mathrm{FT1}$$
$$\mathrm{SA} = (\mathrm{SVS}/D)\cdot(\mathrm{FT2})^{1/2}$$
$$\mathrm{SD} = (\mathrm{SVD}/D)\cdot(\mathrm{FT2})^{1/2}$$
$$\mathrm{UC} = \mathrm{UC_{max}}\cdot\mathrm{FC1}\cdot\mathrm{FC2}\cdot\mathrm{FRAD}$$
$$\mathrm{UN} = \mathrm{UN_{max}}\cdot\mathrm{FN1}\cdot\mathrm{FN2}$$
$$\mathrm{UP} = \mathrm{UP_{max}}\cdot\mathrm{FP1}\cdot\mathrm{FP2}$$

Level III, limiting factors:

$$\mathrm{FCA} = \mathrm{CC}/\mathrm{PHYT}$$
$$\mathrm{FC1} = (\mathrm{FCA_{max}} - \mathrm{FCA})/(\mathrm{FCA_{max}} - \mathrm{FCA_{min}})$$
$$\mathrm{FC2} = \mathrm{C}/(\mathrm{KC} + \mathrm{C})$$
$$\mathrm{FC3} = 1 - \mathrm{CC_{max}}/\mathrm{CC}$$
$$\mathrm{FC4} = (\mathrm{CC}/\mathrm{CC_{min}})^{2/3}$$
$$\mathrm{FNA} = \mathrm{NC}/\mathrm{PHYT}$$
$$\mathrm{FN1} = (\mathrm{FNA_{max}} - \mathrm{FNA})/(\mathrm{FNA_{max}} - \mathrm{FNA_{min}})$$
$$\mathrm{FN2} = \mathrm{NS}/(\mathrm{NS} + \mathrm{KN})$$
$$\mathrm{FN3} = 1 - \mathrm{NC_{min}}/\mathrm{NC}$$
$$\mathrm{FPA} = \mathrm{PC}/\mathrm{PHYT}$$
$$\mathrm{FPH} = \max(0, (\mathrm{PHYT} - 0.5)/(\mathrm{PHYT} + \mathrm{KA}))$$
$$\mathrm{FP1} = (\mathrm{FPA_{max}} - \mathrm{FPA})/(\mathrm{FPA_{max}} - \mathrm{FPA_{min}})$$
$$\mathrm{FP2} = \mathrm{PS}/(\mathrm{PS} + \mathrm{KP})$$
$$\mathrm{FP3} = 1 - \mathrm{PC_{min}}/\mathrm{PC}$$
$$\mathrm{FRAD} = \lg[(\mathrm{RAD} + \mathrm{KL})/(\mathrm{RAD}\cdot\mathrm{BEER} + \mathrm{KL})]/\Omega$$
$$\mathrm{BEER} = \exp(-\Omega)$$
$$\Omega = (\alpha + \beta\cdot\mathrm{PHYT})D$$
$$\mathrm{FT1} = \exp[-2.3|t - 16.5|/15]$$
$$\mathrm{FT2} = \theta^{t-20}$$
$$\mathrm{FT3} = \phi^{t-10}$$
$$\mathrm{FT4} = (t + 273)/280$$
$$\mathrm{FT5} = \exp(0.151t)$$
$$\mathrm{FT6} = \exp(0.203t)$$
$$\mathrm{FZ} = \max(0, (\mathrm{ZOO} - \mathrm{KS})/(\mathrm{ZOO} + \mathrm{KZ}))$$
$$\mathrm{FZP} = \mathrm{ZOO}/\mathrm{PHYT}$$

Level IV, other equations:

$$\mathrm{FISH} = \mathrm{FISH_0}\cdot\{1 + 0.8\sin[0.017453\,(\mathrm{DAY} + 150)]\}$$
$$\mathrm{NWAT} = \mathrm{NC} + \mathrm{ND} + \mathrm{NS} + \mathrm{NZOO}$$
$$\mathrm{PWAT} = \mathrm{PC} + \mathrm{PD} + \mathrm{PS} + \mathrm{PZOO}$$
$$\mathrm{NTOT} = \mathrm{NC} + \mathrm{ND} + \mathrm{NS} + \mathrm{NZOO} + \mathrm{NFISH} + \mathrm{AE}\cdot\mathrm{NSED}$$
$$\mathrm{PTOT} = \mathrm{PC} + \mathrm{PD} + \mathrm{PS} + \mathrm{PZOO} + \mathrm{PFISH} + \mathrm{AE}\cdot\mathrm{PE} + \mathrm{AI}\cdot\mathrm{PI} + \mathrm{AB}\cdot\mathrm{PB}$$
$$\mathrm{NZOO} = \mathrm{FNZ}\cdot\mathrm{ZOO}$$
$$\mathrm{NFISH} = \mathrm{FNF}\cdot\mathrm{FISH}$$
$$\mathrm{PZOO} = \mathrm{FPZ}\cdot\mathrm{ZOO}$$
$$\mathrm{PFISH} = \mathrm{FPF}\cdot\mathrm{FISH}$$
$$\mathrm{PROD} = \mathrm{CDR}\cdot\mathrm{PHYT}$$
$$\mathrm{AB} = (\mathrm{DB}/D)\cdot\mathrm{DMU}$$
$$\mathrm{AE} = \mathrm{LUL}\cdot\mathrm{DMU}/D$$
$$\mathrm{AI} = \mathrm{LUL}\cdot(1 - \mathrm{DMU})D$$

Table V Parameter Values.

Symbol	Definition	Unit	Values		Values Based Upon:
			Model I	Model II	
α	Extinction coefficient of water	m^{-1}	0.27	0.27	Chen and Orlob (1975)
β	Specific extinction coefficient of phytoplankton	$m^2\ g^{-1}$	0.18	0.18	Chen and Orlob (1975)
C	Concentration of inorganic carbon	$mg\ l^{-1}$	100	100	Measurements
CDR_{max}	Maximum growth rate of phytoplankton	day^{-1}	2.3	2.53	Calibration
D	Depth	m	1.8	1.8	Measurements
DB	Depth of biologically very active layer	m	—	$2 \cdot 10^{-3}$	Measurements
DENIT	Denitrification coefficient	day^{-1}	0.02	0.03	Nitrogen balance
DMU	Dry matter in sediment	—	0.07	0.07	Measurements
$FISH_0$	Concentration of fish	$mg\ l^{-1}$	0.55	0.3	Calibration
KA	Michaelis constant for zooplankton grazing on phytoplankton	$mg\ l^{-1}$	0.5	2.0	Chen and Orlob (1975)
KC	Michaelis constant for carbon uptake	$mg\ l^{-1}$	0.5	0.5	Chen and Orlob (1975)
KDN_{10}	Decomposition rate of detritus nitrogen at 10°C	day^{-1}	0.1	0.1	Calibration
KDP_{10}	Decomposition rate of detritus phosphorus at 10°C	day^{-1}	0.25	0.4	Calibration
KE_{20}	Decomposition rate of PE at 20°C	day^{-1}	$2.5 \cdot 10^{-4}$	$2.2 \cdot 10^{-4}$	Calibration
KL	Michaelis constant for light	$kcal\ m^{-2}\ day^{-1}$	400	400	Gargas (1975)
KN	Michaelis constant for nitrogen uptake	$mg\ l^{-1}$	0.2	0.2	Lehman *et al.* (1975), Chen and Orlob (1975)
KP	Michaelis constant for phosphorus uptake	$mg\ l^{-1}$	0.02	0.02	Lehman *et al.* (1975), Chen and Orlob (1975)
KREL	Rate constant for release of nitrogen	day^{-1}	0.0040	0.0040	Jacobson and Jørgensen (1975)
KS	Theshold zooplankton biomass	$mg\ l^{-1}$	0.75	0.75	Steele (1974)
KZ	Michaelis constant for fish feeding on zooplankton	$mg\ l^{-1}$	0.35	0.35	Calibration
LUL	Unstable layer of sediment	m	0.1	0.1	Measurements
DS	$LUL \cdot (1 - DMU)$	m	—	—	—
MA	Mortality of phytoplankton	day^{-1}	0.09	—	Calibration

(*continued over*)

Table V (*continued*)

Symbol	Definition	Unit	Values		Values Based Upon:
			Model I	Model II	
MYZ_{max}	Maximum growth rate of zooplankton	day^{-1}	0.2	0.188	Calibration
MZ	Mortality of zooplankton	day^{-1}	0.025	0.033	Calibration
NH_4P	Ammonia concentration in rainwater	$mg\ l^{-1}$	0.2	0.2	V. Jørgensen (1972)
NO_3P	Nitrate concentration in rainwater	$mg\ l^{-1}$	0.16	0.16	V. Jørgensen (1972)
ϕ	Temperature coefficient for degradation of detritus	—	1.072	1.072	
$PRED_{max}$	Maximum feeding rate of fish on zooplankton	day^{-1}	0.06	0.012	Calibration
PP	Phosphorus concentration in rainwater	$mg\ l^{-1}$	0.0015	0.0015	V. Jørgensen (1972)
RC_{max}	Maximum respiration rate of phytoplankton	day^{-1}	0.13	0.088	Calibration
RZ_{max}	Maximum respiration rate of zooplankton	day^{-1}	0.035	0.028	Calibration
SVD	Settling rate of detritus	$m\ day^{-1}$	0.002	0.0019	Jørgensen (1976)
SVS	Settling rate of *Scenedesmus*	$m\ day^{-1}$	0.06	0.19	Jørgensen (1976)
θ	Temperature coefficient for decomposition of PE	day^{-1}	1.03	1.03	Chen and Orlob (1975)
UC_{max}	Maximum rate of carbon uptake	day^{-1}	0.65	0.55	Calibration
UN_{max}	Maximum rate of nitrogen uptake	day^{-1}	0.03	0.015	Calibration
UP_{min}	Maximum rate of phosphorus uptake	day^{-1}	0.003	0.0014	Calibration
FCA_{min}	Minimum kg C per kg phytoplankton biomass	—	0.15	0.15	Jørgensen (1976)
FCA_{max}	Maximum kg C per kg phytoplankton biomass	—	0.6	0.6	Jørgensen (1976)
FNA_{min}	Minimum kg N per kg phytoplankton biomass	—	0.015	0.015	Jørgensen (1976)
FNA_{max}	Maximum kg N per kg phytoplankton biomass	—	0.10	0.10	Jørgensen (1976)
FPA_{min}	Minimum kg P per kg phytoplankton biomass	—	0.001	0.001	Jørgensen (1976)
FPA_{max}	Maximum kg P per kg phytoplankton biomass	—	0.013	0.013	Jørgensen (1976)
V	Volume of the lake	m^3	420000	420000	
Y	Yield of feeding zooplankton and fish	—	0.63	0.63	Chen and Orlob (1975)
L	$1/Y - 1$	—	0.59	0.59	—

Table VI Forcing Functions (all given as tables).

QTRI	$m^3\ day^{-1}$ tributaries
NTOTRI	Total N ($mg\ l^{-1}$) tributaries
NH_4TRI	mg NH_4^+-N l^{-1} tributaries
NO_3TRI	mg NO_3^--N l^{-1} tributaries
PTOTRI	Total P ($mg\ l^{-1}$) tributaries
PTRI	mg PO_4^{3-}-P l^{-1} tributaries
T	Temperature of lake water
QWAS	$m^3\ day^{-1}$ wastewater
NTOWAS	Total N ($mg\ l^{-1}$) wastewater
NH_4WAS	mg NH_4^+-N l^{-1} wastewater
NO_3WAS	mg NO_3^--N l^{-1} wastewater
PWAS	Total P ($mg\ l^{-1}$) wastewater
RAD	Global irradiance (kcal $m^{-2}\ day^{-1}$)
Q	Outflow ($m^3\ day^{-1}$)
QPREC	Precipitation ($m^3\ day^{-1}$)

Table VII Additional Equations, Model III.

$$\mathrm{dZOO1}/\mathrm{d}t = (\mathrm{MYZ1} - \mathrm{RZ1} - \mathrm{MZ1} - Q/V)\cdot \mathrm{ZOO1} - \mathrm{PRED1}\cdot \mathrm{FISH}/Y$$
$$\mathrm{PRED1} = \mathrm{PRED1}_{max}\cdot \mathrm{FT1}\cdot \mathrm{FZ}$$
$$\mathrm{MYZ1} = \mathrm{MYZ1}_{max}\cdot \mathrm{FD}\cdot \mathrm{FT1}$$
$$\mathrm{RZ1} = \mathrm{RZ1}_{max}\cdot \mathrm{FT1}$$
$$\mathrm{FD} = \mathrm{DET}/(\mathrm{DET} + 2.0)$$
$$\mathrm{DET} = (\mathrm{ND}/\mathrm{FNA} + \mathrm{PD}/\mathrm{FPA})/2$$

MZ1 is the mortality
ZOO1 is the concentration of zooplankton species 1.

(5) Model by Baca and Arnett

$$\frac{\partial \mathrm{PS}}{\partial t} = \underbrace{-\ G_p P A_{pp}}_{\text{phytoplankton uptake}} + \{\underbrace{[I_3\cdot \mathrm{PE}]}_{\text{sediment release}} \underbrace{-\ I_1\cdot \mathrm{PS}}_{\text{sediment uptake}}\} + \underbrace{I_2\cdot \mathrm{PD}}_{\text{gain by organic decay}}$$

and

$$\frac{\partial \mathrm{PD}}{\partial t} = \underbrace{(D_p - C_g Z) P A_{pp}}_{\text{recycled from dead phytoplankton}} + \underbrace{R_z Z A_{pz}}_{\text{recycled from zooplankton}} \underbrace{-\ \{I_4\cdot \mathrm{PD}\}}_{\text{sediment trapping}} \underbrace{-\ I_2\cdot \mathrm{PD}}_{\text{decay}},$$

where

PS is the concentration of inorganic phosphorus,
PD is the concentration of organic phosphorus,
PE is the concentration of inorganic phosphorus in sediment,
P is the phytoplankton concentration,
A_{pp}, A_{pz} are yield coefficients,
G_p is the net phytoplankton growth rate,

D_p is the phytoplankton death rate,
C_g is the zooplankton grazing rate,
R_z is the zooplankton respiration rate,
I_1 is the sediment uptake rate,
I_2 is the organic phosphorus decay rate,
I_3 is the sediment release rate,
I_4 is the sediment trapping rate,
Z is the zooplankton concentration.

The terms in braces apply to the hypolimnion and the term in square brackets designates processes that depend on anaerobic conditions.

For the nitrogen submodel, the equations are:

$$\frac{\partial C_1}{\partial t} = \underbrace{-J_1C_1}_{\text{ammonia oxidation}} - \underbrace{PG_pA_{np}\frac{C_1}{C_1 + C_3}}_{\text{phytoplankton uptake}} + \underbrace{J_4C_4}_{\text{organic nitrogen recycled}} + \underbrace{\{J_5C_5\}}_{\text{anaerobic sediment release}},$$

$$\frac{\partial C_2}{\partial t} = \underbrace{J_1C_1}_{\text{ammonia oxidation}} - \underbrace{J_2C_2}_{\text{nitrite oxidation}},$$

$$\frac{\partial C_3}{\partial t} = \underbrace{J_2C_2}_{\text{nitrite oxidation}} - \underbrace{PG_pA_{np}\frac{C_3}{C_1 + C_3}}_{\text{phytoplankton uptake}} = \underbrace{[J_3C_3]}_{\text{anaerobic denitrification}},$$

$$\frac{\partial C_4}{\partial t} = \underbrace{-J_4C_4}_{\text{organic decay}} + \underbrace{(D_p - C_gZ)PA_{np}}_{\text{recycled from dead phytoplankton}} + \underbrace{R_zZA_{nz}}_{\text{recycled from zooplankton}} - \underbrace{\{J_6C_4\}}_{\text{sediment uptake}},$$

where

C_1 is the ammonia nitrogen concentration,
C_2 is the nitrite nitrogen concentration,
C_3 is the nitrate nitrogen concentration,
C_4 is the organic nitrogen concentration,
C_5 is the sediment nitrogen concentration,
J_1 is the ammonia oxidation rate,
J_2 is the nitrite oxidation rate,
J_3 is the denitrification rate,
J_4 is the organic nitrogen decay rate,
J_5 is the sediment nitrogen decay rate,
J_6 is the sediment uptake rate,
A_{np}, A_{nz} are the nitrogen: carbon ratios for phytoplankton and zooplankton.

In these formulations the rates of phytoplankton uptake of ammonia (C_1) and nitrate (C_3) are governed by their relative concentrations in the water column. Denitrification and release of nitrogen (ammonia) from the sediment depend on the existence of anaerobic conditions.

REFERENCES

Ahlgren, I. (1973a). Limnologiska studier av Sjön Norrviken. III. Avlastningens effekter. *Scripta Limnologica Upsaliencia* No. 333.

Ahlgren, I. (1973b). Limnological studies of Lake Norrviken, a eutrophicated Swedish lake. Nutrients and phytoplankton before and after sewage diversion. *Scripta Limnologica Upsaliencia* No. 334.

Baca, R. G., and R. C. Arnett (1976). A finite-element water quality model for eutrophic lakes. *Finite Elements in Water Resources* vol. 4, eds. W. G. Gray, G. Pinder, and C. Brebbia (Plymouth, Devon: Pentech), pp. 125–147.

Beck, M. B. (ed.) (1978). Mathematical modeling of water quality. Summary Report of a IIASA Workshop, September 13–16, 1977. *Collaborative Proceedings* CP-78-10 (Laxenburg, Austria: International Institute for Applied Systems Analysis).

Bennett, J. R. (1974). On the dynamics of wind-driven lake currents. *Journal of Physical Oceanography* 4(3):400–414.

Bierman, V. J. (1976). Mathematical model of the selective enhancement of blue-green algae by nutrient enrichment. *Modeling Biochemical Processes in Aquatic Ecosystems*, ed. R. P. Canale (Ann Arbor, MI: Ann Arbor Science), pp. 1–32.

Blake, G., and S. Gentil (1979). A discrete lake model and its sensitivity study. *State of the Art in Ecological Modelling*, ed. S. E. Jørgensen (Copenhagen: International Scoiety for Ecological Modelling), pp. 527–544.

Bloomfield, J. A., R. A. Park, D. Scavia, and C. S. Zahorcak (1974). Aquatic modeling in the eastern deciduous forest biome, US International Biological Program. *Modeling the Eutrophication Process*, eds. E. Middlebrooks, D. H. Falkenberg, and T. E. Maloney (Ann Arbor, MI: Ann Arbor Science), pp. 139–158.

Canale, R. P., L. M. DePalma, and A. H. Vogel (1976). A plankton-based food web model for Lake Michigan. *Modeling Biochemical Processes in Aquatic Ecosystems*, ed. R. P. Canale (Ann Arbor, MI: Ann Arbor Science), pp. 33–74.

Caperon, J., and J. Meyer (1972a). Nitrogen-limited growth of marine phytoplankton—I. Changes in population characteristics with steady-state growth rate. *Deep Sea Research* 19:601.

Caperon, J., and J. Meyer (1972b). Nitrogen-limited growth of marine phytoplankton—II. Uptake kinetics and their role in nutrient limited growth of phytoplankton. *Deep Sea Research* 19:619.

Chen, C. W., and G. T. Orlob (1975). Ecologic simulation of aquatic environments. *Systems Analysis and Simulation in Ecology* vol. 3, ed. B. C. Patten (New York, NY: Academic Press), pp. 476–588.

Chen, C. W., M. Lorenzen, and D. J. Smith (1975). A comprehensive water-quality–ecologic model for Lake Ontario. *Report to Great Lakes Environment Research Laboratory, National Oceanic and Atmospheric Administration by Tetra Tech, Inc.*, 202 pp.

Desormeau, C. J. (1978). Mathematical modeling of phytoplankton kinetics with application to two alpine lakes. *Center for Ecological Modeling, Rensselaer Polytechnic Institute, Troy, NY Report* 4, 21 pp.

Di Toro, D. M., D. J. O'Connor, and R. V. Thomann (1971). Nonequilibrium systems in natural water chemistry. *American Chemical Society, Advanced Chemistry Series* 106:131–180.

Di Toro, D. M., D. J. O'Connor, R. V. Thomann, and J. L. Mancini (1975). Phytoplankton–zooplankton–nutrient interaction model for western Lake Erie. *Systems Analysis and Simulation in Ecology* vol. 3, ch. 11, ed. B. C. Patten (New York, NY: Academic Press), pp. 423–474.

Di Toro, D. M., R. V. Thomann, D. J. O'Connor, and J. L. Mancini (1977). Estuarine phytoplankton biomass models—Verification analyses and preliminary applications. *The Sea* vol. 6 Marine modeling, eds. E. D. Goldberg *et al.* (New York, NY: Wiley), pp. 969–1020.

Droop, M. R. (1973). Some thoughts on nutrient limitation in algae. *Journal of Phycology* 9:264.

Droop, M. R. (1974). The nutrient status of algal cells in continuous culture. *Memoranda of the Marine Biological Association, UK* 54:825–855.

Dubois, D. M. (1975). A model of patchiness for prey–predator populations. *Ecological Modelling* 1:67–80.

Dubois, D. M. (1979). Catastrophe theory applied to water quality regulation of rivers. *State of the Art in Ecological Modelling*, ed. S. E. Jørgensen (Copenhagen: International Society for Ecological Modelling), pp. 751–759.

Duckstein, L., J. Casti, and J. Kempf (1977). Modeling phytoplankton growth in small eutrophic ponds with catastrophe theory. *Proceedings of 13th American Water Resources Conference, Tucson, AZ.*

Dugdale, R. C. (1967). Nutrient limitation in the sea: Dynamics, identification, and significance. *Limnology and Oceanography* 12:685.

Eppley, R. W., and W. H. Thomas (1969). Comparison of half-saturation constants for growth and nitrate uptake of marine phytoplankton. *Journal of Phycology* 5:375.

Fuhs, G. W. (1969). Phosphorus content and rate of growth in the diatoms *Cyclotella nana* and *Thalossiosira fluviatilis*. *Journal of Phycology* 5:312.

Fuhs, G. W., S. D. Demmerle, E. Canelli, and M. Chen (1971). Characterization of phosphorus-limited planktonic algae, nutrients and eutrophication: The limiting nutrient controversy. *Proceedings of Symposium of American Society of Limnology and Oceanography and Michigan State University, East Lansing, MI, February 11–12*, p. 113.

Gargas, E. (1975). *A Manual for Phytoplankton. Primary Production Studies in the Baltic* (Baltic Marine Biologists in cooperation with the Danish Agency of the Environment).

Groden, T. W. (1977). Modeling temperature and light adaptation of phytoplankton. *Center for Ecological Modeling, Rensselaer Polytechnic Institute, Troy, NY Report* 2, 17 pp.

Halfon, E., and D. C. L. Lam (1978). The effects of advection–diffusion processes on the eutrophication of large lakes. A hypothetical example: Lake Superior. *Ecological Modelling* 4: 119–132.

Halmann, M., and M. Stiller (1974). Turnover and uptake of dissolved phosphate in freshwater. A study in Lake Kinneret. *Limnology and Oceanography* 19:774.

Ikeda, S., Y. Inoue, and S. Iwai (1979). Multispecies of planktons and nutrients model of lake eutrophication. A simulation study in Lake Biwa. *State of the Art in Ecological Modelling*, ed. S. E. Jørgensen (Copenhagen: International Society for Ecological Modelling), pp. 501–526.

Imboden, D. M. (1974). Phosphorus model for lake eutrophication. *Limnology and Oceanography* 19:297–304.

Imboden, D. M. (1979). Modelling of vertical temperature distribution and its implication on biological processes, in lakes. *State of the Art in Ecological Modeling*, ed. S. E. Jørgensen (Copenhagen: International Society for Ecological Modelling), pp. 545–561.

Imboden, D. M., and R. Gächter (1978). A dynamic lake model for trophic state prediction. *Ecological Modelling* 4:77–98.

Jacobson, O. S., and S. E. Jørgensen (1975). A submodel for nitrogen release from sediments. *Ecological Modelling* 1:147–151.

Jolankai, G., and A. Szöllösi-Nágy (1978). A simple eutrophication model for the Bay of Keszthely, Lake Balaton. *Modeling the Water Quality of the Hydrological Cycle. Proceedings of Baden Symposium, September. International Association of Hydrological Sciences Publication* 125, pp. 137–150.

Jørgensen, S. E. (1976). A eutrophication model for a lake. *Ecological Modelling* 2:147–165.

Jørgensen, S. E. (1978). Focus on ecological modeling. *Ecological Modelling* 4:1–2.

Jørgensen, S. E. (ed.) (1979). State of the art of eutrophication models. *State of the Art in Ecological Modelling* (Copenhagen: International Society for Ecological Modelling), pp. 293–298.

Jørgensen, S. E., L. Bengtsson, and M. Friis (1982). A case study of lake modeling, Södra Bergundasjön. *International Society for Ecological Modelling Journal* 4.

Jørgensen, S. E., and D. R. F. Harleman (eds.) (1978). Hydrophysical and ecological modelling of deep lakes and reservoirs. Summary Report of a IIASA Workshop, December 12–15, 1977. *Collaborative Proceedings* CP-78-7 (Laxenburg, Austria: International Institute for Applied Systems Analysis).

Jørgensen, S. E., L. Kamp-Nielsen, and O. S. Jacobsen (1975). A submodel for anaerobic mud–water exchange of phosphate. *Ecological Modelling* 1:133–146.

Jørgensen, S. E., and H. F. Mejer (1977). Ecological buffer capacity. *Ecological Modelling* 3:39–61.

Jørgensen, S. E., and H. F. Mejer (1979). A holistic approach to ecological modelling. *Ecological Modelling* 7:169–189.

Jørgensen, S. E., H. F. Mejer, and M. Friis (1978). Examination of a lake model. *Ecological Modelling* 4:253–278.

Jørgensen, S. E., H. F. Mejer, L. A. Jørgensen, and L. Kamp-Nielsen (1981). Parameter estimation for a eutrophication model. *Ecological Modelling* 13:111–129.

Jørgensen, S. E., O. S. Jacobsen, and I. Høi (1973). Prognosis for a lake. *Vatten* 29:383–404.

Jørgensen, V. (1972). Nedbarens indhold of plantluaringsstof. *Statens Plantlavlsmode* 22 November, 15pp.

Kamp-Nielsen, L. (1975). A kinetic approach to the aerobic sediment–water exchange of phosphorus in Lake Esrom. *Ecological Modelling* 1:153–160.

Kinnunen, K., J. S. Niemi, and J. Eloranta (1978). Adaptation of the EPAECO model to a lake in central Finland. *Modeling the Water Quality of the Hydrological Cycle. Proceedings of Baden Symposium, September. International Association of Hydrological Sciences Publication* 125, pp. 115–127.

Kremer, J. N., and S. W. Nixon (1978). *A Coastal Marine Ecosystem Simulation and Analysis* (Berlin: Springer), 217 pp.

Lappalainen, K. M. (1975). Phosphorus loading capacity of tubes and a mathematical model for water quality prognoses. *Proceedings of 10th Nordic Symposium on Water Research, Vœrløse, May 20–22, 1974* (Helsinki: "Entrofierung" NORFORSK).

Larsen, D. P., H. T. Mercier, and K. W. Malveg (1974). Modeling algal growth dynamics in Shagawa Lake, Minnesota. *Modeling the Eutrophication Process*, eds. E. J. Middlebrooks, D. H. Falkenberg, and T. E. Maloney (Ann Arbor, MI: Ann Arbor Science), pp. 15–33.

Lehman, J. T., D. B. Botkin, and G. E. Likens (1975). The assumptions and rationales of a computer model of phytoplankton population dynamics. *Limnology and Oceanography* 3:343–364.

Lorenzen, M. W., O. J. Smith, and L. V. Kimmel (1976). A long-term phosphorus model for lakes: Application to Lake Washington. *Modeling Biochemical Processes in Aquatic Ecosystems*, ed. R. P. Canale (Ann Arbor, MI: Ann Arbor Science), pp. 75–92.

Mauersberger, P. (1979). On the role of entropy in water quality modelling. *Ecological Modelling* 7: 191–199.

Mejer, H. F., and S. E. Jørgensen (1979). Energy and ecological buffer capacity. *State of the Art in Ecological Modelling*, ed. S. E. Jørgensen (Copenhagen: International Society for Ecological Modelling), pp. 829–846.

Nyholm, N. (1975). Kinetics studies of phosphate-limited algae growth. *Thesis, Technical University of Copenhagen.*

Nyholm, N. (1978). A simulation model for phytoplankton growth and nutrient cycling in eutrophic, shallow lakes. *Ecological Modelling* 4:279–310.

O'Melia, C. R. (1974). Phosphorus cycling in lakes. *North Carolina Water Resources Research Institute, Raleigh Report* 97, 45 pp.

Park, R. A., C. Collins, and C. Desormeau (1979a). Modeling combined effects of environmental factors on phytoplankton. *Center for Ecological Modeling, Rensselaer Polytechnic Institute, Troy, NY Report.*

Park, R. A., T. W. Groden, and C. J. Desormeau (1979b). Modification to model CLEANER, requiring further research. *Perspectives on Lake Ecosystem Modeling*, eds. D. Scavia and A. Robertson (Ann Arbor, MI: Ann Arbor Science), pp. 87–108.

Park, R. A., *et al.* (1974). A generalized model for simulating lake ecosystems. *Simulation* 23:33–50.

Parker, R. A. (1976). The influence of eddy diffusion and advection of plankton population systems. *International Journal of Systems Science* 7:957–962.

Parker, R. A. (1978). Spatial patterns in some nutrient–plankton models. *Ecological Modelling* 6: 78–90.

Patten, B. C., D. A. Egloff, and T. H. Richardson (1975). Total ecosystem model for a cove in Lake Texoma. *Systems Analysis and Simulation in Ecology* vol. 3, ed. B. C. Patten (New York, NY: Academic Press), pp. 206–423.

Rhee, G. Yu. (1978). Effects of N:P atomic ratios and nitrate limitation on algal growth, cell composition, and nitrate uptake. *Limnology and Oceanography* 23:10–25.

Richey, J. E. (1977). An empirical and mathematical approach toward the development of a phosphorus model of Castle Lake. *Ecosystem Modeling in Theory and Practice*, eds. C. A. S. Hall and I. W. Day, Jr. (New York, NY: Wiley–Interscience), pp. 267–287.

Scavia, D., B. J. Eadie, and A. Robertson (1976). An ecological model for the Great Lakes. *Great Lakes Environmental Research Laboratory, National Oceanic and Atmospheric Association, Ann Arbor, MI Contribution* 64.

Scavia, D., and R. A. Park (1975). A user's manual for the aquatic model CLEANER. *Federal Water Industry, Troy, NY Report* 75-16.

Smith, E. L. (1936). Photosynthesis in relation to light and carbon dioxide. *Proceedings of National Academy of Sciences* 22:504–511.

Smith, M. T. (1973). *Lyngby Kommune Report on Lyngby Lake, Copenhagen*, 165 pp.

Snodgrass, W. J., and C. R. O'Melia (1975). Predictive model for phosphorus lakes. Sensitivity analysis and applications. *Environmental Science and Technology* 9(10):937–944.

Steel, J. A. (1972). The application of fundamental limnological research in water supply system design and management. *Journal of Zoology* 29:41–67.

Steele, J. H. (1962). Environmental control of photosynthesis in the sea. *Limnology and Oceanography* 7:137–150.

Steele, J. H. (1974). *The Structure of Marine Ecosystems* (Oxford: Blackwell), pp. 74–91.

Straškraba, M. (1976). Development of an analytical phytoplankton model with parameters empirically related to dominant controlling variables. *Umweltbiophysik*, eds. R. Glaser, K. Unger, and M. Koch (Berlin: Akademie-Verlag), pp. 33–65.

Straškraba, M. (1979). Natural control mechanisms in models of aquatic ecosystems. *Ecological Modelling* 6:305–322.

Tapp, J. S. (1978). Eutrophication analysis with simple and complex models. *Journal of Water Pollution Control Federation* 50:484–492.

Thomann, R. V., D. M. Di Toro, R. P. Winfield, and D. J. O'Connor (1975). Mathematical modelling of phytoplankton in Lake Ontario. *US Environmental Protection Agency, National Environmental Research Center, Corvallis, OR Report* EPA 660/3-75-005.

Vollenweider, R. A. (1965). Calculation models of photosynthesis–depth curves and some implications regarding day rate estimates in primary production. *Memorie dell Istituto Italiano di Idrobiologia* 18 Suppl.:425–457.

Vollenweider, R. A. (1969). Möglichkeiten und Grenzen elementarer Modelle der Stoffbilanz von Seen. *Archiv Hydrobiologie* 66:1–36.

Vollenweider, R. A. (1975). Input–output models with special reference to the phosphorus loading concept in limnology. *Schweizerische Zeitschrift für Hydrologie* 37:53–83.

Youngberg, B. A. (1977). Application of the aquatic model CLEANER to a stratified reservoir system. *Center for Ecological Modeling, Rensselaer Polytechnic Institute, Troy, NY Report* 1, 22 pp.

CHAPTER 9: NOTATION

AREA	lake surface area
c	concentration
C_j	concentration of constituent in water element j
CONR1	rate constant for conversion of algal to particulate phosphorus
CONR2	rate constant for conversion of particulate to soluble phosphorus
E	eddy diffusion coefficient
K_1	specific rate of phosphorus transfer to sediment
K_2	specific rate of phosphorus transfer from sediment
K_3	fraction of total phosphorus input to sediment that is not available for exchange
KFLM	food level
KMFM	minimum filtering rate multiplier
KP	half-saturation constant for phosphorus
KZ	half-saturation constant for grazing
l_P	phosphorus supply rate
LIGHT	fractional reduction in MYMAX(T) in epilimnion due to availability of light
MYMAX(T)	maximum specific growth rate of phytoplankton as a function of temperature
MYZMAX	maximum specific growth rate of zooplankton
[P]	phosphorus concentration in lake
PA	concentration of algal phosphorus
PC	phosphorus concentration per cell (PCA plus internal storage)
PCA	internal dissolved phosphorus concentration
PCAMIN	minimum value of PCA
PCM	external dissolved phosphorus concentration
PHYT	concentration of algal phytoplankton
PKI	uptake constant
PO	minimum stoichiometric level of phosphorus per cell
PP	concentration of particulate (non-algal) phosphorus
PPIN	rate of supply of particulate phosphorus to epilimnion
PS	concentration of soluble phosphorus
PSED	concentration of soluble phosphorus in sediment
PSIN	rate of supply of soluble phosphorus to epilimnion
Q	outflow rate
SETTL1	settling rate constant for algal phosphorus
SETTL2	settling rate constant for non-algal particulate phosphorus
V_j	volume of water element j
VE	volume of epilimnion
VL	lake volume
VS	sediment volume

x_m	measured value
x_t	predicted value
Y	objective function
$\alpha_k^{\bar{\bar{}}}$	electivity coefficient
ρ_w	hydraulic washout coefficient
σ_P	sedimentation rate constant.

10 Modeling the Distribution and Effect of Toxic Substances in Rivers and Lakes

S. E. Jørgensen

10.1. INTRODUCTION

Only a few models attempting to describe quantitatively the distribution and effect of toxic substances have been published. However, since the impact of toxic substances on man and aquatic ecosystems is well recognized, it is important to develop such models and use them as environmental management tools.

Passage of the Toxic Substances Act of 1976 in the US, unprecedented fines, and continual development of data on lethal and sublethal effects attest to the expansion of control on the production and discharge of such substances. As a result, considerable effort has been devoted in recent years to the development of predictive schemes that would permit a judgment of the effects of a given compound on the environment.

Many water constituents are toxic at certain concentrations and interact directly with the biota of the ecosystem, causing death or severe stress and limiting the use of water resources. With growing industrialization and a steadily increasing number of new toxic compounds, there is indeed a great need for development of water quality models that can be used for predicting safety levels and establishing water quality criteria. This modeling effort should take account of: (i) heavy metals; (ii) oils and chlorinated hydrocarbons; (iii) pesticides; (iv) other organic toxic compounds; and (v) radionuclides. The basis for modeling the distribution and effect of toxic substances in an aquatic ecosystem is knowledge of the processes in the ecosystem. Section 10.2 is devoted to a survey of our present knowledge of these processes and the related coefficients.

The few models in use can be divided into three classes.

I. Food chain or food web models, similar to the eutrophication models presented in Chapter 9, with additional state variables to describe the concentrations of toxic substances, such as compartment models, provide great flexibility and incorporate reasonably complicated food webs. However, such models often become unwieldy and require comprehensive sets of data for their calibration and validation. If there are *m* ecological variables (or components) at *n* spatial locations, and if we consider *p* toxic substances, then there are *mnp* equations (differential or algebraic) to be solved. Thousands of equations can easily result.

II. Other modeling approaches attempt to simplify the procedure. Such models require less data and can be applied more generally, but are less accurate or give less information. Often, for instance, the model does not describe seasonal variations. However, this is not significant when the model is used as a management tool, as the relevant problem most often is to map the worst-case situation.

III. The simplifications are determined by the scope of the model. A typical environmental management problem is the relation between the amount of toxic substance discharged and the concentration in fish. A model focusing on this problem does not need to include all trophic levels, but it might be possible to solve the problem by setting up a model for the concentration of a toxic compound in fish and considering only the processes shown in Figure 10.1.

The three classes of model are surveyed in section 10.3, and their advantages and disadvantages discussed.

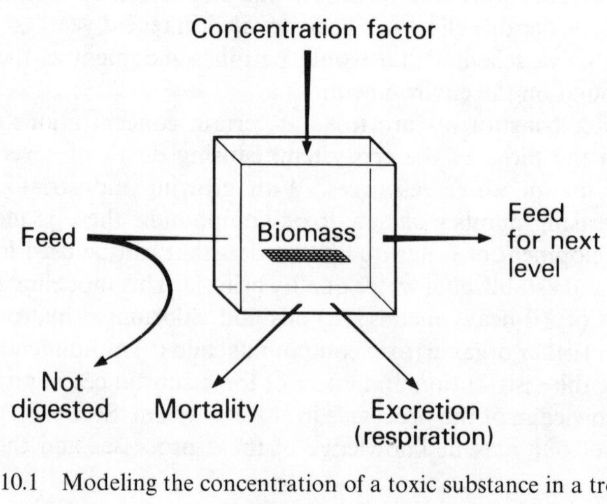

FIGURE 10.1 Modeling the concentration of a toxic substance in a trophic level.

The model should predict a concentration or level of toxic compound in water, fish, phytoplankton, etc. that can be related to the effect, such as increased mortality or decreased growth rate. Since both the model and our knowledge of effects of toxic compounds are uncertain it is necessary to apply high safety factors for the assessment of water quality criteria. The relation of this type of model to other ecological models was mentioned in section 4.2.10, where the modification of mortality and growth rate by toxic substances was demonstrated.

10.2. PROCESSES AND RELATED PARAMETERS

10.2.1. General Considerations

Ecosystem processes in which toxic substances occur can be divided into three groups:

(1) physical processes, such as adsorption and volatilization;
(2) chemical processes, such as oxidation, photolysis, hydrolysis, ionization, and complexation; and
(3) biological processes, such as biodegradation, uptake, and excretion.

The processes are included in the total models. This section contains some general ideas about these processes and the parameters used to describe them. Detailed information is available for a limited number of toxic compounds (Jørgensen, 1979b); when it is not obtainable, general rules have to be used.

10.2.2. Volatilization from Water

The theory of this process has been developed by several people (Liss and Slater, 1974; MacKay and Cohen, 1976; Smith *et al.*, 1977). Figure 10.2 illustrates the major features of a two-film model of mass transfer, which is generally applied in chemical engineering. The water phase is assumed to be well mixed so that any volatile compound is at a uniform concentration C_S except in the vicinity of the interface. A stagnant liquid film of thickness δ_l separates the bulk of the water phase from the interface. A volatile component moves through this film by diffusion. The concentration decreases across the film from C_S to C_{Si}, and the rate at which the component is transported across this film, N_S, is given by

$$N_S = K_l^S(C_S - C_{Si}), \qquad (10.1)$$

where K_l^S is the liquid film mass transfer coefficient (m h^{-1}). A stagnant gas film of thickness δ_g is on the air side. The partial pressure P_{Si} on this side is related

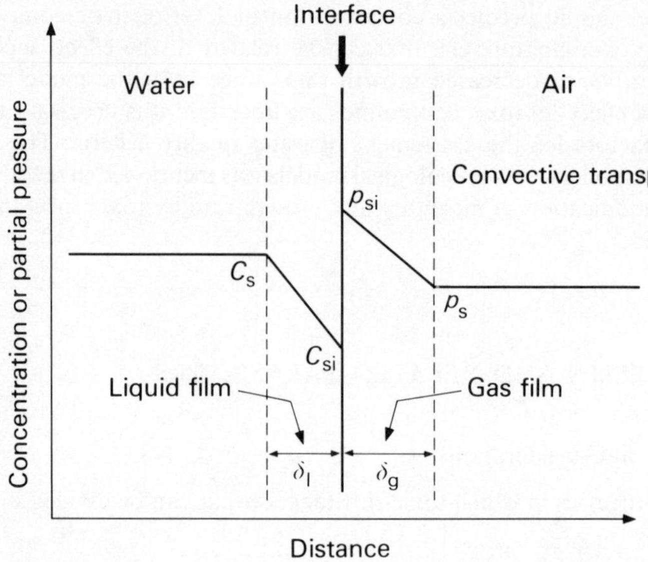

FIGURE 10.2 The two-film model of volatilization from the surface of a water body.

to C_{Si} (molar concentration on the water side) in accordance with Henry's law:

$$P_{Si} = H_c C_{Si} = H x_{Si}, \tag{10.2}$$

where H_c and H are Henry's law constants expressed in moles and as a mole fraction, respectively:

$$H_c = 1.8 \cdot 10^{-3} H. \tag{10.3}$$

The rate of transport across the gas film may be expressed by

$$N_S = \frac{K_g^S}{RT}(P_{Si} - P_S), \tag{10.4}$$

where K_g^S is the gas film mass transfer coefficient (m h^{-1}). By continuing this equation, we obtain

$$K_v^S = \frac{A}{V}\left(\frac{1}{K_l^S} + \frac{RT}{H_c^S K_g^S}\right)^{-1}, \tag{10.5}$$

where

K_v^S is the overall transfer coefficient (h^{-1}),
A is the interfacial area (m^2),
V is the liquid volume (m^3),
T is the temperature (°C).

A similar equation can be used for oxygen transport. We assume that

$$K_l = \frac{D}{\delta_l}, \tag{10.6}$$

where D is the molecular diffusion coefficient, and similarly

$$K_g = \frac{D}{\delta_g}. \tag{10.7}$$

It has been shown that, if molecules are spherical, molecular diffusion coefficients in solution are inversely proportional to molecular diameters d, so that

$$\frac{K_v^S}{K_v^O} = \frac{D^S}{D^O} = \frac{d^O}{d^S}, \tag{10.8}$$

where S indicates the toxic substance and O oxygen. If data on the diffusion coefficients or molecular diameter for the component are not obtainable, the molecular diameter can be estimated from the critical volume V_c, since

$$\frac{\pi d^3}{6} = \frac{V_c}{2N} \quad \text{or} \quad \frac{V_c}{3N}, \tag{10.9}$$

where N is the Avogadro number; the molecular diameter for oxygen, d^O, is 0.298 nm.

H_c^S can be estimated from solubility and vapor pressure:

$$H_c^S = \frac{P_S}{S_{wo}}, \tag{10.10}$$

where P_S is the vapor pressure of S in pure form and S_{wo} is the solubility in water. When data for the considered component are not available, data for a related component can be used.

10.2.3. Sorption

Sorption of toxic components on to suspended matter, sediment, and biota is a very important process. Available data might fit either the Langmuir or Freundlich adsorption isotherms. The latter are more generally used (Smith *et al.*, 1977; and for heavy metals, Reimer and Krenkel, 1973):

$$S_s = KS_w^{1/n}, \tag{10.11}$$

where

S_s is the weight of the component sorbed per gram of sorbent,
S_w is the weight in each liter or milliliter of solution,
K, n are constants.

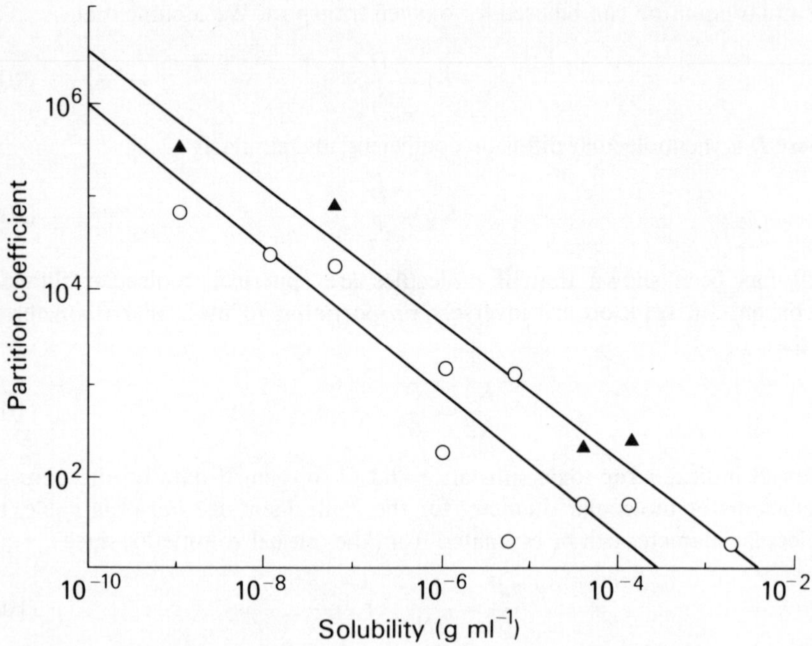

FIGURE 10.3 Partition coefficient plotted against solubility of Coyote Creek sediments (K_p, ○ measured) and of a mixed population of bacteria (K_b, ▲ measured).

At low substrate concentrations, n is often close to unity and K becomes a partition coefficient.

Smith *et al.* (1977) have shown, in a limited number of case studies, that for a given sorbent the logarithm of the partition coefficient and the logarithm of the solubility are linearly related (Figure 10.3). Although this relationship seems to be generally valid, compounds that interact by ion exchange probably would not fit this plot.

10.2.4. Chemical Oxidation

Oxidation of toxic organic components may be important under some environmental conditions. Where it is of importance, a first-order reaction scheme seems to give an acceptable, accurate description.

10.2.5. Photolysis

Photochemical transformation is a significant process for many toxic components (Wolfe *et al.*, 1976; Zepp *et al.*, 1977). The rate of absorption of light,

I_a, by a chemical is determined from

$$I_a = \epsilon I_\lambda S = k_a S, \tag{10.12}$$

where

S is the concentration of the chemical (mol l^{-1}),
ϵ is the molecular light extinction coefficient,
I_λ is the intensity of the incident light,
k_a $= \epsilon I_\lambda$.

By multiplying I_a by the quantum yield ϕ, which is the efficiency of converting absorbed light into chemical energy, we find the rate of direct photolysis:

$$-\frac{dS}{dt} = k_a \phi S = k_p S. \tag{10.13}$$

The photochemical transformation is a first-order reaction, where k_p is dependent on the intensity of the incident light. Zepp *et al.* (1977) have demonstrated how the half-lives for photolysis vary with the season because of variation in I_λ. Wolfe *et al.* (1976) have suggested that ϵ and ϕ be measured in laboratory experiments so that $k_p = f(I_\lambda)$ can be calculated from these values as a function of the hour, day, season, and latitude.

10.2.6. Hydrolysis

Hydrolysis of organic compounds usually results in the introduction of a hydroxyl (–OH) group:

$$RX + H_2O \longrightarrow ROH + HX. \tag{10.14}$$

The kinetics of hydrolysis can be expressed as:

$$R_h = k_h[S] = k_B[OH^-][S] + k_A[H^+][S] + k_N[H_2O][S], \tag{10.15}$$

where k_h, k_B, k_A, and k_N are rate constants. Few data on k_h as a function of pH are available. However, Wolfe *et al.* (1977) give kinetic data for methoxychlor and DDT and other information can be found in Wolfe *et al.* (1976).

10.2.7. Ionization and Complexation

These processes are rapid and can only be included in the model by use of equilibrium expressions. The equilibrium constant can often be found in one of the several handbooks containing the relevant data.

10.2.8. Biodegradation

As mentioned in section 4.1.2, the rate of biodegradation can be described by means of the Monod kinetic equation:

$$\mu = \mu_{\max} \frac{S}{K_S + S} \tag{10.16}$$

$$-\frac{dS}{dt} = \frac{\mu}{Y} X = \frac{\mu_{\max}}{Y} \frac{SX}{K_S + S} \tag{10.17}$$

$$\frac{dX}{dt} = \mu X, \tag{10.18}$$

where

S is the concentration of substrate,
μ is the specific growth rate,
Y is the cell yield,
X is the biomass per unit volume,
K_S is the half-saturation constant.

For many common substrates, K_S is of the order of 10^{-1} μg ml^{-1}. If this is considerably higher than S, the disappearance of substrate is a first-order process in both X and S. The maximum growth rate $\mu_{\max}$ and K_S are known for some of the more important processes (Jørgensen, 1979b).

10.2.9. Interaction Between Biota and Environment

Pollutant uptake is related to the flow of energy into the organism in the forms of food and oxygen. The balanced energy equation states that the energy of the ingested ration, corrected for fecal and nonfecal losses, $e_f R$, is equal to the metabolic rate RESP plus the growth rate dW/dt:

$$e_f R = \text{RESP} + dW/dt. \tag{10.19}$$

It is generally accepted that

$$\text{RESP} = aW^b, \tag{10.20}$$

where a and b are constants and W is the weight of the organism; b is close to 0.75–0.8 and independent of the level of metabolism (Norstrom *et al.*, 1976). The growth rate can be written as a constant MY(n) times W^g, where g is a constant.

Bioaccumulation of toxic compounds, dTC/dt, is proportional to the total respiration Q (uptake from water) and to the uptake from food:

$$\frac{d\text{TC}}{dt} = e_{pf} C_{pf} R + e_{pw} C_{pw} V, \tag{10.21}$$

where

C_{pf} is the concentration of pollutant in the food,
e_{pf} is the efficiency of pollutant uptake from food,
C_{pw} is the concentration of the toxic compound in the water,
e_{pw} is the efficiency of pollutant transfer from water,
R is the ingestion rate,
V is the respired volume of water.

V is inversely proportional to the oxygen concentration and proportional to RESP.

Body clearance of toxic compounds has been shown to follow a first-order reaction scheme (Matida *et al.*, 1971). The rate of toxicant clearance can be written as:

$$\frac{\mathrm{dTC}}{\mathrm{d}t} = k_{cl} \cdot \mathrm{TC} \cdot W^{f}, \tag{10.22}$$

where the clearance coefficient k_{cl} depends mainly on the type of toxic substance, and f is a constant between -0.2 and -0.8. All the parameters mentioned above are dependent upon temperature and oxygen concentration (Jørgensen, 1976).

Since many parameters are required in this submodel to represent the great number of toxic and biotic substances, some parameter generalization would be very useful where the parameters are not available. For e_f, an average value of 0.67 can generally be used; e_{pf} and e_{pw} are highly dependent on the toxic substance, but only slightly dependent on the biota. The growth rate and metabolic rate per unit weight can be found as functions of weight from Figures 10.4 and 10.5 if there is a lack of data.

The excretion rate as a function of length often follows the metabolic rate, which is confirmed by Thomann (1978) for polychlorobiphenyl (PCB) (Figure 10.6) and by Jørgensen (1979a) for cadmium (Figure 10.7). There seems, furthermore, to be a proportional relationship between k_{cl} and water solubility, although the amount of statistical material to support this observation is not yet sufficiently large to allow generalization.

10.2.10. Migration of Trace Metals Across Water–Sediment Interface

Heavy metals accumulate in sediment, and it is therefore of special interest to include a detailed submodel for migration of such trace metals across interfaces between water and sediment. Brooks *et al.* (1968) studied the equilibrium between sediment and interstitial water in basins of Southern California borderland and found that none of the metals considered could exist in solution in the amounts measured if they were bound as simple sulfides (Table 10.1). It seems

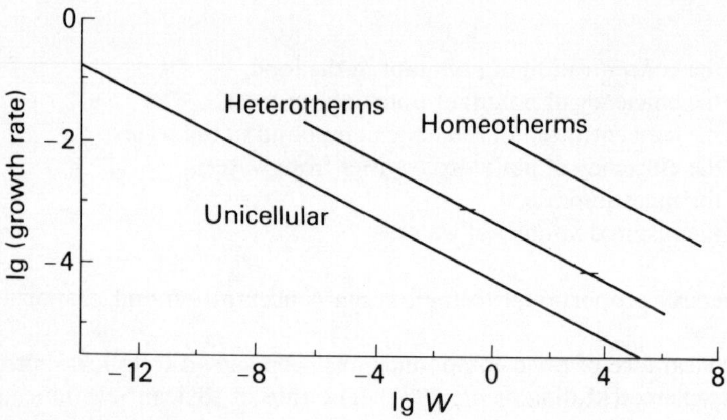

FIGURE 10.4 The relationship between the intrinsic rate of natural growth (cal g^{-1} h^{-1}) and weight (g) for various organisms.

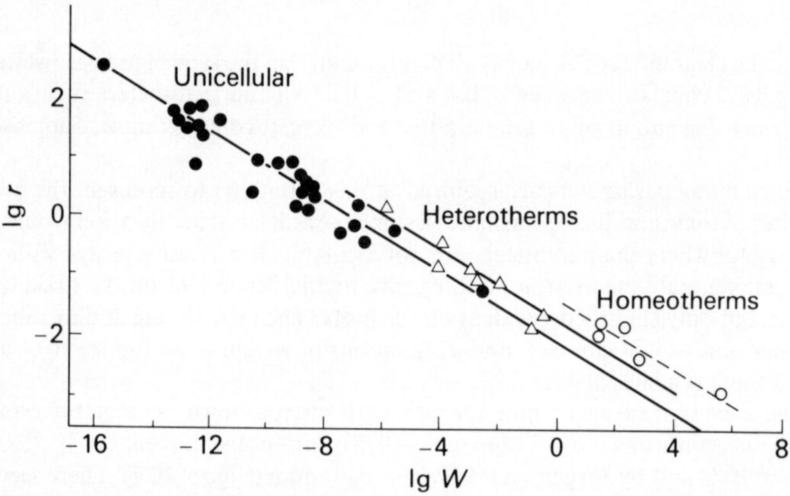

FIGURE 10.5 The relationship between metabolic rate r (day^{-1}) and weight (g) for various organisms.

TABLE 10.1 Solubilities of Metals (Brooks *et al.*, 1968).

Sulfide	K_{sp} ($mol^2\,l^{-2}$)	Solubility in S^{2-}-Free Water (ppb)	Solubility in Seawater with $\Sigma S < 10^{-3}$M (ppb)	Actual Concentration (ppb)
CdS	$3.6 \cdot 10^{-29}$	$8 \cdot 10^{-4}$	$5 \cdot 10^{-12}$	0.2–8
CoS	$3.0 \cdot 10^{-26}$	0.01	$2 \cdot 10^{-9}$	0.4–3
CuS	$8.5 \cdot 10^{-45}$	$6 \cdot 10^{-12}$	$6 \cdot 10^{-23}$	0.7–14
FeS	$3.7 \cdot 10^{-29}$	39	$3 \cdot 10^{-2}$	1–69
NiS	$1.4 \cdot 10^{-24}$	0.08	$1 \cdot 10^{-7}$	2–16
ZnS	$1.2 \cdot 10^{-23}$	0.3	$9 \cdot 10^{-7}$	1–152

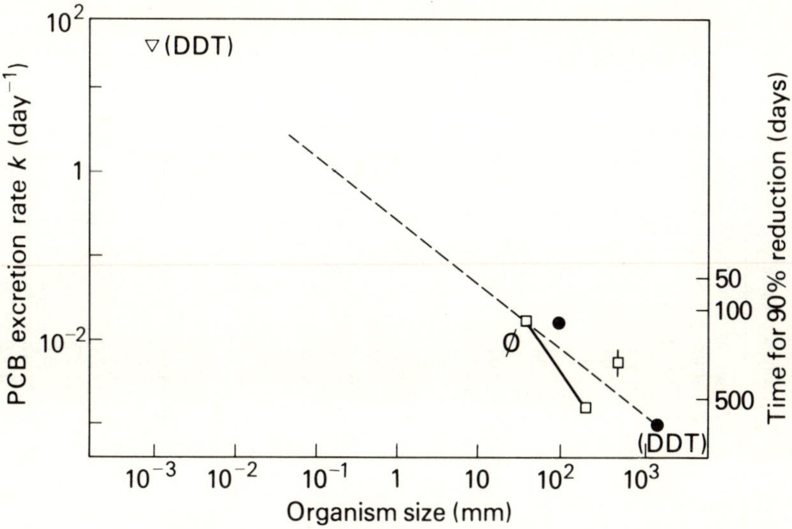

FIGURE 10.6 Variation of PCB excretion rate with organism size.

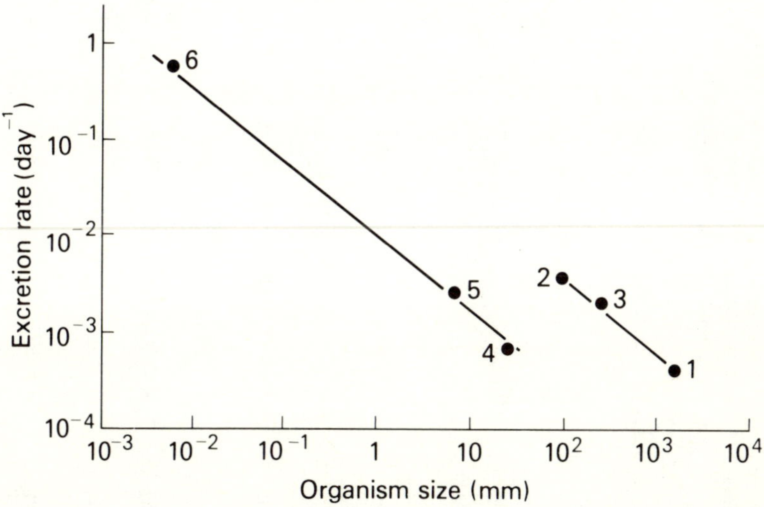

FIGURE 10.7 Rate of excretion of cadmium (day^{-1}) plotted against the length of various organisms. 1, man; 2, mice; 3, dogs; 4, oysters; 5, clams; 6, phytoplankton.

TABLE 10.2 Major Soluble Metal Species in Interfacial Seawater from Equilibrium Calculations.

Trace Metal	Environmental Condition	Assumed Controlling Solid	Solubility Product K_{sp} ($\ln pK_{sp}$)[a]	Major Soluble Complex (at pH 7.5)		
				Major Species	Formation Constant β ($\ln \beta$)[a]	Percentage of Total Soluble Concentration
Cd(II)	Oxidizing	$CdCO_3$	13.59	$CdCl^+$	2.69	56.5
				$CdCl_2^0$	2.69	15.2
				$CdCl_4^{2-}$	2.91	10.0
				$CdCl_6^{4-}$	2.25	9.1
				$CdCl_3^-$	2.78	9.0
Cd(II)	Reducing	CdS	26.96	$Cd(HS)_2^0$	15.21	97.2
				$Cd(HS)_3^-$	17.09	2.2
				$Cd(HS)_4^{2-}$	19.26	0.1
				$Cd(HS)^+$	7.94	<0.1
				$CdCl^+$	2.69	<0.1
Cr(III)	Reducing	$Cr(OH)_3$	33.52	$Cr(OH)_4^-$	33.12	90.2
				$Cr(OH)_2^+$	19.15	9.2
				$Cr(OH)^{2+}$	10.63	0.1
				Cr(citric acid)^0	−5.55[b]	<0.1
				Cr(serine)$^{2+}$	8.79	<0.1
Cu(II)	Oxidizing	$Cu_2CO_3(OH)_2$	33.16	$Cu[B(OH)_4]_2^0$	12.55	33.6
				$CuCO_3^0$	6.77	30.1
				$Cu[B(OH)_4]^+$	7.13	27.7
				$CuCl^+$	1.58	4.6
				$Cu(CO_3)_2^{2-}$	10.01	3.1
Cu(II)	Reducing	CuS	36.38	$Cu(HS)_3^-$	−4.40[b]	99.4
				$CuS(HS)_3^{3-}$	−4.40[b]	0.5
				Cu(histidine)$^+$	11.71	<0.1
				Cu(histidine)$_2^0$	19.73	<0.1
				$Cu[B(OH)_4]_2^0$	12.55	<0.1
Fe(III)	Oxidizing	$Fe(OH)_3$	39.29	$Fe(OH)_4^-$	35.29	97.1
				Fe(citric acid)$^-$	28.14	1.5
				$Fe(OH)_2^+$	20.37	1.2
				Fe(*p*-OH benzoic acid)$^+$	15.2	<0.1
				$Fe[B(OH)_4]_2^+$	15.6	<0.1

Fe(II)	Reducing	FeS	16.90	$Fe(citric\ acid)^{2-}$	18.64	86.6
				$FeHPO_4^0$	7.03	9.0
				Fe^{2+}	—	2.4
				$FeCl^+$	0.38	0.9
				$FeCl_2^0$	0.81	0.7
Hg(II)	Oxidizing	$HgCl_2$	15.10	$HgCl_4^{2+}$	15.6	80.4
		HgO	26.24[b]	$HgCl_3^+$	14.8	15.4
				$HgCl_2^0$	13.9	4.1
				$Hg(OH)_2^0$	22.4	<0.1
				$HgClBr^0$	2.01[b]	<0.1
Hg(II)	Reducing	HgS	53.89	HgS_2^{2-}	0.57[b]	98.5
				$Hg(cysteine)^0$	46.2	1.2
				$HgS(HS)_2^{2-}$	−3.70[b]	0.2
				$Hg(HS)_3^-$	−3.50[b]	<0.1
				$HgS(H_2S)_2^0$	−4.25[b]	<0.1
Mn(II)	Oxidizing	$Mn(OH)_2$	12.19	$MnCl^+$	1.06	38.1
		$MnCO_3$	9.20	$MnCl_2^0$	1.52	34.0
		MnO_2	0.92[b]	Mn^{2+}	—	17.2
				$MnCl_3^-$	1.11	7.4
				$MnHCO_3^+$	1.8	2.1
Mn(II)	Reducing			$MnCl^+$	1.08	38.2
		$MnCO_3$	9.20	$MnCl_2^0$	1.52	34.1
		MnS (pink)	9.70	Mn^{2+}	—	17.2
		MnS (precipitate)	15.70	$MnCl_3^-$	1.11	7.4
				$MnHCO_3^+$	1.8	1.7
Ni(II)	Oxidizing	$NiCO_3$	8.2	Ni^{2+}	—	47.0
		$Ni(OH)_2$ (fresh)	14.81	$NiCl^+$	0.72	38.4
				$NiCl_2^0$	0.70	10.0
		$Ni(OH)_2$ (aged)	17.31	$NiSO_4^0$	2.18	4.2
				$Ni(OH)^+$	4.70	0.3
Ni(II)	Reducing	NiS(α)	18.5	$Ni(cysteine)^0$	10.48	99.0
		NiS(γ)	25.7	$Ni(tyrosine)_2^0$	10.36[b]	<0.1
				$Ni(histidine)^+$	9.79	<0.1
				$Ni(histidine)_2^0$	17.76	<0.1
				Ni^{2+}	—	<0.1

(continued over)

TABLE 10.2 (*continued*)

Trace Metal	Environmental Condition	Assumed Controlling Solid	Solubility Product K_{sp} (ln pK_{sp})[a]	Major Soluble Complex (at pH 7.5)		
				Major Species	Formation Constant β (ln β)[a]	Percentage of Total Soluble Concentration
Pb(II)	Oxidizing	$PbCO_3$	13.30	$PbCO_3^0$	7.4	64.6
				$PbCl_6^{4-}$	3.02	28.7
				$PbCl^+$	1.60	2.4
				$Pb(CO_3)_2^{2-}$	9.89	2.0
				$PbCl_2^0$	1.78	1.0
Pb(II)	Reducing	PbS	27.65	$PbS(HS)^-$	-6.9^b	83.9
				$PbS(H_2S)^0$	-6.8^b	16.0
				$PbCO_3^0$	7.4	<0.1
				$PbCl_6^{4-}$	3.02	<0.1
				$PbCl^+$	1.60	<0.1
Zn(II)	Oxidizing	$ZnCO_3$	10.65	Zn^{2+}	—	38.1
		$ZnSiO_3$	21.03^b	$Zn(OH)_2^0$	12.89	29.6
				$ZnCl^+$	0.43	18.0
				$ZnCl_2^0$	0.61	6.6
				$ZnSO_4^0$	2.27	4.3
Zn(II)	Reducing	ZnS	22.60	$Zn(HS)_3^-$	-3.0^b	99.2
				Zn^{2+}	—	0.2
				$Zn(OH)_2^0$	12.89	0.1
				$ZnCl^+$	0.43	<0.1
				$ZnCl_2^0$	0.61	<0.1

[a] All values of pK_{sp} and β are corrected to the conditions of $I = 0$ and $T = 12°C$ using the Davies modification of the Debye–Huckel expression and the van't Hoff equation. When ΔH_r (enthalpy of reaction) was unknown, K_{sp} and β values for the other temperature (usually $T = 15$–$25°C$) were used.

[b] $HgO(s) + H_2O \rightleftharpoons Hg^{2+} + 2OH^-$; $MnO_2(s) + 2H^+ \rightleftharpoons Mn^{2+} + \frac{1}{2}O_2 + H_2O$; $ZnSiO_3(s) + H_2O \rightleftharpoons Zn^{2+} + 2OH^- + SiO_2(s)$; Cr^{3+} + citric acid $(H_3L) \rightleftharpoons CrL^0 + 3H^+$; $CuS(s) + H_2S(aq) + HS^- \rightleftharpoons Cu(HS)_3^-$; $CuS(s) + 3HS^- \rightleftharpoons CuS(HS)_3^{3-}$; $Hg^{2+} + Cl^- + Br^- \rightleftharpoons HgClBr^0$; $HgS(s) + S^{2-} \rightleftharpoons HgS_2^{2-}$; $HgS(s) + 2HS^- \rightleftharpoons HgS(HS)_2^{2-}$; $HgS(s) + H_2S(aq) + HS^- \rightleftharpoons Hg(HS)_3^-$; $HgS(s) + 2H_2S(aq) \rightleftharpoons HgS(H_2S)_2^0$; Ni^{2+} + tyrosine$^-$ (or HL^-) $\rightleftharpoons Ni(HL)_2^0$; $PbS(s) + H_2S(aq) \rightleftharpoons PbS(H_2S)^0$; $PbS(H_2S)^0 \rightleftharpoons PbS(HS)^- + H^+$; $ZnS(s) + H_2S(aq) + HS^- \rightleftharpoons Zn(HS)_3^-$.

most probable that complexes form and solubilize the metals even in the presence of sulfide.

In a reducing environment Fe and Mn are released to the interfacial seawater in significant quantities, while concentrations of other metals, such as Cd, Cu, Ni, Pb, and Zn, decrease in comparison with the original seawater background. These changes are related to the formation of sulfides. Lu and Chen (1977) suggested that these sulfide solids (except Cr, which might form $Cr(OH)_3$) would regulate the free metal ion concentration. They set up an equilibrium model controlled by the sulfides (except for chromium, which was controlled by $Cr(OH)_3$). The results of the model are shown in Table 10.2 for $T = 12\,°C$, $I = 0.7$, where I is the ionic strength. In the model the total concentrations of ligands in seawater were used to calculate the concentrations of complexes formed. Table 10.3 lists the typical total concentrations of known ligands for which data are available on stability in interstitial water. These concentrations were used in the Lu and Chen model. The model was validated for three types of sediment: silty clay, silty sand, and sandy silt, with different concentrations of heavy metals. After long-term incubation, the concentrations of Cd, Hg, Mn, Ni, Pb, and Fe in the clay sediment were found to be close to the model calculations. Therefore, it seems that the formation of solids and the complex formation control the solubility of trace metals. By use of this model as a base, it has been shown that sulfide complexes are the most important soluble species for Cd,

TABLE 10.3 Concentrations of Ligands in Interstitial Seawater (Reducing Conditions). A factor of 20 was applied because the concentrations of N and P compounds released into interfacial water are about 20 times higher than those in average seawater.

Ligand	Concentration (M)	Ligand	Concentration (M)
Total soluble carbonate	$8 \cdot 10^{-3}$	Glutamic acid	$1.09 \cdot 10^{-6}$
Total soluble borate	$6 \cdot 10^{-4}$	Glycine	$4.0 \cdot 10^{-6}$
Total soluble silicate	$5 \cdot 10^{-4}$	Glycollic acid	$7.89 \cdot 10^{-6}$
Ammonium ion	$4 \cdot 10^{-4}$	Histidine	$2.58 \cdot 10^{-7}$
Nitrite	$7 \cdot 10^{-7}$	*p*-hydroxybenzoic acid	$4.35 \cdot 10^{-7}$
Nitrate	$1.4 \cdot 10^{-6}$	Hydroxyproline	$3.05 \cdot 10^{-7}$
Orthophosphate	$2.5 \cdot 10^{-5}$	Lactic acid	$1.11 \cdot 10^{-7}$
Sulfide	$5 \cdot 10^{-4}$	Lencine	$7.63 \cdot 10^{-7}$
Sulfate	$2.8 \cdot 10^{-2}$	Lysine	$6.85 \cdot 10^{-7}$
Fluoride	$8 \cdot 10^{-4}$	Malic acid	$1.49 \cdot 10^{-5}$
Chloride	$0.5 \cdot 10^{-6}$	Methionine	$1.34 \cdot 10^{-7}$
Bromide	$8 \cdot 10^{-6}$	Ornithine	$8.47 \cdot 10^{-7}$
Iodide	$5 \cdot 10^{-7}$	Proline	$1.74 \cdot 10^{-7}$
Acetic acid	$2 \cdot 10^{-4}$	Serine	$1.90 \cdot 10^{-6}$
Alanine	$1.12 \cdot 10^{-6}$	Threonine	$8.40 \cdot 10^{-7}$
Arginine	$1.15 \cdot 10^{-7}$	Tryptophane	$9.80 \cdot 10^{-8}$
Aspartic acid	$1.2 \cdot 10^{-6}$	Tyrosine	$5.24 \cdot 10^{-7}$
Citric acid	$1.04 \cdot 10^{-6}$	Valine	$5.13 \cdot 10^{-7}$
Cysteine	$1.65 \cdot 10^{-7}$		

Hg, and Pb, organic complexes for Fe and Ni, chloride complexes for Mn, and hydroxide complexes for Cr.

Under oxidizing conditions the controlling solids may change gradually from metallic sulfides to carbonate, hydroxide, oxyhydroxide, oxide, and silicate solids, resulting in a change of the solubility of trace metals. The solubility of Cd, Cu, Ni, Pb, and Zn increases owing to the formation of more soluble solids, while the solubility of Fe and Mn might decrease because of the formation of high oxidation states, less soluble oxides, or hydroxides. Through thermodynamic calculations using one-twentieth of the concentration in the reducing conditions of organic ligands and $[NH_3] = 0$, $[NO_2^-] = 0$, $[S^{2-}] = 0$, [total carbonate] $= 10^{-2}$, [total soluble silicate] $= 4 \cdot 10^{-4}$, and $[\Sigma PO_4^{3-}] = 2 \cdot 10^{-6}$, it was found that only Zn and Cr solubilities are close to the experimental data. Cd, Cu, Ni, and Pb were far below the equilibrium concentrations and Fe and Mn far above. This discrepancy may be explained by adsorption.

From these considerations, it seems that metal migration is mainly controlled by the chemistry of the water, with the redox condition as one of the principal factors, except for Cr and ionic Hg. The released amounts of Cd, Cu, Ni, Pb, and Zn increase as the redox condition becomes more oxidizing, while the opposite is true for Fe and Mn. Equilibrium thermodynamic calculations seem to hold under reducing concentrations, but equilibrium constants for formation of humic complexes must be included for Cu, Zn, and Fe.

TABLE 10.4 Stability Constants ($-pK$) of Complexes in Water (Zitko and Carson, 1976)

Cation	Ligand			
	Glycine	ATP	Glutathione	Acetic Acid
Ca^{2+}	1.31	3.60		0.39
Mg^{2+}	3.44	4.00		0.82
Zn^{2+}	5.52	4.85	8.30	1.57
Cd^{2+}	4.80		10.50	1.70
Cu^{2+}	8.62	6.13		2.24

If data are lacking, typical seawater concentrations for ligands might be used (Table 10.3). Furthermore, these calculations can be used under oxidizing conditions for Zn and Cr, while further examination is necessary for other metals under these conditions to find the equilibrium description, as sorption phenomena constitute a major factor. Lu and Chen (1977) considered only seawater, but other ligand formations may play a role for other aquatic ecosystems. Other possible complex formations are listed in Table 10.4.

10.3. MODELS

10.3.1. Food Chain/Web Models

The literature on the transport of hazardous substances generally begins by assigning to the ecological system a series of compartments positioned in space and time. The concept of a compartment arises from a grouping of ecological properties, species, and types (e.g. phytoplankton, zooplankton, fish). The continuum of the environment is replaced by finite, discrete, interacting trophic levels. The details of each compartment must be specified, and attention is often directed toward a part of the ecosystem, depending on the aims of the model. The ecological concepts of compartment analyses have been reviewed by Patten (1971) and have already been applied in conjunction with eutrophication models. Figure 4.2 (p. 118) is based upon these concepts, used here for modeling the processes of nitrogen in lake systems.

Thomann *et al.* (1974) modeled the distribution of cadmium in western Lake Erie, applying a food chain model with a basic structure similar to that of eutrophication models. The equations are mass balances around each discrete trophic level at some position in space. The mass of toxicant per unit biomass at the considered level is used as the state variable. The rates of uptake of heavy metals directly from water by zooplankton and fish are not considered. The model includes, in addition to the model for cadmium, a general eutrophication model.

Miller's (1979) model for mercury in water considers six compartments: water, sediment, suspended matter, invertebrates, plants, and fish. Other compartment models have been constructed in food chain studies of hazardous substances. Gillett *et al.* (1974) modeled the movement of pesticides; Hill *et al.* (1976) considered the dynamic behavior of vinyl chloride in aquatic ecosystems; Lassiter *et al.* (1976) focused on the fate of mercury in aquatic systems; and Lassiter (1978) considered the dynamics of methyl parathion and benzothiophene. These models take into account the following processes (having rates V).

(A) Volatilization V_v is described as a first-order reaction:

$$V_v = k_v[\mathrm{COMP}],$$

where k_v is the rate constant and COMP the concentration of the component.

(B) Photolysis V_p is described as a first-order reaction:

$$V_{ph} = k_{ph}[\mathrm{COMP}],$$

where k_{ph} is the rate constant.

(C) Oxidation V_o is described often as a second-order reaction:

$$V_o = k_o[O_2][\mathrm{COMP}],$$

where k_o is the rate constant and $[O_2]$ the oxygen concentration.

(D) Hydrolysis V_H can proceed with hydrogen ions, water, and hydroxyl ions:

$$V_H = (k_H[H^+] + k_{H_2O} + k_{OH}[OH^-])[COMP],$$

where k_H, k_{H_2O}, and k_{OH} are rate constants.

(E) Microbial biodegradation V_b is described by means of a Michaelis–Menten expression:

$$V_b = \frac{k_b[COMP]}{k_S + [COMP]}[M], \qquad (10.23)$$

where k_b is a rate constant, [M] the activity of the microbiota, and k_S a half-saturation constant.

(F) Adsorption is considered a fast reaction. The definition may vary according to the situation as well as to the chemical substance. Fast reactions can be considered to be equilibria for which the computations are algebraic, whereas slow reactions (A to E) are best described by differential equations. A Freundlich adsorption isotherm is often used to indicate the equilibrium:

$$[CAD] = a[COMP]^b, \qquad (10.24)$$

where [CAD] is the concentration (e.g. mg per kg dry matter) of the compound adsorbed on suspended matter or sediment, and a and b are constants.

(G) Ionization and complexation are fast chemical reactions, for which the computations are algebraic. Equilibrium is described by a mass equation. As an example (Lassiter, 1978), let us consider a chemical that can exist in three ionic states: uncharged, and singly and doubly positively charged. In contact with particles (P) (suspended or sedimentary) the material can adsorb physically, or either charged form can attach to particles by cation exchange. Thus, the chemical CE can exist in six states:

		Equilibrium constant
$CE + H^+ \rightleftharpoons$	CEH^+	K_1
$CE + P \rightleftharpoons$	CEP	K_5
$CEH^+ + H^+ \rightleftharpoons$	CEH_2^{2+}	K_2
$CEH^+ + P \rightleftharpoons$	$CEH\text{-}P$	K_{51}
$CEH_2^{2+} + P \rightleftharpoons$	$CEH_2\text{-}P$	K_{52}

(10.25)

An algebraic equation can be written to represent equilibrium conditions for each reaction. An equation representing conservation of mass is also used, so that an algebraic computation for the six forms is possible. One convenient form for computation is the expression of each state as a fraction of the total chemical present; e.g., the expression for thc unionized fraction α_0 is

$$\alpha_0 = 1 + K_5[P] + K_1[H^+] + K_1K_{51}[H^+][P] + K_1K_2[H^+]^2 + K_1K_2K_{52}[H^+]^2[P], \qquad (10.26)$$

and so on for the other five fractions. Most chemicals do not present as difficult a problem as the one in the example. The importance of computing the ionized species is that kinetic reactions vary with species. Processes A to G contain explicit, quantifiable reference to the environment, such as pH, concentration of suspended matter, microbial activity, water velocity, dispersive mixing, and geometry of the water body.

(H) Interaction between the biota and the environment must also be considered. Generally, the models are structured in trophic levels 2–4, but it is also possible to consider a food web.

The processes determining the concentrations of toxic substances in an organism are shown in Figure 10.1. The organisms take up toxic substances through feed and directly from the water. The first process is described as $\mu_{max} f(\text{temp}) \cdot \text{YT} \cdot f(\text{feed})$, where μ_{max} is the maximum growth rate, $f(\text{temp})$ is a temperature function, YT is the utility coefficient for the toxic compound in feed, and $f(\text{feed})$ represents the concentration of feed. $f(\text{temp})$ and $f(\text{feed})$ were described in detail in Chapter 4. YT is strongly dependent on the toxic component.

The uptake directly from water is often described by a concentration factor, which is, however, dependent on the size of the organism. The excretion rate V_e can be described by a first-order reaction:

$$V_e = k_e[\text{COMP}]_o, \qquad (10.27)$$

where the rate constant k_e depends on the size of the organism and $[\text{COMP}]_o$ is the concentration of the toxic component in the organism, e.g. based upon dry weight. The other processes will not change the concentration of the toxic component in the organism, and the equations mentioned in Chapter 4 can be applied directly.

A compartment model requires, as demonstrated, detailed knowledge of the processes of transportation of toxic components in an aquatic environment. We know that the equations shown give a good description of the individual processes, but we know the equation parameters for only a limited number of processes. However, some general rules can be set up, as shown in section 10.2.

The state of the art is that, by means of this type of model, we can give an approximate description of the fate of a toxic component in an aquatic ecosystem, although more needs to be known about parameter values, and especially about the relationships between environmental factors and these values. Furthermore, it is noticeable that no feedback mechanisms have been taken into consideration, as our present knowledge of these is rather limited.

10.3.2. Simplifications

As mentioned in the introduction, the models described above are unwieldy and can only be calibrated and validated against comprehensive sets of data, which are available only in few cases. Consequently, a management model needs a higher degree of simplicity, in one way or another.

Thomann (1978) reduced the number of equations by discretizing ecological space. He considered the trophic position as one-dimensional:

$$\psi = Sf(L, t), \qquad (10.28)$$

where ψ is measured in milligrams of toxicant per liter and per unit length of organism, S is in milligrams of toxicant per liter, L is the trophic length, and t is time. This modeling approach uses up to three spatial dimensions plus one for the ecological dimension. The governing equation for ψ in a completely mixed water volume is

$$V\left(\frac{\partial \psi}{\partial t} + \frac{\partial V_L \psi}{\partial L}\right) = Q\psi_{\text{in}} - Q\psi + W'(L) - S'(L). \qquad (10.29)$$

We then make several assumptions. (1) The uptake of toxicant is proportional to the water concentration C_{pw} and (2) to the biomass along L; (3) excretion of toxicant occurs from the entire food chain according to first-order kinetics on ψ; and (4) $\psi_{\text{in}} = 0$, i.e. no input of toxicant mass is associated with the food chain. The governing equation then becomes:

$$V\left(\frac{\partial \psi}{\partial t} + \frac{\partial V_L \psi}{\partial L}\right) = -Q(L)\psi - K(L)V\psi + k_{\text{u}}(L)m(L)VC_{\text{pw}} \qquad (10.30)$$

and for the water phase:

$$V\frac{\mathrm{d}C_{\text{pw}}}{\mathrm{d}t} = W - QC_{\text{pw}} - \lambda VC_{\text{pw}} + V\int_{L_1}^{\infty} K(L)\psi \,\mathrm{d}C_{\text{pw}} - VC_{\text{pw}}\int_{L_1}^{\infty} k_{\text{u}}(L)m(L)\mathrm{d}L. \qquad (10.31)$$

This model seems able to simplify models of toxic substance distribution in aquatic ecosystems.

Baccini and Imboden (1977) developed a model using a thermodynamic approach by considering formation of certain stable metal species in a set of possible inorganic and organic ligands. The proposed model treats the plankton, allochthonous particles, and chemically undefined dissolved substances as uniform ligands in a given set of conditions.

A third simplification was suggested by Jørgensen (1979a). This approach includes the processes shown in Figure 10.1 for the considered trophic levels. The following differential equations for biomass and the concentration of a toxic substance in the biomass can be evaluated (the notation is in Table 4.13, pp. 140–142) (taken from Jørgensen, 1979a).

$$\frac{\mathrm{dBIO}(n)}{\mathrm{d}t} = \mathrm{BIO}(n)(\mathrm{MY}(n) \cdot \mathrm{YF}(n) - \mathrm{MORT}(n) - \mathrm{RESP}(n) - \mathrm{MY}(n+1)) \tag{10.32}$$

$$\frac{\mathrm{dTOX}(n)}{\mathrm{d}t} = \mathrm{BIO}(n)(\mathrm{MY}(n) \cdot \mathrm{YT}(n) \cdot \gamma(n-1) - \mathrm{MORT}(n) \cdot \gamma(n) - \mathrm{EXCR}(n) \cdot \gamma(n) - \mathrm{MY}(n+1) \cdot \gamma(n) + \mathrm{UT}(n) \cdot \mathrm{TOX}(0)). \tag{10.33}$$

YF is the efficiency for uptake of food and UT (day^{-1}) is the rate of uptake from water. Since

$$\gamma(n) = \frac{\mathrm{TOX}(n)}{\mathrm{BIO}(n)} \tag{10.34}$$

and

$$\gamma'(n) = \frac{\mathrm{TOX}'(n) \cdot \mathrm{BIO}(n) - \mathrm{BIO}'(n) \cdot \mathrm{TOX}(n)}{(\mathrm{BIO}(n))^2}, \tag{10.35}$$

we have

$$\frac{\mathrm{d}\gamma(n)}{\mathrm{d}t} = \mathrm{MY}(n)(\gamma(n-1) \cdot \mathrm{YT}(n) - \gamma(n) \cdot \mathrm{YF}(n)) + \gamma(n)(\mathrm{RESP}(n) - \mathrm{EXCR}(n)) + \mathrm{UT}(n) \cdot \mathrm{TOX}(0). \tag{10.36}$$

Only a few data, however, are available on the uptake rate, while many references give information about the concentration factor CF at steady state. $\mathrm{dBIO}(n)/\mathrm{d}t$, $\mathrm{dTOX}(n)/\mathrm{d}t$, and $\mathrm{d}\gamma(n)/\mathrm{d}t$ are all equal to zero, and under the circumstances of the experiments on which the CF value is based, MORT(n), MY($n+1$), and $\gamma(n-1)$ are zero as well:

$$\mathrm{MY}(n) \cdot \mathrm{YF}(n) - \mathrm{RESP}(n) = 0 \tag{10.37}$$

and

$$-\mathrm{EXCR}(n) \cdot \gamma(n) + \mathrm{UT}(n) \cdot \mathrm{TOX}(0) = 0 \tag{10.38}$$

or

$$\frac{\gamma(n)}{\mathrm{TOX}(0)} = \frac{\mathrm{UT}(n)}{\mathrm{EXCR}(n)} = \mathrm{CF}. \tag{10.39}$$

$n = 0$ corresponds to the water phase in these equations. From eqn. 10.39, UT(n) or EXCR(n) can be found if EXCR(n) or UT(n) and CF are known.

One purpose of modeling the distribution of toxic substances is to find a relationship between the input of toxic substances to the aquatic ecosystem and the approximate concentrations at different trophic levels. However, the concentration will show seasonal variations, probably reaching a maximum at the highest growth rate (summer) (Gallegos and Whicker, 1972; Betzer and Pilson, 1974). As the object of the model is to find the maximum concentration rather than to simulate the seasonal variations, it is suggested that, in addition to modeling the conditions of maximum growth, the concentration should be determined for different growth rates.

The following state variables are used.

(1) The concentration of soluble toxic substances in water, TOX(0): the differential equation is based upon:

$$\frac{\mathrm{dTOX}(0)}{\mathrm{d}t} = \sum \text{sources} - \sum \text{sinks} + \text{inflow} - \text{outflow}. \tag{10.40}$$

(2–5) The concentration of toxic substances in trophic levels 1–4: the equations are as indicated above (see eqn. 10.36). Jørgensen (1979a) discussed the dependence of concentration on temperature.

(6) The equilibrium between water and the six state variables: the concentration of adsorbed material on suspended matter, $\gamma(0)$, can be described by use of an adsorption coefficient KA, which must be found for each case study:

$$\gamma(0) = \mathrm{KA} \cdot \mathrm{TOX}(0), \tag{10.41}$$

although some indication can be given on the basis of the literature. In this context, $\gamma(n-1)$ for filter feeders (see eqn. 10.36) should be a weighted average of $\gamma(1)$ and $\gamma(0)$. An estimated ratio of the concentrations of phytoplankton and suspended matter should be used for weighting.

(7, 8) The last two state variables cover the sediment: the concentrations $\gamma(s)$ in the sediment and $\gamma(i)$ in the interstitial water. $\gamma(s)$ and $\gamma(i)$ are related by equilibrium equations, which must be established in every case study, although general expressions are available for some heavy metals (Lu and Chen, 1977). If it is critical whether the conditions are oxidizing or reducing, an oxygen model must be superimposed on the heavy metal model. $\gamma(s)$ is found on the basis of:

$$\frac{d\gamma(s)}{dt} = \frac{\gamma(s) + rat(0)\cdot\gamma(0) + rat(1)\cdot\gamma(1)}{BIOL} - SOL, \qquad (10.42)$$

where rat(0) is the mass ratio of suspended matter that has settled to active sediment and rat(1) the mass ratio of settled phytoplankton to active sediment, based upon dry matter. BIOL is the amount of dry matter in the sediment. It is not constant, since the organic matter decomposes at a rate dependent on the composition of the sediment and the temperature. SOL is the rate at which the toxic compound dissolves. The expression for SOL depends on the equilibrium equation relating $\gamma(s)$ to $\gamma(i)$. Finally, $\gamma(i)$ and TOX(0) must be related by a diffusion expression. However, the submodel for the water–sediment exchange of toxic substances requires further studies before a satisfactory description of the processes in the sediment can be set up.

10.3.3. Models Considering a Single Trophic Level

As mentioned in the introduction, it may often be possible to answer essential management questions by considering only the trophic level in focus, e.g. fish. The model for methyl mercury by Fagerstrøm and Åsell (1973) illustrates this modeling approach. The principles of this model are summarized below.

(1) Roach and pike are recruited to the model at an age of one year with a prescribed body burden of methyl mercury (MM). At a prescribed age the fate of the fish is no longer taken into account.

(2) Methyl mercury is gained via food according to:

$$\begin{pmatrix}\text{food}\\ \text{intake}\end{pmatrix}\cdot\begin{pmatrix}\text{concentration}\\ \text{of MM in food}\end{pmatrix}\cdot\begin{pmatrix}\text{assimilation}\\ \text{efficiency of MM}\end{pmatrix},$$

and via respired water according to:

$$\begin{pmatrix}\text{volume of}\\ \text{respired water}\end{pmatrix}\cdot\begin{pmatrix}\text{concentration}\\ \text{of MM in water}\end{pmatrix}\cdot\begin{pmatrix}\text{withdrawal}\\ \text{efficiency of MM}\end{pmatrix}.$$

Food intake and volume of respired water are in turn calculated from body weight and growth rate according to Winberg (1960).

(3) MM is lost by excretion, which is assumed proportional to body burden and specific metabolic rate.

(4) All metabolic rates are adjusted with respect to temperature.

(5) A pike is assumed to have a maximum and a minimum length of prey and both are linear functions of the length of the pike (Domanewski,

1962; Lawler, 1965). The optimum length is assumed to be the arithmetic mean of the two limiting lengths and the frequency distribution of prey is assumed normal, with the interval between maximum and minimum lengths equaling four standard deviations. Superimposed on this preference function is the survival function of the prey, so the final sampling function is skewed. A prescribed fraction of the prey is assumed to be pike, i.e. a certain amount of cannibalism is presupposed.

(6) The driving force for the model is the growth of the fish. Parameters for both growth and survival are treated as constants, i.e. stable size and age distributions are assumed. Deviations from this situation can be brought about by letting the parameters be, for example, time-dependent, temperature-dependent, or affected by feedback from the MM concentrations.

It was concluded that the model is able to mimic the static picture that is common in nature: (a) MM concentrations in fish increase with age; and (b) MM concentrations in predatory species exceed those in prey species by a factor of 2–5.

Aoyama *et al.* (1978) set up a model for predicting heavy metal concentrations in fish. The model considers the concentration factor (water/fish), the excretion rate, and bioaccumulation through the food chain. The model has been calibrated and validated; laboratory data were applied, with good results.

Leung (1978) developed a model of accumulation of pesticides in fish. This model considers that the following processes increase the concentration of pesticides in fish: (a) uptake associated with prey, which is ingested, (b) pesticide entering through gills, and (c) pesticide adsorbed on to the body. The model contains the following processes that decrease the concentration: (1) defecation of pesticide, (2) excretion of pesticide, (3) loss due to release of products, (4) transformation to a daughter product, (5) loss due to ingestion of fish by higher predators, and (6) loss due to nonpredatory death. The model has been applied successfully for DDT and methoxychlor.

Seip (1978) modeled the uptake of heavy metals by algae. The basis of his model is the following equation for the amount of zinc, Z_i, accumulated in age class i during the time interval Δt:

$$Z_i = [u_i C_{\mathrm{w}}^q - S_i(C_{\mathrm{A}} - C_{\mathrm{w}})]N_i \Delta t, \tag{10.43}$$

where

u_i is the uptake rate in age class i,
S_i is the secretion rate from age class i,
C_{A} is the concentration of the heavy metal in algae,
C_{w} is the concentration of the heavy metal in water,

N_i is the biomass of age class i,
q is a constant.

The concentration of zinc in the algae is given by:

$$C_A = \frac{\sum_i Z_i}{\sum N_i} + C_I, \tag{10.44}$$

where C_I is the initial concentration of the heavy metal in algae. Data from literature and from observations at a polluted and an almost unpolluted locality were used in parameterization. The model was used to predict concentration factors along a natural gradient of zinc concentrations in a Norwegian fjord. The calculated values were in good agreement with observations.

Orlob *et al.* (1980) developed a model to simulate the fate of copper in a marine environment. The model is based on the fact that the ionic form has the toxic effect. As little as 10 $\mu g\ l^{-1}$ or even less of ionic copper has sublethal effects on sensitive marine biota. Consequently, the model includes a quantitative description of the processes that determine the concentration of ionic copper: (1) formation of copper ion complexes, (2) adsorption of ionic copper on suspended matter, and (3) adsorption of complexed copper on suspended matter. The three forms, ionic copper, adsorbed copper, and complexed copper, are in an equilibrium described by three equations.

A model developed for description of the chromium distribution in a Danish firth (Mogensen, 1979) also demonstrates the use of simplifications based on a knowledge of the actual processes in the ecosystem. Since Cr(III) has a very low solubility, most of the chromium is accumulated in the sediment. The model is able to describe the distribution of chromium in the water on the basis of the analysis of several sediment cores. This pattern of distribution can again be used to relate the discharge of chromium to the sediment as a function of the distance to the point of discharge (Mogensen and Jørgensen, 1979). The accumulations of chromium in algae and in benthic animals were included in the model; other biological processes with chromium were considered insignificant.

Lam and Simons (1976) applied a model of lead nitrate spill to Lake Ontario, using an approach similar to the two mentioned above. They represent the free ion concentration, the concentration of precipitated lead, and the ability of water to precipitate lead by state variables. The distribution is described by means of a hydrodynamic model. The concentration of free lead ions is related to the toxicities of algae, invertebrates, and fish.

Gromiec and Gloyna (1973) considered the transport of radionuclides in water. The governing equation is the same as that used by Mogensen and Jørgensen (1979):

$$\frac{\partial C}{\partial t} = E\frac{\partial^2 C}{\partial x^2} - U\frac{\partial C}{\partial x} - \sum_{i=1}^{n} f_i K_i(G(C) - C_1), \tag{10.45}$$

where

C is the concentration of the toxic substance,
E is the eddy diffusion coefficient,
U is the advection flow,
G is decay, a function of C.

The first two terms defined mixing characteristics and dilution, and the third term covers uptake and release by various aquatic surfaces. This approach excludes chemical processes such as those mentioned in points A to E and G and simplifies the processes F and H. These simplifications are often allowed in transport models of heavy metals, since the governing processes for heavy metals are sorption on plants, including algae, on suspended matter, and on sediment. Gromiec and Gloyna demonstrated how such a model can be applied to the transport of radionuclides. A laboratory investigation of the sorption processes is used as the basis for parameter estimation. However, generally a degree of knowledge of the release of heavy metals from the sediment seems necessary, since a substantial part of the heavy metals will accumulate in the sediment.

The case studies by Gromiec and Gloyna and by Mogensen and Jørgensen can be used to predict concentrations in the considered species but can be included as submodels in total models as well. No case was studied by the use of two or more models, so no result of comparison is available. However, from the survey of the three classes of model it is apparent that the more comprehensive models, class I, will often require so much data and such detailed knowledge of the individual processes in the environment that this type of model will very often be omitted because of limited resources.

Whether class II or class III models should be selected for a given problem cannot be answered generally. If the problem is limited to one trophic level, or to one species or group of species, a class III model will suffice and will give more details too, while a problem that requires the entire ecosystem to be considered demands a class II model.

10.4. CONCLUDING DISCUSSION

We are today able to model the distribution and effect of only a small number of the overwhelmingly many compounds that have or might have impact on the environment. Workable models have been set up for some pesticides (DDT, methoxychlor), PCB, and some heavy metals (mercury, lead, and chromium). The models are able to describe discharge in the worst situation with acceptable accuracy. The concentrations of toxic substances in fish and the food chain accumulation can also be predicted fairly well. In other words, models of toxic substances can be used as management tools for a limited number of compounds, although some of the processes are not known in sufficient detail. However, research has intensified during the last few years, so that more knowledge has been gained about increasing numbers of compounds and processes, as well as the parameters used to describe them.

Although a general model cannot be developed for all compounds, more research must be devoted to general principles, if the task of modeling the distribution and effects of the more important toxic substances is not to be insuperable. This chapter has demonstrated some basic procedures, but more experience in the application of models of toxic substances is required before we can develop models for classes of components (the same type of equation for all components, but with different parameters that depend on physico-chemical data, such as solubility). It is not too optimistic to expect that this point will soon be reached, but simultaneous efforts in the development of total models and research into the individual processes are needed.

REFERENCES

Aoyama, I., Yos. Inoue, and Yor. Inoue (1978). Simulation analysis of the concentration process of trace heavy metals by aquatic organisms from the viewpoint of nutrition ecology. *Water Research* 12:837–842.

Baccini, R., and D. Imboden (1977). A copper model for an aquatic ecosystem. *Paper presented at SIL Conference, Copenhagen.*

Betzer, S. B., and M. E. Q. Pilson (1974). The seasonal cycle of copper concentration in *Busycon canaliculatum*. *Biological Bulletin* 142:165–175.

Brooks, R. R., B. J. Presley, and I. R. Kaplan (1968). Trace elements in the interstitial waters of marine sediment. *Geochimica et Cosmochimica Acta* 32:397–414.

Domanewski, L. N. (1962). *Bjulleten Instituta Biologii Vodochranilischtsch* 12:50 (German translation no. 160 from Swedish Natural Science Research Council).

Fagerstrøm, T., and B. Åsell (1973). Methyl mercury accumulation in an aquatic food chain. A model and implications for research planning. *Ambio* 2:164–171.

Gallegos, A. F., and F. W. Whicker (1972). Radio cesium retention by rainbow trout as affected by temperature and weight. *Report of National FNF Series* 100, 115642:1–25.

Gillett, J. W., *et al.* (1974). A conceptual model for the movement of pesticides through the environment. *National Environmental Research Center, US Environmental Protection Agency, Corvallis, OR Report* EPA 660/3-74-024, p. 79.

Gromiec, M. J., and E. F. Gloyna (1973). Radioactivity transport in water. *Final Report No. 22 to US Atomic Energy Commission, Contract* AT(11-1)-490.

Hill, J., *et al.* (1976). Dynamic behavior of vinyl chloride in aquatic ecosystems. *Environmental Research Laboratory, US Environmental Protection Agency, Athens, GA Report* EPA 600/3-76-001, p. 64.

Jørgensen, S. E. (1976). A model of fish growth. *Ecological Modelling* 2:303–313.

Jørgensen, S. E. (1979a). Modelling the distribution and effect of heavy metals in an aquatic ecosystem. *Ecological Modelling* 6:199–223.

Jørgensen, S. E. (ed.) (1979b). *Handbook of Environmental Data and Ecological Parameters* (Copenhagen: International Society for Ecological Modelling).

Lam, D. C. L., and T. J. Simons (1976). Computer model for toxicant spills in Lake Ontario. *Environmental Biogeochemistry* vol. 2 Metals transfer and ecological mass balances, ed. J. O. Nriago (Ann Arbor, MI: Ann Arbor Science), pp. 537–549.

Lassiter, R. R. (1978). Principles and constraints for predicting exposure to environmental pollutants. *US Environmental Protection Agency, Corvallis, OR Report* EPA 118-127519.

Lassiter, R. R., J. L. Malanchuk, and G. L. Baughman (1976). Comparison of processes determining the fate of mercury in aquatic systems. *Proceedings of Conference on Environmental Modeling and Simulation. US Environmental Protection Agency, Washington, DC Report* EPA 600/9-76-016, pp. 619–623.

Lawler, H. G. (1965). Toxic substances in fish. *Journal of Fisheries Research Board of Canada* 23:1358–1377.

Leung, D. K. (1978). Modeling the bioaccumulation of pesticides in fish. *Center for Ecological Modeling, Rensselaer Polytechnic Institute, Troy, NY Report* 5.

Liss, P. S., and P. G. Slater (1974). Flux of gases across the air–sea interface. *Nature* 247:181–184.

Lu, J. C. S., and K. Y. Chen (1977). Migration of trace metals in interface of seawater and polluted surficial sediments. *Environmental Science and Technology* 11:174–182.

MacKay, D., and Y. Cohen (1976). Prediction of volatilization rate of pollutants in aqueous systems. *Proceedings of Symposium on Nonbiological Transport and Transformation of Pollutants on Land and Water, 11–13 May, National Bureau of Standards, Gaithersburg, MD.*

Matida, Y., H. Kumada, S. Kimura, Y. Saiga, T. Nose, M. Yokote, and H. Kawatsu (1971). Toxicity of mercury compounds to aquatic organisms and accumulation of the compounds by the organisms. *Bulletin of Freshwater Fisheries Research Laboratory, Tokyo* 21:197–225.

Miller, D. R. (1979). Models for total transport. *Principles of Ecotoxicology Scope* vol. 12, ed. G. C. Butler (New York, NY: Wiley), pp. 71–90.

Mogensen, B. (1979). Chromium pollution in a Danish fjord. *Licentiate Thesis, Royal Danish School of Pharmacy, Copenhagen.*

Mogensen, B., and S. E. Jørgensen (1979). Modelling the distribution of chromium in a Danish firth. *Proceedings of 1st International Conference on State of the Art in Ecological Modelling, Copenhagen, 1978*, ed. S. E. Jørgensen (Copenhagen: International Society for Ecological Modelling), pp. 367–377.

Norstrom, R. J., *et al.* (1976). A bioenergetics-based model for pollutant accumulation by fish. Simulation of PCB and methyl-mercury residue levels in Ottawa River yellow perch. *Journal of Fisheries Research Board of Canada* 33:248–267.

Orlob, G. T., D. Hrovat, and F. Harrison (1980). Mathematical model for simulation of the fate of copper in a marine environment. *American Chemical Society, Advances in Chemistry Series* 189:195–212.

Patten, B. C. (ed.) (1971). *Systems Analysis and Simulation in Ecology* vol. 1 (New York, NY: Academic Press), 607 pp.

Reimer, R. S., and P. A. Krenkel (1973). Sorption phenomenon in the organics of bottom sediments. *Heavy Metals in the Aquatic Environment*, ed. P. A. Krenkel (Oxford: Pergamon), pp. 117–132.

Seip, K. L. (1978). Mathematical model for uptake of heavy metals in benthic algae. *Ecological Modelling* 6:183–198.

Smith, J. H., *et al.* (1977). Environmental pathways of selected chemicals in freshwater systems, Part I. *US Environmental Protection Agency, Corvallis. OR Report* EPA 600/7-77-113.

Thomann, R. V. (1978). Size-dependent model of hazardous substances in aquatic food chain. *US Environmental Protection Agency, Corvallis, OR Report* EPA 600/3-78-036.

Thomann, R. V., *et al.* (1974). A food chain model of cadmium in western Lake Erie. *Water Research* 8:841–851.

Winberg, G. G. (1960). *Fisheries Research Board of Canada Translation Series* No. 194 (*Nauchnye Trudy Belorusskovo Gosudarstvennovo Universiteta imeni V.I. Lenina*, pp. 1–253 (1956)).

Wolfe, N. L., R. G. Zepp, G. L. Baughman, R. C. Fincher, and J. A. Gordon (1976). Chemical and photochemical transformation of selected pesticides in aquatic systems. *US Environmental Protection Agency, Corvallis, OR Report* EPA 600/3-76-067.

Wolfe, N. L., *et al.* (1977). Methoxychlor and DDT degradation in water rates and products. *Environmental Science and Technology* 11:1077–1081.

Zepp, R. G., *et al.* (1977). Photochemical transformation of DDT and methoxychlor degradation products DDE and DMDE by sunlight. *Archives of Environmental Contamination and Toxicology* 6:305–314.

Zitko, V., and W. G. Carson (1976). The mechanism of the effect of water hardness on the lethality of heavy metals to fish. *Chemisphere* 11:299–303.

CHAPTER 10: NOTATION

A	interfacial area
a, b	constants (adsorption isotherm)
$(C_I)\ C_A$	(initial) concentration of heavy metal in algae
C_{pf}	concentration of pollutant in food
C_{pw}	concentration of pollutant in water
C_S	concentration in the liquid phase
C_{Si}	concentration at interface
C_w	concentration of heavy metal in water
CAD	concentration of adsorbed component
COMP	concentration of component
d^O	diameter of oxygen molecule
D^S, D, D^O	molecular diffusion coefficients
e_f	correction for fecal and nonfecal losses
e_{pf}	efficiency of pollutant uptake from food
e_{pw}	efficiency of pollutant uptake from water
E	eddy diffusion coefficient
$G(C)$	decay, a function of concentration
H, H_c	Henry's law constants
I	ionic strength
I_a	rate of light absorption
I_λ	intensity of light
k_{cl}	clearance coefficient
k_e	rate constant for excretion
k_o	rate constant for oxidation
k_{ph}	rate constant for photolysis
k_v	rate constant for volatilization
K_g^S	gas film mass transfer coefficient
K_l^S	liquid film mass transfer coefficient
K_S	half-saturation constant
K_v^S	overall film mass transfer coefficient

L	trophic length
[M]	activity of microbiota
MY(n)	growth rate coefficient
n	constant
N	Avogadro number
N_i	biomass of age class i
N_S	transportation rate
P_S	vapor pressure of S
P_{Si}	partial pressure at interface
Q	uptake from water
R	ingestion rate
RESP	respiration rate
S	concentration of chemical or substrate S
S_i	secretion rate
S_s	weight of component sorbed per unit of sorbent
S_w	concentration in aqueous solution
S_{wo}	solubility in water
TC	toxic compound concentration
u_i	uptake rate in age class i
U	flow velocity
UT	rate of uptake from water
V_b	microbial biodegradation rate
V_c	critical volume
V_e	excretion rate
V_H	rate of hydrolysis
V_o	rate of oxidation
V_{ph}	rate of photolysis
V_v	rate of volatilization
W	weight
x	distance
X	biomass per unit volume
Y	cell yield
YF	efficiency for uptake of food
YT	utility coefficient
Z_i	amount of heavy metal (zinc) in age class i taken up during time interval Δt
δ_g	thickness of gas film
δ_l	thickness of liquid film
ϵ	molecular extinction coefficient
μ	biodegradation rate (growth rate)
ϕ	quantum yield
ψ	mass of toxicant per liter and per unit length of organism.

11 Sensitivity Analysis, Calibration, and Validation

M. B. Beck

11.1. INTRODUCTION

In Chapter 2 a procedure for modeling was introduced. The chapter showed that the modeling procedure divided essentially into two parts: first, the development of a model from existing general theory and basic principles; and then the evaluation of the model against observations of the behavior of the field system. Throughout the intervening chapters of the book much has been said about model development, but there has been little discussion of model evaluation. The objectives of this chapter are to expand upon some of the topics introduced in Chapter 2, for example, sensitivity analysis, calibration, and validation, and thus to complete the discussion of model development with a discussion of model evaluation. This chapter, therefore, is strongly methodological in content. However, it is difficult to discuss model calibration, for instance, in the absence of experimental data, and the value of any method is best judged by its application in practice. Accordingly, we shall use illustrative case studies to support the principal themes of the chapter.

Section 11.2 deals with (*a priori*) sensitivity analysis; it resumes the discussion of section 2.4.1 of Chapter 2. The mathematical treatment of section 11.2 is introduced using the Streeter–Phelps model of BOD–DO interaction and then followed by a case study of a more complex model for stream quality of the Berkel River in the Netherlands (van Straten and de Boer, 1979). Sensitivity analysis can, of course, be carried out without the constraints of having to use field observations. In that sense, as indicated in Chapter 2, it may technically belong to the *a priori* phase of the modeling procedure. Section 11.3 examines the problem of model calibration, which covers the individual subproblems of model structure identification, parameter estimation, and verification (see section 2.6). This section provides a review of previous applications of parameter estimation algorithms in water quality modeling. It also discusses briefly the development of one particular estimation algorithm, the extended Kalman filter (EKF) (see, for example, Jazwinski, 1970), and gives two examples of the use of this algorithm: for the River Cam in England (Beck and Young, 1976) and for the Jordan River, Utah, USA (Bowles and Grenney, 1978a,b). Section 11.4 deals with model validation, using a case study of Lyngby Lake in Denmark to illustrate the long-term prediction of the response of a lake to substantially changed nutrient loading conditions (Jørgensen *et al.*, 1978).

11.2. SENSITIVITY ANALYSIS

Sensitivity analysis addresses the problem of examining the relative magnitudes of changes in the model predictions with respect to changes in the values of the model parameters $\boldsymbol{\beta}$. The main distinction between *a priori* sensitivity analysis (section 2.4.1) and *a posteriori* sensitivity analysis (section 2.7.1) is that the latter requires knowledge of the variance–covariance structure of the calibrated parameter estimation errors. One can easily apply *a priori* sensitivity analysis (as will be discussed in this chapter) to a calibrated model; the Berkel River study is one example of such an application. Hence the distinction of "*a priori*" is dropped from the title of this section, although it should be noted that (*a posteriori*) sensitivity analysis in the sense of section 2.7.1 will not be discussed.

Equation 2.1 of section 2.4.1 defined a sensitivity coefficient s_{ij} for the change Δc_i in the ith state variable of the model resulting from a change $\Delta\beta_j$ in the value of the jth parameter:

$$s_{ij} = \frac{\Delta c_i/\bar{c}_i}{\Delta\beta_j/\bar{\beta}_j}, \tag{11.1}$$

where we have normalized the relationship by including a nominal reference value $\bar{\beta}_j$ for the parameter, which would give a nominal reference value $\bar{c}_i$ for the state variable. In general, Δc_i and $\Delta\beta_j$ are understood as small changes in the neighborhoods of $\bar{c}_i$ and $\bar{\beta}_j$. As stated earlier, a definition of the type given by (11.1) enables the analyst to investigate whether a certain percentage change in a parameter has no real significance ($s_{ij} \approx 0$), whether β_j is a dominant parameter, or whether a small change in β_j induces instability in the model structure.

We can develop this intuitive notion somewhat further in order to discuss *parameter influence coefficients* or *parameter sensitivity functions* (Tomovic, 1964; Eykhoff, 1974). Let us take the example of stream BOD distribution from the linear classical Streeter–Phelps model (see also Rinaldi and Soncini–Sessa, 1978):

$$\frac{\mathrm{d}c(\tau)}{\mathrm{d}\tau} = -K_1 c(\tau); \qquad c_0 = c(\tau) \quad \text{for} \quad \tau = 0, \tag{11.2}$$

in which $c(\tau)$ is the BOD of the river at a point corresponding to a travel time τ, and K_1 is the first-order kinetic rate constant for BOD decay (stream flow is assumed to be constant with time and distance). Our objective is to compute the difference between the solution $c(\tau)$ of (11.2), given the initial condition $(\bar{c}_0 + \Delta c_0)$ and the parameter value $(K_1 = \bar{K}_1 + \Delta K_1)$, and the nominal solution $\bar{c}(\tau)$ of (11.2), given $\bar{c}_0$ and $\bar{K}_1$. Therefore, the method that we shall develop allows for analysis of the sensitivity of the model performance to changes in the initial conditions; each case is treated independently here.

First, we shall be more specific and define $c(\tau)$ to be a function not only of time of travel but also of the parameter K_1, i.e. $c(\tau; K_1)$. The solution of (11.2) can now be written as a function of the reference solution $\bar{c}(\tau; \bar{K}_1)$ and of small perturbations in the vicinity of the reference solution by taking a first-order Taylor series expansion:

$$c(\tau; K_1) = \bar{c}(\tau; \bar{K}_1) + \left[\frac{\partial c(\tau; K_1)}{\partial K_1}\right]_{\bar{K}_1} \Delta K_1 + \cdots. \tag{11.3}$$

Here, we shall define the *sensitivity coefficient* as

$$s(\tau; \bar{K}_1) \triangleq \left[\frac{\partial c(\tau; K_1)}{\partial K_1}\right]_{\bar{K}_1}. \tag{11.4}$$

The notation $[\cdot]_{\bar{K}_1}$ represents evaluation of the partial derivative at the point $K_1 = \bar{K}_1$; thus this derivative is clearly the "instantaneous" gradient of the solution c with respect to K_1, to which gradient $\Delta c/\Delta K_1$ would be an approximation—compare, therefore, (11.1) and (11.4). The significance of (11.3) is that it shows how the family of solutions $c(\tau; K_1)$ in the neighborhood of the reference solution may be quickly computed once the sensitivity coefficient of (11.4) is known. Partly for this reason we wish to obtain an expression for generating $s(\tau; \bar{K}_1)$.

If we differentiate the model relationship of (11.2) with respect to K_1 and evaluate the result for $K_1 = \bar{K}_1$, then

$$\left[\frac{\partial^2 c(\tau; K_1)}{\partial K_1 \, \partial \tau}\right]_{\bar{K}_1} = -\bar{K}_1 \left[\frac{\partial c(\tau; K_1)}{\partial K_1}\right]_{\bar{K}_1} - [c(\tau; K_1)]_{\bar{K}_1}. \tag{11.5}$$

After interchanging the order of differentiation in the second partial derivative on the left-hand side of (11.5) and applying the definition of (11.4), we have

$$\frac{ds(\tau; \bar{K}_1)}{d\tau} = -\bar{K}_1 s(\tau; \bar{K}_1) - \bar{c}(\tau; \bar{K}_1); \qquad s(0; \bar{K}_1) = 0, \tag{11.6}$$

where the initial condition for this equation has been derived from (11.4) and the analytical solution of (11.2). We have, therefore, a differential equation whose solution provides the variations of the sensitivity coefficient with respect to the independent variable for a chosen nominal reference value of the parameter, i.e. $\bar{K}_1$. We shall discuss later why this might be important.

To complete the picture, however, it is also possible, given a small change Δc_0 in the initial condition of (11.2), to obtain likewise a differential equation for an associated sensitivity coefficient, namely

$$\frac{ds(\tau; \bar{c}_0)}{d\tau} = -\bar{K}_1 s(\tau; \bar{c}_0); \qquad s(0; \bar{c}_0) = 1, \tag{11.7}$$

in which, once again, the initial condition is derived through differentiation of the analytical solution to (11.2).

Inspection of (11.7) shows that the sensitivity of the model response (i.e. the BOD along the river) to changes in the upstream BOD decays exponentially as τ, the time of travel, increases. In other words, as expected, for points very distant from the reference point (of discharge) the equilibrium BOD is essentially independent of the in-stream BOD at the reference point. The parameter sensitivity function of (11.6) is rather more interesting. Figure 11.1 shows the evolution of this equation for a nominal BOD decay rate coefficient ($\bar{K}_1$) of 0.3 day^{-1} and a BOD at the reference point (τ_0) of 10.0 g m^{-3}. The sensitivity coefficient is always negative—an increase in BOD decay rate coefficient decreases the remaining BOD—and shows a peak negative value. The minimum is the result of a balance between two opposing effects: (i) a change in K_1 has a greater effect on the remaining BOD, the more the BOD has decayed, i.e. at points further from the reference point; (ii) a change in K_1 has a relatively greater effect, the higher the absolute value of BOD, i.e. at points close to the reference point. We may say that the model response is most sensitive to the chosen value of K_1 at $\tau = 3$ (days); alternatively, when $\tau = 1$ (day) a change of $+0.05$ (day^{-1}) in K_1, i.e. $K_1 = 0.35$ (day^{-1}), would lead to an approximate change of -0.37 (g m^{-3}) in the remaining BOD at that location (from eqn. 11.3). However,

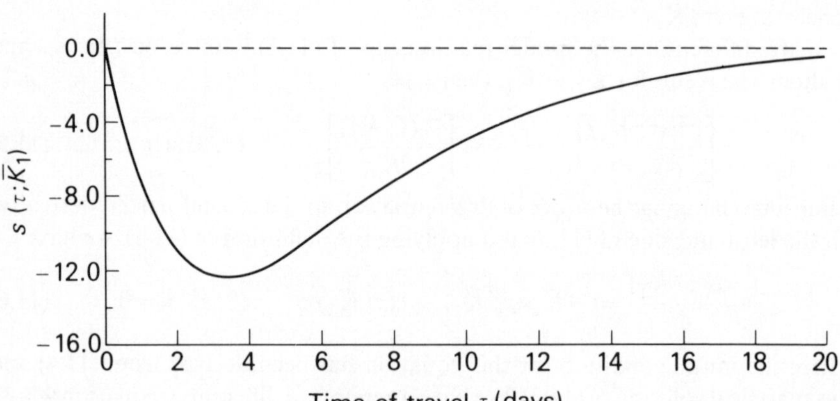

FIGURE 11.1 Coefficient $s(\tau; \bar{K}_1)$ expresses the sensitivity of the stream BOD to changes in the BOD decay rate constant K_1 (see also eqn. 11.6). The units of $s(\tau; \bar{K}_1)$ are g m^{-3} day^{-1}.

such results are only approximate, and are only valid in the locality of the nominal reference solution because of the linearization inherent in the analysis.

What can be concluded from the preceding analysis of parameter sensitivity functions? Probably the most important factor would be the concise representation of how the model solution is influenced by its parameter values *at each point in space* (or time), i.e. how model solutions differ from the nominal solution at different spatial locations. In fact, if our intuitive definition of the coefficient of sensitivity in eqn. 11.1 is recalled, it is seen to be ambiguous because it does not state how Δc_i is defined: for example, is it a difference in peak response or a difference in equilibrium response, and so on? Thus relationships of types (11.6) and (11.7) will demonstrate where the model solution is most sensitive to the parameter value, whether at a nonequilibrium (or transient, or near-field) position for $\tau \to 0$, or whether at an equilibrium (steady state, or far-field) position, $\tau \to \infty$. It would, of course, be possible to generate equivalent results by recomputing and comparing the model solutions of (11.2) for each change of the parameter value or initial condition. In this context we could also approximate these same model solutions by solving (11.6) or (11.7) simultaneously with (11.3), though it is not obvious whether there would generally be any computational advantage offered by such an alternative.

The method is not confined to representations of linear systems. In general, for the nonlinear function

$$\frac{dc_i(\tau)}{d\tau} = f_i\{c_1, c_2, \ldots, c_n, \tau; \beta_1, \beta_2, \ldots, \beta_q\}, \tag{11.8}$$

with n state variables and q parameters, the equation for the sensitivity $s_{ij}(\tau)$ of state c_i to parameter β_j is

$$\frac{ds_{ij}(\tau)}{d\tau} = \left[\frac{\partial f_i\{\cdot\}}{\partial c_1}\right]_{\bar{c}} s_{1j} + \left[\frac{\partial f_i\{\cdot\}}{\partial c_2}\right]_{\bar{c}} s_{2j} + \cdots + \left[\frac{\partial f_i\{\cdot\}}{\partial c_n}\right]_{\bar{c}} s_{nj}$$

$$+ \left[\frac{\partial f_i\{\cdot\}}{\partial \beta_j}\right]_{\bar{c}} \qquad \text{for } i = 1, 2, \ldots, n; j = 1, 2, \ldots, q. \tag{11.9}$$

The notation $[\cdot]_{\bar{c}}$ indicates that all partial derivatives are computed by substitution of those values for the state vector $\boldsymbol{c}$ and parameter vector $\boldsymbol{\beta}$ that are defined for the nominal reference solution of the model. Inspection of the subscripts i and j indicates that in principle there are q such equations as (11.9)—where $\boldsymbol{\beta}$ is a vector of q parameters—to be solved simultaneously for each of the n state variable equations. The initial conditions of (11.9) are given by $s_{ij}(0) = 0$, provided that the initial conditions of the state vector $\boldsymbol{c}_0$ are not considered as model parameters.

We shall further generalize the relationships of the type given by (11.9). Let us suppose that an n-state-variable (lumped-parameter) model is described by the following vector differential equation:

$$\frac{\mathrm{d}\boldsymbol{c}(\tau)}{\mathrm{d}\tau} = \boldsymbol{f}\{\boldsymbol{c}, \tau; \boldsymbol{\beta}\}, \tag{11.10}$$

in which $\boldsymbol{f}\{\boldsymbol{c}, \tau; \boldsymbol{\beta}\}$ is an n-element, vector-valued function. If we define an n-element sensitivity vector $\boldsymbol{s}(\tau; \beta_j)$ for the jth model parameter β_j, then the resulting n parameter sensitivity functions may be written concisely as:

$$\frac{\mathrm{d}\boldsymbol{s}(\tau; \beta_j)}{\mathrm{d}\tau} = \left[\frac{\partial \boldsymbol{f}\{\boldsymbol{c}, \tau; \bar{\boldsymbol{\beta}}\}}{\boldsymbol{c}}\right]_{\bar{\boldsymbol{c}}} \boldsymbol{s}(\tau; \beta_j) + \left[\frac{\partial \boldsymbol{f}\{\bar{\boldsymbol{c}}, \tau; \boldsymbol{\beta}\}}{\partial \beta_j}\right]_{\bar{\boldsymbol{\beta}}}. \tag{11.11}$$

The (Jacobian) matrix

$$\left[\frac{\partial \boldsymbol{f}\{\boldsymbol{c}, \tau; \bar{\boldsymbol{\beta}}\}}{\partial \boldsymbol{c}}\right]_{\bar{\boldsymbol{c}}}$$

is an $n \cdot n$ matrix with elements i, l:

$$\left[\frac{\partial f_i\{\boldsymbol{c}, \tau; \bar{\boldsymbol{\beta}}\}}{\partial c_l}\right] \quad \text{for} \quad i, l = 1, 2, \ldots, n.$$

The vector

$$\left[\frac{\partial \boldsymbol{f}\{\bar{\boldsymbol{c}}, \tau; \boldsymbol{\beta}\}}{\partial \beta_j}\right]_{\bar{\boldsymbol{\beta}}}$$

is an n-element column vector. Hence, for models more complex than the simple example of eqn. 11.2, eqn. 11.11 gives the simultaneous solution for the sensitivity of each state variable equation to changes in the value of β_j.

Eykhoff (1974), quoting Meissinger (1960), lists several further possible applications of parameter sensitivity functions. One of these applications, specifically based on the interpretation of the sensitivity coefficient as a gradient, and implicitly as the *gradient* of a model error function with respect to a parameter value, will become important later when we discuss algorithms for parameter estimation in section 11.3. In section 2.4.1 the notion of *parameter identifiability* was introduced in relation to the insensitivity of the model responses to the assumed parameter values. When one is working with large, complex models, in which it is not necessarily self-evident how each parameter affects each state variable response, sensitivity analysis may yield important insights into the properties of the model and into the likelihood of successful model calibration.

11.2.1. A Case Study: Berkel River, Netherlands

Van Straten and de Boer (1979) carried out a sensitivity analysis of a water quality model for the Berkel River in the Netherlands. This section reports some of their results.

Although only a 10 km reach of the river was chosen for the study, the reach has three weirs and, at the time of the experiment, the time of travel along the reach was 108 h. A discharge of mechanically treated sewage is located approximately 2.5 km downstream of the upstream boundary of the reach. As is typical of such lowland rivers, the growth of floating algae during low-flow, summer conditions can be considerable. For this reason a model was developed to characterize the relationships between five state variables: dissolved oxygen (DO), carbonaceous BOD (C-BOD), nitrogenous BOD (N-BOD), algae (expressed as chlorophyll *a*), and soluble reactive phosphorus (SRP). The model assumes that dispersive properties of the river can be neglected so that the partial differential equations may be solved along the characteristic trajectories of stream flow (e.g. Di Toro, 1969; see also Chapter 6). De Boer (1979) calls this a "moving cell model," since it idealizes the behavior of the system as a sequence of batch reactors moving with the flow. Thus the model has five equations written in the form of (11.10). Estimates $\hat{\boldsymbol{\beta}}$ of the parameter values for the model were obtained largely by trial-and-error fitting of the model to field observations. The model combines expressions for the various source and sink terms that are frequently found in stream dissolved oxygen models (e.g. O'Connor and Di Toro, 1970) and in lake phytoplankton models (e.g. Di Toro *et al.*, 1971).

A sample of the solutions to the sensitivity functions of eqn. 11.11 is given in Figures 11.2 and 11.3. Here the sensitivity of the state variable responses to

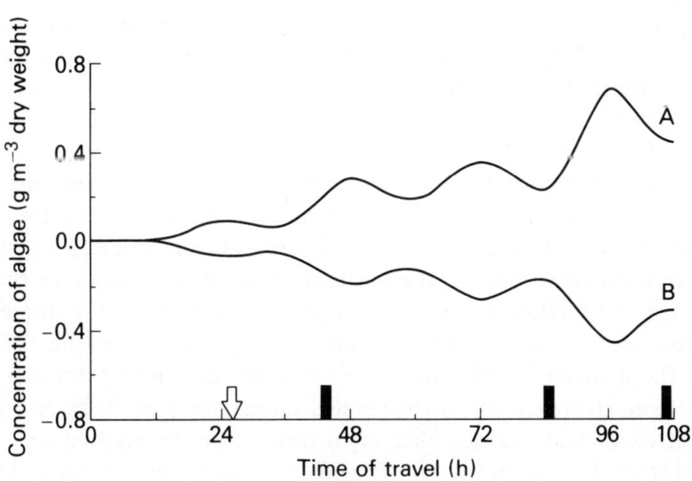

FIGURE 11.2 Solutions of the sensitivity functions of eqn. 11.11 for the Berkel River model (adapted from van Straten and de Boer, 1979). The curves represent (for a given time of travel, τ) the change in algal concentration for: A, a 10% increase in the value of the algal growth rate coefficient; B, a 10% increase in the value of the light extinction coefficient.

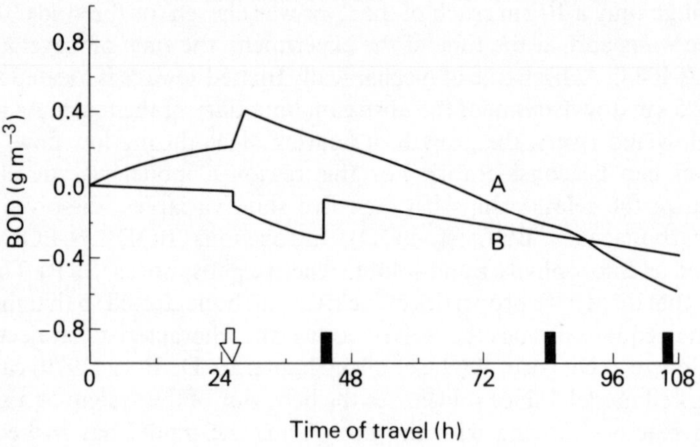

FIGURE 11.3 Solutions of the sensitivity functions of eqn. 11.11 for the Berkel River model (adapted from van Straten and de Boer, 1979). The curves represent (for a given time of travel, τ) the change in carbonaceous BOD for: A, a 10% increase in the value of the algal death rate coefficient; B, a 10% increase in the value of the carbonaceous BOD decay rate coefficient.

changes in the estimated parameter values is expressed as an absolute change in the state variable for a relative increase of 10% in the parameter value. The two trajectories of Figure 11.2 represent the sensitivity of the algal concentration to changes in: A, the algal growth rate coefficient; and B, the light extinction coefficient. A diurnal pattern is clearly evident and the growth rate coefficient is seen to have a more dominant effect than the extinction coefficient. Toward the end of the reach the 10% increase in the growth rate coefficient would alter the simulated algal concentration by 0.7 g m^{-3} (dry weight), an increase over the nominal concentration of about 35%. Figure 11.3 shows two trajectories for the sensitivity of the carbonaceous BOD to changes in the values of: A, the algal death rate coefficient; and B, the carbonaceous BOD decay rate coefficient. The sensitivity function for the former parameter is especially interesting. A higher algal death rate coefficient would initially lead to a higher BOD, as a result of the increased production of BOD from the greater amount of dead algal material. However, as τ increases less algae remain in the system and thus at downstream locations a higher algal death rate means that less BOD is produced from dead algal matter. The effect of increasing the algal death rate coefficient is, therefore, to decrease BOD levels at the downstream end of the reach. Trajectory B in Figure 11.3 shows that BOD is particularly sensitive to the value of the BOD decay rate coefficient between the point of wastewater discharge and the first weir. In fact it was assumed that a higher rate of BOD decay would occur in this reach because of the easily degradable substances present in an effluent that has not received any biological treatment. Curve B, therefore, reflects a 10% increase in a relatively higher value for the BOD decay rate coefficient over that stretch of river.

All the trajectories of Figures 11.2 and 11.3 exhibit a trend toward higher sensitivity of the model responses as the distance downstream increases. One may conclude, therefore, that: (a) if the estimated parameter values are not correct, then predictions from the model become increasingly unreliable at higher travel times; and (b) a greater experimental sampling effort should be allocated to the lower reaches of the system, for it is here that the model suggests the behavior of the system will be most sensitive with regard to subsequent calibration of the parameter values.

11.3. CALIBRATION AND VERIFICATION

The subject of model calibration is subdivided into the two problems of (*a posteriori*) model structure identification and parameter estimation. In Chapter 2 we pointed out that the application of a parameter estimation algorithm, particularly a recursive estimation algorithm, is frequently a part of the process of solving the model structure identification problem (see also Beck, 1979a). In this section, therefore, methods of model structure identification will not be discussed as a separate issue in itself. Several methods of model structure identification can be found in the literature (e.g. Box and Jenkins, 1970; Akaike, 1974; Van den Boom and Van den Enden, 1974; Unbehauen and Göhring, 1974; Wellstead, 1976, 1978), although almost invariably they address the problem in terms of black box model representations. It was also suggested in Chapter 2 that model structure identification, at least for the internally descriptive models that are the principal subject matter of this book, is the crucial problem in calibration and technically very difficult to solve. It will be apparent that one of the case studies used to illustrate the application of a parameter estimation algorithm is, strictly speaking, a study of the problem of model structure identification.

In general, it is beyond the scope of this book to discuss estimation algorithms in detail. The texts by Eykhoff (1974) and Gelb (1974) provide suitably instructive introductions to the subject; the book by Mehra and Lainiotis (1976) gives a somewhat more specialized treatment of later developments. The review section of this chapter, section 11.3.4, will in addition give references to several other illustrative applications of estimation algorithms. The discussion of parameter estimation begins, however, with the recall of some basic introductory concepts from Chapter 2. This is followed by a brief summary of the principal components of off-line estimation algorithms and then a more detailed treatment of recursive parameter estimation algorithms.

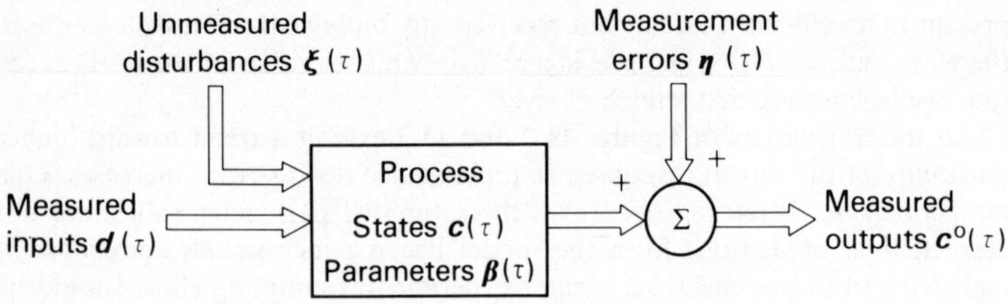

FIGURE 11.4 Definition of the system and its associated variables.

11.3.1. Parameter Estimation

Figure 11.4 gives the general definition of the (water quality) system and its associated variables, as discussed in Chapter 2. Already, however, we have restricted the discussion to *lumped-parameter models* (either dynamic or steady state) by specifying only one independent variable τ (meaning either time or time of travel). This is partly for clarity and simplicity, although to a great extent this is indicative of the limitations of currently published studies of parameter estimation in water quality models. Our treatment of parameter estimation focuses, therefore, on the problem as specified by the following model form.

State vector variations:

$$\dot{\boldsymbol{c}}(\tau) = \boldsymbol{f}\{\boldsymbol{c}(\tau), \boldsymbol{\beta}(\tau), \boldsymbol{d}(\tau)\} + \boldsymbol{\xi}(\tau). \tag{11.12a}$$

Output response observations:

$$\boldsymbol{c}^{o}(\tau) = \boldsymbol{m}\{\boldsymbol{c}(\tau), \boldsymbol{\beta}(\tau)\} + \boldsymbol{\eta}(\tau). \tag{11.12b}$$

The dot notation in (11.12a) denotes differentiation with respect to τ.

- $\boldsymbol{d}(\tau)$ is a vector of measured input forcing variables,
- $\boldsymbol{c}(\tau)$ is a vector of state variables (the symbol $\boldsymbol{c}$ is used because, in general, the state variables are here represented by the concentrations of dissolved and suspended materials in the water body),
- $\boldsymbol{c}^{o}(\tau)$ is a vector of measured output response variables,
- $\boldsymbol{\beta}(\tau)$ is a vector of model parameters,
- $\boldsymbol{\xi}(\tau)$ is a vector of unmeasured system disturbance variables,
- $\boldsymbol{\eta}(\tau)$ is a vector of measurement errors,
- $\boldsymbol{f}, \boldsymbol{m}$ are nonlinear vector-valued functions.

The basic requirement of parameter estimation is that *estimated model output responses* $\hat{\boldsymbol{c}}^{o}(\tau)$, computed when the input measurements $\boldsymbol{d}(\tau)$ are given, i.e.

$$\hat{\boldsymbol{c}}^{o}(\tau) = \boldsymbol{f}'\{\hat{\boldsymbol{c}}(\tau), \hat{\boldsymbol{\beta}}(\tau), \boldsymbol{d}(\tau)\}, \tag{11.13}$$

can be compared with observed values for the outputs $\boldsymbol{c}^{\mathrm{o}}(\tau)$. Estimates of the outputs are required because, in general, the measured outputs are not straightforward error-corrupted observations of the state variables. Minimization of some function of the errors between observed and estimated outputs, i.e.

$$\boldsymbol{\epsilon}\{\hat{\boldsymbol{\beta}}\} = \boldsymbol{c}^{\mathrm{o}} - \hat{\boldsymbol{c}}^{\mathrm{o}}\{\hat{\boldsymbol{\beta}}, \hat{\boldsymbol{c}}\}, \tag{11.14}$$

leads to a set of "best" estimates $\hat{\boldsymbol{\beta}}$ for the parameter values according to the chosen error function. In (11.14) the functional arguments $\hat{\boldsymbol{\beta}}$ and $\hat{\boldsymbol{c}}$ have been selectively retained in order to illustrate two points. First, since the objective is *parameter estimation* the errors $\boldsymbol{\epsilon}$ are indicated as being dependent only upon the parameters. Second, however, the errors depend also upon $\hat{\boldsymbol{c}}$ since the estimated outputs $\hat{\boldsymbol{c}}^{\mathrm{o}}$ are functions of the state variable estimates $\hat{\boldsymbol{c}}$, which may not necessarily be uniquely determined by the choice of parameter estimates and input measurements alone. This implies, then, solution of the related problem of *state estimation* (as we shall see in section 11.3.2). Essentially, the input–output field data $\boldsymbol{d}$ and $\boldsymbol{c}^{\mathrm{o}}$ are the fixed basis for model calibration and parameter estimation; they are, as it were, the fixed external description of the system behavior. The objective of calibration is to make plausible hypotheses about the *internal* description ($\hat{\boldsymbol{c}}$ and $\hat{\boldsymbol{\beta}}$) of the behavior (Figure 11.4) according to a chosen criterion and to certain assumptions about the random processes $\boldsymbol{\xi}$ and $\boldsymbol{\eta}$.

To reiterate the discussion of section 2.6.4, let us start with the method of least-squares parameter estimation, which minimizes the error (loss) function,

$$J = \sum \boldsymbol{\epsilon}^{\mathrm{T}}\{\hat{\boldsymbol{\beta}}\}\boldsymbol{\epsilon}\{\hat{\boldsymbol{\beta}}\}. \tag{11.15}$$

The fundamental role of least-squares estimation as a solution to the problem of evaluating parameters by reference to field data is undisputed; it is nearly always quoted as the basis for further development of more intricate methods (for example, Draper and Smith, 1966; Åström and Eykhoff, 1971; Young, 1974; Söderström *et al.*, 1978). Least-squares estimation does, however, suffer from an important disadvantage, namely the problem of *bias in the estimates*, i.e.

$$E\{\hat{\boldsymbol{\beta}}\} \neq \boldsymbol{\beta}, \tag{11.16}$$

where $E\{\cdot\}$ is the expectation operator. In other words, since we are estimating the parameters on the basis of field data that are a sample realization of randomly distributed variables, the parameter estimates can also be characterized as having probability distributions. Furthermore, if Gaussian distributions are assumed, the properties of the parameter estimates can be specified by the values of their means and covariances. Thus (11.16) states that the mean or most probable values of the parameter estimates are not equal to the true values of the parameters. Least-squares estimators, however, are not always biased, although they are in many cases of practical interest. Such a discrepancy between the true parameter values and the least-squares estimates will occur when, as frequently, the statistical properties of the random sequences $\boldsymbol{\xi}$ and $\boldsymbol{\eta}$ do not conform to those of zero-mean sequences that are not correlated with each other and that are not correlated with themselves in time, i.e. $\boldsymbol{\xi}$ and $\boldsymbol{\eta}$ are not *white noise* sequences.

There are many other estimation algorithms, few of which, however, depart essentially from the principle of least squares. We shall mention only two other approaches. The first is the maximum likelihood method (e.g. Åström and Bohlin, 1966; Box and Jenkins, 1970; Källström *et al.*, 1976; van Straten, 1983). In this case, if the loss function of (11.15) is modified to give a weighted form of the squared loss function,

$$J^* = \sum (\boldsymbol{\epsilon}^T\{\hat{\boldsymbol{\beta}}\}\mathbf{W}\boldsymbol{\epsilon}\{\hat{\boldsymbol{\beta}}\}), \tag{11.17}$$

where $\mathbf{W}$ is a matrix of weighting coefficients, then maximum likelihood estimation usually corresponds to minimizing the loss function (11.17) with a particular choice of $\mathbf{W}$. This choice is that $\mathbf{W}$ is set equal to the inverse of the variance–covariance matrix of the associated output response observation errors. That is to say, maximum likelihood estimation assumes some statistical knowledge about the system under investigation and weights the squared errors in accordance with the accuracy of the system response measurements. If the measurement errors are assumed to be independent of each other, in which case $\mathbf{W}$ becomes a diagonal matrix with nonzero elements on its leading diagonal only, the minimization of (11.17) is sometimes referred to as weighted least-squares estimation.

The second method, known as the method of instrumental variables (e.g. Kendall and Stuart, 1961; Johnston, 1963; Young, 1976), seeks to generate a set of variables, the instrumental variables, which, it is hoped, conform to certain statistical requirements that ensure the prevention of bias in the parameter estimates. Unlike the method of maximum likelihood, an instrumental variable estimator does not necessarily require the analyst to qualify formally the structure of the correlated properties of the sequences $\boldsymbol{\xi}$ and $\boldsymbol{\eta}$.

These are thus some of the more important approaches to the problem of parameter estimation. It is now appropriate to recall from Chapter 2 the distinction between off-line and recursive methods of parameter estimation.

11.3.2. Off-Line Estimation Algorithms

The basis of an off-line method is that, during each iteration, the model output responses are compared with the field observations, while the parameter estimates are held constant, and at the *end* of each iteration a value for the error function, for example (11.15), is computed. The algorithm assesses the shape of the error function surface for the particular values of the parameters and error function at each iteration, and then attempts to descend toward the minimum of the error function by specifying a corrected set of parameter values for the next iteration. The form of this iterative search for the "best" set of parameter values may be defined, for example, by the following class of *gradient algorithms*:

$$\hat{\boldsymbol{\beta}}^i = \hat{\boldsymbol{\beta}}^{i-1} - \lambda \boldsymbol{\Psi} \nabla_{\boldsymbol{\beta}}(J). \tag{11.18}$$

Here superscript i denotes the revised parameter estimates at the end of the ith iteration; λ is a positive scalar step length, $\nabla_{\boldsymbol{\beta}}(J)$ denotes a vector of gradients of the loss function J with respect to the parameters $\boldsymbol{\beta}$, and $\boldsymbol{\Psi}$ is a matrix of coefficients to be determined. It is not always necessary to solve the problem numerically. For instance, some models may have a form that, when substituted into the error function (11.15), allows differentiation of J with respect to the parameters to yield explicit analytical solutions for the best estimates. It may also be observed from (11.18) that when $\nabla_{\boldsymbol{\beta}}(J) = \mathbf{0}$, $\hat{\boldsymbol{\beta}}^i = \hat{\boldsymbol{\beta}}^{i-1}$; in other words, the iterative numerical algorithm has converged on the set of parameter estimates that minimizes the error function.

Without going into the extensive field of methods for finding the extremum of a function, i.e. optimization, we may note from (11.18) that in implementing an off-line estimation algorithm the following three factors require consideration.

(i) There is the determination of the gradients $\nabla_{\boldsymbol{\beta}}(J)$ and, most probably, the computation of numerical approximations thereof. If we look at the form of J in (11.15), and thence recall (11.14), we can see how the gradients required for (11.18) will be dependent upon the sensitivity of the model responses $\hat{\mathbf{c}}^{\mathrm{o}}\{\hat{\boldsymbol{\beta}}\}$ to the parameter values. More specifically, J is a function of the errors $\boldsymbol{\epsilon}$, which are a function of the estimated output responses $\hat{\mathbf{c}}^{\mathrm{o}}$, which in turn are a function of the state estimates $\hat{\mathbf{c}}$. Differentiation of J with respect to $\hat{\boldsymbol{\beta}}$, therefore, implies differentiation of $\mathbf{c}$, the state vector, with respect to $\boldsymbol{\beta}$, i.e.

$$\left[\frac{\partial \mathbf{c}}{\partial \boldsymbol{\beta}}\right]_{\hat{\mathbf{c}}}.$$

This Jacobian matrix is clearly the matrix of sensitivity coefficients, as defined in (11.4), but evaluated for the current set of state and parameter estimates $\hat{\mathbf{c}}$, $\hat{\boldsymbol{\beta}}$. It can now be seen, through (11.18), how sensitivity coefficients are directly linked with the performance of parameter estimation algorithms.

(ii) There is also the choice of the matrix $\boldsymbol{\Psi}$: to name but two basic alternatives, we have the method of *steepest descent* in which, simply,

$$\boldsymbol{\Psi} = \mathbf{I},$$

that is,

$$\psi_{ij} = \begin{cases} 1 & \text{for } i = j \\ 0 & \text{for } i \neq j, \end{cases}$$

where $\mathbf{I}$ is the unit identity matrix; or the *Newton–Raphson method*, where

$$\psi_{ij} = \frac{\partial^2 J}{\partial \beta_i \, \partial \beta_j},$$

i.e. elements that are second partial derivatives of the error function with respect to the parameters.

(iii) Lastly, after determination of the direction in which to reduce the error function surface toward its minimum, there is the problem of specifying the magnitude λ of the step to be taken in that direction.

From these basic ingredients arises a variety of methods that attempt to combine and exploit the better features of each individual algorithm in terms of its stability and convergence properties (for example, Wilde, 1964; Wilde and Beightler, 1967; Shastry *et al.*, 1973).

The off-line schemes of parameter estimation are, therefore, seen to be essentially developed from the explicit solution of an optimization problem. The routine that searches for the optimum, however, need not be coupled to an intelligent appraisal of the shape of the loss function surface, as it was in the example algorithm of (11.18). A much easier routine to implement would be one that searches in a random fashion for the minimum of the errors between the model and the field data (e.g. Halfon, 1979). For such a *random search* routine the parameter values must usually be specified *a priori* to lie between certain bounds:

$$\boldsymbol{\beta}_{\min} \leq \hat{\boldsymbol{\beta}} \leq \boldsymbol{\beta}_{\max}. \tag{11.19}$$

For a large number of iterations i through the field data, the combination of parameter values that yields the lowest value of the loss function is chosen for the best parameter estimates.

11.3.3. Recursive Estimation Algorithms: An Example

We have chosen to discuss recursive estimation algorithms in greater detail than off-line algorithms and to discuss, in particular, the extended Kalman filter (EKF) algorithm (for example, Jazwinski, 1970). The application of this algorithm will then be illustrated by two case studies in section 11.3.5. The presentation is somewhat limited in its breadth, therefore, although there will be advantages in terms of demonstrating the estimation of parameters that may vary in time and space.

The Extended Kalman Filter

Unfortunately there is no concise derivation of the EKF algorithm since it is implicit that one is first familiar with the form of the linear Kalman filter (LKF)

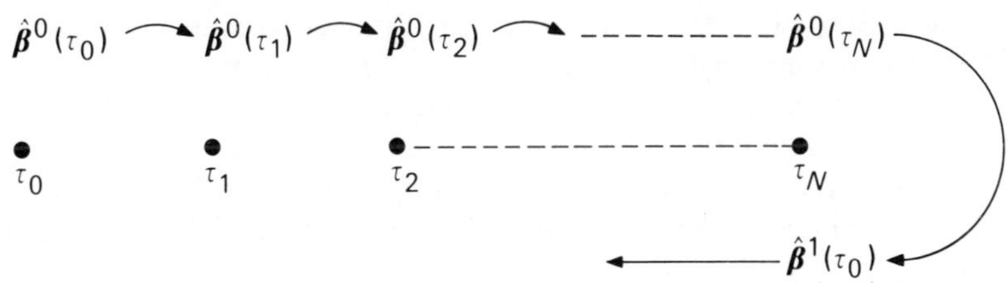

FIGURE 11.5 Conceptual picture of a recursive estimation scheme. The notation τ_k denotes the kth discrete sampling instant in a time series with N samples.

(Kalman, 1960; Kalman and Bucy, 1961; Gelb, 1974). This in turn implies a reasonable appreciation of the notion of state estimation, which is especially important for the present purposes because the key link between the LKF and EKF will be the interpretation of the parameter estimation problem as a problem of state estimation. We may recall, therefore, the ambivalent attitude toward use of the terms "state" and "parameter," which was suggested in section 2.3. It will be helpful in the discussion here to visualize a parameter as merely a special case of a constant state variable.

In Chapter 2 the basic idea of a recursive parameter estimator was introduced. Figure 11.5 repeats the diagram shown before as Figure 2.8(b). Attention will be confined to a revised form of the output observations of (11.12b), that is,

$$\boldsymbol{c}^o(\tau_k) = \boldsymbol{m}\{\boldsymbol{c}(\tau_k), \boldsymbol{\beta}(\tau_k)\} + \boldsymbol{\eta}(\tau_k), \tag{11.20}$$

in which τ_k is the kth discrete point in time (or time of travel). This merely reflects the more realistic situation in which discretely sampled observations of water quality are available. A close study of Figure 11.5 allows one almost intuitively to write down the structure of a recursive estimator: thus the newly revised estimate of the parameter vector at the current time, say $\hat{\boldsymbol{\beta}}(\tau_k)$, would be a function of its immediate past estimate $\hat{\boldsymbol{\beta}}(\tau_{k-1})$ and a correction term that is based on the error between the observation and model prediction at time τ_k, i.e.

$$\hat{\boldsymbol{\beta}}(\tau_k) = \hat{\boldsymbol{\beta}}(\tau_{k-1}) + \mathbf{G}(\tau_k)\boldsymbol{\epsilon}(\tau_k). \tag{11.21}$$

$$\begin{matrix}\text{new} \\ \text{estimate}\end{matrix} = \begin{matrix}\text{old} \\ \text{estimate}\end{matrix} + \begin{matrix}\text{weighting} \\ \text{factor}\end{matrix} \begin{matrix}\text{prediction} \\ \text{error}\end{matrix}$$

The error vector $\boldsymbol{\epsilon}(\tau_k)$ is the error between a model prediction (estimate) of the system response at time τ_k and the noise-corrupted measurements of that output response. The gain matrix of weighting factors, $\mathbf{G}(\tau_k)$, is essentially a function of the available field data. It may be noted in passing that (11.21) is structurally similar to the off-line algorithm of (11.18); Young (1974) has indicated how some forms of recursive algorithm are themselves also gradient algorithms. It is not difficult to see that a recursive *state* estimator can be constructed along lines exactly analogous to (11.21), that is,

$$\hat{\boldsymbol{c}}(\tau_k^+) = \hat{\boldsymbol{c}}(\tau_k^-) + \mathbf{G}(\tau_k^-)\boldsymbol{\epsilon}(\tau_k^-). \tag{11.22}$$

In this case $\mathbf{G}(\tau_k^-)$ is a function partly of the model parameter values; $\hat{c}(\tau_k^+)$ denotes the newly updated state vector estimate *after* receipt of all the measured information at time τ_k, whereas $\hat{c}(\tau_k^-)$ represents a best forward extrapolation (in time) of the state c *prior* to receipt of the current measurements. $\epsilon(\tau_k^-)$ is likewise based upon this same extrapolation $\hat{c}(\tau_k^-)$. The terms "prior," "current," and "after" are intended here to be indicative of the sequential procedure of the algorithm.

An important difference between (11.21) and (11.22) lies in the arguments of $\hat{c}$ and $\hat{\boldsymbol{\beta}}$. As one would expect, the state of water quality will change over the interval from τ_{k-1} to τ_k; it is therefore prudent to use a model to make some form of extrapolated prediction over this interval for comparison with the next measurements obtained at τ_k. In contrast, the assumed model of parameter variations would be that, in fact, they remain constant. Thus the best prediction of the parameter values at a later instant is that they have the same values as estimated at present. With this in mind, it is appropriate now to develop a *conceptual picture* of the EKF algorithms; this is shown in Figure 11.6. A model of "reality" is embedded in the filter. Predictions, of the kind $\hat{c}(\tau_k^-)$ in (11.22), are computed from the model by using the measured input disturbances

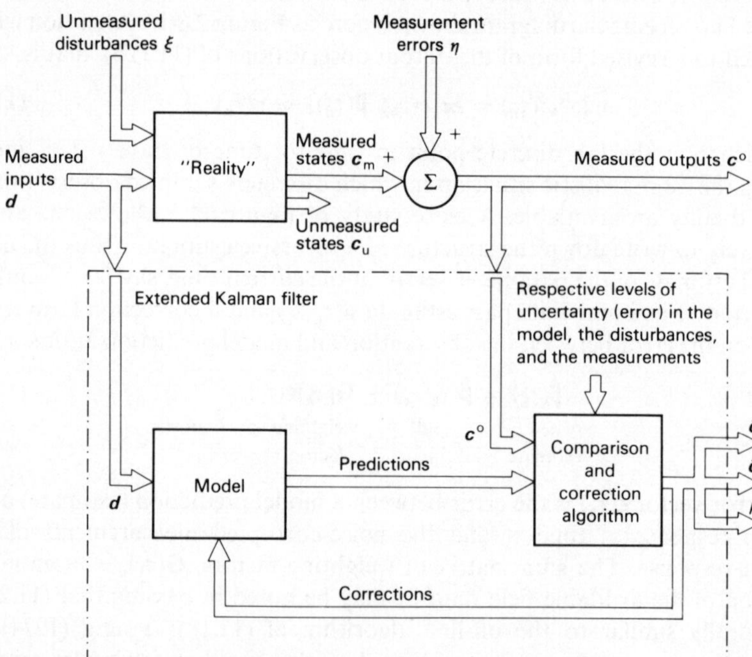

FIGURE 11.6 Conceptual picture of the extended Kalman filter. $\hat{c}_m$, $\hat{c}_u$, and $\hat{\boldsymbol{\beta}}$ are, respectively, estimates of the measured state variables, the unmeasured state variables, and the parameters.

of the system, $\boldsymbol{d}(\tau_k^-)$, and fed forward to the corrector algorithm together with the current observations $\boldsymbol{c}^o(\tau_k)$ of the system output response. For the corrector algorithm, i.e. (11.22), it is apparent that additional (and parallel) algorithms are required that compute the evolution of the uncertainty, or error bounds, associated with the state estimates. The results from the corrector algorithm computation are the updated estimates; these in turn are fed back to the model for revision of subsequent predictions.

Two qualitative interpretations of the filter may be helpful at this point. First, the filter can be seen as an algorithm for "translating" information about the observed input–output behavior of the real system into model-related estimates of the state variables and parameters. Second, the name "filter" also suggests the intuitive idea that the algorithm has the objective of filtering from the given field data the unwanted influences of measurement noise and uncertain disturbances.

The following development of the EKF depends on two key items:

(i) the formulation and outline solution of the *combined state–parameter estimation* problem; and
(ii) the method of *computation of the gain matrix* $\mathbf{G}(\tau_k^-)$, and the dependence of this matrix on specified measures of the levels of uncertainty in the model as a representation of reality, in the input system disturbances, and in the output response observations.

A formal derivation of the EKF is given in the source reference, Jazwinski (1970). Alternatively, Young (1974) provides an outline of how the EKF algorithms can be obtained from an extension of linear regression analysis; this same outline is treated in depth by Beck (1979b).

Combined state–parameter estimation. Let us assume that the type of model to be used is the *linear* form of (11.12a) and (11.20):

$$\dot{\boldsymbol{c}}(\tau) = \mathbf{A}\boldsymbol{c}(\tau) + \mathbf{B}\boldsymbol{d}(\tau) + \boldsymbol{\xi}(\tau), \qquad (11.23a)$$

with discretely sampled, noise-corrupted output response measurements $\boldsymbol{c}^o$:

$$\boldsymbol{c}^o(\tau_k) = \mathbf{H}\boldsymbol{c}(\tau_k) + \boldsymbol{\eta}(\tau_k), \qquad (11.23b)$$

in which the dot notation refers to differentiation with respect to time τ. The matrices $\mathbf{A}$, $\mathbf{B}$, and $\mathbf{H}$ have the dimensions $n \cdot n$, $n \cdot m$, and $p \cdot n$, respectively; n, m, and p correspond to the dimensions of the state ($\boldsymbol{c}$), input ($\boldsymbol{d}$), and output ($\boldsymbol{c}^o$) vectors. Equation 11.23b expresses the fact that in general the outputs are not necessarily simple error-corrupted observations of the state variables (the outputs may be linear combinations of the state variables). $\boldsymbol{\xi}$ is an n-dimensional vector of unmeasured, random process disturbances, $\boldsymbol{\eta}$ is a p-dimensional vector of chance measurement errors, and both $\boldsymbol{\xi}$ and $\boldsymbol{\eta}$ will be assumed to approximate zero-mean, white, Gaussian sequences (formal definitions are given below). For the linear system of (11.23) the linear Kalman filter would thus provide recursive estimates $\hat{\boldsymbol{c}}(\tau_k|\tau_k)$ of the state vector $\boldsymbol{c}(\tau_k)$, where the argument of $\hat{\boldsymbol{c}}$ indicates that these are estimates conditioned upon all available measurements up to and including those at the current time τ_k. The restriction of specifying (11.23) as linear is primarily a convenience for illustrative purposes; the actual problem at hand will soon become nonlinear in any case.

Let us suppose now that some of the unknown, or imprecisely known, elements of the matrices in (11.23), that is a vector of parameters $\boldsymbol{\beta}$, say, are required to be estimated simultaneously with the state vector. One approach to realizing a simultaneous state–parameter estimator is to augment the state vector $\boldsymbol{c}$ with the parameter vector and accordingly to postulate a set of additional differential equations representing the parameter "dynamics." (In this sense the method of quasilinearization is similar to the EKF since it, too, sets up the parameter estimation problem by interpreting the parameters as additional state variables (Bellman and Kalaba, 1965; Lee, 1968).) If the augmented state vector $\boldsymbol{c}^*$ is defined by

$$\boldsymbol{c}^* \triangleq \begin{bmatrix} \boldsymbol{c} \\ \hdashline \boldsymbol{\beta} \end{bmatrix}$$

the state–parameter dynamics and observation equation are given in the following general nonlinear form:

$$\dot{\boldsymbol{c}}^*(\tau) = \boldsymbol{f}^*\{\boldsymbol{c}^*(\tau), \boldsymbol{d}(\tau)\} + \boldsymbol{\xi}^*(\tau), \tag{11.24a}$$

$$\boldsymbol{c}^o(\tau_k) = \boldsymbol{m}^*\{\boldsymbol{c}^*(\tau_k)\} + \boldsymbol{\eta}(\tau_k). \tag{11.24b}$$

The functions $\boldsymbol{f}^*\{\cdot\}$ and $\boldsymbol{m}^*\{\cdot\}$ are vector-valued; they are nonlinear because of the product terms involving elements of $\boldsymbol{\beta}$ with elements of $\boldsymbol{c}$ and $\boldsymbol{d}$. $\boldsymbol{\xi}^*(\tau)$ denotes that the vector of stochastic disturbances in (11.24a) is now of a different order to that defined for $\boldsymbol{\xi}(\tau)$ in (11.23a).

Let us consider the problem of specifying the dynamics of the parameters $\boldsymbol{\beta}$. Of particular importance to the subsequent discussion are two such specifications: (a) we might assume that the parameters are constant, that is, time-invariant:

$$\dot{\boldsymbol{\beta}}(\tau) = \boldsymbol{0}; \tag{11.25}$$

or (b) it might be proposed that they vary in an unknown "random walk" fashion:

$$\dot{\boldsymbol{\beta}}(\tau) = \boldsymbol{\zeta}(\tau). \tag{11.26}$$

in which $\boldsymbol{\zeta}$ is a vector of zero-mean, white, Gaussian disturbances. Were there to be more *a priori* information on the parameter variations, it would be appropriate, for instance, to define the dynamics as oscillatory in accordance with some diurnal or seasonal fluctuation.

The algorithms and computation of the gain matrix. The EKF is a linear approximation of the nonlinear filter that would ideally be needed to provide estimates of c^* in (11.24). Briefly, there are three main steps in its derivation.

(i) Linearization of the nonlinear augmented state equations by means of a first-order Taylor series expansion about a deterministic nominal reference trajectory for the state variables. From this a set of *linear* equations is obtained for the dynamic variations of small perturbations about the reference trajectory.

(ii) Application of a linear Kalman filter for estimation of the small-perturbations vector.

(iii) Substitution of the current augmented state–parameter vector estimates as the choice of nominal reference trajectory. If this specific choice is not made, the state estimates can be obtained by combining the filter estimates of the small perturbations with the known deterministic reference trajectory. However, as a consequence of such a choice it is possible to formulate the EKF directly in terms of the augmented state–parameter vector c^*, rather than in terms of the small perturbations; and by this choice it is also more probable that the perturbations about the reference trajectory are in fact sufficiently small to justify linearization as a valid approximation. Figure 11.7 gives further explanation of this outline; a more complete interpretation of the derivation is given in Beck (1979b).

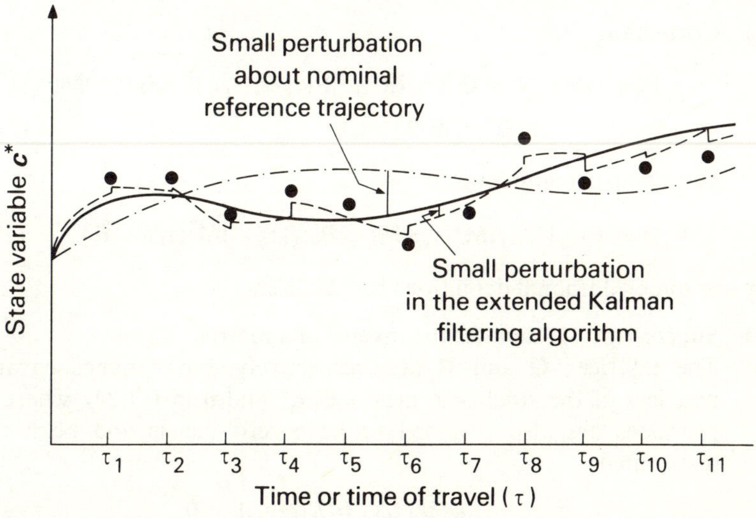

FIGURE 11.7 An example showing how the relinearization procedure of the extended Kalman filter is capable of preserving only *small* perturbations about the reference trajectory (in the EKF the current state estimates are substituted for the reference trajectory). Continuous line: the true state; ● observed values of the state; chain line: a nominal reference trajectory; broken line: the state estimates from the filtering algorithm.

We have lastly to introduce some precise notation. Thus, let us define the state–parameter estimation error variance–covariance matrix as

$$\mathbf{P}(\tau|\tau_k) \triangleq E\{(\boldsymbol{c}^*(\tau) - \hat{\boldsymbol{c}}^*(\tau|\tau_k))(\boldsymbol{c}^*(\tau) - \hat{\boldsymbol{c}}^*(\tau|\tau_k))^{\mathrm{T}}\} \tag{11.27}$$

in which $E\{\cdot\}$ is the expectation operator and the argument $(\tau|\tau_k)$ denotes an estimate, or error variance, of the state variable (or parameter) at time τ based upon all past measured information up to and including that available at τ_k.

The EKF algorithms can now be stated summarily in the following form.

State estimates

(i) For *prediction* between sampling instants,

$$\hat{\boldsymbol{c}}^*(\tau_k|\tau_{k-1}) = \hat{\boldsymbol{c}}^*(\tau_{k-1}|\tau_{k-1}) + \int_{\tau_{k-1}}^{\tau_k} \boldsymbol{f}^*\{\hat{\boldsymbol{c}}^*(\tau|\tau_{k-1}), \boldsymbol{d}(\tau)\}\mathrm{d}\tau. \tag{11.28a}$$

(ii) For correction at the sampling instant,

$$\hat{\boldsymbol{c}}^*(\tau_k|\tau_k) = \hat{\boldsymbol{c}}^*(\tau_k|\tau_{k-1}) + \mathbf{G}^*(\tau_k)[\boldsymbol{c}^{\circ}(\tau_k) - \boldsymbol{m}^*\{\hat{\boldsymbol{c}}^*(\tau_k|\tau_{k-1})\}]. \tag{11.28b}$$

Error covariances

(i) Prediction,

$$\mathbf{P}(\tau_k|\tau_{k-1}) = \boldsymbol{\Phi}(\tau_k, \tau_{k-1})\mathbf{P}(\tau_{k-1}|\tau_{k-1})\boldsymbol{\Phi}^{\mathrm{T}}(\tau_k, \tau_{k-1}) + \mathbf{Q}. \tag{11.28c}$$

(ii) Correction,

$$\mathbf{P}(\tau_k|\tau_k) = [\mathbf{I} - \mathbf{G}^*(\tau_k)\mathbf{M}(\tau_k)]\mathbf{P}(\tau_k|\tau_{k-1})[\mathbf{I} - \mathbf{G}^*(\tau_k)\mathbf{M}(\tau_k)]^{\mathrm{T}} + \mathbf{G}^*(\tau_k)\mathbf{R}\mathbf{G}^{*\mathrm{T}}(\tau_k). \tag{11.28d}$$

Gain matrix computation,

$$\mathbf{G}^*(\tau_k) = \mathbf{P}(\tau_k|\tau_{k-1})\mathbf{M}^{\mathrm{T}}(\tau_k)[\mathbf{M}(\tau_k)\mathbf{P}(\tau_k|\tau_{k-1})\mathbf{M}^{\mathrm{T}}(\tau_k) + \mathbf{R}]^{-1}. \tag{11.28e}$$

There are some additional definitions for (11.28).

(a) Superscript -1 denotes the inverse of a matrix.
(b) The matrices $\mathbf{Q}$ and $\mathbf{R}$ are, respectively, the variance–covariance matrices of the stochastic processes $\boldsymbol{\xi}^*$ and $\boldsymbol{\eta}$ in (11.24), where these processes are also assumed to have zero means and be normally distributed:

$$E\{\boldsymbol{\xi}^*(\tau_k)\} = E\{\boldsymbol{\eta}(\tau_k)\} = \mathbf{0}$$

and

$$E\{\boldsymbol{\xi}^*(\tau_k)\boldsymbol{\xi}^{*\mathrm{T}}(\tau_l)\} = \mathbf{Q}\delta_{kl}; \qquad E\{\boldsymbol{\eta}(\tau_k)\boldsymbol{\eta}^{\mathrm{T}}(\tau_l)\} = \mathbf{R}\delta_{kl},$$

with

$$\delta_{kl} = \begin{cases} 0 & \text{for } k \neq l \\ 1 & \text{for } k = l. \end{cases}$$

(c) The model-predicted response errors (residual errors) are given by

$$\boldsymbol{\epsilon}(\tau_k) = \boldsymbol{c}^{\text{o}}(\tau_k) - \boldsymbol{m}^*\{\boldsymbol{c}^*(\tau_k|\tau_{k-1})\} = \boldsymbol{c}^{\text{o}}(\tau_k) - \hat{\boldsymbol{c}}^{\text{o}}(\tau_k|\tau_{k-1}). \quad (11.29)$$

(d) The gain matrix $\mathbf{G}^*(\tau_k)$ refers to the combined state–parameter vector.

(e) The matrices $\mathbf{\Phi}(\tau_k, \tau_{k-1})$ and $\mathbf{M}(\tau_k)$ refer (implicitly) to the (augmented) *state transition matrix* (e.g. Dorf, 1965) and the observations matrix for the linearized small-perturbations system obtained at the first step of the derivation shown above, i.e.

$$\mathbf{\Phi}(\tau_k, \tau_{k-1}) \triangleq \exp\{\mathbf{F}(\tau_{k-1})[\tau_k - \tau_{k-1}]\}$$

with

$$\mathbf{F}(\tau_{k-1}) \triangleq [\partial f_i^*\{\boldsymbol{c}^*(\tau), \boldsymbol{d}(\tau)\}/\partial c_j^*]\big|_{\substack{\boldsymbol{c}^*(\tau) = \boldsymbol{c}^*(\tau_{k-1}|\tau_{k-1}) \\ \boldsymbol{d}(\tau) = \boldsymbol{d}(\tau_{k-1})}}$$

and

$$\mathbf{M}(\tau_k) \triangleq [\partial m_i^*\{\boldsymbol{c}^*(\tau_k)\}/\partial c_j^*]\big|_{\boldsymbol{c}^*(\tau_k) = \boldsymbol{c}^*(\tau_k|\tau_{k-1})}$$

in which f_i^* and m_i^* are individual elements of the nonlinear vector-valued functions $\boldsymbol{f}^*$ and $\boldsymbol{m}^*$ given by (11.24).

Since the EKF algorithms are recursive it is clear that a set of initial values at time τ_0 must be specified for the state–parameter estimates $\hat{\boldsymbol{c}}^*(\tau_0|\tau_0)$ and their associated error variances $\mathbf{P}(\tau_0|\tau_0)$. Also implied by the algorithms is the fact that values have to be specified for the covariance matrices $\mathbf{Q}$ and $\mathbf{R}$, which, as indicated elsewhere (Beck, 1979a), is an especially difficult task with respect to $\mathbf{Q}$. The importance of algorithms (11.28), however, is not to be able to recall their exact form but to notice the following two items.

(i) Even though the EKF is a linear approximation designed to treat the system of nonlinear equations 11.24, the consequences of the linearization procedure appear only as the matrices $\mathbf{\Phi}$ and $\mathbf{M}$ in the error covariance and gain matrix computation algorithms. For the state estimation computation, (11.28a) and (11.28b), the original nonlinear functions are preserved.

(ii) Apart from their use as a measure of the confidence bounds on the accuracy of the estimates,† the covariance algorithms essentially

† However, in view of the linearization approximation one should be very cautious in making this interpretation for the parameter estimates, although such caution is not necessary with a recently proposed modified form of the EKF (Ljung, 1979).

serve the purpose of making an intelligent choice of gain matrix. One can observe that (qualitatively) when the measurement error variance, i.e. $\mathbf{R}$, is large, the gain matrix tends to be small. In other words, the filtering algorithm will ignore large errors $\epsilon(\tau_k)$ between predicted and observed responses since it attributes these to measurement error. Conversely, when the filter is uncertain of the model performance, i.e. $\mathbf{P}(\tau_k|\tau_{k-1})$ is large, the gain matrix is relatively large and significant correction of the estimates will follow if large prediction errors are perceived.

Figure 11.8 shows some additional schematic features of the EKF algorithms; further details of the operation of the algorithms, especially with respect to the solution of the model structure identification problem, are given in Beck (1979a).

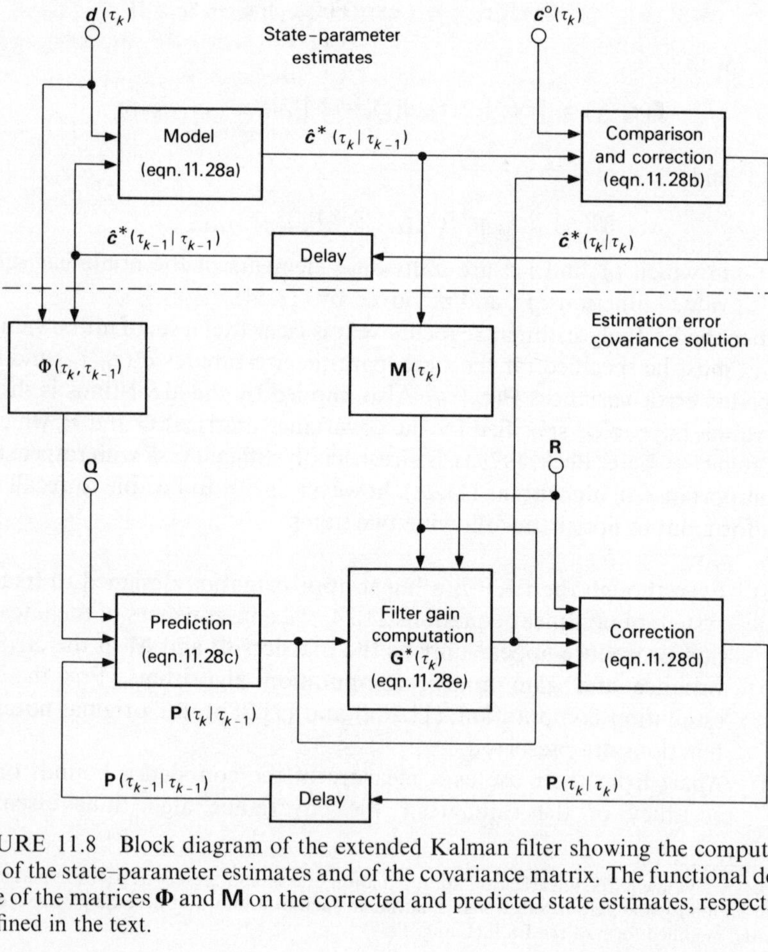

FIGURE 11.8 Block diagram of the extended Kalman filter showing the computation both of the state–parameter estimates and of the covariance matrix. The functional dependence of the matrices $\mathbf{\Phi}$ and $\mathbf{M}$ on the corrected and predicted state estimates, respectively, is defined in the text.

11.3.4. Some Recent Applications of Parameter Estimation

Table 11.1 summarizes the literature on recent applications of parameter estimation algorithms in the development and assessment of stream and lake water quality models. The applications can be broadly classified according to whether they used off-line (O) or recursive (R) estimation schemes in the manner described previously. The models are classified according to the choice of:

(i) independent variable(s), i.e. time and/or one-dimensional space;
(ii) quality characteristics, i.e. state variables; and
(iii) model equation form, i.e. differential or difference equations.

A few remarks are necessary to qualify the contents of Table 11.1. For instance, the paper by Ivakhnenko *et al.* (1977) is primarily concerned with the problems of model discrimination and model structure identification rather than with the problem of parameter estimation. In fact in the Group Method of Data Handling (GMDH) the implicit problem of parameter estimation is treated with a least-squares estimator. Other references (Shastry *et al.*, 1973; Beck and Young, 1976; Jolankai and Szöllösi-Nágy, 1978; Halfon *et al.*, 1979) are similarly oriented toward model structure identification.

Four papers, those of Huck and Farquhar (1974), Beck (1975), Gnauck *et al.* (1976), and Halfon *et al.* (1979), deal with models and estimation procedures that are representative of input–output (i.e. black box) time-series analysis techniques. Halfon *et al.* (1979) also make use of frequency response methods (see section 2.4). Earlier examples of similar approaches are given by Thomann (1967), Fuller and Tsokos (1971), and Edwards and Thornes (1973). Other contemporary studies along the same lines include the use of: correlation analysis (Schurr and Ruchti, 1975); a recursive instrumental variable (IV) approach (Beck, 1978a); and further applications of the Box and Jenkins (1970) maximum likelihood methods (e.g. Mehta *et al.*, 1975). Thus, although Table 11.1 focuses on internally descriptive model calibration, this does not suggest that there has been any lack of attention given to input–output model identification.

Lastly, we may observe that only one study (Thé, 1978) has addressed the particularly difficult problem of parameter estimation in distributed-parameter models with the use of field data.

11.3.5. Two Case Studies

The two case studies in this section (Beck and Young, 1976; Bowles and Grenney, 1978a,b) illustrate the application of the extended Kalman filter algorithm discussed in section 11.3.3. Both examples assume a lumped-parameter form of model; one of them represents a steady state condition, while the other treats

TABLE 11.1 Summary of Recent Applications of Parameter Estimation Algorithms in Water Quality Modeling. Model description includes a definition of independent and dependent variables. O and R denote off-line and recursive estimation algorithms, respectively.

Authors	Field Data	Algorithm	Type of Model
Stream water quality modeling			
Koivo and Phillips (1971)	—	Stochastic approximation (least squares); R	Time and space; BOD, DO; analytical solution to first-order partial differential equation.
Koivo and Phillips (1972)	—	Least squares; O	Space; BOD, DO; steady state analytical solution to first-order partial differential equation.
Koivo and Phillips (1976)	—	Linear Kalman filter; R	Time and space; BOD, DO; difference equations.
Koivo and Koivo (1978)	—	Least squares; R (state estimation only)	Time and space; BOD, DO; first-order partial differential equation.
Lee and Hwang (1971)	—	Quasilinearization (least squares); O	Space; BOD, DO; ordinary differential equation.
Shastry *et al.* (1973)	Sacramento River (1962)	Weighted least squares; maximum likelihood; O	Space; BOD, DO; ordinary differential equation.
Huck and Farquhar (1974)	St. Clair River (1971)	Maximum likelihood; O	Single-point spatial location, time variations; DO, chloride; black box time-series model.
Beck (1975)	River Cam (1972)	Maximum likelihood; O	Time; BOD, DO; ordinary differential equation; also black box time-series model.
Beck and Young (1976)	River Cam (1972)	Extended Kalman filter; R	Time; BOD, DO; ordinary differential equation.
Whitehead and Young (1975)	Bedford–Ouse River (1973)	Multivariable instrumental variable—approximate maximum likelihood (MIVAML); R	Time; BOD, DO; difference equations.

Young and Whitehead (1977)	River Cam (1972), Bedford-Ouse River (1973)	MIVAML; R	Time; BOD, DO; difference equations.
Lettenmaier and Burges (1976)	—	Extended Kalman filter; R	Space; BOD, DO; ordinary differential equations.
Erni and Ruchti (1977)	Aare River	Differential approximation method; O	Single-point spatial location, time variations; DO; difference equations.
Ivakhnenko *et al.* (1977)	River Cam (1972)	Group Method of Data Handling (GMDH); O	Single-point spatial location, time variations; BOD, DO; difference equations.
Stehfest (1978)	River Rhine (1971)	Quasilinearization (least squares); O	Space; BOD, DO; ordinary differential equations.
Stehfest (1978)	River Rhine (1971)	Quasilinearization (least squares); O	Space; easily degradable organic matter, slowly degradable organic matter, bacterial mass, protozoan mass, DO; ordinary differential equations.
Bowles and Grenney (1978a)	Jordan River, Utah	Extended Kalman filter; R	Space; BOD, DO, NH_3-N, NO_3-N, algal N, organic N; ordinary differential equations.
Moore and Jones (1978)	River Cam (1972)	Coupled Bayesian–Kalman filter; R	Time; BOD, DO; ordinary differential equations.
Rinaldi *et al.* (1979)	Bormida River	Least squares; O	Space; BOD, DO; analytical solution to first-order ordinary differential equations.
Tamura (1979)	—	Linear Kalman filter (and others); R	Time and space; BOD, DO; difference equations.
Thé (1978)	River Rhine	Linear Kalman filter; R	Time and space; conductivity; second-order partial differential equation (finite-difference approximation solution).
Whitehead (1980a)	River Cam (1972)	Instrumental variable; R	Time; BOD, DO; ordinary differential equations.
Whitehead (1980b)	Bedford-Ouse River (1978)	Extended Kalman filter; R	Time; DO (in four river reaches); NH_3-N (in two river reaches); ordinary differential equations.

(*continued over*)

TABLE 11.1 (*continued*)

Authors	Field Data	Algorithm	Type of Model
Lake water quality modeling			
Di Cola *et al.* (1976)	Leopold's Park Pond, Brussels (1973–75)	Least squares; O (solved as an optimal control problem)	Time; autotrophs, herbivores, carnivores; ordinary differential equations.
Gnauck *et al.* (1976)	Saidenbach Reservoir, GDR (1966–70); Kliçava Reservoir, CSSR (1963–72)	Least squares; R	Time; DO, chlorophyll *a*, particulate organic matter; regression relationship.
Jolankai and Szöllösi-Nágy (1978)	Lake Balaton, Hungary (1971–77)	Maximum likelihood; R	Time; soluble reactive phosphorus, chlorophyll *a*, exchangeable phosphorus in sediment; ordinary differential equations.
Lewis and Nir (1978)	Greifensee, Switzerland (1973)	Weighted least squares; O	Time; soluble reactive phosphorus, particulate phosphorus; ordinary differential equations.
Halfon *et al.* (1979)	Small lake ecosystem	Least squares; O (also frequency-domain analysis)	Time; soluble phosphorus, particulate phosphorus, a low-molecular-weight form of phosphorus, colloidal phosphorus; ordinary differential equations.
Benson (1979)	Lake Placid, British Columbia, Canada	Least squares; O	Time; phytoplankton biomass; ordinary differential equations.
Di Toro and van Straten (1979), van Straten (1983)	Lake Ontario (1972)	Maximum likelihood; weighted least squares; O	Time; 16 state variables divided between epilimnion and hypolimnion layers; ordinary differential equations.
Scavia (1980)	Saginaw Bay, Lake Huron	Extended Kalman filter; R	Time; eight state variables; ordinary differential equations.

a dynamic situation. These examples reflect problems and constraints that are typical of current studies in model calibration: problems of too few data; data with too high levels of error and uncertainty; and hence the technical difficulties that have in general restricted the application of estimation algorithms to small models with only a few state variables.

Case Study 1: *River Cam, England*

Like the Berkel River discussed in section 11.2.1, the River Cam is a slow, lowland river that is susceptible to significant growth of algae during summer. This particular study concerns a field experiment carried out in summer 1972 on a 4.5 km stretch just downstream from Cambridge. The field data and a complete set of results for the model development exercise are given in Beck (1978a).

The dynamic behavior of BOD–DO interaction in a reach may be approximated by a lumped-parameter, ordinary differential equation form of model. The model assumes that the mixing properties of the reach can be idealized as those of a continuously stirred tank reactor (CSTR) (see also Thomann, 1963 and Chapter 6, section 6.3). If the assumptions of Dobbins (1964) are adopted for the definition of BOD–DO interaction, the model takes the form:

$$\begin{bmatrix} \dot{c}_1(\tau) \\ \dot{c}_2(\tau) \end{bmatrix} = \begin{bmatrix} -(K_1 + Q(\tau)/V) & 0 \\ -K_1 & -(K_2 + Q(\tau)/V) \end{bmatrix} \begin{bmatrix} c_1(\tau) \\ c_2(\tau) \end{bmatrix} + \begin{bmatrix} Q(\tau)/V & 0 \\ 0 & Q(\tau)/V \end{bmatrix} \begin{bmatrix} d_1(\tau) \\ d_2(\tau) \end{bmatrix} + \begin{bmatrix} L_A(\tau) \\ K_2 C_s(\tau) + D_B(\tau) \end{bmatrix} + \begin{bmatrix} \xi_1(\tau) \\ \xi_2(\tau) \end{bmatrix}. \tag{11.30a}$$

This equation for the state vector dynamics corresponds to (11.23a) of section 11.3.3. The output observations equation for the error-corrupted measurements of downstream BOD and DO, corresponding to (11.23b), is given by

$$\begin{bmatrix} c_1^{\circ}(\tau_k) \\ c_2^{\circ}(\tau_k) \end{bmatrix} = \begin{bmatrix} c_1(\tau_k) \\ c_2(\tau_k) \end{bmatrix} + \begin{bmatrix} \eta_1(\tau_k) \\ \eta_2(\tau_k) \end{bmatrix}, \tag{11.30b}$$

where τ_k represents the kth day of the experiment. Other terms in (11.30) are defined as follows.

- $c_1(\tau), c_2(\tau)$ are state variables representing concentrations of BOD and DO, respectively, at the downstream end of the reach (g m^{-3}),
- $d_1(\tau), d_2(\tau)$ are measured input variables representing concentrations of BOD and DO, respectively, at the upstream end of the reach (g m^{-3}),
- $Q(\tau)$ is the stream discharge in the reach (m^3 day^{-1}),
- V is the (constant) volume of water in the reach (m^3),
- K_1 is the BOD decay rate constant (day^{-1}),

K_2 is the reaeration rate constant (day^{-1}),
$C_s(\tau)$ is the saturation concentration of DO (g m^{-3}),
$L_A(\tau)$ is the rate of addition of BOD to the reach by, for example, local surface runoff (g m^{-3} day^{-1}),
$D_B(\tau)$ is the net rate of addition of DO to the reach by the combined effects of photosynthesis, respiration, and decomposition of mud deposits (g m^{-3} day^{-1}),
$\xi_1(\tau), \xi_2(\tau)$ are unknown stochastic disturbances of BOD and DO, respectively (g m^{-3} day^{-1}),
$\eta_1(\tau_k), \eta_2(\tau_k)$ are random measurement errors associated with the downstream BOD and DO observations, respectively (g m^{-3}).

In this example, given input–output data for d_1, d_2, c_1^o, and c_2^o, we wish to estimate the parameters $K_1, K_2, L_A(\tau)$, and $D_B(\tau)$ in addition to the state variables $c_1(\tau)$ and $c_2(\tau)$. The augmented state–parameter vector is thus defined by

$$\mathbf{c}^* = [c_1, c_2, K_1, K_2, L_A, D_B]^T \tag{11.31}$$

and the augmented state–parameter dynamics become

$$\begin{bmatrix} \dot{c}_1^*(\tau) \\ \dot{c}_2^*(\tau) \\ \dot{c}_3^*(\tau) \\ \dot{c}_4^*(\tau) \\ \dot{c}_5^*(\tau) \\ \dot{c}_6^*(\tau) \end{bmatrix} = \begin{bmatrix} -(c_3^*(\tau)+Q(\tau)/V)c_1^*(\tau)+(Q(\tau)/V)d_1(\tau)+c_5^*(\tau) \\ \left(\begin{array}{c} -(c_4^*(\tau)+Q(\tau)/V)c_2^*(\tau)-c_3^*(\tau)c_1^*(\tau)+(Q(\tau)/V)d_2(\tau) \\ + c_4^*(\tau)C_s(\tau) + c_6^*(\tau) \end{array}\right) \\ 0 \\ 0 \\ 0 \\ 0 \end{bmatrix} + \begin{bmatrix} \xi_1^*(\tau) \\ \xi_2^*(\tau) \\ 0 \\ 0 \\ \xi_5^*(\tau) \\ \xi_6^*(\tau) \end{bmatrix}. \tag{11.32a}$$

Clearly the observations equation (11.30b) is modified to

$$\mathbf{c}^o(\tau_k) = [\mathbf{I} \,\vdots\, \mathbf{0}]\mathbf{c}^*(\tau_k) + \boldsymbol{\eta}(\tau_k), \tag{11.32b}$$

in which $\mathbf{0}$ is the null matrix. Equations 11.32 are now in a form that corresponds directly with the general representation of the combined state–parameter estimation problem in (11.24). For estimation purposes K_1 and K_2 are assumed to be time-invariant, while $L_A(\tau)$ and $D_B(\tau)$ are idealized as random walk parameters. The specifications of the covariance matrices $\mathbf{P}(\tau_0|\tau_0)$, $\mathbf{Q}$, and $\mathbf{R}$ (section 11.3.3) for this example are given in Table 11.2.

The results of processing the field data with the extended Kalman filtering algorithms are shown in Figure 11.9; the field data cover the period from 6 June to 25 August 1972, inclusive. In Figure 11.9(a) the peak estimate of downstream BOD at day τ_{58} is a consequence of both a high stream discharge and high upstream BOD caused by a thunderstorm. Inspection of Figure 11.9(e)

TABLE 11.2 Specification of Leading Diagonal Elements for the Covariance Matrices $\mathbf{P}(\tau_0|\tau_0)$, $\mathbf{Q}$, and $\mathbf{R}$ for the River Cam Case Study. All other elements in the matrices are assumed to be zero.

State/Parameter	*A priori* Estimation Error Covariance $\mathbf{P}(\tau_0\|\tau_0)$	Unmeasured Input Disturbance Covariance $\mathbf{Q}$	Measurement Error Covariance $\mathbf{R}$	Units
c_1	1.0	0.4	0.4	$(\mathrm{g\,m^{-3}})^2$
c_2	1.0	0.4	0.2	$(\mathrm{g\,m^{-3}})^2$
K_1	0.005	0	—	$(\mathrm{day^{-1}})^2$
K_2	0.005	0	—	$(\mathrm{day^{-1}})^2$
L_A	2.0	0.05	—	$(\mathrm{g\,m^{-3}\,day^{-1}})^2$
D_B	2.0	0.05	—	$(\mathrm{g\,m^{-3}\,day^{-1}})^2$

indicates that a significant addition of BOD to the reach occurs over the period from τ_{40} to τ_{50} and from about τ_{65} onward, even though this particular section of the River Cam receives no direct local surface runoff and these intervals were periods of warm, sunny weather. The estimates of D_B (Figure 11.9(f)) likewise suggest a net addition of DO to the stream during τ_{40}–τ_{50}; but for the initial 20–25 days of the experiment there exists the opposite apparent effect of a net removal of DO from the water body (a result primarily, it is thought, of biased BOD measurements). The recursive estimation trajectories for the parameters K_1 and K_2 in Figures 11.9(c) and (d) are more or less stationary, although the estimates $\hat{K}_2$ of the reaeration rate constant undergo substantial modification between τ_{40} and τ_{50}.

In section 2.6.3 of Chapter 2 we tried to illustrate the qualitative features of model structure identification. Figure 2.9 attempted in a conceptual fashion to show how recursive parameter estimation algorithms can yield useful diagnostic information about this problem. Figure 11.9(d) is a specific realization of that earlier conceptual picture. It was also stated earlier that "for much of the time ... model structure identification ... is confronted with the need to offer plausible hypotheses about 'unexplained' relationships in a set of field data." Why, for example, is there an apparent addition of BOD to the reach at certain times of the experiment? The primary hypothesis to emerge from the analysis is that significant growth of a floating algal population is responsible for part of the observed dynamic BOD–DO interaction. It is further postulated that the growth rate of algae is governed by the prevailing sunlight conditions, an assumption that thus requires further systematic evaluation (Beck and Young, 1976; Beck, 1978b). The results of Figure 11.9 are representative merely of a part of the difficult process of model structure identification.

Case Study 2: Jordan River, Utah

The second case study is concerned with a steady state water quality modeling problem (Bowles and Grenney, 1978a). We shall see how it illustrates different

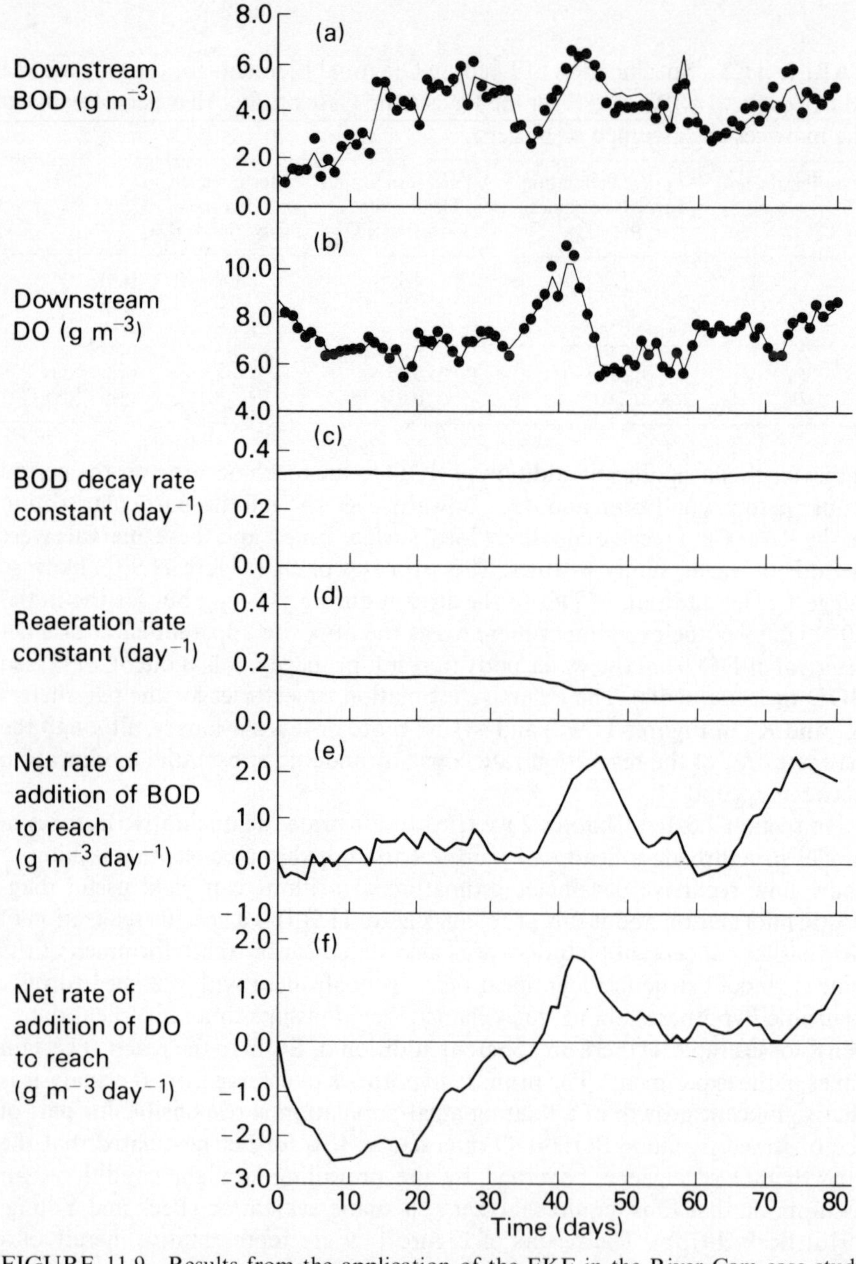

FIGURE 11.9 Results from the application of the EKF in the River Cam case study (adapted from Beck and Young, 1976): (a) observations and state estimates, $\hat{c}_1^*$, for downstream BOD; (b) observations and state estimates, $\hat{c}_2^*$, for downstream DO; (c) estimates, $\hat{c}_3^*$, for the BOD decay rate constant (K_1); (d) estimates, $\hat{c}_4^*$, for the reaeration rate constant (K_2); (e) estimates, $\hat{c}_5^*$, for the net rate of addition of BOD to the reach (L_A); (f) estimates, $\hat{c}_6^*$, for the net rate of addition of DO to the reach (D_B).

aspects of the EKF algorithm, principally at the stage of formulating the system equations. The state vector variations are again provided in the form of lumped-parameter, ordinary differential equations, though here distance along the river (or time of travel) is the independent variable:

$$\begin{bmatrix} \dot{c}_1(\tau) \\ \dot{c}_2(\tau) \\ \dot{c}_3(\tau) \\ \dot{c}_4(\tau) \\ \dot{c}_5(\tau) \\ \dot{c}_6(\tau) \end{bmatrix} = \begin{bmatrix} K^*(L_1(\tau)-c_1(\tau))-K_1c_1(\tau) \\ K^*(L_2(\tau)-c_2(\tau))+K_{52}c_5(\tau)-K_{23}c_2(\tau)-f_2\{c_2(\tau),\, c_3(\tau)\}c_4(\tau) \\ K^*(L_3(\tau)-c_3(\tau))+K_{23}c_2(\tau)-f_3\{c_2(\tau),\, c_3(\tau)\}c_4(\tau) \\ K^*(L_4(\tau)-c_4(\tau))-K_{45}c_4(\tau)+f_4\{c_2(\tau),\, c_3(\tau)\}c_4(\tau) \\ K^*(L_5(\tau)-c_5(\tau))+K_{45}c_4(\tau)-K_{52}c_5(\tau) \\ K^*(L_6(\tau)-c_6(\tau))-D'_B+K_2(C_s(\tau)-c_6(\tau))-K_1c_1(\tau)-\gamma K_{23}c_2(\tau) \end{bmatrix} + \begin{bmatrix} \xi_1(\tau) \\ \xi_2(\tau) \\ \xi_3(\tau) \\ \xi_4(\tau) \\ \xi_5(\tau) \\ \xi_6(\tau) \end{bmatrix}. \tag{11.33}$$

$c_1(\tau)$ to $c_6(\tau)$ are the respective stream concentrations of carbonaceous BOD, ammonia nitrogen, nitrate nitrogen, algal nitrogen, organic nitrogen, and dissolved oxygen (g m^{-3}),

K^* is a rate constant for the addition of components by lateral inflow (day^{-1}, in travel time),

K_1, K_2, $C_s(\tau)$ are as defined for (11.30),

$L_1(\tau)$ to $L_6(\tau)$ are the respective concentrations of each component in lateral inflow (g m^{-3}),

K_{23} is the first-order rate constant for conversion of ammonia nitrogen to nitrate nitrogen (day^{-1}),

K_{45} is the death rate constant for algae (day^{-1}),

K_{52} is the rate constant for decomposition of organic nitrogen into ammonia nitrogen (day^{-1}),

D'_B is the rate of oxygen consumption by bottom mud deposits (g m^{-3} day^{-1}),

γ is a coefficient for the rate of oxygen consumption by nitrification (dimensionless),

$f_2\{\cdot\}$, $f_3\{\cdot\}$, $f_4\{\cdot\}$ are nonlinear functions describing, respectively, the rates of uptake of ammonia nitrogen and nitrate nitrogen by algae and the growth rate of algae (day^{-1}).

For this second case, therefore, the state vector behavior is *nonlinear*. Moreover, the observations are not as straightforward as in the River Cam study: one of the observations is available as the linear sum of two of the state variables, organically bound nitrogen and nitrogen bound as algal cell material. The noise-corrupted output observations are thus given by

$$\mathbf{c}^{o}(\tau_k) = \begin{bmatrix} 1 & 0 & 0 & 0 & 0 & 0 \\ 0 & 1 & 0 & 0 & 0 & 0 \\ 0 & 0 & 1 & 0 & 0 & 0 \\ 0 & 0 & 0 & 1 & 1 & 0 \\ 0 & 0 & 0 & 0 & 0 & 1 \end{bmatrix} \mathbf{c}(\tau_k) + \boldsymbol{\eta}(\tau_k), \tag{11.34}$$

where these are measurements defined at each spatial location τ_k.

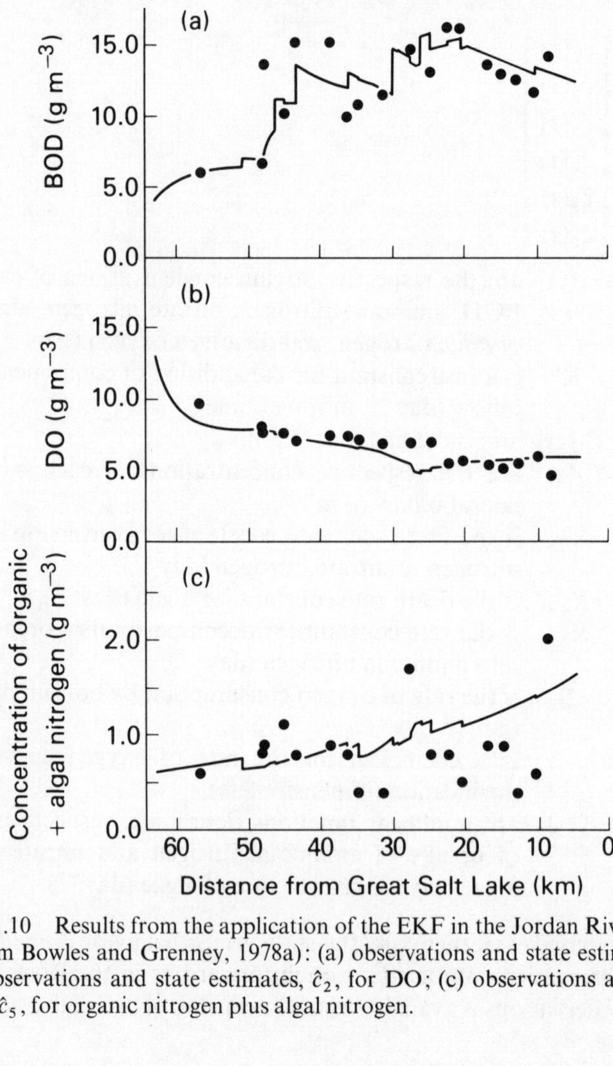

FIGURE 11.10 Results from the application of the EKF in the Jordan River case study (adapted from Bowles and Grenney, 1978a): (a) observations and state estimates, $\hat{c}_1$, for BOD; (b) observations and state estimates, $\hat{c}_2$, for DO; (c) observations and state estimates, $\hat{c}_4 + \hat{c}_5$, for organic nitrogen plus algal nitrogen.

Some of the results quoted by Bowles and Grenney are shown in Figures 11.10 and 11.11. Figure 11.10 represents a typical set of estimates for BOD, DO, and algal and organically bound nitrogen concentration when the EKF is applied to the problem of state estimation *only* with eqns. 11.33 and 11.34. In contrast to the trajectories of Figure 11.9, Figure 11.10 gives the patterns of both predicted, $\hat{c}(\tau_k|\tau_{k-1})$, and corrected, $\hat{c}(\tau_k|\tau_k)$, state estimates generated by the prediction and correction algorithms of the filter (eqns. 11.28a and b, respectively). Some of the changes evident in the estimated states are, however, due to the effects of point-wise addition of constituents from discharges coincident with the measurement location. The persistently poor correspondence between the estimates and observations for algal and organically bound nitrogen is described by Bowles and Grenney as a consequence of the large errors attributed to these measurements. This, then, is a case of the filter "believing" it has an accurate model for this part of the system (recall the discussion of section 11.3.2) so that large errors are ignored as the spurious consequences of chance.

Lastly, Figure 11.11 provides estimates for the parameter $L_2(\tau)$, the ammonia nitrogen concentration in lateral inflow, when the augmented state vector is defined as

$$\boldsymbol{c}^* = [c_1, c_2, c_3, c_4, c_5, c_6, L_2]^{\mathrm{T}}$$

and when $L_2(\tau)$ is idealized as a random walk parameter. The principal objective of this analysis, therefore, is not model calibration, in the sense of model structure identification or parameter estimation. Rather, such an application of the EKF is directed toward the combined use of model and field data for *reconstruction* of information about variables that may be of interest from the point of view of water quality management. Similar and more extensive results for the estimation of spatially varying parameters in this case study can be found in Bowles and Grenney (1978b).

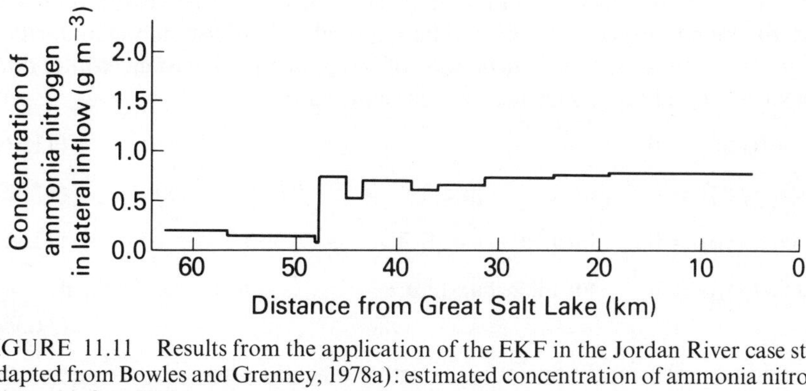

FIGURE 11.11 Results from the application of the EKF in the Jordan River case study (adapted from Bowles and Grenney, 1978a): estimated concentration of ammonia nitrogen in lateral inflow (L_2).

Concluding Remarks

The two case studies have illustrated some of the potential benefits of using recursive estimation algorithms in solving the problem of model calibration. There are, however, limitations on the performance of these algorithms, and of the extended Kalman filter in particular. For example, the choices of variance–covariance matrices $\mathbf{P}(\tau_0|\tau_0)$, $\mathbf{Q}$, and $\mathbf{R}$, which are unavoidably subjective choices, determine in part the behavior of the filtering algorithms. It is also necessary to have reasonable *a priori* estimates $\hat{\boldsymbol{\beta}}(\tau_0|\tau_0)$ of the model parameters, since the EKF does not guarantee convergence to a globally "optimum" set of estimates. Nevertheless, if the analyst is aware of these and other limitations, then the analysis will be all the more effective, and less likely to lead to erroneous conclusions. However in this respect a recent development of particular significance is a modified form of the EKF, proposed by Ljung (1979), which overcomes the difficulties of convergence associated with the original EKF.

11.3.6. Verification

Most of what needs to be stated about verification was introduced in section 2.6.5. The essence of model verification, as understood here, is the problem of checking the statistical properties of the predicted model response errors, i.e.

$$\boldsymbol{\epsilon}(\tau_k) = \mathbf{c}^{o}(\tau_k) - \hat{\mathbf{c}}^{o}(\tau_k), \tag{11.35}$$

where $\hat{\mathbf{c}}^{o}(\tau_k)$ is a model-related estimate of the vector of measured outputs (see also Figure 2.10). Typically, according to the definition of the system and its variables in Figure 11.4, assumptions are made about the statistical properties of the stochastic sequences $\boldsymbol{\xi}$ and $\boldsymbol{\eta}$ (system disturbances and measurement errors, respectively). Usually these assumptions require $\boldsymbol{\xi}$ and $\boldsymbol{\eta}$ to be zero-mean, white noise sequences, or to have been generated by simple manipulations from such sequences. It is customary also to assume that the sequences are drawn from random variables with Gaussian distributions. If all these assumptions are valid, the model response errors will be required to conform, in the majority of cases, with the statistical properties of zero-mean, Gaussian, white noise sequences. Stated more specifically, it is required that:

$$E\{\boldsymbol{\epsilon}(\tau_k)\} = \mathbf{0} \tag{11.36a}$$

$$E\{\epsilon_i(\tau_k)\epsilon_j(\tau_l)\} = 0 \qquad \text{for all } k, l \text{ and for } i \neq j;\ i, j = 1, 2, \ldots, p \tag{11.36b}$$

$$E\{\epsilon_i(\tau_k)\epsilon_i(\tau_l)\} = 0 \qquad \text{for } k \neq l;\ i = 1, 2, \ldots, p \tag{11.36c}$$

$$E\{\epsilon_i(\tau_k)d_j(\tau_l)\} = 0 \qquad \text{for all } k, l \text{ and for } i = 1, 2, \ldots, p;\ j = 1, 2, \ldots, m. \tag{11.36d}$$

Taken in turn, the conditions of (11.36) state that: the errors have sample mean values of zero (11.36a); are not cross-correlated among themselves (11.36b);

are not autocorrelated (11.36c); and are not correlated with the measured input forcing functions (11.36d). Results illustrative of this kind of analysis are given in Beck (1978b).

11.4. VALIDATION

Findeisen *et al.* (1978) define validation in the following terms: "a model can never be completely validated; we can never prove that its results conform to reality in all respects; it can only be invalidated." Models are working hypotheses about the nature of the behavior of a system. While the analyst may seek confirmation of his hypotheses in the process of validation, the basic purpose of validation is in fact to find invalid hypotheses; and knowledge of invalid hypotheses should ultimately lead to revised and better approximations (models) of reality. That, we presume, is a goal of every modeling exercise.

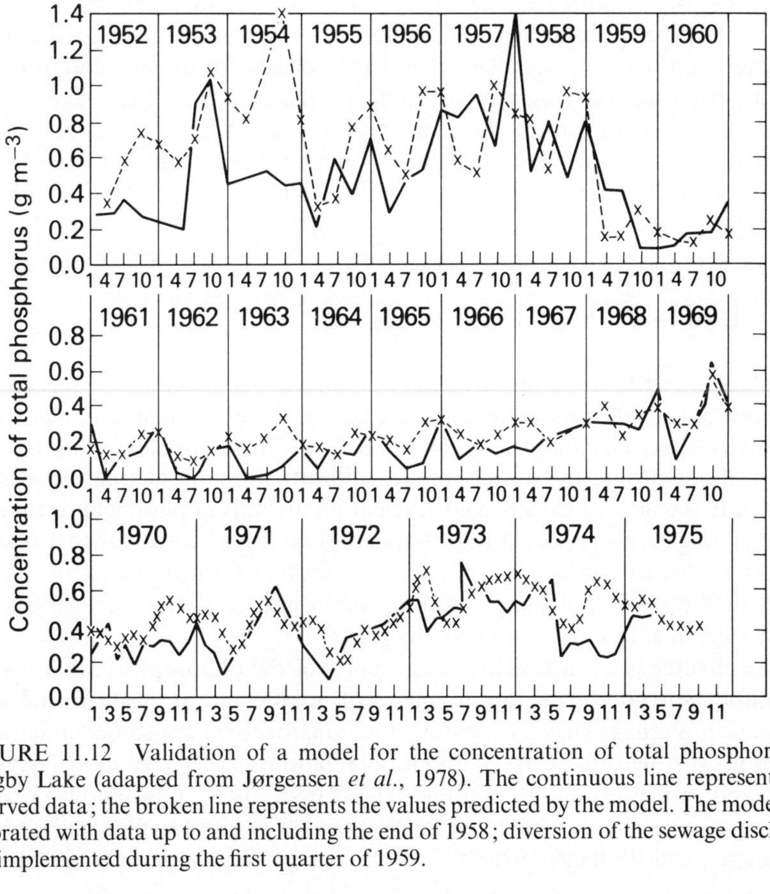

FIGURE 11.12 Validation of a model for the concentration of total phosphorus in Lyngby Lake (adapted from Jørgensen *et al.*, 1978). The continuous line represents the observed data; the broken line represents the values predicted by the model. The model was calibrated with data up to and including the end of 1958; diversion of the sewage discharge was implemented during the first quarter of 1959.

In precise mathematical terms no further remarks on model validation will be made here. Suffice it to say that verification, as discussed in the previous section, is not a particularly rigorous test of whether a model is "good" or "bad." It is possible to fit a model to a set of field data with almost arbitrarily small error, provided that there is a sufficient number of parameters in the model. Such a model would probably be of trivial value in terms of solving problems or in terms of understanding system behavior. It is of greater importance to learn whether a model that has been calibrated and verified with one set of data gives a reasonable approximation of behavior observed in a second (and independent) set of field data. As an example of this latter type of validation procedure, let us take the case of Lyngby Lake in Denmark (Jørgensen *et al.*, 1978). A model was calibrated and verified for the period 1952–58, during which time the lake was receiving an effluent of wastewater that had been treated both mechanically and biologically. After 1958 the wastewater discharge was diverted to a coastal location. Figure 11.12 shows a comparison of the field data for 1952–75 with the estimated responses of the model. Having calibrated the model for the period 1952–58, Jørgensen *et al.* made no further adjustment of the model parameter estimates. Thus 1959–75 is a period used for model validation; it is also a period in which we may assume that the behavior of the lake was significantly different from the observed behavior (1952–58) used for calibration. The reader is left to draw his own conclusions about the results of Figure 11.12—validation, like many other topics of this chapter, is a matter of personal judgment.

11.5. SUMMARY AND CONCLUSIONS

Chapters 2 and 11 are complementary chapters. The objective of this chapter has been to provide a more detailed discussion of the problems and methods of sensitivity analysis, calibration, and validation. There are clearly limits to the power of our current methods of model calibration, as this chapter has demonstrated. It is true that elegant and efficient algorithms of parameter estimation are available in theory, but at the interface between practice and theory in water quality model calibration it is a robust, workable algorithm that the analyst really requires. When dealing with field data it is rare, if ever, that the analysis yields elegant solutions with great efficiency.

The chapter has not touched upon many of the philosophical matters that inevitably surround the development of mathematical models and their calibration. Perhaps this is a serious omission, for there are some fundamental questions that need to be asked about the relationship between model development and model evaluation (Young, 1978; Beck, 1981; Lewandowski, 1981). Thomann, for example, urges that more attention be given to model verification (Thomann and Winfield, 1976): "... it is no longer of great moment if hundreds of sets of nonlinear equations are successfully solved on a large computer. What is of significance, however, is whether the numerical computations are reasonable representations of the real world."

Thomann's concern over the rapidly increasing size of water quality models without a consistent increase in the verified capabilities of these models is a concern to which others would probably subscribe. But what is "reasonable" to one person may well be quite different from what is "reasonable" to another person. Some recent attempts at a formal resolution—by reference to experimental observations—of these vexed questions of model verification and sufficient model complexity provide sobering evidence for the builder of large models (Maciejowski, 1978, 1979). On the other hand, a small model calibrated accurately against limited historical field data may not be capable of predicting responses to substantially altered future input disturbances. No doubt there will always be a gap between what can be simulated, in theory, and what can be verified in practice. One hopes that the size of the gap will not increase.

REFERENCES

Akaike, H. (1974). A new look at statistical model identification. *IEEE Transactions on Automatic Control* AC-19: 716–722.

Åström, K. J., and T. Bohlin (1966). Numerical identification of linear dynamic systems from normal operating records. *Theory of Self-adaptive Control Systems*, ed. P. H. Hammond (New York, NY: Plenum), pp. 96–111.

Åström, K. J., and P. Eykhoff (1971). System identification—A survey. *Automatica* 7: 123–162.

Beck, M. B. (1975). The identification of algal population dynamics in a nontidal stream. *Computer Simulation of Water Resources Systems*, ed. G. C. Vansteenkiste (Amsterdam: North-Holland), pp. 483–494.

Beck, M. B. (1978a). A comparative case study of dynamic models for DO–BOD–algae interaction in a freshwater river. *Research Report* RR-78-19 (Laxenburg, Austria: International Institute for Applied Systems Analysis).

Beck, M. B. (1978b). Random signal analysis in an environmental sciences problem. *Applied Mathematical Modelling* 2(1): 23–29.

Beck, M. B. (1979a). Model structure identification from experimental data. *Theoretical Systems Ecology*, ed. E. Halfon (New York, NY: Academic Press), pp. 259–289.

Beck, M. B. (1979b). System identification, estimation and forecasting of water quality—Part 1: Theory. *Working Paper* WP-79-31 (Laxenburg, Austria: International Institute for Applied Systems Analysis).

Beck, M. B. (1981). Hard or soft environmental systems? *Ecological Modelling* 11: 233–251.

Beck, M. B., and P. C. Young (1976). Systematic identification of DO–BOD model structure. *Proceedings of American Society of Civil Engineers, Journal of Environmental Engineering Division* 102(EE5): 909–927.

Bellman, R. E., and R. E. Kalaba (1965). *Quasilinearization and Nonlinear Boundary Value Problems* (New York, NY: American Elsevier).

Benson, M. (1979). Parameter fitting in dynamic models. *Ecological Modelling* 6: 97–115.

de Boer, B. (1979). A moving cell method of the dissolved oxygen and phytoplankton dynamics in rivers. *Hydrological Sciences Bulletin* 24(1): 199–211.

Bowles, D. S., and W. J. Grenney (1978a). Steady-state river quality modeling by sequential extended Kalman filters. *Water Resources Research* 14:84–95.

Bowles, D. S., and W. J. Grenney (1978b). Estimation of diffuse loading of water quality pollutants by Kalman filtering. *Applications of Kalman Filter to Hydrology, Hydraulics and Water Resources*, ed. C.-L. Chiu (Pittsburgh, PA: University of Pittsburgh, Stochastic Hydraulics Program), pp. 581–597.

Box, G. E. P., and G. M. Jenkins (1970). *Time-series Analysis, Forecasting and Control* (San Francisco, CA: Holden–Day).

Di Cola, G., L. Guerri, and H. Verheyden (1976). Parametric estimation in a compartmental aquatic ecosystem. *Preprints of Fourth IFAC Symposium on Identification and System Parameter Estimation, Institute of Control Sciences, Moscow*, Part 2, pp. 157–165.

Di Toro, D. M. (1969). Stream equations and method of characteristics. *Proceedings of American Society of Civil Engineers, Journal of Sanitary Engineering Division* 95(SA4):699–703.

Di Toro, D. M., D. J. O'Connor, and R. V. Thomann (1971). A dynamic model of the phytoplankton population in the Sacramento–San Joaquin Delta. *American Chemical Society, Advances in Chemistry Series* 106:131–180.

Di Toro, D. M., and G. van Straten (1979). Uncertainty in the parameters and predictions of phytoplankton models. *Working Paper* WP-79-27 (Laxenburg, Austria: International Institute for Applied Systems Analysis).

Dobbins, W. E. (1964). BOD and oxygen relationships in streams. *Proceedings of American Society of Civil Engineers, Journal of Sanitary Engineering Division* 90:53–78.

Dorf, R. C. (1965). *Time-domain Analysis and Design of Control Systems* (Reading, MA: Addison–Wesley).

Draper, N. R., and H. Smith (1966). *Applied Regression Analysis* (New York, NY: Wiley).

Edwards, A. M. C., and J. B. Thornes (1973). Annual cycle in river water quality: A time-series approach. *Water Resources Research* 9:1286–1295.

Erni, P. E., and J. Ruchti (1977). Der Sauerstoffhaushalt von Fliessgewässern: Kritische Prüfung eines mathematischen Sauerstoffmodells anhand der Identifikation seiner Parameter in Flüssen und künstlichen Rinnen. *Schweizerische Zeitschrift für Hydrologie* 39(2):261–276.

Eykhoff, P. (1974). *System Identification–Parameter and State Estimation* (Chichester: Wiley).

Findeisen, W., A. Iastrebov, R. Lande, J. Lindsay, M. Pearson, and E. S. Quade (1978). A sample glossary of systems analysis. *Working Paper* WP-78-12 (Laxenburg, Austria: International Institute for Applied Systems Analysis).

Fuller, F. C., and C. P. Tsokos (1971). Time-series analysis of water pollution data. *Biometrics* 27:1017–1034.

Gelb, A. (ed.) (1974). *Applied Optimal Estimation* (Cambridge, MA: MIT Press).

Gnauck, A., J. Wernstedt, and W. Winkler (1976). On the use of real-time estimation methods for the mathematical modelling of limnological ecosystems. *Preprints of Fourth IFAC Symposium on Identification and System Parameter Estimation, Institute of Control Sciences, Moscow*, Part 2, pp. 124–133.

Halfon, E. (1979). On the parameter structure of a large-scale ecological model. *Contemporary Quantitative Ecology and Related Ecometrics*, eds. G. P. Patil and M. L. Rosenzweig (Fairland, MD: International Cooperative Publishing House), pp. 279–293.

Halfon, E., H. Unbehauen, and C. Schmid (1979). Model order estimation and system identification theory and application to the modeling of ^{32}P kinetics within the trophogenic zone of a small lake. *Ecological Modelling* 6:1–22.

Huck, P. M., and G. T. Farquhar (1974). Water quality models using the Box–Jenkins method. *Proceedings of American Society of Civil Engineers, Journal of Environmental Engineering Division* 100(EE3):733–752.

Ivakhnenko, A. G., G. I. Krotov, V. I. Cheberkus, and V. N. Vysotskiy (1977). Identification of dynamic equations of a complex plant on the basis of experimental data using self-organization of models, Part I—One-dimensional problems. *Soviet Automatic Control* 10(2):24–30.

Jazwinski, A. H. (1970). *Stochastic Processes and Filtering Theory* (New York, NY: Academic Press).

Johnston, J. (1963). *Econometric Methods* (New York, NY: McGraw-Hill).

Jolankai, G., and A. Szöllösi-Nágy (1978). A simple eutrophication model for the Bay of Keszthely, Lake Balaton. *Modeling the Water Quality of the Hydrological Cycle. Proceedings of Baden Symposium, September. International Association of Hydrological Sciences Publication* 125, pp. 137–150.

Jørgensen, S. E., H. Mejer, and M. Friis (1978). Examination of a lake model. *Ecological Modelling* 4:253–278.

Källström, C. G., T. Essebo, and K. J. Åström (1976). A computer program for maximum likelihood identification of linear multivariable stochastic systems. *Preprints of Fourth IFAC Symposium on Identification and System Parameter Estimation, Institute of Control Sciences, Moscow*, Part 2, pp. 508–521.

Kalman, R. E. (1960). A new approach to linear filtering and prediction problems. *Transactions of American Society of Mechanical Engineers, Series D: Journal of Basic Engineering* 82:35–45.

Kalman, R. E., and R. S. Bucy (1961). New results in linear filtering and prediction theory. *Transactions of American Society of Mechanical Engineers, Series D: Journal of Basic Engineering* 83:95–108.

Kendall, M. G., and A. Stuart (1961). *Advanced Theory of Statistics* (London: Griffin).

Koivo, A. J., and G. R. Phillips (1971). Identification of mathematical models for DO and BOD concentrations in polluted streams from noise corrupted measurements. *Water Resources Research* 7(4):853–862.

Koivo, A. J., and G. R. Phillips (1972). On determination of BOD and parameters in polluted stream models from DO measurements only. *Water Resources Research* 8(2):478–486.

Koivo, A. J., and G. R. Phillips (1976). Optimal estimation of DO, BOD and stream parameters using a dynamic discrete time model. *Water Resources Research* 12(4):705–711.

Koivo, H. N., and A. J. Koivo (1978). Least-squares estimator for polluted stream variables in a distributed parameter model. *Advances in Water Resources* 1:191–194.

Lee, E. S. (1968). *Quasilinearization and Invariant Imbedding* (New York, NY: Academic Press).

Lee, E. S., and I. Hwang (1971). Stream quality modeling by quasilinearization. *Journal of Water Pollution Control Federation* 43(2):306–317.

Lettenmaier, D. P., and S. J. Burges (1976). Use of state estimation techniques in water resource system modeling. *Water Resources Bulletin* 12(1):83–99.

Lewandowski, A. (1981). Issues in model validation. *Working Paper* WP-81-32 (Laxenburg, Austria: International Institute for Applied Systems Analysis).

Lewis, S., and A. Nir (1978). A study of parameter estimation procedures of a model for lake phosphorus dynamics. *Ecological Modelling* 4:99–117.

Ljung, L. (1979). Asymptotic behavior of the extended Kalman filter as a parameter estimator for linear systems. *IEEE Transactions on Automatic Control* AC-24:36–50.

Maciejowski, J. M. (1978). The modeling of systems with small observation sets. *Lecture Notes in Control and Information Sciences* vol. 10 (Berlin: Springer).

Maciejowski, J. M. (1979). Model discrimination using an algorithmic information criterion. *Automatica* 15:579–593.

Mehra, R. K., and D. G. Lainiotis (eds.) (1976). *System Identification: Advances and Case Studies* (New York, NY: Academic Press).

Mehta, B. M., R. C. Ahlert, and S. L. Yu (1975). Stochastic variation of water quality in the Passaic River. *Water Resources Research* 11(2):300–308.

Meissinger, H. F. (1960). The use of parameter influence coefficients in computer analysis of dynamic systems. *Proceedings of Western Joint Computer Conference, San Francisco*, pp. 181–192.

Moore, R. J., and D. A. Jones (1978). Coupled Bayesian–Kalman filter estimation of parameters and states of dynamic water quality model. *Application of Kalman Filter to Hydrology, Hydraulics and Water Resources*, ed. C.-L. Chiu (Pittsburgh, PA: University of Pittsburgh, Stochastic Hydraulics Program), pp. 599–635.

O'Connor, D. J., and D. M. Di Toro (1970). Photosynthesis and oxygen balance in streams. *Proceedings of American Society of Civil Engineers, Journal of Sanitary Engineering Division* 96(SA2):547–571.

Rinaldi, S., and R. Soncini-Sessa (1978). Sensitivity analysis of generalized Streeter–Phelps models. *Advances in Water Resources* 1:141–146.

Rinaldi, S., R. Soncini-Sessa, H. Stehfest, and H. Tamura (1979). *Modeling and Control of River Quality* (New York, NY: McGraw-Hill).

Scavia, D. (1980). Uncertainty analysis of a lake eutrophication model. *Ph.D. Dissertation, University of Michigan, Ann Arbor, MI.*

Schurr, J. M., and J. Ruchti (1975). Kinetics of oxygen exchange, photosynthesis and respiration in rivers determined from time-delayed correlations between sunlight and dissolved oxygen. *Schweizerische Zeitschrift für Hydrologie* 37(1):144–174.

Shastry, J. S., L. T. Fan, and L. E. Erickson (1973). Nonlinear parameter estimation in water quality modeling. *Proceedings of American Society of Civil Engineers, Journal of Environmental Engineering Division* 99(EE3):315–331.

Söderström, T., L. Ljung, and I. Gustavsson (1978). A theoretical analysis of recursive identification methods. *Automatica* 14:231–244.

Stehfest, H. (1978). On the monetary value of an ecological river quality model. *Research Report* RR-78-1 (Laxenburg, Austria: International Institute for Applied Systems Analysis).

van Straten, G. (1983). Maximum likelihood estimation of parameters and uncertainty in phytoplankton models. *Uncertainty and Forecasting of Water Quality*, eds. M. B. Beck and G. van Straten (Berlin: Springer), pp. 157–171.

van Straten, G., and B. de Boer (1979). Sensitivity to uncertainty in a phytoplankton–oxygen model for lowland streams. *Working Paper* WP-79-28 (Laxenburg, Austria: International Institute for Applied Systems Analysis).

Tamura, H. (1979). On some identification techniques for modeling river quality dynamics with distributed lags. *Handbook of Large-scale Systems Engineering Applications*, eds. M. G. Singh and A. Titli (Amsterdam: North-Holland), pp. 274–294.

Thé, G. (1978). Parameter identification in a model for the conductivity of a river based on noisy measurements at two locations. *Modeling, Identification and Control in Environmental Systems*, ed. G. C. Vansteenkiste (Amsterdam: North-Holland), pp. 823–830.

Thomann, R. V. (1963). Mathematical model for dissolved oxygen. *Proceedings of American Society of Civil Engineers, Journal of Sanitary Engineering Division* 89(SA5):1–30.

Thomann, R. V. (1967). Time-series analysis of water quality data. *Proceedings of American Society of Civil Engineers, Journal of Sanitary Engineering Division* 93(SA1):1–23.

Thomann, R. V., and R. P. Winfield (1976). On the verification of a three-dimensional phytoplankton model of Lake Ontario. *Proceedings of Conference on Environmental Modeling and Simulation. US Environmental Protection Agency, Washington, DC Report* EPA 600/9-76-016, pp. 568–572.

Tomovic, R. (1964). *Sensitivity Analysis of Dynamic Systems* (New York, NY: McGraw-Hill).

Unbehauen, H., and B. Göhring (1974). Tests for determining model order in parameter estimation. *Automatica* 10:233–244.

Van den Boom, A. J. W., and A. W. M. Van den Enden (1974). The determination of the orders of process and noise dynamics. *Automatica* 10:245–256.

Wellstead, P. E. (1976). Model order testing using an auxiliary system. *Proceedings of Institution of Electrical Engineers* 123:1373–1379.

Wellstead, P. E. (1978). An instrumental product moment test for model order estimation. *Automatica* 14:89–91.

Whitehead, P. G. (1980a). An instrumental variable method of estimating differential-equation models of dispersion and water quality in nontidal rivers. *Ecological Modelling* 9:1–14.

Whitehead, P. G. (1980b). Real-time monitoring and forecasting of water quality in the Bedford Ouse river system. *Hydrological Forecasting. Proceedings of Oxford Symposium. International Association of Hydrological Sciences Publication* 129, pp. 333–342.

Whitehead, P. G., and P. C. Young (1975). A dynamic–stochastic model for water quality in part of the Bedford Ouse river system. *Computer Simulation of Water Resources Systems*, ed. G. C. Vansteenkiste (Amsterdam: North-Holland), pp. 417–438.

Wilde, D. J. (1964). *Optimum Seeking Methods* (Englewood Cliffs, NJ: Prentice-Hall).

Wilde, D. J., and C. S. Beightler (1967). *Foundations of Optimization* (Englewood Cliffs, NJ: Prentice-Hall).

Young, P. C. (1974). A recursive approach to time-series analysis. *Bulletin of the Institute of Mathematics and Its Application* 10:209–224.

Young, P. C. (1976). Some observations on instrumental variable methods of time-series analysis. *International Journal of Control* 23:593–612.

Young, P. C. (1978). General theory of modelling for badly defined systems. *Modelling, Identification and Control in Environmental Systems*, ed. G. C. Vansteenkiste (Amsterdam: North-Holland), pp. 103–135.

Young, P. C., and P. G. Whitehead (1977). A recursive approach to time-series analysis for multivariable systems. *International Journal of Control* 25:457–482.

CHAPTER 11: NOTATION

$\mathbf{A}$	system matrix for state variable relationships in a linear system
$\mathbf{B}$	input matrix for relationships between input variables and state variables in a linear system
$\boldsymbol{c}$	vector of state variables
$\boldsymbol{c}^*$	augmented vector of state variables and parameters
$\boldsymbol{c}^{o}$	vector of measured output response variables
$\hat{\boldsymbol{c}}$	vector of state estimates
$\hat{\boldsymbol{c}}^{o}$	vector of model-predicted values for the output response variables
$\bar{\boldsymbol{c}}$	nominal deterministic reference trajectory for the state vector
C_s	saturation concentration of dissolved oxygen
$\boldsymbol{d}$	vector of measurable input disturbances
D_B	net rate of addition of DO to reach of river by the combined effects of photosynthesis, respiration, and decomposition of mud deposits
D'_B	rate of oxygen consumption by mud deposits
$E\{\cdot\}$	expectation operator
$\boldsymbol{f}\{\cdot\}$	nonlinear vector-valued function for variations in the state variables
$f_2\{\cdot\}, f_3\{\cdot\}, f_4\{\cdot\}$	nonlinear functions for, respectively, the uptake of ammonia nitrogen and nitrate nitrogen by algae, and the growth rate of algae
$\mathbf{F}(\tau_k)$	system matrix for relationships between small perturbations in a linearized system

$\mathbf{G}(\tau_k)$	gain matrix in the correction procedure of the Kalman filter
$\mathbf{G}^*(\tau_k)$	gain matrix in the correction procedure of the extended Kalman filter
$\mathbf{H}$	matrix for the relationships between observations in a linear system
$\mathbf{I}$	unit identity matrix
J	(squared error) loss function
K_1	BOD decay rate constant
K_2	reaeration rate constant
K_{23}	first-order rate constant for conversion of ammonia nitrogen to nitrate nitrogen
K_{45}	rate constant for death of algae
K_{52}	rate constant for decomposition of organic nitrogen to ammonia nitrogen
K^*	rate constant for the addition of components by lateral inflow
L_A	rate of addition of BOD to reach by, for instance, local surface runoff
L_i	concentration of water quality component i in lateral inflow
m	number of input variables
$\boldsymbol{m}\{\cdot\}$	nonlinear vector-valued function for relationships between state and output variables
$\mathbf{M}(\tau_k)$	matrix for the relationships between observations of a linearized small-perturbation system
n	number of state variables
p	number of output variables
$\mathbf{P}(\tau \mid \tau_k)$	(state–parameter) estimation error covariance at time τ, given all measured information up to and including that available at time τ_k
q	number of parameters
$\mathbf{Q}$	system noise covariance matrix
$Q(\tau)$	stream discharge
$\mathbf{R}$	measurement error covariance matrix
s	sensitivity coefficient
V	volume of water in reach of river
$\mathbf{W}$	matrix of weighting coefficients
$\boldsymbol{\beta}$	vector of parameters
$\bar{\boldsymbol{\beta}}$	nominal reference vector of parameter values
$\hat{\boldsymbol{\beta}}$	vector of parameter estimates
γ	coefficient for rate of oxygen consumption by nitrification
δ_{kl}	Kronecker delta function

$\boldsymbol{\epsilon}$	predicted model response errors (or residual errors)
$\boldsymbol{\zeta}$	vector sequence of disturbances for random walk parameters
$\boldsymbol{\eta}$	vector of (stochastic) measurement errors
λ	scalar step length in an off-line parameter estimation algorithm
$\boldsymbol{\xi}$	vector of (stochastic) unmeasured input disturbances of system
τ	independent variable of time or of time of travel
τ_k	kth discrete sample in time or in time of travel
$\boldsymbol{\Phi}(\tau_k, \tau_{k-1})$	state transition matrix of a linearized small-perturbation system
$\boldsymbol{\Psi}$	matrix of coefficients in an off-line parameter estimation algorithm.

12 Models for Management Applications

D. P. Loucks

12.1. INTRODUCTION

Water quality management planning involves the identification and evaluation of various management alternatives for achieving economic and water quality goals. Economic goals are often expressed in terms of cost effectiveness (cost minimization) and in terms of the distribution of the cost among those who should pay. Water quality goals are usually expressed as wastewater effluent standards or water quality standards in waste-receiving water bodies, or both. The effectiveness of any management alternative may be measured in terms of how well it accomplishes these goals. Water quality management models can assist planners in identifying and evaluating possible management alternatives in order to determine which alternative is best.

Water quality management models are usually extensions of some of the simulation or predictive models discussed in the previous chapters. In addition to the predictive equations, management models include as unknowns the design and operating policy variables of each management alternative. Relationships are included that describe the resulting effluent or water quality, and the cost of each management alternative as a function of the design and operating policy variables. Also included in these models are the constraints defining the desired effluent and/or water quality standards.

Most water quality management models are optimization models. As such, the relationships that define economic costs and the resulting effluent or water quality as functions of the unknown design or operating variables must conform to what is required for model solution using particular optimization solution procedures. If simplifications or modifications are necessary purely for model solution, then, as emphasized throughout this chapter, the resulting solution should be checked using more accurate water quality simulation models. Management models are used for a preliminary evaluation of various alternatives and for identifying what data are important and needed prior to the implementation of a more expensive data collection and simulation study.

12.2. MANAGEMENT ALTERNATIVES FOR WATER QUALITY CONTROL

The first and most obvious method of water quality control is to limit the amount of waste discharged into water bodies. This type of control can take on numerous forms, some of which are described below.

(1) Each waste producer is required to discharge less waste, for example through process changes or removal of at least some minimum specified fraction of the waste prior to releasing the remainder into natural waters or land waste disposal areas. Removal of waste can be accomplished by a variety of physical, biological, and chemical processes.
(2) The portion of the treated wastewater effluent that, if released into the natural water body, would result in a lower water quality than desired is stored instead. Ponds or tanks can be used for effluent storage. The quantity and timing of discharge of stored effluent to land or water should depend in part on the assimilative capacity of the receiving body.
(3) Waste is piped, either prior to or following some treatment, to areas within or outside of the region for additional treatment and/or disposal at land or water sites having greater assimilative capacities. This alternative also permits the processing of wastes at larger regional facilities that benefit from economies of scale in construction and operating costs, as well as from increased operating efficiencies.
(4) In-stream quality can be improved by artificial aeration or flow augmentation. The deficit of dissolved oxygen can be decreased by injection of air into the water. Increasing the stream flow in periods of low flows by releasing water from upstream reservoirs may also improve the stream quality by dilution and by changing velocities and temperatures, which, in turn, affect the reaction rates of various quality constituents.

Each of these means of water quality management will be discussed in greater detail later in this chapter. Prior to this discussion some remarks on management objectives and the criteria that are used for evaluating alternative combinations of the various central options are appropriate.

12.3. MANAGEMENT OBJECTIVES AND QUALITY STANDARDS

Water quality management objectives are multiple and conflicting. Those in control of activities that generate wastes would naturally prefer to dispose of their wastes at no cost to themselves and, if possible, to others as well. This policy leads to higher profits if income is being derived from the waste-making activities, or to less taxes if the wastes are derived from human settlements such as cities and municipalities.

However, if the discharge of wastes does result in added costs elsewhere, i.e. in environmental damages, those who incur these costs and damages would also prefer not to incur them. They can argue that dischargers of waste into water bodies, for example, should pay for the environmental damage. Yet because those who discharge waste are usually not affected by the damage caused by that waste, there is no economic incentive to control that discharge. Water pollution is said to be an externality, i.e. it is imposed on individuals other than those who cause it. This is the central conflict involving water quality management in river basins throughout the world.

Because the private market system fails to charge each polluter an amount equal to the damages resulting from his waste discharge, regulatory action is often required. The types of incentive that water quality regulatory agencies have used to compensate for the failure of individual polluters to consider the damages they impose on others are: (1) legislative, including direct regulation, the establishment of effluent or stream quality standards, licensing, and zoning; (2) legal, including compensation for damages and fines for violation of law; and (3) economic, including effluent charges or taxes, subsidies, accelerated depreciation allowances, and the like. Whatever the methods used, the objective should be to achieve a more efficient and equitable allocation of natural resources from the standpoint of society as a whole.

One of the difficulties in finding a plan that is both efficient and equitable is the problem of quantifying water quality benefits or damages. This problem is similar to that of attributing a monetary benefit to such things as aesthetics and clean air. There is also the problem of determining equitable distributions of costs and benefits. Thus, the selection of the desired water quality and the determination of who will pay for it often become political decisions. This political aspect is reflected in both the water quality management objectives and the quality standards. Political systems have clearly demonstrated their sensitivity to these multiobjective aspects of water quality management problems.

Those who develop regional water quality management models usually assume the actual or potential existence of some governmental institution that has the authority to control water quality within its region, by economic incentives such as effluent charges and/or by legal means such as effluent standards. The main purpose of most regional water quality models is to examine alternatives that will reduce both the private costs and public damages resulting from water pollution.

To begin quantifying a rather general objective for regional water quality management, let us consider a river basin in which there are numerous individuals or groups s that discharge pollutants into the natural water courses. Included among those individuals or groups of individuals are organizations, such as state and federal pollution control agencies, that have financial as well as political interests in the quality of the natural water within the basin. Water quality control alternatives such as waste water treatment and effluent storage impose costs on private agencies and, because of cost-sharing programs, on public or governmental bodies as well. Quality control alternatives such as flow augmentation and artificial aeration may only add to the cost paid by one or more public agencies. Regardless of who pays, the cost to each individual or group s can, for the moment, be denoted as a function of the scale of all alternatives used for water quality control.

If S_i is the scale of some waste reduction alternatives at site i costing $C_i^s(S_i)$ for each group s, a cost-effective objective without regard to cost distribution or to the political influence of each group can be written as:

$$\text{minimize} \sum_i \sum_s C_i^s(S_i). \tag{12.1}$$

Such an objective may or may not result in an acceptable solution. This depends in part on the quality standards imposed. Usually the minimum allowable quality standards are not intended to represent the desired quality. Environmental protection goals and allowances for future growth and uncertainties often result in planned or desired qualities that are higher than specified by minimum quality standards. One way to achieve a quality that comes closer to the target is through the proper allocation of effluent taxes T_i^s to the group s discharging wastes at site i. The tax could be dependent on the waste released at each site i.

The fraction $P_i = P_i(S_i)$ of the waste reduced or removed at each site i is, of course, a function of the scale S_i of the waste reduction measures employed at each site. If W_i is the constant quantity of waste available at site i prior to the implementation of any waste reduction measures, then $W_i(1 - P_i)$ is the remaining quantity that will be discharged. Ideally any tax on the amount of waste discharged should reflect the external damages attributable to that discharge. The purpose of the tax or subsidy is to provide an economic incentive for reducing the external damages, if any, that result from the discharge of wastes.

Effluent charges typically are paid to the appropriate river basin authority or other public agency responsible for water quality management. The task of such an agency in establishing and implementing effluent charges is to set them in an equitable manner so as to cover the agency cost, not paid from other sources, of measures taken by the agency to achieve a desired water quality and for agency administration and operation. The dischargers themselves are interested in minimizing their costs of waste treatment or reduction plus the tax they must pay for the wastes that they discharge.

Let $C_i(S_i)$ represent the annual agency costs of measures S_i taken at site i to improve water quality, and let $C_i^s(S_i)$ be the individual or private costs, as before. The aim of the control agency is to establish effluent tax rates T_i^s per unit of waste discharge $W_i(1 - P_i)$ in such a way as to minimize the total cost of water quality control:

$$\text{minimize} \sum_i \Bigg(\underbrace{C_i(S_i)}_{\substack{\text{agency}\\ \text{costs}}} + \underbrace{\sum_s C_i^s(S_i)}_{\substack{\text{other private}\\ \text{and/or public}\\ \text{costs}}} \Bigg) \tag{12.2}$$

while ensuring that

$$\underbrace{\sum_s \sum_i T_i^s W_i(1 - P_i(S_i))}_{\substack{\text{total effluent}\\ \text{tax income}}} \geq \underbrace{\sum_i C_i(S_i)}_{\substack{\text{total agency}\\ \text{cost}}} \tag{12.3}$$

and that the desired quality is maintained at all sites j:

$$\underbrace{\bar{Q}_j(\bar{S}, \bar{W})}_{\text{quality at site } j} \geq \underbrace{\bar{Q}_j^{\min}}_{\substack{\text{minimum desired}\\ \text{quality at } j}} . \tag{12.4}$$

The individuals who must pay the cost of waste reduction or treatment and/or an effluent tax on the waste $W_i(1 - P_i)$ discharged at sites i are of course interested in minimizing their total costs:

$$\text{minimize} \sum_i [\underbrace{C_i^s(S_i)}_{\substack{\text{waste reduction}\\ \text{cost}}} + \underbrace{T_i^s W_i(1 - P_i)}_{\text{effluent charge}}] \quad \forall s. \tag{12.5}$$

These objectives do not attempt to quantify the benefits or damages associated with the resulting water quality, except through the establishment of effluent charges and, perhaps, quality standards. The charges and standards would have to be defined prior to their incorporation into water quality models. One of the advantages, however, of model construction and solution prior to the final establishment of charges and standards is the ability to estimate the costs of each group s and the resulting water quality associated with various proposed combinations of charges and standards.

This problem, defined by eqns. 12.2–12.5, is an example of a multilevel–multiobjective planning problem. The controlling agency objective (12.2) is to be minimized subject to the minimization of a number of other objectives (12.5), which are not controlled by the agency except through the establishment of the effluent tax rates T_i^s in those objectives. Research has only begun to identify some approaches to solving the overall multilevel–multiobjective problem. In the absence of satisfactory solution procedures, the agency-level objective by itself (the minimization of total cost of all measures implemented for water quality control, eqn. 12.2) is typically chosen for a preliminary evaluation of water quality management alternatives.

Yet with or without effluent taxes, the agency objective of cost minimization has not been generally accepted. This is in part because there are, indeed, other water quality management objectives at other levels of planning and decision making, and it is not obvious how these objectives should be combined to simulate, in a specific situation, what might be the actual response to a decision at the central agency level.

The political process of establishing effluent charges and minimum acceptable qualities, in the form of either effluent or stream quality standards, involves the participation of each group of interested individuals within the river basin. Some groups have more political influence than others. This depends not only on their political skills but also on how strongly they feel about certain issues. To include the effect of this influence in water quality models, it is often assumed that relative weights can be defined and used in the objective function. Each weight w_s reflects the relative influence that group s exerts compared with all other groups defined by the model. If the proper weights, charges, and standards are used, a socially or politically equitable and efficient water quality management policy might result from the following objective:

$$\text{minimize} \sum_s w_s \sum_i [C_i^s(S_i) + T_i^s W_i(1 - P_i)]. \tag{12.6}$$

The difficulty here, of course, is that the relative political weights are unknown, even to the decision makers, until the final decision is made. By varying the relative weights, however, an analyst can define some of the efficient alternatives from the infinite set of possible alternatives. If the objective is piecewise linear, the number of efficient alternatives defined by this procedure will be smaller than the number of possible efficient alternatives. Efficient alternatives can also be defined by setting upper bounds on all but one of the terms within the square brackets in (12.5) and minimizing the other. Clearly, both the former weighting approach and this latter target approach, and even other more efficient iterative multiobjective approaches, merely define possible solutions, not necessarily a best solution.

If all the relative weights are assigned values of unity, the objective function (12.6) will represent the minimization of total costs and effluent charges without regard to any redistribution of the costs and charges among various polluters. As the relative weights or target levels change, so will the alternatives associated with those weights or targets. Relatively high weights or low targets will correspond to those polluters having a relatively strong political position and interest, which in effect will reduce their share of the total cost. Varying the weights or targets permits an examination of the stream quality that is likely to be associated with various cost distributions.

Other means have been used to incorporate considerations of equity into otherwise strictly cost-effective models (e.g. those having objective functions of the type (12.2)). These include constraints specifying equal scales of various alternatives or some function of these scales, such as equal treatment efficiencies or equal costs per capita contributing to the total waste at various sites. If S_i is the scale of quality control alternatives employed at site i, then constraints requiring equal scales of control could be written as

$$S_i = S_k \qquad \forall i \in Z_k, \tag{12.7}$$

where Z_k is the set of sites in zone k of the region. Zones within a region may be defined geographically or by types of polluter.

Although numerous analysts have included equity within the constraint set of their models, it could be argued that equity is an objective—one of many that water quality planners consider. The relative weight given to an equity objective depends in part on the economic costs of achieving it, as well as on the administrative and political costs of not achieving it, i.e. the unquantifiable cost associated with implementing a plan that calls for a wide range of quality control requirements within a region and that minimizes only the total economic costs. Equity within any zone of a river basin can be expressed as an objective by defining and then minimizing the absolute difference between the minimum and maximum scales of water quality control within the zone. Let $S_k^{\min}$ and $S_k^{\max}$ denote variable lower and upper bounds on the scale S_i of a single control alternative within zone k so that

$$S_k^{\min} \le S_i \le S_k^{\max} \qquad \forall i \in Z_k. \tag{12.8}$$

Part of the equity objective can involve minimizing the sum of the weighted differences between $S_k^{\min}$ and $S_k^{\max}$ over all zones k:

$$\text{minimize} \sum_k w_k(S_k^{\max} - S_k^{\min}). \tag{12.9}$$

Constraints on water quality can be defined as (a) effluent standards restricting the waste $W_i(1 - P_i)$ released at site i to be no greater than the maximum allowable quantity:

$$W_i(1 - P_i) \le W_i^{\max}, \tag{12.10}$$

or as (b) receiving-water quality standards requiring the quality $Q_i(\bar{S})$ at a site or reach i within the water body to be no less than some minimum allowable quality:

$$Q_i(\bar{S}) \ge Q_i^{\min}. \tag{12.11}$$

These objectives and constraints will be defined in greater detail below.

12.4. WATER QUALITY CONTROL ALTERNATIVES

In each of the models to be described, wastewater treatment and/or reduction through process changes (i.e. the removal of some fraction of the total waste load prior to discharging the remainder into the receiving water bodies) will be included as a method of controlling water quality. Models have been developed for assisting in the design of wastewater treatment facilities and in the design of treatment processes within these facilities (Loucks, 1967; Lawrence and McCarty, 1970; Fan *et al.*, 1974; Middleton and Lawrence, 1974, 1976; Van Note *et al.*, 1975; Grady, 1977). Such models are useful for defining the costs of wastewater treatment. Since the design and cost of alternative treatment facilities are fairly well defined and known, the discussion in this section will begin with an examination of how to dispose of the treated wastewater effluent.

12.4.1. Wastewater Disposal on Land

There are two alternatives for disposing of wastewater effluent from a treatment plant. These consist of discharging the effluent either into a receiving water body or on to land for further waste reduction prior to drainage into a water body. Wastewater disposal on land, or land application as it is commonly called, may be attractive in certain areas where land is available. Land application permits further removal of nitrogen, phosphorus, organics, pathogens, and traces of heavy metals from the wastewater effluent. Of all these constituents, nitrogen is the most mobile in soils, and hence its removal usually controls the rate of land application (Haith, 1973; Koenig and Loucks, 1977). Nitrogen in the drainage waters from well aerated soils exists mostly in the form of nitrate nitrogen, NO_3-N.

Outlined in this section is a simple simulation model to assist in evaluating irrigation land application alternatives that do not generate more than the predefined maximum allowable NO_3-N concentrations in the drainage waters. Effluent disposal on land is commonly accomplished using spray irrigation methods. Nitrogen is removed from the soil mostly by plant uptake and subsequent harvest or consumption by grazing.

Figure 12.1 illustrates a typical land application system. Of interest are the storage lagoon capacity V, the land application area A, the irrigation volumes Q_{2t} in each period t (of length Δt) within a year, and the maximum discharge rate Q_2^*,

$$Q_2^* = \max_t (Q_{2t}/\Delta t), \tag{12.12}$$

that minimize the total annual cost $C(V, A, Q_2^*)$. This cost can then be compared with the additional cost of advanced wastewater treatment required to meet the same NO_3-N and other constituent effluent standards.

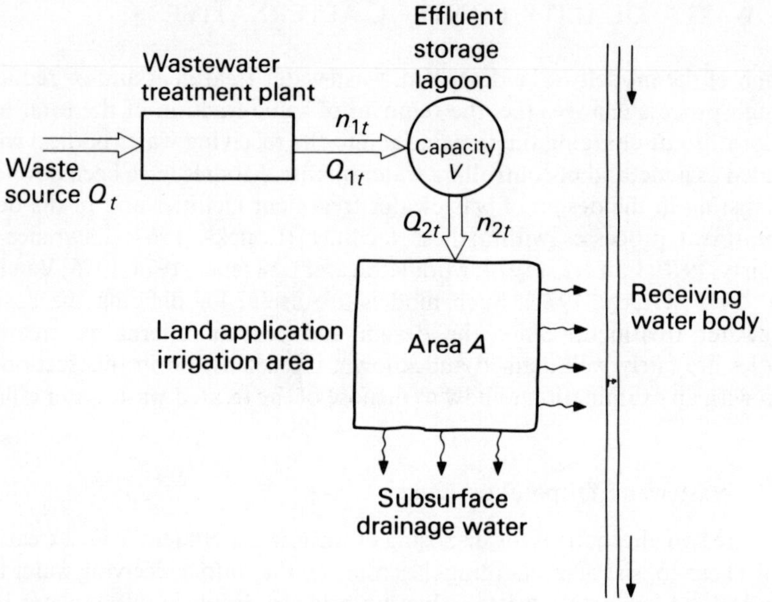

FIGURE 12.1 Components of the land application system of wastewater disposal.

The relationships between the components of the land application system can be defined by a series of mass balance equations. The wastewater volume mass balance for the storage lagoon equates the final storage volume S_{t+1} to the initial storage volume S_t plus the difference between the inflow volume Q_{1t} and the outflow volume Q_{2t} in each period t of the year:

$$S_{t+1} = S_t + Q_{1t} - Q_{2t} \qquad \forall t. \tag{12.13}$$

If t is the last period within the year, then $t + 1 = 1$. The lagoon storage capacity V equals the maximum of all the storage volumes S_t:

$$V = \max_t (S_t) \qquad \forall t. \tag{12.14}$$

The inflow volumes are known, but the outflow volumes Q_{2t} and lagoon capacity V are unknown decision variables. Equation 12.13 can be modified to include evaporation losses if desired. Such losses could be based on the average lagoon volume $\bar{S}_t$ in each period:

$$\bar{S}_t = (S_t + S_{t+1})/2 \qquad \forall t. \tag{12.15}$$

Nitrogen in the lagoon may be removed by ammonia volatilization, denitrification, algal uptake, and settling. If a first-order decay process is assumed, the mass of nitrogen (the product of concentration $n_{2,t+1}$ and volume S_{t+1}) in the lagoon at the end of each period t equals the initial lagoon nitrogen mass $S_t n_{2t}$

plus the difference between the mass input $Q_{1t}n_{1t}$ and the combined mass outflow $Q_{2t}(n_{2,t+1} + n_{2t})/2$ and decay $k_t\bar{S}_t(n_{2,t+1} + n_{2t})/2$, where k_t is the temperature-dependent nitrogen removal rate in period t $[\mathrm{T}^{-1}]$:

$$\underset{\substack{\text{mass at end}\\ \text{of period } t}}{S_{t+1}n_{2,t+1}} = \underset{\substack{\text{mass at}\\ \text{beginning}\\ \text{of period } t}}{S_t n_{2t}} + \underset{\substack{\text{mass in}\\ \text{influent}}}{Q_{1t}n_{1t}} - \underset{\text{mass in effluent}}{Q_{2t}(n_{2,t+1} + n_{2t})/2}$$

$$- \underset{\substack{\text{mass decay in}\\ \text{period } t}}{k_t\bar{S}_t(n_{2,t+1} + n_{2t})/2}. \tag{12.16}$$

This completes the equations involving the storage lagoon. What remains to be described is the spray irrigation site.

If it is assumed that the soil moisture content, expressed as a depth [L], is maintained at field capacity M throughout the year (since otherwise additional lagoon storage capacity may be required), the water balance for the irrigated area is defined by equating the irrigation rate Q_{2t}/A with the evapotranspiration rate E_t plus the drainage rate d_t less the average precipitation rate P_t in period t:

$$Q_{2t}/A = E_t + d_t - P_t \qquad \forall t. \tag{12.17}$$

Each of these terms is expressed in units of length. To prevent surface runoff, the irrigation rate, allowing for precipitation and loss by evaporation, should not exceed the maximum drainage capacity d, or

$$d_t \le d \qquad \forall t. \tag{12.18}$$

Drainage occurs, i.e. $d_t > 0$, when the application rate exceeds that required to just maintain soil moisture content at field capacity.

Soil nitrogen relationships can be approximated by separately defining mass balance equations for organic and inorganic nitrogen. Average organic nitrogen levels in the soil must reach an equilibrium value F $[\mathrm{M\,L}^{-2}]$ if the waste disposal system is to be operated at a steady state. This value may be determined by the native fertility of the soil, or if an objective of land application is to build up soil productivity, F will be a desired equilibrium value. Soil organic nitrogen levels will deviate from the equilibrium value during any period owing to mineralization of some fraction m_t of the organic nitrogen and addition of organic nitrogen XO_t $[\mathrm{M\,L}^{-2}]$ from wastewater irrigation during the period. If O_t is the deviation $[\mathrm{M\,L}^{-2}]$ at the beginning of period t then the total organic nitrogen level at the beginning of period $t + 1$ is $F + O_{t+1}$:

$$F + O_{t+1} = F + O_t - m_t(F + O_t) + XO_t$$

or

$$O_{t+1} = (1 - m_t)O_t - m_t F + XO_t. \tag{12.19}$$

The nitrogen addition per unit land area is some fraction α of the nitrogen in the lagoon effluent $Q_{2t}(n_{2,t+1} + n_{2t})/2$ that is in the organic form, divided by the total land area A:

$$XO_t = \alpha Q_{2t}(n_{2,t+1} + n_{2t})/2A \qquad \forall t. \tag{12.20}$$

The soil inorganic nitrogen content per unit land area, I_{t+1} [M L^{-2}], at the end of each period t equals the sum of the initial inorganic nitrogen content I_t, the fraction m_t of organic nitrogen that was mineralized in period t, and the inorganic nitrogen addition XI_t in the wastewater effluent, less that leached from the soil by drainage, L_t, and that removed from the soil by plant growth, N_t, during the period:

$$I_{t+1} = I_t + m_t(F + O_t) + XI_t - L_t - N_t \qquad \forall t. \tag{12.21}$$

The addition of inorganic nitrogen is that contained in the lagoon effluent divided by the whole irrigation area A:

$$XI_t = (1 - \alpha)Q_{2t}(n_{2,t+1} + n_{2t})/2A \qquad \forall t. \tag{12.22}$$

In the preceding equations, soil nitrogen losses from ammonia volatilization, denitrification, and surface runoff were assumed insignificant. The effect of this assumption will be conservative. The mineralization fraction m_t will depend on the average soil temperature during each period t. Of course, when the soil is frozen, essentially no mineralization or drainage takes place ($d_t \approx 0$).

The inorganic nitrogen loss from leaching, L_t [ML^{-2}], depends on the average inorganic nitrogen concentration $(I_{t+1} + I_t)/2M$ in the soil water when leaching or drainage occurs, i.e. when $d_t > 0$:

$$L_t = d_t(I_t + I_{t+1})/2M \qquad \forall t. \tag{12.23}$$

Plant uptake of inorganic nitrogen, N_t, will depend on the type of cover crop grown and harvested or consumed as well as on the available inorganic nitrogen in the soil. If $N_t^{\max}$ is the upper limit of the nitrogen uptake (which will depend on the type of plant) and if up to 70% of the soil nitrogen is available to the irrigated crops, then N_t equals the smaller of these two maxima:

$$N_t = \min(0.7[I_t + m_t(F + O_t) + XI_t], N_t^{\max}] \qquad \forall t. \tag{12.24}$$

The final constraint applies to the quality of the drainage water. If $n_t^{\max}$ is the maximum allowable nitrate nitrogen concentration, then

$$(I_t + I_{t+1})/2M \leq n_t^{\max} \qquad \forall t. \tag{12.25}$$

Model Solution Procedure

A simulation procedure can be defined to solve this model. Input data include wastewater inflows Q_{1t} to the storage lagoon and their nitrogen concentrations

n_{1t}, the nitrate nitrogen decay rate k_t, the hydrological parameters P_t and E_t, and the soil data parameters M, d, and m_t. Then a particular set of lagoon volume discharges Q_{2t} can be selected whose total equals the sum of the lagoon inflows Q_{1t}. The maximum of all Q_{2t} determines the capacity Q_2^* of the pumps and pipes required to connect the lagoon to the irrigation area. The outflows Q_{2t} together with the known inflows Q_{1t} determine the storage capacity V of the lagoon as found from the simultaneous solutions of eqns. 12.13 and then solution of eqn. 12.14. (Since each of eqns. 12.13 is linearly dependent on the others, each S_t in the solution to (12.13) can be adjusted by a constant amount to ensure any desired minimum storage volume for increased detention times and nitrogen removal.)

When each lagoon storage volume S_t has been determined, and since the influent nitrate nitrogen concentrations n_{1t} are known, the simultaneous solutions of eqns. 12.16 yield the nitrate nitrogen concentrations n_{2t} of the lagoon effluent. Finally, (12.17) and (12.18) can be used to compute the minimum irrigation area required for the effluent discharges Q_{2t}.

Equation 12.18 ensures that for any particular land area A, the drainage d_t cannot exceed the drainage capacity d. Knowing the effluent nitrate nitrogen concentrations n_{2t} permits the solution of (12.20) and (12.22) for the nitrogen additions per unit land area, XO_t and XI_t. It is then possible to solve (12.19).

The simultaneous solutions of (12.19) can provide an estimate of deviations in soil organic matter per unit area, O_t, at the beginning of each period t associated with particular values of A and Q_{2t}. Knowledge of each O_t permits the simultaneous solutions of (12.21), (12.23), and (12.24), if it is assumed that the plant uptake of nitrogen, N_t, equals $0.7[I_t + m_t(F + O_t) + XI_t]$. After the solution of these equations for each inorganic nitrogen mass per unit area, I_t, if any N_t exceed $N_t^{\max}$, those N_t are set equal to $N_t^{\max}$ and the equations must be solved again. This procedure is continued until eqns. 12.24 are satisfied.

It only remains to check that the constraint eqns. 12.25 limiting the nitrate nitrogen concentration in the drainage water are satisfied. If not, the irrigation area A must be increased if any actual nitrate nitrogen concentration exceeds $n_t^{\max}$. The area can be decreased if all nitrate nitrogen concentrations are less than $n_t^{\max}$. Once the minimum area has been found, the total annual cost $C(V, A, Q_2^*)$ can be determined (Haith *et al.*, 1977).

This simulation procedure can be repeated for different combinations of volume discharges Q_{2t} in an effort to identify the least-cost design. As described above, the simulation procedure involves nothing more complex than the simultaneous solution of several sets of linear equations, some of which may not even be necessary if, say after a period when the soil is frozen, the initial storage volume and nitrate nitrogen concentration in the lagoon are known. When the simulation equations are used for any particular problem, some unit conversion coefficients may be necessary to maintain the desired units.

12.4.2. Regional Treatment and Transport

An alternative to numerous separate treatment facilities is the transport of partially treated wastewater to one or more regional advanced wastewater treatment facilities for further removal of constituents prior to discharge into a water body or on to a land area. This usually results in increased treatment efficiencies and reliabilities and possibly lower total costs because of economies of scale in wastewater treatment. Nevertheless, added to the cost of advanced treatment at any regional facility is the cost of wastewater transport.

Figure 12.2 illustrates a possible situation in which a regional treatment facility might be considered. Each existing treatment plant needs to be upgraded to meet new effluent standards and increasing volumes of wastewater flow. Let us assume that the types and concentrations of wastes in the effluent of each plant are approximately the same. Therefore, the annual cost $C_i(Q_i^T)$ of increased waste removal capacity at each treatment site i can be defined as a function of the treated wastewater volume Q_i^T at that site.

An alternative to increasing the efficiency of each plant is to transport wastewater effluent from one or more treatment plants to one or more regional advanced waste treatment facilities that could be located at various existing or new treatment plant sites. The wastewater volume Q_i^T that is to be treated at each site i will be the difference between the total wastewater inflow to that site and the total wastewater outflow. The inflow is the volume Q_i^0 collected at

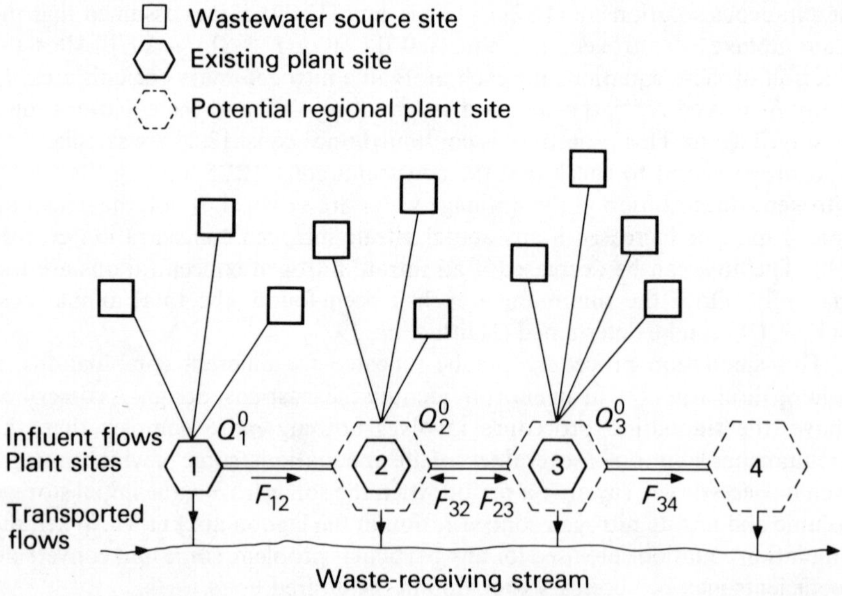

FIGURE 12.2 Existing and potential wastewater treatment sites.

the site plus the volume F_{ji} transported to the site from adjacent sites j. The outflow from any site i is the volume F_{ij} transported to adjacent sites j. Hence at each of sites 1, 2, 3, and 4 shown in Figure 12.2 the total inflow must equal the total outflow plus the volume treated, if any.

$$Q_1^0 = Q_1^T + F_{12} \tag{12.26}$$

$$Q_2^0 + F_{12} + F_{32} = Q_2^T + F_{23} \tag{12.27}$$

$$Q_3^0 + F_{23} = Q_3^T + F_{32} + F_{34} \tag{12.28}$$

$$F_{34} = Q_4^T. \tag{12.29}$$

Because site $i = 1$ is not a potential regional wastewater treatment site, but can be expanded to meet the required effluent standards for the existing influent volume Q_1^0, the treated flow Q_1^T cannot exceed Q_1^0. This condition is satisfied by eqn. 12.26.

The annual pipeline construction and pumping costs $C_{ij}(F_{ij})$ associated with the transport of wastewater from site i to adjacent site j will be some function of the flow F_{ij}. Because of the difference in elevation between those sites, these costs may depend on the direction of flow. The capacity of each pipe segment between sites i and $i + 1$ must equal at least the flow volume $F_{i,i+1}$ or $F_{i+1,i}$. If flow in either direction is possible, as between sites 2 and 3, one or both of these flows will undoubtedly be zero, as will its cost, in the model solution. In the example illustrated in Figure 12.2, the capacity of the pipeline between sites $i = 1$ and 2 must equal F_{12}. The pipeline capacity between sites 2 and 3 will equal $F_{23} + F_{32}$ (since one will be zero), and the pipeline capacity between sites 3 and 4 will equal F_{34}.

Let us assume a cost minimization objective:

$$\text{minimize} \sum_{i=1}^{4} \left(C_i(Q_i^T) + \sum_j C_{ij}(F_{ij}) \right). \tag{12.30}$$

The two components of the objective include annual wastewater treatment, pipeline construction, and pumping costs. The sum over sites j includes only those adjacent to site i.

If the cost functions exhibit fixed costs and economies of scale, i.e. decreasing marginal costs for certain ranges of Q_i^T and F_{ij}, then the problem can be formulated as a linear mixed-integer programming model. In this case it is probable that only one regional plant, if any, will be in the solution, even though the model constraints (defined by eqns. 12.26–12.29) allow for more than one regional facility. When the treatment plant cost functions $C_i(Q_i^T)$ are being derived, the waste concentration in the influent as well as the maximum allowable concentrations in the effluent must be known and considered. More detailed analyses can be found in Roman (1970) and Chi (1972).

12.4.3. Multiple Point Source Waste Reduction to Meet Water Quality Standards

The models presented so far have assumed the existence of quality standards specifying the maximum allowable constituent concentrations in the wastewater effluent that can be discharged into a water body. There may also exist stream quality standards specifying the maximum allowable constituent concentrations within the water body. These maxima may vary with the location within the water body, e.g. along a stream or estuary. Numerous models have been proposed for use in estimating the degree of waste removal at various point source sites along a water body that will meet both effluent and water quality standards. The most common of these models apply to the management of dissolved oxygen concentrations in streams and estuaries.

The oxygen required for the decomposition or assimilation of any particular quantity of biodegradable water is expressed as biochemical oxygen demand, BOD. The oxygen demand of a waste can be separated into two components, the amount required for the assimilation of the carbonaceous waste material, BOD^c, and that required for the assimilation of the nitrogenous waste material, BOD^n. This division permits a more accurate description of the oxygen demand at any point in the stream, lake, or estuary than would the total BOD, because the rates of deoxygenation associated with the two components differ. Another reason for explicitly considering the nitrogenous component of BOD is that as the percentage of carbonaceous BOD that is removed increases, say to 80 or 90%, the percentage of the nitrogenous component in the remaining wastewater effluent increases (Loucks and Jacoby, 1972). As water quality standards require increasingly high waste removals or treatment efficiencies, the nitrogenous wastes discharged into natural waters become increasingly important for the prediction of dissolved oxygen concentrations.

The depletion of dissolved oxygen by the metabolic processes of waste-consuming organisms, plant respiration, benthic deposits, and the like is offset by the absorption of oxygen from the atmosphere, from plant photosynthesis, and possibly from other natural and artificial means. Differential equations describing the processes of oxygen depletion and replacement were described in Chapter 6. The solutions of these differential equations, subject to the appropriate boundary and initial conditions, represent the temporal and longitudinal distributions of BOD^c, BOD^n, and dissolved oxygen concentration along a water course. If both natural and wastewater flows are constant, steady state can be assumed. For water quality control alternatives that are inflexible with respect to time, it is often reasonable to base the scale of these alternatives on some critical steady state conditions that can occur at specified locations during certain times of the year.

Let each waste source site along a nondispersive river be denoted by the index i and each quality-monitoring site in the water body by the index j. The oxygen demand BOD_j [$M L^{-3}$] at any quality site j resulting from the discharge of a mass of BOD_i^c and BOD_i^n per unit time [$M T^{-1}$] at all source sites i can be predicted by using equations such as

$$BOD_j = \frac{1}{Q_j} \sum_i (b_{ij}^c BOD_i^c + b_{ij}^n BOD_i^n) + \hat{BOD}_j, \tag{12.31}$$

where each parameter b_{ij} is the BOD mass at site j resulting from unit mass of BOD discharged at site i and $\hat{BOD}_j$ is the BOD [$M L^{-3}$] at site j resulting from all sources other than at sites i. Similarly, the dissolved oxygen deficit D_j at any quality site j can be written as

$$D_j = \frac{1}{Q_j} \sum_i (d_{ij}^c BOD_i^c + d_{ij}^n BOD_i^n) + \hat{D}_j \tag{12.32}$$

if the water body has a flow volume Q_j at site j. In (12.32), $\hat{D}_j$ is the dissolved oxygen deficit [$M L^{-3}$] at site j resulting from all BOD sources other than at sites i. Each parameter d_{ij} is the mass of dissolved oxygen at site j resulting from unit BOD discharge at site i.

In the two equations above, the masses of BOD_i^c and BOD_i^n discharged into the water body at each site i in each period may be unknown decision variables. If W_i^c and W_i^n [$M T^{-1}$] denote the total masses of carbonaceous and nitrogenous oxygen-demanding waste produced per unit time at site i, and if P_i^c is the fraction of the carbonaceous waste removed by treatment, then

$$BOD_i^c = W_i^c(1 - P_i^c). \tag{12.33}$$

The amount of nitrogenous BOD removed is some function $f_i(P_i^c)$ of the carbonaceous BOD removal efficiency. Hence,

$$BOD_i^n = W_i^n(1 - f_i(P_i^c)). \tag{12.34}$$

Equations 12.31–12.34 can be combined to form a mathematical model whose solution can identify various combinations of wastewater treatment efficiencies along a river or estuary that will satisfy both effluent and water quality standards. The planning problem illustrated in Figure 12.3 involves four point sources of waste and numerous quality-monitoring sites. The problem is to determine the degree of treatment, P_i^c, at each site i that satisfies effluent standards BOD_i^{max} at wastewater discharge sites i and stream quality standards BOD_j^{max} and DO_j^{min} for both BOD and dissolved oxygen concentration at various sites j. Since there are usually many alternative combinations of wastewater treatment efficiencies P_i^c that will meet the standards, the objective of the analysis will be to identify those that minimize the sum of the costs of wastewater treatment $C_i(P_i^c)$ at all sites i:

$$\text{minimize} \sum_{i=1}^{4} C_i(P_i^c). \tag{12.35}$$

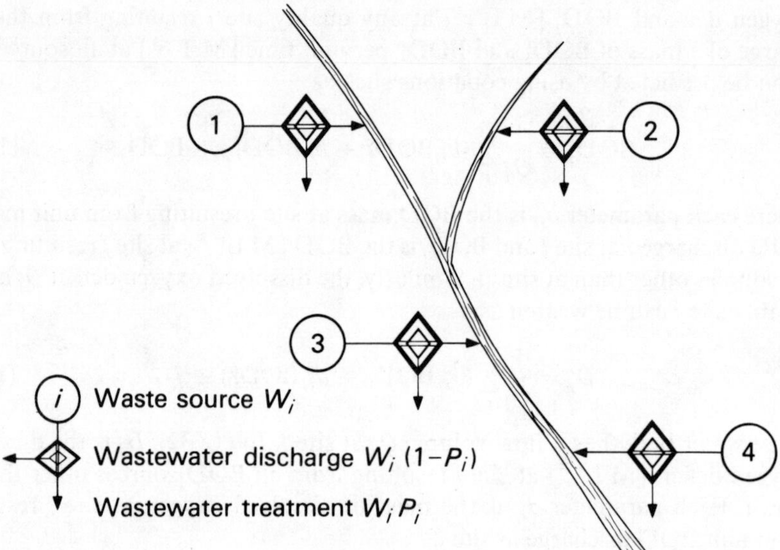

FIGURE 12.3 Wastewater discharge sites along a river system.

Constraint equations associated with the BOD effluent standards $\mathrm{BOD}_i^{\max}$ [M L^{-3}] at each site i are

$$\frac{1}{Q_i^{\mathrm{w}}}(\mathrm{BOD}_i^{\mathrm{c}} + \mathrm{BOD}_i^{\mathrm{n}}) \leq \mathrm{BOD}_i^{\max}, \tag{12.36}$$

where Q_i^{w} is the wastewater volume discharged from site i. For any water quality site j in the river the quality standards for BOD and DO can be expressed as

$$\frac{1}{Q_j}\sum_i (b_{ij}^{\mathrm{c}}\mathrm{BOD}_i^{\mathrm{c}} + b_{ij}^{\mathrm{n}}\mathrm{BOD}_i^{\mathrm{n}}) + \widehat{\mathrm{BOD}}_j \leq \mathrm{BOD}_j^{\max} \tag{12.37}$$

$$\mathrm{DO}_j^{\mathrm{S}} - \frac{1}{Q_j}\sum_i (d_{ij}^{\mathrm{c}}\mathrm{BOD}_i^{\mathrm{c}} + d_{ij}^{\mathrm{n}}\mathrm{BOD}_i^{\mathrm{n}}) + \hat{D}_j \geq \mathrm{DO}_j^{\min}. \tag{12.38}$$

The term $\mathrm{DO}_j^{\mathrm{S}}$ in (12.38) is the saturation dissolved oxygen concentration at site j. An alternative formulation of this model is given elsewhere (Loucks *et al.*, 1981, ch. 10).

A sufficient number of quality sites j must be selected to ensure that the maximum BOD or dissolved oxygen deficit within the entire river section of interest is not greater than the maximum acceptable. An alternative to the initial selection of numerous quality sites is to select a few such sites, solve the model, determine where the maximum BOD and DO deficit occur, and, if these concentrations are unacceptable, constrain the concentrations at these critical sites and repeat the procedure. For long river systems this iterative trial-and-error procedure may be less costly than solving a model that includes a large number of quality sites. Two or three iterations are usually all that is required.

Equations 12.35–12.38 can be made piecewise linear for solution by linear programming or linear mixed-integer programming algorithms. The unknown variables are the waste removal efficiencies P_i^c of the treatment plants at the waste source sites i. The coefficients of the model, b_{ij} and d_{ij}, must reflect the design flow conditions between sites i and j. They can be determined by methods discussed in Chapter 6.

Some of the more detailed nonlinear models, discussed in Chapter 6, that simulate many water quality constituents can also be converted into economic management models. For example, the application of finite-difference or finite-element techniques permits the conversion of the QUAL II simulation model to an optimization or management model. A large number of water quality constituent concentrations can then be considered. However, to do this, the relationship between, say, carbonaceous BOD removal efficiencies P_i^c and the removal efficiencies for nitrogen, phosphorus, and all other constituents in the wastewater influent must be defined. The removal efficiencies (and costs) are not independent, just as the nitrogenous BOD removal efficiency $f_i(P_i^c)$ was not independent of the carbonaceous BOD removal efficiency P_i^c in the BOD–DO model described above.

The interdependence (or joint-product effect) of removal efficiencies may be defined by dividing the range of carbonaceous BOD removal efficiencies P_i^c into several segments h. This permits the consideration of piecewise linear cost functions,

$$C_i(P_i^c) \approx \sum_h C_{ih} P_{ih}^c, \tag{12.39}$$

as well as piecewise linear efficiency functions $f_i^k(P_i^c)$ for any waste constituent k:

$$f_i^k(P_i^c) \approx \sum_h f_{ih}^k P_{ih}^c. \tag{12.40}$$

In (12.40), coefficient f_{ih}^k is the fraction of constituent k removed per unit fraction of carbonaceous BOD removed in segment h at site i. If linear programming is used to solve the resulting model, one must be sure that each P_{ih}^c is at its maximum value if $P_{i,h+1}^c > 0$. Otherwise, the cost and efficiency functions will be in error. Mixed-integer programming or separable programming algorithms that ensure this condition may be necessary. Other ways of modeling these joint-product effects are also available (Loucks *et al.*, 1981).

In addition to the linear programming models just discussed, the literature contains numerous examples of water quality management models based on other optimization algorithms. Dynamic programming (Liebman and Lynn, 1966), along with linear programming, is perhaps one of the most commonly used approaches (for example, see Thomann and Sobel, 1964; Sobel, 1965; Loucks *et al.*, 1967; Deininger, 1969, 1970; Gunther, 1970; Hass, 1970; Chia Shun Shih, 1970; Lehmann, 1971, 1974; Dorfman *et al.*, 1972; Milaszewski and Roman, 1972, 1978; Roman, 1974; Hock, 1978). A very comprehensive study, using dynamic programming, was reported by Newsome (1972) and Warn (1978) for the River Trent in England. Others have used nonlinear programming (Bayer, 1977) and geometric programming (Ecker, 1975), to name only some of the numerous techniques proposed and applied.

12.4.4. Flow Augmentation

The models discussed above are all based on a critical design flow volume or cross-sectional area, denoted as Q_j or A_j at each quality site j. If the discharge rate for BOD and other wastes from one or more sources is constant, the waste concentration and dissolved oxygen deficit at various sites may either decrease or increase, depending on the magnitude of the stream flow.

Increasing the stream flow has four primary effects on water quality. First, the volume of water increases. If the increased flow is of higher quality than the base flow, the increased flow dilutes the waste constituent concentrations and increases the minimum dissolved oxygen concentration. Second, the water velocity increases and, in turn, usually increases the reaeration rate and lengthens the distance over which an oxygen-demanding pollutant causes an oxygen deficit. Third, if base flows are augmented with cooler water, the deoxygenation rate decreases and the saturation concentration of dissolved oxygen increases. Conversely, if higher-temperature waters are used for augmentation, the deoxygenation rate increases and the saturation dissolved oxygen concentration decreases. Finally, increased stream flow may increase the BOD addition through runoff and scour of benthic deposits. All of these factors may well result in flow augmentation being beneficial at some sites and detrimental at others (Jaworski *et al.*, 1970; Loucks and Jacoby, 1972).

Let us consider the example of a single waste discharge site upstream from some quality sites. If the dissolved oxygen deficit and BOD of the flow upstream of the discharge site are less than those in the discharged wastewater, and if the increased stream flows do not pick up too much additional BOD from increased runoff and scour, then the additional dilution of the wastewater flow will decrease the BOD and, therefore, will increase the minimum dissolved oxygen concentration downstream from the waste discharge site. Though the minimum dissolved oxygen concentration is increased, these same conditions may lower the actual dissolved oxygen concentration at one or more specific quality sites, as illustrated in Figure 12.4.

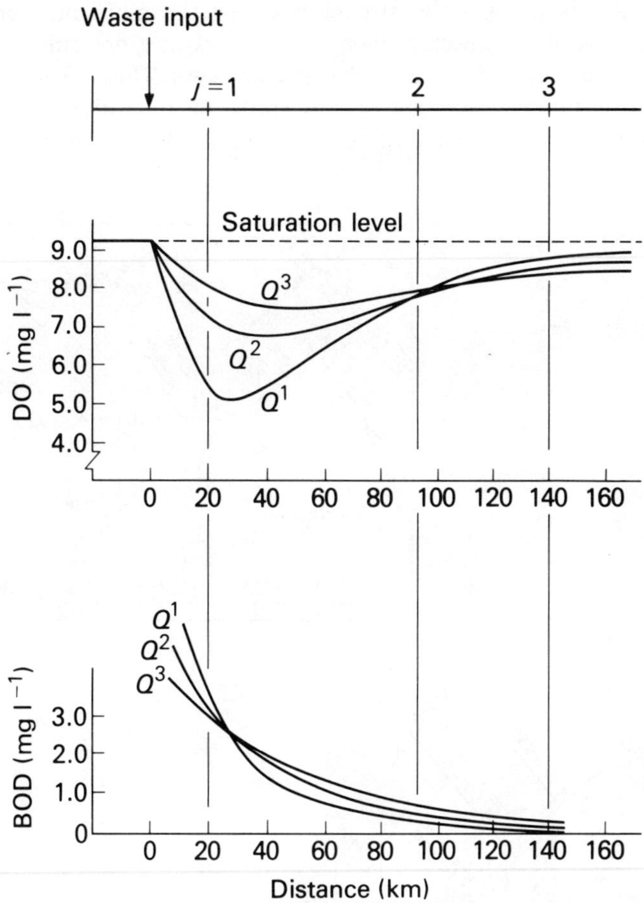

FIGURE 12.4 BOD and dissolved oxygen concentration downstream from a single waste source under increasing flow conditions: $Q^1 < Q^2 < Q^3$.

In Figure 12.4 each pair of functions corresponds to the oxygen "sag" concentration and the BOD resulting from a single waste source at site $i = 0$ for three stream flows $Q^1 < Q^2 < Q^3$. For the conditions stated above, the minimum dissolved oxygen concentration increases with the stream flow. At a site $j = 1$, an increase in flow results in a decrease in the dissolved oxygen deficit and BOD, i.e. the water quality increases with flow. The opposite may occur at quality sites $j = 2$ and 3, where the quality decreases as the stream flow increases, at least up to a certain flow volume. Hence there can exist situations, as illustrated in Figure 12.4, in which the waste removal efficiencies at upstream treatment facilities designed to meet both dissolved oxygen and BOD stream quality standards at low design stream flows (e.g. the minimum average seven-day consecutive flow expected once in ten years) are not sufficient to meet these same quality standards at higher stream flows. This was illustrated very clearly in an optimization–simulation study of the water quality of the Saint John River in the United States and Canada (H.G. Acres Ltd., 1971). As

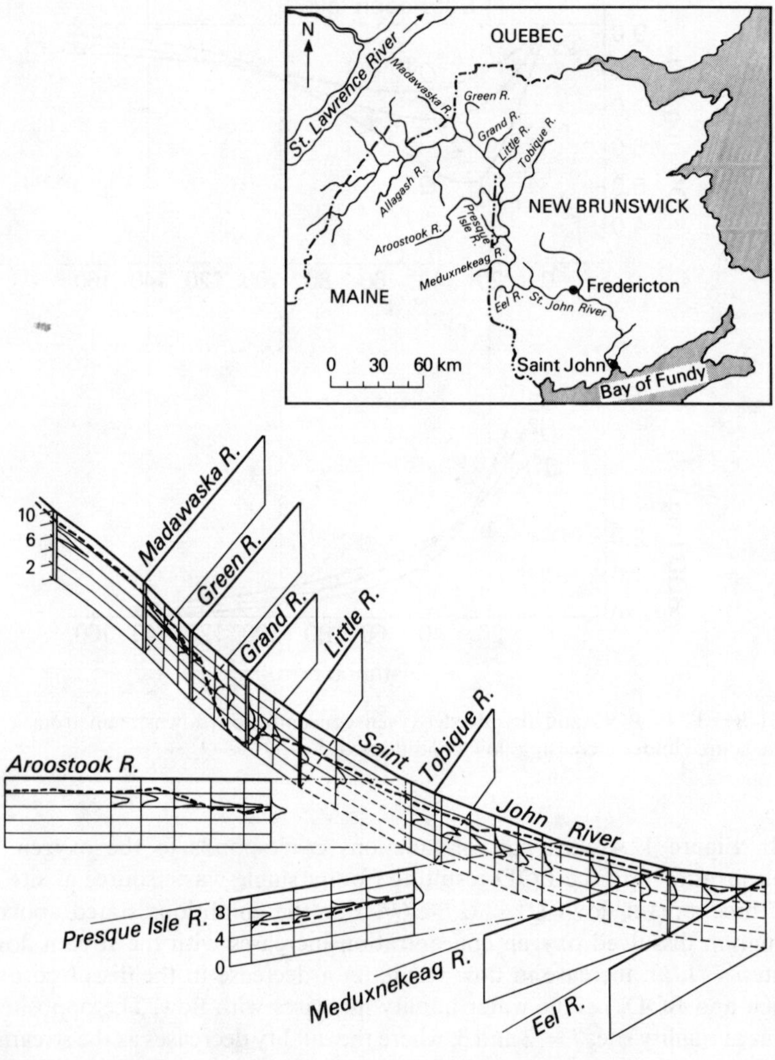

FIGURE 12.5 Comparison of dissolved oxygen concentrations predicted by analytical model for low-flow conditions and by a daily simulation model for the Saint John River in the United States and Canada (after H. G. Acres Ltd., 1971). Vertical scale, DO (mg l^{-1}).

seen in Figure 12.5, daily flows typically higher than the seven-day, ten-year low design flow condition resulted in lower predicted DO from increased runoff and scour from a large point source of BOD near Grand River.

Since stream flows in excess of the commonly chosen low design flow occur much more frequently, for a considerable portion of the time the stream quality at particular downstream sites may be less than what would occur in a critical low design flow condition. The selection of the critical design flow Q_j, therefore, may become an important consideration in the determination of waste removal efficiencies at treatment facilities.

The critical stream flow Q_j can be augmented by the release of additional waters from reservoirs or by reduction of water withdrawals. The increased flow may be a means of improving the critical flow conditions (e.g. by increasing the volume of dilution water, reducing temperatures, or increasing the reaeration rates), thereby reducing the required treatment capacities.

The net cost of flow augmentation can be defined as the minimum cost necessary to increase the flows during the period of low flows for the sole purpose of water quality management. In other words, the cost of augmentation is that required over and above the costs needed to maximize the net benefits of flow regulation for uses other than for quality management. If such conditions do not exist at the time flow augmentation is being considered, then the benefits derived from flow augmentation, apart from improved quality, should be subtracted from the gross costs of augmentation. This net cost $C_q(\Delta Q)$ of the flow increase ΔQ can then be included in a cost minimization objective function (Loucks and Jacoby, 1972).

If the temperature and velocity of the flow between any pair of sites i and j change because of flow augmentation, the parameters and variables that define the two transfer coefficients b_{ij} and d_{ij}, and $\widehat{\mathrm{BOD}}_j$ and $\hat{D}_j$ will change. Hence $b_{ij}(Q_{ij})$, $d_{ij}(Q_{ij})$, $\widehat{\mathrm{BOD}}_j(Q_j)$, and $\hat{D}_j(Q_j)$ are functions of the stream flow Q_{ij} or Q_j. These functions would take the place of the constants b_{ij}, d_{ij}, $\widehat{\mathrm{BOD}}_j$, and $\hat{D}_j$; in addition, $Q_j + \Delta Q$ would replace the term Q_j in (12.37) and (12.38).

12.4.5. Artificial In-Stream Aeration

In-stream aeration is another method of increasing the dissolved oxygen concentration in rivers. This is usually accomplished by injecting air into water through a network of perforated pipes or by rotating devices that cause surface turbulence, thereby increasing the area over which oxygen transfer can occur. These methods may be particularly efficient for the temporary improvement of near-anaerobic conditions, i.e. at sites where the dissolved oxygen deficits are relatively high, but they require energy and may cause excessive noise.

The oxygen transfer rate due to artificial aeration varies directly with the dissolved oxygen deficit D, with the water quality and temperature, and with the flow. If it is assumed that the rate of oxygen transfer per unit of power input is constant over a period t and that the aerators are operating at full capacity, the rate of oxygen transfer O_j [M T^{-1}] at site j is a function of aerator power capacity PA_j [L M T^{-1}] and dissolved oxygen deficit D_j [M L^{-3}] at site j:

$$O_j = O_j(PA_j, D_j). \tag{12.41}$$

Aeration devices are often the most cost-effective means of meeting dissolved oxygen standards that would otherwise be violated during rarely encountered extreme low-flow, high-temperature conditions (Pinaldi *et al.*, 1979). The typical dissolved oxygen control problem consists of determining the least expensive number of aeration units to be used, their location along the stream, and their design capacity measured in terms of power.

The cost of aerators operated at full capacity can be expressed as a function of their capacity. The function can account for the expected down time necessary for maintenance and repair. If the aerators are not operated at capacity, additional equations will be needed to define separately the annual capacity and the operating costs (Ortolano, 1972). Let $C_j(PA_j)$ be the cost of each aerator unit of power capacity PA_j at location j along a stream having a specified extreme design flow of Q_j and known upstream oxygen-demanding waste discharges. Such costs, together with equations for predicting the oxygen deficit and oxygen addition from artificial aeration, can be used to develop an optimization model for finding the optimum capacities PA_j at various sites j. If the objective is one of total cost minimization, this can be written as:

$$\text{minimize} \sum_j C_j(PA_j). \tag{12.42}$$

This objective is to be met subject to the maintenance of stream quality standards for dissolved oxygen. If a stream is divided into homogeneous reaches and all reach junctions and potential aerator sites j are numbered successively in the downstream direction, then the dissolved oxygen deficit D_j just upstream of site j, prior to any artificial addition of oxygen at that site, is a function of the BOD mass B_{j-1}, the initial deficit mass $D_{j-1}Q_{j-1}$, and the mass of added oxygen O_{j-1} at the next upstream site $j-1$:

$$D_j = B_{j-1}d_{j-1,j} + (D_{j-1}Q_{j-1} - O_{j-1})e_{j-1,j}. \tag{12.43}$$

The parameter $d_{j-1,j}$ is the dissolved oxygen deficit at site j resulting from unit BOD mass at site $j-1$, and $e_{j-1,j}$ is the dissolved oxygen deficit at site j resulting from a unit deficit mass at site $j-1$.

Dissolved oxygen standards would apply to each D_j prior to any artificial reaeration. For a given dissolved oxygen saturation concentration DO_j^S at site j,

$$DO_j^S - D_j \geq DO_j^{min} \quad \forall j, \tag{12.44}$$

where DO_j^{min} is the minimum allowable dissolved oxygen concentration at site j. The final set of constraints defines the production of oxygen at each site j as a function of the power capacity of the aerator (eqn. 12.41).

Of course, there may also be a constraint on the total number of aerators permitted, but this constraint tends to be redundant, or infeasible, if the sites j are chosen at locations of maximum deficit and the cost functions exhibit economies of scale. To estimate the annual cost functions $C_j(PA_j)$ some estimates of operating times are needed. These estimates have to be based on an analysis of the hydrological record of stream flow conditions, some assumptions regarding the oxygen-demanding waste discharges during critical flow conditions, and the cost of energy.

The control problem defined by (12.42–12.44) can be structured for solution by discrete dynamic programming in which the stages, the possible locations of the aerators, are variable and dependent in part on the location and capacity of aerators installed upstream (Rinaldi *et al.*, 1979).

12.5. LAKE QUALITY MANAGEMENT

There are a number of alternatives for managing lake water quality. For man-made lakes these include the selection of the impoundment site if it is not already fixed, the design of the outlet structure and the release policy, the control of constituents in the inflow, artificial destratification by such means as diffused air or mechanical pumping, dredging, and other ways of altering the normal physical, biological, and chemical processes that affect water quality (Symons, 1969). Simulation models that are able to predict with any reasonable accuracy the impact on water quality of any of these management alternatives are indeed just beginning to appear. Optimization models that incorporate water quality prediction together with various management alternatives have not yet been developed for lakes and reservoirs, with the exception of fully mixed impoundments. However, just as multiparameter water quality simulation models such as QUAL II (Chapter 6) are being adapted for optimization (management) modeling, so will the multiparameter simulation models for stratified lakes and reservoirs.

12.5.1. Constant-Volume Well Mixed Lakes

For reasonably well mixed lakes of constant volume, the one-dimensional models discussed in Chapter 7 (section 7.2) can be developed and include as unknown decision variables the extent and cost of waste discharge reduction. If it is assumed that a lake has a constant volume V $[L^3]$, an outflow rate Q_t $[L^3\,T^{-1}]$, and a nutrient or waste input rate N_t $[M\,T^{-1}]$ having a net decay rate constant K $[T^{-1}]$, the concentration C_{Nt} $[M\,L^{-3}]$ of N at the end of period t after an initial concentration $C_{N,t-1}$ can be found by integrating the differential equation that defines the rate of change in C_N:

$$\frac{dC_N}{dt} = \frac{N_t}{V} - \frac{Q_t C_N}{V} - KC_N. \tag{12.45}$$

This yields

$$C_{Nt} = \frac{N_t}{Q_t + KV} - \left(\frac{N_t}{Q_t + KV} - C_{N,t-1}\right)\exp\left[-\Delta t\left(\frac{Q_t}{V} + K\right)\right]. \tag{12.46}$$

The term Δt is a particular interval within period t. Hence if Δt is zero, the concentration of N is $C_{N,t-1}$, i.e. what it was at the end of period $t - 1$ or at the beginning of period t. If the inflow rates N_t and Q_t are constant for a long time, i.e. as Δt approaches infinity, the concentration C_N approaches the equilibrium concentration, $N_t/(Q_t + KV)$.

To illustrate the use of (12.46) in a lake quality management model, let us consider the discharge N_{ti} [M T^{-1}] of a mass of nutrient in each period t at various sites i along a lake. The rate of discharge is assumed constant over each period (having a time interval Δt equal to 1), but the mass of nutrient discharged can be altered by removing a fraction X_i at a cost $C_i(X_i)$. Thus the mass of nutrient discharged is $N_{ti}(1 - X_i)$ at each site i. Equation 12.46 can now be written for each period within a year:

$$C_{Nt} = \frac{\Sigma_i N_{ti}(1 - X_i)}{Q_t + KV}(1 - e_t) + C_{N,t-1}e_t \qquad \forall t, \tag{12.47}$$

where each known e_t can be expressed by

$$e_t = \exp\left[-\left(\frac{Q_t}{V} + K\right)\right] \qquad \forall t \tag{12.48}$$

and the variables X_i are bounded:

$$X_i^{\min} \leq X_i \leq X_i^{\max} \qquad \forall i. \tag{12.49}$$

The variables C_{Nt} and X_i are unknown. If the rates N_t and Q_t repeat themselves over each cycle of periods t, the initial concentration C_{N0} can be set equal to the concentration C_{NT} at the end of the final period T to yield a steady state solution of concentrations C_{Nt} for all periods $t = 1, 2, \ldots, T$. This eliminates the need to specify arbitrary or observed initial concentrations C_{N0}, unless of course a dynamic model is desired for a predefined number of periods t. In addition, the model could be made more realistic by the inclusion of changes in the decay rates K_t in each period that could result from changes in water temperature.

A simple economic objective might be to minimize the total cost:

$$\text{minimize} \sum_i C_i(X_i), \qquad (12.50)$$

subject to (12.47) and (12.49) and to constraints ensuring that the concentrations C_{Nt} do not exceed some acceptable level $C_N^{\max}$:

$$C_{Nt} \leq C_N^{\max} \qquad \forall t. \qquad (12.51)$$

Once again, it should be emphasized that these models are extremely simplified for most lake quality management problems. Nevertheless, with some judgment as to the appropriate values of Q_t and V, they can sometimes be used as a preliminary means of establishing average concentrations of nutrients or pollutants in well mixed lakes and of identifying the mix of water quality control alternatives that should be further analyzed using more detailed simulation models (Russell, 1975; Canale, 1976; Pavoni, 1977).

12.6. PREDICTING ALGAL BLOOM POTENTIAL

One of the major concerns of lake management in nutrient-rich areas is the potential for algal blooms. Large algal populations or blooms can seriously depress dissolved oxygen concentrations at night when algal respiration continues but oxygen production has stopped. Particularly disastrous can be the sudden collapse of a bloom, resulting in a tremendous oxygen demand for decomposition of the dead algal cells. In either case the depression of dissolved oxygen levels can kill fish and other aquatic life and produce bad odors and appearance. In addition, large algal populations can clog filters in water supply systems and often produce substances toxic to fish, shellfish, and animals that drink the water.

A set of optimization models can be used to predict the potential for algal blooms and thus to help estimate the effectiveness of alternative management strategies (diversion of nutrient-bearing waste and thermal discharges or lake destratification). This is done by determining the largest probable bloom that could occur under the conditions placed on algal growth by temperature and by nutrient and light availability.

Such models were developed and applied by Bigelow *et al.* (1977) to saltwater lakes and estuaries in the Netherlands. At any given temperature, the maximum potential algal bloom is limited by nutrient and light availability. The nutrient constraints in the model limit the total concentration of critical nutrients to that actually available. Let x_j be the unknown concentration of living algae in species j [M L^{-3}], a_{ij} the mass of nutrient i contained in unit mass of species j [M M^{-1}], y_i the concentration [M L^{-3}] of nutrient i that is temporarily held in nonliving matter (dead algae and decomposable organic matter) and that will become available for algal growth, and w_i the unknown concentration $[\mathrm{M\,L^{-3}}]$ of dissolved nutrient i immediately available for algal growth. The total concentration N_i $[\mathrm{M\,L^{-3}}]$ of nutrient i potentially available for algal growth is expressed as:

$$N_i = w_i + y_i + \sum_j a_{ij} x_j. \tag{12.52}$$

Upon the death of an algal cell, the nutrients in the cell enter the pool (y_i) of nutrients bound up in nonliving material. Upon decay of the nonliving material, the nutrients become available for further algal growth. The rate of change of the nutrient concentration in the nonliving material, $\mathrm{d}y_i/\mathrm{d}t$, is the difference between the rate at which algae die and the rate at which nutrient i is released from nonliving material into the water column and joins the available nutrient (w_i). Let d_j be the death rate $[\mathrm{T^{-1}}]$ of species j and u_i the rate of release of nutrient i from nonliving material (the mineralization rate); then

$$\frac{\mathrm{d}y_i}{\mathrm{d}t} = \sum_j a_{ij} d_j x_j - u_i y_i. \tag{12.53}$$

If algal blooms build up rather slowly compared with the rate at which nutrients are released from nonliving material, the concentration of nutrients, y_i, will be reasonably close to equilibrium ($\mathrm{d}y/\mathrm{d}t \approx 0$) at the peak of the algal bloom. Then, a reasonable estimate of y_i is the equilibrium concentration y_i^*, obtained by setting (12.53) to zero:

$$y_i^* = \frac{1}{u_i} \sum_j a_{ij} d_j x_j. \tag{12.54}$$

From eqns. 12.52–12.54 it follows that the equilibrium concentration N_i^* of nutrient i is

$$N_i^* = w_i + \frac{1}{u_i} \sum_j a_{ij} d_j x_j + \sum_j a_{ij} x_j = w_i + \sum_j a_{ij} \left(1 + \frac{d_j}{u_i}\right) x_j. \tag{12.55}$$

This relationship specifies how the total biomass in all algal species is constrained by the available nutrients if (12.54) applies.

In warm summer months, algal blooms can build up very rapidly such that the concentration of nonliving material lags behind the buildup of the algal populations. If, at the peak of the bloom, the amount of nutrient i in nonliving matter is approximately $k_i y_i^*$ for some empirically determined k_i, $0 \le k_i \le 1$, then

$$N_i = w_i + \sum_j a_{ij} \left(1 + k_i \frac{d_j}{u_i}\right) x_j. \tag{12.56}$$

Of course, $k_i = 0$ corresponding to $y_i = 0$ would result in the restriction that the total mass of nutrient i in all living algae must be less than N_i, the total nutrient available. While this yields the physically possible upper bound on algal biomass, it may not be sufficiently restrictive to allow determination of the maximum probable or realistically achievable algal population density. The latter is the more important quantity from the point of view of lake management.

The constraints on light availability affect the capacity of an algal species to grow and compete with other algal species. The amount of light that algae receive depends on the incident solar radiation throughout the day, $I_0(t)$, the fraction of the radiation transmitted across the air–water interface, β, and the light extinction rate η [L^{-1}] in the water column. In general the light intensity at depth z is given by

$$I_z(t) + \beta I_0(t)e^{-\eta z}. \tag{12.57}$$

An algae population can grow if its individuals receive light of an adequate intensity throughout the day as they move throughout the epilimnion or upper layers of a lake or estuary so that production exceeds respiration and mortality. If $P_j[I_z(t)]$ is the rate of production of algal species j at light intensity $I_z(t)$, for species j to participate in a bloom the average production rate of algal biomass throughout the 24 h must be no less than the minimum required for growth, $P_j^{\min}$:

$$\frac{1}{24}\int_0^{24}\frac{1}{z_{\max}}\int_0^{z_{\max}} P_j[\beta I_0(t)e^{-\eta z}]\,dz\,dt \geq P_j^{\min}, \tag{12.58}$$

where $z_{\max}$ is the maximum mixing depth. For any temperature, the production rate $P_j[I]$ peaks at some characteristic value for which the algal species is best adapted. Thus for each species j there will be a feasible range,

$$\eta_j^{\min} \leq \eta \leq \eta_j^{\max}, \tag{12.59}$$

within which species j can thrive and outside of which the average light intensity is either too low or too high.

The extinction coefficient η depends on the natural color or turbidity of the water, resulting in a background extinction rate η_0, and on the densities of algae (self-shading) and nonliving matter in the water column. If the contribution to the total extinction coefficient of a unit concentration of algal species j is η_j, then the extinction coefficient resulting from background and living algae is:

$$\eta_0 + \sum_j \eta_j x_j. \tag{12.60}$$

To this must be added the contribution from nonliving algae. This is primarily an effect of the chlorophyll remaining in dead algae because reflection is not nearly as important as the absorption of light. If the decay rate [T^{-1}] of the light-absorbing capacity of dead algae is v, then the differential equation describing the behavior of the contribution η_d of dead algae to the total extinction coefficient η is

$$\frac{d\eta_d}{dt} = \sum_j \eta_j d_j x_j - \nu\eta_d. \qquad (12.61)$$

Again, if the bloom grows slowly so that the concentration of chlorophyll in dead algae and hence its contribution η_d to the extinction coefficient stay in relative equilibrium, then $d\eta_d/dt \approx 0$, and the equilibrium extinction coefficient contribution from dead algae is approximately

$$\eta_d^* \approx \sum_j \frac{\eta_j d_j}{\nu} x_j. \qquad (12.62)$$

However, if blooms grow rapidly, η_d may lag behind the value given in (12.62) and the total extinction coefficient at the peak of the bloom is best given by:

$$\begin{aligned}\eta &= \eta_0 + \sum_j \eta_j x_j + k\eta_d^* \\ &= \eta_0 + \sum_j \eta_j\left(1 + \frac{kd_j}{\nu}\right)x_j \qquad (12.63)\end{aligned}$$

for some empirically determined k between 0 and 1.

For incorporation into the linear programming model, the acceptable extinction coefficient interval $(\eta_j^{\min}, \eta_j^{\max})$ for each species is used to determine a set of subintervals $(\eta_s^{\min}, \eta_s^{\max})$. Typically a number of subintervals s will be contained within any particular species interval. Each subinterval between $\eta_s^{\min}$ and $\eta_s^{\max}$ will include acceptable extinction coefficient values for a set S_s of algal species j.

A set of linear programming models for finding the maximum probable or reasonable biomass or chlorophyll concentration of an algal bloom can be developed using these nutrient and light availability constraints. To find the maximum potential biomass or chlorophyll concentration one would maximize for each s either biomass:

$$B^s = \sum_{j \in S_s} x_j \qquad (12.64)$$

or chlorophyll:

$$C^s = \sum_{j \in S_s} C_j x_j, \qquad (12.65)$$

where C_j is the chlorophyll content $[\mathrm{M\,M^{-1}}]$ per unit mass of algal species j. This maximization would be subject to nutrient constraints,

$$N_i = w_i + \sum_{j \in S_s} a_{ij}\left(1 + k_i \frac{d_j}{u_i}\right)x_j, \qquad (12.66)$$

to the light limitations,

$$\eta_s^{\min} \le \eta_0 + \sum_{j \in S_s} \eta_j \left(1 + \frac{kd_j}{v}\right) x_j \le \eta_s^{\max}, \tag{12.67}$$

and to nonnegativity,

$$x_j \ge 0, \qquad w_i \ge 0.$$

Solving the model for each set S_s with the corresponding extinction coefficient interval gives a set of potential concentrations of biomass $\{B^s\}$ or chlorophyll $\{C^s\}$. The largest of these,

$$B^{\max} = \max_s B^s \tag{12.68}$$

and

$$C^{\max} = \max_s C^s, \tag{12.69}$$

provide estimates of the potential or maximum probable values of these parameters.

The peak biomass of a bloom could be estimated with a dynamic ecosystem simulation model. This requires considerably more information than needed for this simple optimization model, as well as the ability to identify the environmental conditions that would give rise to the maximum possible bloom.

This optimization model makes no attempt to describe the time dynamics of the aquatic system. Rather it uses the available nutrient constraints and information about light requirements and self-shading characteristics of algal species to predict the maximum biomass or chlorophyll concentration that might reasonably be achieved during a particular week or month. Using a management period of this length should ensure that sufficient time is available for potential blooms to materialize, as long as algal populations are already present in reasonable densities in natural environments.

There may be no feasible solution to the optimization model for some s because an algal bloom may be unable to develop at a particular temperature and in corresponding light conditions. In winter and early spring, when the water is cold, there may be no feasible solution for any value of s. Feasibility could also be affected by small changes in some of the parameters. From the dual variables of the nutrient and light availability constraints, one can obtain some indication of the impact on bloom potential caused by a change in the available nutrient concentrations N_i and algal extinction coefficient limits.

As formulated, the model considers only the effect of a bloom in terms of total biomass or chlorophyll. The oxygen demand from decomposition of dead algal cells, which would occur over a short time were the maximum bloom to collapse suddenly and all the algae die, can also be determined. The effect of zooplankton has been omitted from the model but could be incorporated in an approximate way. The model is relatively simple and serves as an approximate description of a very complex system. Its potential value lies in its simplicity, its ability to estimate answers to some important management questions, and its relatively modest data requirements compared with more complex and realistic dynamic algal models.

12.7. CONCLUDING REMARKS

It would be ideal if one could say that water quality models such as those discussed in this chapter could be used directly to identify optimum, or at least improved, solutions to water quality management problems. Perhaps they can in situations where there exist sufficient data and sufficient modeling expertise to develop, calibrate, and verify models and their solutions and where the objectives of those responsible for water quality planning and management are clearly defined. Where ideal conditions do not exist, these models can serve another purpose.

Water quality management models, such as those outlined in this chapter, are most commonly used for developing an understanding of the relative impacts on water quality of alternative management practices and for determining the significance or importance of having more accurate or more detailed data. These insights can guide the development of effective plans and decisions, if not actually identify the best plan or decision.

Water quality modeling, if done well, can give an understanding of why some management alternatives are better than others for a particular river basin. Modeling can provide estimates of how the river system will respond, at least in a relative sense, to different waste discharges. In addition, models can be used to help identify preferred management plans, based on various management objectives and assumptions concerning future resource costs, technology, and social and legal requirements.

In acknowledging the role that water quality models can and should play in the planning process, one must recognize the inherent limitations of models as representatives of any real problem. The input data, including management objectives and assumptions concerning the physical, biological, and chemical processes in the water body, may be controversial or uncertain. Of course, the input affects the output. While the input data and model may be the best available, one's knowledge about the actual water body and about how future events may alter its behavior will always be limited. In addition, since public water quality objectives change, water quality models must be viewed as flexible tools that can adapt to changing circumstances as they are perceived and to changing data as they become available.

REFERENCES

Acres, H. G., Ltd. (1971). Water quality management methodology and its application to the Saint John River. Appendix G. *H. G. Acres Ltd., Niagara Falls, Ontario Report* (unpublished).

Bigelow, J. H., J. G. Bolten, and J. C. DeHaven (1977). Protecting an estuary from floods—A policy analysis of the Oosterschelde. Vol. IV, Assessment of algal blooms, a potential ecological disturbance. *Rand Corporation, Santa Monica, CA Report* R-2121/4-NETH.

Bayer, M. B. (1977). A modeling method for evaluating water quality policies in nonserial river systems. *Water Resources Bulletin* 13(6):1141–1151.

Canale, R. P. (ed.) (1976). *Modeling Biochemical Processes in Aquatic Ecosystems* (Ann Arbor, MI: Ann Arbor Science).

Chi, T. (1972). Water conveyance models. *Models for Managing Regional Water Quality*, eds. R. Dorfman, H. D. Jacoby, and H. A. Thomas, Jr. (Cambridge, MA: Harvard University Press), ch. 8.

Chia Shun Shih (1970). System optimization for river basin water quality management. *Journal of Water Pollution Control Federation* 42(10):1792–1804.

Deininger, R. A. (1969). Über die Planung eines wirtschaftlich optimalen Systems von Kläranlagen. *Gas- und Wasserfach* 110(52):1443–1445.

Deininger, R. A. (1970). Les calculatrices et la recherche opérationelle pour la planification des installations d'épuration. *Centre Beige d'Etude et de Documentation des Eaux Rapport* 314.

Dorfman, R., H. D. Jacoby, and H. A. Thomas, Jr. (eds.) (1972). *Models for Managing Regional Water Quality* (Cambridge, MA: Harvard University Press).

Ecker, J. G. (1975). A geometric programming model for optimal allocation of stream dissolved oxygen. *Management Science* 21:658–668.

Fan, L. T., M. T. Kuo, and L. E. Erickson (1974). Effect of suspended wastes on system design. *Proceedings of American Society of Civil Engineers, Journal of Environmental Engineering Division* 100(EE6):1231–1247.

Grady, C. P. L., Jr. (1977). Simplified optimization of activated sludge process. *Proceedings of American Society of Civil Engineers, Journal of Environmental Engineering Division* 103(EE3):413–429.

Gunther, W. (1970). Optimierungsmodelle zur Planung und Betreibung von Anlagensystemen zur Abwasserbehandlung in Flussgebieten. *Wasserwirtschaft–Wassertechnik* 20(3): 81–84.

Haith, D. A. (1973). Optimal control of nitrogen losses from land disposal areas. *Proceedings of American Society of Civil Engineers, Journal of Environmental Engineering Division* 99(EE6):923–937.

Haith, D. A., A. Koenig, and D. P. Loucks (1977). Preliminary design of wastewater land application systems. *Journal of Water Pollution Control Federation* 49(12):2371–2379.

Hass, J. E. (1970). Optimal taxing for the abatement of water pollution. *Water Resources Research* 6(2):353–365.

Hock, B. (1978). Water quality modeling as a tool for decision making in Hungary. *Modeling the Water Quality of the Hydrological Cycle. Proceedings of Baden Symposium, September. International Association of Hydrological Sciences Publication* 125, pp. 312–321.

Jaworski, N. A., W. J. Weeber, and R. A. Deininger (1970). Optimal reservoir releases for water quality control. *Proceedings of American Society of Civil Engineers, Journal of Sanitary Engineering Division* 96(SA3):727–742.

Koenig, A., and D. P. Loucks (1977). A management model for wastewater disposal on land. *Proceedings of American Society of Civil Engineers, Journal of Environmental Engineering Division* 103(EE2):181–196.

Lawrence, A. W., and P. L. McCarty (1970). Unified basis for biological treatment design and operation. *Proceedings of American Society of Civil Engineers, Journal of Sanitary Engineering Division* 96(SA3):757–778.

Lehmann, H. (1971). Ein Beitrag zur Schaffung optimaler Beschaffenheitsverhältnisse in Fliessgewässern unter Berücksichtigung der Abwasserbehandlung, der Wasseraufbereitung und sonstiger Nutzungen. *Wasserwirtschaft–Wassertechnik* 21(12):404–410.

Lehmann, H. (1974). Planung der Wasserbeschaffenheit in Fliessgewässern bei Berücksichtigung ökonomischer Kriterien. *Wasserwirtschaft–Wassertechnik* 24(8):262–267.

Liebman, J. C., and W. R. Lynn (1966). The optimal allocation of stream dissolved oxygen. *Water Resources Research* 2(3):581–591.

Loucks, D. P. (1967). Risk evaluation in sewage treatment plant design. *Proceedings of American Society of Civil Engineers, Journal of Sanitary Engineering Division* 93(SA1):25–39.

Loucks, D. P., and H. D. Jacoby (1972). Flow regulation for water quality management. *Models for Managing Regional Water Quality*, eds. R. Dorfman, H. D. Jacoby, and H. A. Thomas, Jr. (Cambridge, MA: Harvard University Press), ch. 9.

Loucks, D. P., C. S. Revelle, and W. R. Lynn (1967). Linear programming models for water pollution control. *Management Science* 14(4):B-166–B-181.

Loucks, D. P., J. R. Stedinger, and D. A. Haith (1981). *Water Resource Systems Planning and Analysis* (Englewood Cliffs, NJ: Prentice-Hall).

Middleton, A. C., and A. W. Lawrence (1974). Cost optimization of activated sludge systems. *Biotechnology and Bioengineering* 16:807–826.

Middleton, A. C., and A. W. Lawrence (1976). Least cost design of activated sludge systems. *Journal of Water Pollution Control Federation* 48(5):889–905.

Milaszewski, R., and M. Roman (1972). Optymalizacja stopnia oczyszczania ścieków dla systemu oczyszcralni. *Gospodarka Wodna* 36(1):23–29.

Milaszewski, R., and M. Roman (1978). Some physical and economic aspects of optimization of the degree of wastewater and water treatment. *Modeling the Water Quality of the Hydrological Cycle. Proceedings of Baden Symposium, September. International Association of Hydrological Sciences Publication* 125, pp. 373–382.

Newsome, D. H. (1972). The Trent River model—An aid to management. *Modeling of Water Resource Systems* vol. II, ed. A. K. Biswas (Montreal: Harvest House), pp. 490–509.

Ortolano, L. (1972). Artificial aeration as a substitute for wastewater treatment. *Models for Managing Regional Water Quality*, eds. R. Dorfman, H. D. Jacoby, and H. A. Thomas, Jr. (Cambridge, MA: Harvard University Press), ch. 7.

Pavoni, J. L. (ed.) (1977). *Handbook for Water Quality Management Planning* (New York, NY: Van Nostrand Reinhold).

Rinaldi, S., R. Soncini-Sessa, H. Stehfest, and H. Tamura (1979). *Modeling and Control of River Quality* (New York, NY: McGraw-Hill).

Roman, M. (1970). Podstawy stosowania grupowych oczyszcralni ścieków. *Nowa Technika w Inzynierii Sanitarnej, Seria Wodociegi i Kanalizacja* (Warsaw: Arkady).

Roman, M. (1974). Kompleksowa optymalizacja stopnia oczyszczania ścieków i wody. *Prace Naukowa Politechniki Warszawskeij, Seria Budo Wnictwo* No. 33 (Warsaw: Warsaw Polytechnic), pp. 47–60.

Russell, C. S. (ed.) (1975). *Ecological Modeling in a Resource Management Framework. Resources for the Future Working Paper* QE-1 (Baltimore, MD: Johns Hopkins University Press).

Sobel, M. J. (1965). Water quality improvement programming problems. *Water Resources Research* 1(4):477–487.

Symons, J. M. (1969). Water quality behavior in reservoirs. *US Department of Health, Education and Welfare, Cincinnati, OH, Public Health Service Publication* 1930.

Thomann, R. V., and M. J. Sobel (1964). Estuarine water quality management and forecasting. *Proceedings of American Society of Civil Engineers, Journal of Sanitary Engineering Division* 90(SA5):9–36.

Van Note, R. H., P. V. Herbert, R. M. Patel, C. Chupek, and L. Feldman (1975). A guide to the selection of cost-effective wastewater treatment systems. *US Environmental Protection Agency, Washington, DC Report* EPA 430/9-75-002.

Warn, A. E. (1978). The Trent mathematical model. *Mathematical Models in Water Pollution Control*, ed. A. James (New York, NY: Wiley), ch. 7.

CHAPTER 12: NOTATION

a_{ij}	mass of nutrient i in unit mass of algal species j
A	land area
b_{ij}	BOD mass at site j resulting from unit mass of BOD discharged at site i
B^s	maximum potential biomass concentration
$\hat{\text{BOD}}_j$	BOD at site j resulting from all sources other than at sites i
$C_i(S_i)$	agency cost of waste reduction
$C_i^s(S_i)$	private cost of waste reduction for group s
C^s	maximum potential chlorophyll concentration
C_j	chlorophyll content per unit mass of algal species j
d_{ij}	mass of dissolved oxygen at site j resulting from unit BOD discharge at site i
d_j	death rate of algal species j
$d_{j-1,j}$	dissolved oxygen deficit at site j resulting from unit BOD mass at site $j-1$
d_t	drainage rate
D_j	dissolved oxygen deficit at site j
$\hat{D}_j$	dissolved oxygen deficit at site j resulting from all other BOD sources other than at sites i
DO_j^S	saturation dissolved oxygen concentration at site j
$e_{j-1,j}$	dissolved oxygen deficit at site j resulting from unit deficit mass at site $j-1$
E_t	evapotranspiration rate
f_{ih}^k	fraction of constituent k removed per unit fraction of carbonaceous BOD removed in segment h at site i
F	equilibrium level of organic nitrogen in soil
F_{ij}	wastewater volume transferred from site i to adjacent sites j
I_t	initial inorganic nitrogen content
$I_z(t)$	light intensity at water depth z
k_t	temperature-dependent nitrogen removal rate in period t
K	nutrient or waste decay rate constant
L_t	inorganic nitrogen leached by drainage
m_t	mineralized fraction of organic nitrogen
M	field capacity
n	concentration of nitrogen
N_t	inorganic nitrogen removed by plant growth during period t; nutrient or waste input or discharge rate
O_j	rate of oxygen transfer at site j
O_t	deviation of level of organic nitrogen at start of period t
$P_i(S_i)$	fraction of waste reduced at site i
P_j	rate of production of algal species j

P_t	average precipitation rate
PA_j	aerator power capacity at site j
Q_i^w	wastewater volume discharged from site i
Q_j	flow volume at site j
$\bar{Q}_i$	quality of water at site i
Q_2^*	maximum discharge rate
Q_{2t}	irrigation volume in each period t (of length Δt) within a year
s	subinterval of algal species interval
S_i	scale of waste reduction measures at site i
S_s	set of algal species j
S_t	storage volume at start of period t
T_i^s	tax rate per unit of waste discharge
u_i	rate of release of nutrient i from nonliving material (mineralization rate)
V	storage lagoon capacity
w	weighting constant
w_i	concentration of dissolved nutrient i immediately available for algal growth
W_i	constant quantity of waste available at site i
W_i^c, W_i^n	total masses of carbonaceous and nitrogenous oxygen-demanding waste produced per unit time at site i
x_j	concentration of living algae in species j
X	fraction of nutrient or waste
XI_t	inorganic nitrogen addition
XO_t	organic nitrogen addition
$(y_i^*)y_i$	(equilibrium) concentration of nutrient i temporarily held in nonliving matter
Z_k	set of sites in zone k of a region
α	fraction of nitrogen in lagoon effluent
β	fraction of radiation transmitted across air–water interface
η	light extinction rate
ν	decay rate of light-absorbing capacity of dead algae.

13 Future Directions

G. T. Orlob

13.1. ACHIEVEMENTS IN MODELING WATER QUALITY

There is little doubt that mathematical models have reached a level of acceptance in the scientific and engineering communities that makes them viable in research and problem solving. Over roughly two decades, from the early 1960s, modeling of water quality has moved forward—albeit, with difficulty at times—from mere intellectual exercise of the mathematically inclined to creative synthesis of scientific principles and empirical observations aimed at quantification of natural aquatic phenomena. One has only to skim the most recent journals of science and engineering to appreciate that "model," "modeling," "systems," "systems analysis," etc. are solidly embedded in the lexicon of today's researchers and practitioners. The hardware to facilitate the use of models has advanced so rapidly, especially in the design of inexpensive high-capability minicomputers, that the machine seems less of a restriction than ever. Moreover, to an important degree models of hydraulic, hydrological, water quality, and ecological behavior of the earth's water resources have become useful tools in problem solving, not merely ends in themselves. It is the development of such models that we have tried to chronicle here.

If we consider that the "modern era" in modeling of water quality commenced with the now classic Streeter–Phelps formulation of the DO sag, then the chronology of our accomplishments since 1960 has been roughly as follows.

1960–65

The Streeter–Phelps approach was extended by discretization of the stream system to allow spatial variability in load patterns, geometry, and kinetic coefficients. Temperature was introduced as a state variable in one-dimensional stream and primitive reservoir models that considered heat exchanges at the air–water interface. Temporal variability was accommodated by providing independently derived solutions of the hydrodynamic equations for one-dimensional systems. The advection–diffusion equation was adopted as the basic description for water quality transport processes. The first simple models, essentially one-dimensional and steady state, began their use in water quality management.

1965–70

Discretized one-dimensional models of stream and reservoir systems were extended to include additional sources and sinks, e.g. nitrogenous oxygen demands and photosynthesis. One-dimensional networks were used to represent two-dimensional vertically mixed systems. Thermal stratification was simulated for deep reservoirs and lakes that could be represented as one-dimensional continua. Nutrient budgets and primary productivity of lakes were simulated for simple, fully mixed systems. Water quality models began to figure prominently in assessments of the impacts of pollution.

1970–75

Multilayered two-dimensional models of wind-driven circulation were developed, providing a basis for description of water quality in large lakes. Water-quality–ecological interactions were incorporated in multisegment models. The kinetics of primary productivity and higher trophic levels were developed for model application. The Michaelis–Menten–Monod limiting nutrient concepts became prominent in modeling of eutrophying aquatic systems. Finite-element models were developed for two-dimensional systems, adding flexibility to the already well accepted finite-difference techniques. Sensitivity testing assumed greater importance in assessing model reliability. Increased dimensionality of models heightened concern for adequacy of data for calibration and validation.

1975–80

During the last five years or so, increasing attention has been given to improvements in model reliability and assessment of capability. Reassessments were made of trade-offs between model detail and the required data base. Automatic calibration and parameter estimation techniques were devised. Improvements were made in the description of important hydrodynamic processes essential to the characterization of water quality, for example, the effects of internal mixing and surface wind shear. The coupling of hydrodynamics and water quality, as in the case of thermal stratification, was given increased attention with the development of improved two- and three-dimensional lake models. Complexities of the aquatic ecosystem were the subject of an increasing number of biologically oriented modelers, leading to improved descriptions of the ecological relationships and the process of eutrophication. A general proliferation of models gave rise to concern for lack of documentation of useful models and practical mechanisms for technology exchange.

These two decades, which account for virtually all of the history of water quality modeling as we know it, have not been without some negative aspects. Where these developed, they were attributed generally to overzealousness on the part of practitioners of the modeling "art" and to lack of effective communication between the modeler and the potential user of his product. There was a general tendency in the mid-1960s to oversell the potentials of modeling, especially in the field of water quality management. This was encouraged by the rapid advances in the sixties of systems technology and computer sciences. Also, there was a natural tendency in academia to rise to the intellectual challenges of modeling, leading to the overproduction of models, as evidenced in the literature. However, despite these general difficulties, water quality modeling is apparently a viable part of our technology, still offering challenges to the modeling enthusiast.

13.2. AREAS FOR IMPROVEMENT

At the end of each of the preceding chapters the author has endeavored to summarize the major accomplishments within the frame of his topic, at the same time indicating the direction toward needed improvement. Certain improvements refer specifically to a particular relationship or model and need not be repeated here, but there are some areas of concern that seem to apply to modeling generally, at least to the modeling of water quality. To bring our book to a conclusion on an optimistic note, it seems worth while to identify a few areas where the efforts of future modelers may be productive.

13.2.1. Model Reliability

Among the most perplexing of questions is "How good is the model?" A clear and compelling need exists to answer this question in explicit and quantitative terms. However, this must be done in the context of the intended use of the model and the person or persons who will interpret the results.

In the context of water quality management, it appears generally to be the case that most models are far too complex to serve as useful aids in the decision process. Yet, simplification of the model (and concomitant reduction in the data base required) may be at the sacrifice of credibility to the decision maker. One solution may be the utilization of hierarchical models, which at one extreme of detail preserve some essential attributes of the water body, while at the other assure rapid, inexpensive, and more informative exposition of the consequences of alternatives.

Quantitative measures of reliability include statistical measures of the processes of calibration, verification, and validation. Each of these processes is much in need of formalization in the modeling process, perhaps even of standardization to some degree. It is apparent that much of the difficulty with model credibility stems from lack of accepted criteria for ascertaining reliability. Development of such criteria should be beneficial to the modeling art.

13.2.2. Data and Data Management

An almost universal complaint by modelers and users of model output is aimed directly at the quantity and quality of data. Generally, the data are not sufficient and often they are of the wrong kind. There is a need for improved methods for structuring the data collection program that supports modeling. The data management activity should be developed in parallel with the model if possible or, if an existing model is to be employed, it should be developed with the aid of the model.

Design of the data collection scheme depends on the question of model reliability, discussed above. Of course, if data collection is more costly than model development (usually the case) then it would seem prudent, if not imperative, that model reliability assessment and data system design be carried out simultaneously. Research aimed at improving the relationship of these two crucial modeling activities should have a high priority.

13.2.3. Models and Submodels

Water quality models described in the previous chapters are largely descriptive and deterministic in structure, of the distributed-parameter type. In addition, many of these are dynamic, an added complexity that exacerbates the problems of parameter estimation, reliability estimation, calibration, etc. While such models often have considerable merit as management tools or aids in research, they may prove demanding of data, costly to operate, and even conducive, because of their complexity, to increasing human error. There is a need to obviate these weaknesses, perhaps with supporting models or "submodels" of simpler design.

Possibilities for simplification include:

(1) reducing spatial segmentation;
(2) changing from dynamic to steady state;
(3) reducing the number of parameters, i.e. "lumped" instead of distributed parameters;
(4) introducing inputs stochastically instead of deterministically.

Development of hierarchical models, compatible submodels, or complementary models of different conceptual form (e.g. lumped versus distributed, or stochastic versus deterministic) should increase the utility of modeling generally.

13.2.4. First Principles Versus Empiricism

The nature of modeling suggests that there will always be some dependence on empiricism. However, to the extent practical, empirical representation should be supplanted by first principles (see Chapter 3). The argument here is for a fundamental understanding of the water body whenever possible rather than succumbing to the temptation to mount increasingly demanding programs to collect data in support of empirical modeling.

A case in point is the development of an improved hydrodynamic model to describe wind-induced mixing in the hypolimnion of a stratified lake. This comparatively recent innovation in lake modeling promises to improve substantially the description of internal mixing processes that previously were treated almost entirely as analogous to molecular diffusion.

13.2.5. Hydrodynamics in Relation to Water Quality

There has been a tendency, especially among the more biologically oriented, to concentrate on rigorous development of ecological relationships in water quality modeling, occasionally at the expense of comparable rigor in characterizing the hydromechanical behavior of the water body. While there exist cases where hydrodynamics is of small consequence, it is more often the case that the major mechanisms of water quality and ecological change are closely linked to movement of the water and internal mixing processes. A successful modeling effort will depend on sound description of these motions, notwithstanding an elegant development of the pertinent water-quality–ecological relationships. Examples of the importance of giving appropriate weight to both aspects of water quality modeling are seen in efforts to characterize eutrophication of large lakes, e.g. the Great Lakes, where wind-driven circulations determine the fate of nutrients discharged in shoreline areas.

13.2.6. One, Two, and Three Dimensions

Capability to model the water quality changes in relatively small, deep impoundments that are strongly stratified seems to be well developed, although some improvements are possible yet in describing the processes of internal mixing over the annual cycle. Also, two-dimensional vertically mixed systems, i.e. shallow lakes and reservoirs without strong stratification, are accommodated fairly well by existing models. However, long, narrow reservoirs or fjord-type lakes and large, deep lakes are still prime candidates for modeling effort. A first essential step is a rigorous model of two-dimensional stratified flow in which the internal mixing is represented in terms of the traditional parameters describing the flow.

Attempts to design such a model using the finite-element method appear promising, primarily because of the flexibility the method provides for dealing with irregular physical boundaries. Incorporation in such a model of an improved description of turbulent mixing, such as the Launder K–ϵ representation, is a necessary direction of future research.

13.2.7. Kinetics and Stoichiometry

The long tradition in water quality analysis of the universality of certain kinetic coefficients and stoichiometric constants has encouraged modelers to extend these notions into the realm of ecological modeling. While there are some indications that this confidence is justified for very simple or lumped-parameter models, there remains considerable reservation (especially among the aquatic biologist modelers). Additional research and modeling experience are required to assure the reasonable extension of existing models to new situations.

13.2.8. Documentation

Lack of complete documentation is probably the greatest obstacle to user acceptance, despite the existence of a sizable number of proven water quality models. Since model development, itself, is more or less an "art," there is considerable variability in the quality of in-program comment and user manuals. There is a need, as yet not addressed, to standardize or to provide some measure of consistency in this important step in model development.

13.2.9. Register for Software

An initial motivation for the development of this book was the apparent lack of an effective mechanism for technology transfer in the field of mathematical modeling of water resource systems. Although there has been substantial progress to improve this situation by just such organizations as IIASA, as this volume itself attests, there is still a need to formalize an institutional structure for technology exchange. This structure is envisioned as an international register of software (perhaps with counterparts at a national level), which would document proven models and, if feasible, serve as a clearing house for dissemination of programs and essential documentation. This is a role already assumed by IIASA on a limited basis. In the future it is expected that the scope of this effort will be expanded and that other projects, such as that represented by this volume, will continue to enhance technology transfer in the field of mathematical modeling.

13.3. CONCLUSION

During the course of preparing this book, a product of the collective effort of an international team of dedicated experts in the field of water quality modeling, it became painfully apparent that we had little hope of capturing that elusive "state of the art." Modeling is itself dynamic, often changing so rapidly that a manuscript completed one week can be out of date the next. Moreover, we continued to be plagued by the lag between creation of the model and its appearance in the technical literature, most often a year or more. Finally, we were continually reminded that the art we were seeking was not really found in the technical press after all, but appeared most often in reports of very limited distribution. If we were lucky we found such a document in the files of one of our group, if not we had to depend on the assistance of our friends and advisors, whose help was generously given.

Considering all these difficulties in gathering sufficient useful information to justify this book, we decided that we could not in good conscience consider our effort as actually characterizing the state of the art. However, we hope we have come fairly close to the target within our own frame of reference. This has called for identification and discussion of a selected group from among the many models we have reviewed. Our criteria for inclusion, admittedly arbitrary on our part, embraced such subjective determinations as "practical," "well documented," "widely used," "transferable," etc. We recognize that some important models may have been excluded, either inadvertently or by our subjective judgment. Nevertheless, we hope that our sampling of the literature and available reports has been sufficiently comprehensive to have stimulated the reader to examine our treatise carefully, to point out to us its deficiencies (which we acknowledge before the fact), and, above all, to contribute to further advancing the "state of the art."

Index